# THE PULL OF HISTORY

## Human Understanding of Magnetism and Gravity through the Ages

# THE PULL OF HISTORY

## Human Understanding of Magnetism and Gravity through the Ages

**Yoshitaka Yamamoto**
Sundai Preparatory School, Japan

World Scientific

NEW JERSEY • LONDON • SINGAPORE • BEIJING • SHANGHAI • HONG KONG • TAIPEI • CHENNAI • TOKYO

*Published by*

World Scientific Publishing Co. Pte. Ltd.
5 Toh Tuck Link, Singapore 596224
*USA office:* 27 Warren Street, Suite 401-402, Hackensack, NJ 07601
*UK office:* 57 Shelton Street, Covent Garden, London WC2H 9HE

**British Library Cataloguing-in-Publication Data**
A catalogue record for this book is available from the British Library.

**THE PULL OF HISTORY**
**Human Understanding of Magnetism and Gravity through the Ages**



ISBN 978-981-3223-76-9

For any available supplementary material, please visit
http://www.worldscientific.com/worldscibooks/10.1142/10540#t=suppl

Desk Editor: Christopher Teo

# Preface

This book was originally published in 2003 in Japanese as *Jiryoku to juryoku no hakken* (Discovering magnetism and gravity). For the English edition, however, I have updated and substantially rewritten certain sections. To that extent, the present work represents a revised edition rather than a faithful translation of the original.

In this work, I trace the evolution of man's understanding of action at a distance (also known as noncontact force) from antiquity to the modern era and consider how this intellectual evolution shaped the rise of modern science. From ancient times through the Middle Ages, magnetic force was identified as the primary example of action at a distance. In modern times, universal gravitation was added to the category of action at a distance, a step that in many respects marks the birth of modern physics.

My investigation of this theme began almost thirty years ago when I first began studying the history of science. It struck me as curious that Johannes Kepler frequently described gravitational force as magnetic. Although I searched a variety of sources, nowhere could I find a clear, satisfactory answer to the question of why he identified gravitation with magnetic force. My curiosity was further piqued by the fact that, whereas it was precisely Kepler's and Newton's positing of gravitational force that laid the foundation of modern cosmology, continental thinkers, particularly Cartesians and other mechanists, emphatically rejected the Newtonian idea of gravitation on the grounds that it was "occult."

As I pursued answers to such questions, I gradually came to see that the key concept in the formation of modern physics was force—above all, force in the form of action at a distance. Histories of physics, and especially histories of mechanics, have mainly traced the construction and development of the law of motion. I believe, however, that by shifting the focus to changes and transitions in the concept of force, it is possible to gain new insights into the birth of modern physics. This conviction is what motivated me to write this book.

Several points are developed in this work that, to the best of my knowledge, have never before been discussed in science histories. Marbode's discovery in the twelfth century of the electrostatic force of chalcedony is one example, but others involve more than pointing out specific facts. My perspective on Thomas Aquinas, for example, no doubt differs from the mainstream view. Nonetheless, I believe the argument can be made that Thomas's theory of astrological causality, according to which celestial bodies exert force on terrestrial bodies, may have derived from the discovery around that time of the directionality of the loadstone. As I discuss at the end of Part 1, the first scientific experiments on magnets were conducted by one of Thomas's contemporaries, Petrus Peregrinus.

Part 2 opens with an account of Nicolaus Cusanus and his proposal for the quantitative measurement of magnetic force in the mid-fifteenth century. This represented a decisive turning point in the development of the concept of physical force. It is also underlined that the Renaissance concept of magic changed significantly from the fifteenth century to the sixteenth century. On religious grounds, Marsilio Ficino and other fifteenth-century thinkers rejected demonic magic (black magic), which called upon the power of demons, and differentiated it from natural magic (white magic), which made use of natural forces. Yet they did not deny the existence of black magic. In the sixteenth century, however, Pietro Pomponazzi and some other scholars no longer believed that demonic magic existed. The main focus of scientific inquiry during the latter half of the sixteenth century was natural magic. Natural magic, particularly as practiced by Giambattista della Porta, was essentially experimental, an approach that opened one of the doors to modern science. Natural magic viewed the natural world as a network of sympathy and antipathy, that is, as occult actions between natural things. The remarkable properties of magnets—namely, their powers of attraction and repulsion coupled with their peculiar directionality—were considered typical of these occult actions and were the central focus of study. Thus, action at a distance was explored mainly from the perspective of magic.

This brings us to William Gilbert, a figure who plays a leading role in conventional histories of magnetic science. Studies of Gilbert, however, rarely discuss the connection between his discovery that the Earth is a large magnet and subsequent changes in the prevailing view of the universe—specifically, the rise of the heliocentric model. While Gilbert's experimental methods and his

Earth-magnet discovery are recognized as a milestone in the history of modern science, his argument that the Earth is a living entity with a "soul" has generally been either downplayed or entirely overlooked. Yet his view of the Earth as a living, self-moving entity is precisely what hastened the downfall of Aristotelian natural physics (which viewed the Earth as inanimate and immovable) and opened the way to Keplerian gravitational theory. The main theme of Part 3 is the development of such theory—specifically, how the magnetic philosophy of Gilbert and Kepler led to Hooke's and Newton's elucidation of the "system of the world."

In writing this book, I was able to find answers to the questions that had troubled me for so long. The result, I believe, is in its own way a new and original account of the birth of modern science.

The research necessary for this work required access to a vast amount of literature. I was able to find most sources at the Japan Diet Library, but to access others I was assisted by many of my friends, colleagues, and students. I am also indebted to many people for their invaluable assistance in the preparation of the manuscript. I particularly wish to express my gratitude to Ms. Masae Hayashi, Mr. Shuji Ino, Ms. Satsuki Hayakawa, Ms. Miho Kanno, Ms. Katsumi Kawanishi, Ms. Keiko Kimura, Mr. Takahiro Miyake, Mr. Noritoshi Morinaga, Dr. Koichi Nakamura, Dr. Masao Nomura, Mr. Yasuhiko Oshima, Ms. Ikuko Okai, Dr. Shigeo Sawai, Ms. Mizue Taguchi, Mr. Yoshihiro Takata, Mr. Kazuki Yoshimoto, and my brother, Shigenobu Yamamoto.

With regards to the English translation, I am exceedingly grateful to Mr. Dean Robson, Ms. Cathy Hirano, and Ms. Esther Sanders. Finally, I wish to thank Misuzu Shobo, the publisher of the original

Japanese edition; Mr. Shogo Morita, Ms. Ayako Ishigami, Ms. Misako Nakagawa, and Mr. Takashi Arai of the Misuzu Shobo editorial staff; and World Scientific, the publisher of the present English edition.

Yoshitaka Yamamoto

# Introduction

As a physicist and a historian of physics, I am constantly reminded that modern physics, and indeed modern natural science in general, arose mainly in Europe. The late twentieth century saw much controversy over the way in which the development of natural science unfolded on that continent during the sixteenth and seventeenth centuries. Some focused on the revival of Platonism, others on technological achievements from the late medieval period onward, and still others on the magical beliefs prevalent during the Renaissance. In particular, the contribution of magic to the formation of modern science was the subject of much interest and controversy among scholars. There are some excellent books on science and magic, including those by such eminent scholars as Frances A. Yates and Paolo Rossi. Their writings interpret the methodology, logic, and views of nature of early scientists, philosophers, and so-called magicians and weave all of this into the fabric of science history. But they neither examine the development of individual concepts nor seek answers to specific questions.

The concept for this book arose from my conviction that historical inquiry requires precisely such case studies. Questions about the contribution of magic or technology, if discussed only in general terms, become mired in differing interpretations of historical documents, and definitive conclusions fail to emerge. The sheer vastness of the question "What made the birth of modern science possible?" precludes a simple answer. To delve deeper, we need to focus on key concepts and examine their development in concrete terms.

In physics, the key concept is *force*, and more specifically *gravitational force*. The shift from a geocentric (Ptolemaic) to a heliocentric (Copernican) model was a watershed in the development of modern astronomy. From the point of view of physics, however, merely placing the sun at the center of things was not sufficient to establish an accurate understanding of the solar system. This understanding came only with the discovery of universal gravitation as the force that enabled our sun to keep its planets in their respective orbits. As one scholar puts it, "the clear comprehension of mechanical force and the conscious incorporation of it into the basic structure of physics can be regarded as the beginning of modern science."[1]

Very briefly, the history of physics can be summarized as beginning with the atomistic theories of ancient Greek scholars, who created the notions of the atom and void space. Two thousand years later, in the seventeenth century, came the discovery of universal gravitation, the quintessential example of *action at a distance*. Then, in the nineteenth century, the concept of field was born, and finally this led to the physics of the present day by way of the discovery of the quantum. In modern physics, the notion of action at a distance has been reduced to the concept of field, but it is indisputably true that the introduction of this notion—in the specific form of universal gravitation—marked the starting point of modern physics and played a crucial role in the history of science during the seventeenth and eighteenth centuries.

To fill in some of the details: Nicolaus Copernicus introduced a revolutionary way of understanding the cosmos; Johannes Kepler further refined Copernican thinking with his three laws, which included the statement that orbits are elliptical; and then Isaac Newton introduced the law of universal gravitation as a mathematical explanation of the physical order of the solar system. The shift posed by Copernicus was geometric, but it took Kepler and Newton to establish a basis for this shift on principles of physics and dynamics, relying on gravitational force between celestial bodies as a key concept. In 1600, William Gilbert became the first to say that the Earth was an active magnet rather than an inert lump of soil. Kepler took this idea and carried it further, arguing that the sun exerted on the planets some kind of force analogous to magnetic force. Newton's theory replaced the Aristotelian-Ptolemaic view of the universe and became the point of departure for modern physics, with the concept of gravitational force as the cornerstone of our understanding of the cosmos.

From ancient times, people recognized matter and motion, but this recognition in itself did not lead to the birth of physics. We owe a lot to Rene Descartes and Galileo Galilei for establishing a mechanistic way of viewing matter and for establishing principles of mechanics. However, Descartes's system of mechanics was inadequate in that it allowed for motion to be transmitted only by collision and in no other way. For Galileo's part, he did not have the notion of force, and so he failed to view the solar system dynamically. Moreover, neither man was able to understand the significance of Kepler's discovery. It was only when Robert Hooke and Newton linked the idea of force to the achievements of Copernicus and Kepler that the solar system became a subject of the study of dynamics, and with this the true meaning of Kepler's laws was finally recognized. Thus, the formation and development of the concept of force was the key to the way in which the heliocentric model was completed as part of the discipline of physics and modern physics advanced with its first step toward triumph.

We can assume that people first developed the concept of force through the sensations of exertion and resistance when they lifted or carried heavy loads or when they pushed or pulled one another. The concept of force was thus anthropomorphic, and therefore it was only natural for people to believe that the intervention of some physical intermediary, comparable to an outstretched arm or hand, was required to exert force on a distant object.

Even as far back as ancient Greece, Plato noted that

> as to the flowing of water, the fall of the thunderbolt, and the marvels that are observed about the attraction of amber and the Heraclean stones [i.e., magnets],—in none of these cases is there any attraction.[2]

He asserted that what appeared to be attraction must somehow be explainable by contiguous action of one kind or another. Aristotle, too, thought it obvious that

> that which primarily is moved locally and corporeally must be either in contact with or continuous with that which moves it … That which is the first movement of a thing … is always together with that which is moved by it (by "together" I mean that there is nothing intermediate

> between them). This is universally true wherever one thing is moved by another.[3]

In the thirteenth century, Roger Bacon affirmed the idea that "proximity is required as a necessary condition of action."[4] Later, at the dawn of the modern age, Gilbert claimed that "no action can take place by means of matter unless by contact."[5] In the seventeenth century, Walter Charleton, in a tone typical of the mechanists and atomists of his day, postulated as follows:

> When we see a piece [*sic*] of Amber, Jet, hard Wax, or other Electrique, after sufficient friction, to attract straws, shavings of wood, quils, and other festucous bodies of the same lightness, objected within the orbe of their Alliciency; and that with a cleanly and quick motion: Why should we not conceive, that this Electricity or Attraction may hold a very neer Analogy to that attraction of Gnats, by the exserted and nimbly retracted tongue of a Chamaeleon.[6]

Charleton wrote this in 1654, just a few decades before Newton published his laws of gravity. It is a "general law of nature," said Charleton, that "Nothing can act upon a Distant subject." Thus, even into the early modern period it was considered self-evident that an intermediary agent was required for any force to act on a distant object.

During the first half of the seventeenth century, Descartes and Pierre Gassendi, also mechanistic-atomistic theorists, competed with chemical philosophers of the Paraclesian school, both sides seeking to occupy the position of supremacy long held by the Aristotelians. One heavily disputed issue was "weapon salve," which followers of Paracelsus believed could cure injured soldiers when applied not to their wounds, but to the swords that had injured them: "*Paracelsus* saith," wrote one British Aristotelian critic of the notion in 1631, "if the weapon be annoynted, the wounded partie may be cured, though 20 miles absent." The use of weapon salve was derided not only, of course, by mechanists and atomists, but also by physicians who adhered to the teachings of Aristotle and Galen. The practice was rejected outright as either nonsense or magic. The refusal to accept this magical therapy can be traced back to one of the conventional beliefs of the time: that action without an intermediary agent is impossible. The same British Aristotelian summed up the position succinctly: "*Nullum agens agit in distans* [No action can operate at a distance]."[7]

But the logic that rejected weapon salve also repudiated the possibility that the sun could exert a gravitational force on the Earth, or the Earth on the moon. Even Galileo denied the influence of the moon upon the tidal movements on the Earth. As a result, when Newton introduced his law of universal gravitation, he was subjected to unimaginably harsh criticism at the hands of both progressive scientists, including Cartesian epigones and Gottfried Leibniz, and old-guard Aristotelians. Leibniz wrote: "A body is never moved naturally, except by another body which touches it and pushes it; after that it continues until it is prevented by another body which touches it. Any other kind of operation on bodies is either miraculous or imaginary."[8] It was simply beyond thinkers of both the new and old schools to consider that Newton's ideas could represent anything other than magic or astrology.

It was none other than magnetism that provided the logic to defend the notion of action at a distance. Even as early as the second century A.D., Alexander of Aphrodisias acknowledged magnetism to be the sole instance of action occurring at a distance. And in the fourteenth century, William of Ockham (most famous for Ockham's Razor, a maxim now often paraphrased as "Entities should not be multiplied beyond necessity") asserted that "the lodestone acts at a distance immediately and not through a medium."[9] Gilbert, too, who denied action at a distance in general terms, accepted it in the case of magnets. These thinkers were alike in reaching a common-sense conclusion based on the premise that direct experience is to be accepted at face value as an accurate reflection of reality.

Many shared this conception of magnetism as the singular example of action at a distance. Paracelsus, thought to be the originator of weapon-salve, gave credence to the idea that celestial bodies can influence the Earth. In his treatise *The Diseases that Deprive Man of His Reason*, he puts it thus:

> The stars have the power to hurt and weaken our body and to influence health and illness. They do not fall into us materially or substantially, but influence reason invisibly and insensibly, like a magnet attracting iron.[10]

A century later Francis Bacon called magnetism "an instance of divergence" and designated the motion caused by action at a distance "magnetic motion."[11]

Because of its capacity to act without direct contact, magnetism had, since ancient times, appeared miraculous or enigmatic and was often considered to be a living, spiritual, or even magical force. In the sixteenth century, Georgius Agricola, a German physician known for his discourses on mining and mineralogy, explained that

> theologians attribute the powers this mineral [i.e., lodestone] possesses to divine origin, scientists to natural origins, the nature of which cannot be interpreted.[12]

In 1600 Gilbert made a similar observation:

> Loadstone and also amber do some philosophers invoke when in explaining many secrets their senses become dim and reasoning cannot go further. Inquisitive theologians also would throw light on the divine mysteries set beyond the range of human sense, by means of loadstone and amber.[13]

And even in fairly recent times, several renowned thinkers have associated this force of nature with the inexplicable. In the eighteenth century, for example, Scottish economist Adam Smith wrote that

> when we observe the motion of the iron, in consequence of that of the loadstone, we gaze and hesitate, and feel a want of connection betwixt two events which follow one another in so unusual a train.[14]

Even today, for those who have no knowledge of how it works, magnetism can appear nothing short of wonderful. Twentieth-century physicist Albert Einstein related the following experience from his childhood:

> The development of [our] thought world is in a certain sense a continuous flight from "wonder." A wonder of such nature I experienced as a child of 4 or 5 years, when my father showed me a compass. That this needle behaved in such a determined way did not at all fit into the nature of events, which could find

> a place in the unconscious world of concepts (effect connected with direct "touch").[15]

Were it not for our conditioning to the idea through education, motion occurring without direct contact would remain incomprehensible even today.

The story of how we have come to our present level of understanding of magnetism, historically perhaps the most incomprehensible of all natural phenomena, is a source of endless fascination. Ancient peoples had long used the lodestone in religious ceremonies, medical treatment, and rites of exorcism. Then Greek civilization, and with it the beginnings of natural philosophy, came onto the scene, and two types of explanation for magnetism emerged: on the one hand, mechanists and reductionists believed that invisible particles or some imponderable fluid acted as an intermediary agent to effect the attraction of one object by another; on the other hand, hylozoists believed that magnetism was a living or soul-like force that defied further explanation.

In the Middle Ages, the hylozoistic view became dominant, and magnetism became a symbol for the network of "occult forces" or the workings of "sympathy and antipathy." In the thirteenth century, Saint Thomas Aquinas remarked that

> in the physical order things have certain occult forces, the reason of which man is unable to assign; for instance that the magnet attracts iron.[16]

An organic view of nature, together with belief in an occult force, persisted into the Renaissance and was the basis for the magical philosophy that held sway in certain circles. Expressing ideas that were typical at the time was sixteenth-century philosopher Pietro Pomponazzi:

> The magnet attracts iron, and diamond repels it. Sapphire cures ulcers and improves eyesight. There are a number of other examples of this kind of occult force.[17]

Magnetism was not only typical of occult force but also virtually the only manifestation of such force in the material world.

As if the lodestone's power to attract were not enough to inspire wonder, thinkers of bygone centuries also had to grapple with the observed facts that lodestones pointed north and imparted to iron needles both attractive force and northward directionality. At one time it was believed that lodestones were attracted to the North Star or to a celestial pole and that this property was bestowed upon them by the heavens. It was therefore believed that lodestones embodied some sort of sympathetic relationship between the Earth and the heavens and even offered proof of the validity of astrology. We see an early example of this viewpoint in Thomas Aquinas:

> The magnet, for instance, attracts iron … because it shares in the powers of the heavens.[18]

An even more specific statement is offered by the fifteenth-century Renaissance humanist Marsilio Ficino:

> A balanced needle influenced on its end by a lodestone points towards the Bear; this happens because the lodestone draws it in that direction, since the power of the Bear prevails in this stone; and from there it is transferred into the iron and pulls both toward the Bear.

> Moreover this sort of power both was infused from the beginning and also grows continually by the rays of the Bear.[19]

Once sailors began making use of magnets in the form of compasses for navigation, the way was paved for the discovery of the Earth's magnetic field, which in turn led to the novel idea that the Earth was active and no mere lump of dirt. In 1600, Gilbert published *On the Magnet*, in which he revealed his discovery that the Earth was a giant magnet. He described our planet as a living, soul-like being, and thereafter it became possible to think of the Earth as a moving planet, a premise that was essential to the viability of heliocentrism.[20] Soon after that, in 1605, Flemish engineer Simon Stevin wrote the following about the Earth's movement as proposed by Copernicus:

> Since this motion is simply described by him [i.e., Copernicus], without any natural argument or proof, this supposition long troubled me in my mind ... But thereafter there was published the book about the great terrestrial magnet, described by *Guilelmus Gilbertus* [William Gilbert], in which the natural cause of this motion in my opinion is hit off and revealed.[21]

Kepler shared Gilbert's view of the Earth as a magnet, and by association he conceived of the idea of a gravitational force between astronomical bodies:

> Just as the kinship of their bodies makes the loadstone attract loadstone or iron; so also in the case of the moon it is not unbelievable that she should be moved by the terrestrial body which is akin, although neither in that case nor in this case is there any contact between the bodies.[22]

Although this association between magnetism and gravity has been almost entirely forgotten and strikes us today as rather odd, centuries ago it was crucial in inspiring scientists with the idea that celestial bodies might be exerting a force upon one another. Gilbert himself wrote, in a work published posthumously:

> The Moon is drawn to the Earth by magnetism.[23]

Robert Hooke, who was one of the first to state in the form of conjecture what Newton later proved mathematically as the law of universal gravitation, explained in 1666 the part that magnetism played in the drama of the discovery of gravity:

> GILBERT began to imagine [gravity] a magnetical attractive power, inherent in the parts of the terrestrial globe: the noble VERULAM [Francis Bacon] also, in part, embraced this opinion; and KEPLER (not without good reason) makes it a property inherent in all celestial bodies, sun, stars, planets.[24]

It was none other than the magnetism-related magical discourses and practices of the era preceding modern science that enabled the concept of force—universal gravitation—to take shape and gain acceptance. Without knowing this, we might all too easily explain the birth of modern physics as merely the result of some stroke of genius by Kepler or Newton.

My writing of this book is an attempt to fill what I see as a serious gap in science history. I have sought to explain the genesis and development of modern science, placing importance on the evolution of the concept of force in Europe from antiquity to the beginning of modern times. I have focused especially on magnetic force and gravitational pull, emphasizing the contributions of both magic and technology to their discovery.

This is not to say that there is any lack of books on the history of either magnetism or gravity; on the contrary, there have been many tomes written about these subjects. Most, however, regard Greek philosophy as containing the seeds of modern science and then skip straight to the Renaissance, ignoring an entire millennium. Typically, they then state that modern science was formed during the period from the Renaissance to the early modern era through struggles against Aristotelian philosophy. In the area of mechanics, this type of neglect has been redressed somewhat since the time of French physicist Pierre Duhem, but the history of electromagnetics remains largely unstudied to this day. A case in point is Edmund T. Whittaker's *History of the Theories of Aether and Electricity*: beginning in the thirteenth century, it describes how Aristotelianism spread throughout Western Europe during this period under the influence of Thomas Aquinas, and then how William of Ockham's struggle for liberation from Thomist philosophy in the next century prepared the ground for the flowering of the Renaissance and the successes of Copernicus and Kepler.

Physics is not the only subject where we see this tendency to ignore what is essentially most of the Middle Ages. Thomas Steele Hall's *Ideas of Life and Matter*, for example, is a scholarly study of the history of biology and medicine, a meticulous (for the most part) genealogy of theories of life from ancient Greece to modern times. But following a discussion of Galen, a physician of the second century A.D., the camera fades and the scene suddenly cuts to the Renaissance.

Science history tends to pay only scant attention to anything that has been labeled “absurd” or “reactionary” or has otherwise drawn contempt from the perspective of *modern* science, such as superstition, conjecture, folklore, and religious discourse. In the eleventh century, for instance, Marbode (the Bishop of Rennes) mentioned in a poem extolling the supernatural powers of precious stones that a type of quartz called chalcedony possessed an attractive force. This discovery of a third substance, following amber and jet, capable of drawing other objects with static electricity has to my knowledge been completely overlooked by historians writing about electromagnetics. Consider also Thomas Aquinas’s claim that lodestones were granted their power by the heavens. Despite the importance of this idea in terms of its historical influence, few documents record how it was passed down to the Renaissance. And the contribution of Giambattista Della Porta’s *Natural Magick* to the study of magnets during the sixteenth century may not have been totally neglected but it was certainly underestimated.

So persistent has this type of thinking been that in a detailed treatise on magnetism published in the 1940s, author A. Crichton Mitchell leaves out any and all discussion of Saint Augustine, whose philosophy had an enormous and tangible impact on the understanding of magnetism for some one thousand years. Mitchell explains that he does not mention “legendary matter with which the history of the subject has been diluted”; and further that

> it has also been decided to omit most of those references, chiefly from the early Church Fathers, in which the magnet’s attraction for iron is only used as an illustration of some doctrinal theme, and are obviously not intended as serious contributions to the science of Magnetism.[25]

If Mitchell belittles some contributions as "not serious," it is because he is looking at them with modern eyes. We must remember that in their time these contributions were indeed very serious, and that without acknowledging the contributions and influence of key figures such as Saint Augustine, far too much is missed.

To close this introduction, I'd like to draw attention to the words of a scholar who eloquently points would-be historians toward a better way. In his preface to a work about superstitions in medieval Europe, Jean-Claude Schmitt, a contemporary historian of the Annales school, offers the following advice:

> For a long time, students of the history of Christianity, especially those who were clerics, believed that they could study Church traditions by using concepts passed down by the Church. We need, however, to reexamine the concepts most prominent in the Christian heritage, especially such deep-seated concepts as magic, superstition, and even religion. In order to be objective about Christian culture in historical terms, one must be critically distanced from Christian terminology. In other words, historians need to distinguish between the technical terms that they establish for their study and the existing terms of Christian culture, the very subject of this scholarly exercise.[26]

Here, Schmitt is cautioning Christian historians to remember that among the lower strata of medieval European society Christianity coexisted and competed with indigenous beliefs and other religions known since antiquity. Although Christianity came out the winner and eventually dominated all other forms of religion, the history of the Middle Ages, when this contest was ongoing, cannot be well understood by passing judgments from a contemporary perspective.

I believe that by replacing "Christianity" with a particular science such as "physics," "the Church" with "academia," and "clerics" with "scholars," Schmitt's warning applies with equal validity to the history of medieval science.

Schmitt adds that, "undaunted by the prospect of sounding paradoxical, I daresay that in the long period where devotion to religion was dominant, a true 'history of religion' did not exist." The same caveat applies in cases where the yardstick of modern science is used to measure the scientific developments of an era rife with an assortment of religious and magical beliefs. Again, "history of religion" can be replaced with "history of science." Scholars of the Middle Ages, after all, knew nothing of the fixed notions that we, in hindsight, have come to designate the "history of physics." If we wish to truly learn about the birth of science, we must examine premodern writings about force within the context of their own times.

I am, both by training and by profession, a physicist. I know full well that venturing outside my specialty to write a work of history goes beyond audacity, almost to the point of recklessness—as though I were an unlicensed driver barreling through an unfamiliar neighborhood. Nonetheless, I could not resist the urge to shed light on facts that physics historians have, if not misread, at least ignored or undervalued. It is my sincere hope that this challenge will contribute to a redefining of the foundation of modern physics and, by extension, of modern science in general. It is this, the secret of the birth of science, that remains to be unlocked.

# Contents

# Part 1
# Antiquity and the Middle Ages

Chapter 1

# Ancient Greece: The Science of Magnetism is Born

## Section 1—Early Attempts to Explain Magnetism

Tradition has it that the first ancient Greek scholar ever to discuss the magnet was Thales (c. 624–546 B.C.) of Miletus, a flourishing port city in Ionia on the Aegean Sea. In fact, none of Thales's writings were preserved, so whatever we claim to know about him has come down to us from scholars of later generations. About two hundred years after the time of Thales, for example, Aristotle (384–322 B.C.) noted in *On the Soul* that

> Thales, too, to judge from what is recorded about him, seems to have held soul to be a motive force, since he said that the magnet has a soul in it because it moves the iron.[1]

In *Lives of Eminent Philosophers*, Diogenes Laertius, writing a further five centuries later, in the third century A.D., corroborates the traditional claim regarding magnets:

> Aristotle and Hippias affirm that, arguing from the magnet and from amber, [Thales] attributed a soul or life even to inanimate objects.[2]

We cannot actually confirm whether or not Thales knew of amber's power to attract (due to static electricity), because the writings of Hippias, too, have been lost; nor does Aristotle mention amber in any of his extant works.

According to both Aristotle and Diogenes Laertius, Thales referred to magnets only to explain the action of the soul, mentioning them only in the course of defending his claim that everything has a soul. Nowhere do they suggest that Thales explained what magnets were, nor, certainly, that he claimed magnets as a new discovery of his own. To the contrary, their writings suggest that, by Thales's time, magnets and their action were already well known. Incidentally, both the Greek word for "soul" (*psuche*) and the Latin equivalent (*anima*) are actually much broader in meaning than is conveyed by the modern English word "soul." *Psuche* spans all the nuances of "soul," "life," and "mind"; it refers to all things lifelike and to the principle of life itself[3]. Thales was a proponent of hylozoism—the belief that all material things have life—and, for him, the magnet gave a straightforward demonstration of this doctrine.

We do not know what else Thales may have said about magnets or magnetism, but we do know that he was the first to explain matter in terms of an *original substance* and was, in this sense, the first to explain natural phenomena in scientific terms. He is said to have claimed that "everything is made of water," thereby positing water as a substance whose existence preceded everything else and that could be relied on to remain the same, whatever the vicissitudes of nature. Yet if the original substance is unchanging, why then do objects exist in different states, and how can things undergo change? The first attempt at resolving these questions came—to the best of our knowledge—from a fellow native of Miletus, Anaximenes (c. mid-sixth century B.C.). He held air to be the original substance and believed that changes in its density were responsible for changes seen in all other substances:

> As it [i.e., air] is condensed and rarefied it appears different: when it is diffused into a more rarefied condition it becomes fire; wind, again, is air moderately condensed; cloud is produced from air by compression; when it is yet more condensed it is water, and then earth; and when it is as dense as possible it is stones.[4]

This idea, presumably, came from the everyday experience of seeing water turn to ice when cooled and to vapor when heated. The positing by Heraclitus (c. 540–480 B.C.) of fire as the original substance can be seen as an extension of the same thinking. In any case, Thales's water, Anaximenes's air, and Heraclitus's fire were all lifelike existences with souls. We can surmise that these three substances were all singled out because of their obvious importance in the sustenance of human life. During this period in history, the universe as a whole was thought to be alive. And magnets were seen as "living proof," so to speak, of the life to be found in all natural objects, including those that appeared to be inanimate.

The Milesian philosophers thus accepted the world as it was—that is, as they were able to apprehend it via the senses. But in the first half of the fifth century B.C., across the water in Elea, southern Italy, Parmenides (c. 515–445 B.C.) believed otherwise. He held that reason (*logos*) alone was to be trusted and that the senses deceived. With this deliberate protest against the status quo, Parmenides elevated pure thought, for the first time, to a position of supremacy over sensory perception, counterposing rationalism to empiricism. He argued that it was irrational to discuss the existence of "things that are not" and that therefore there could be no such thing as change or movement: to say that something has changed or moved, we must assume the existence of that which no longer exists (at least not in its former state or at its former position). For this reason, he believed, not only qualitative change but also creation and disappearance as such were mere illusions. The philosophers who came after Parmenides faced the urgent task of determining how to respond to this radical notion of the negation of change.

Those who wished to unseat Parmenides in the latter half of the century included Empedocles (c. 495–c. 435 B.C.), a native of Sicily; Leucippus of Miletus (c. 480–? B.C.); and Democritus of Thracia (460–c. 370 B.C.). While Empedocles originated the so-called four-element theory of matter and Leucippus and Democritus advanced a competing theory called atomism, in fact both camps were attempting to refute Parmenides by advancing a basic strategy, if you will, for understanding nature as something that was ever-changing yet still orderly. Theirs were two separate attempts to establish harmony between, on the one hand, Thales's idea of a changeless original substance as a first principle in explaining natural phenomenon and, on the other, the diversity and ceaseless mutability of objects seen in everyday life. Empedocles parted company with the idea of a single original substance and instead believed that four elements—earth, water, air, and fire—were the roots of all substances. In contrast, the atoms, or particles, spoken of by Leucippus and Democritus were of varied shapes and sizes, but all were formed from a single substance. In this respect, atomism differed from the four-element theory in its interpretation of Thales's basic tenet, yet both theories were alike in being reductionist: both held that all the complexity of the natural world, with all the bewildering change and diversity known to man through the evidence of the senses, could and should be explained by just one substance or, at any rate, a very small number of substances.

Where, then, does the magnet fit into all of this? The first attempts to provide a rational explanation for magnetism came, in fact, from both Empedocles and Democritus, and also from Diogenes of Apollonia (c. 450 B.C.), who believed that air was primordial.

In Empedocles's context, the words "earth," "water," and "air" had much broader meanings than they do in modern times, referring, respectively, to solids, liquids, and gases in general. "Fire," too, denoted not what we would normally think of but rather something akin to what we today would call "energy." Moreover, for Empedocles these elements did not represent different states, or modes, with the possibility of solids or gases turning to liquid and vice versa. Instead, these were elements in the literal sense of the term: they could be neither created nor destroyed, all substances could be reduced to them, and they themselves could be reduced no further. The four elements did not undergo change, but were combined by "love" and separated by "strife" and thus were mixed in various proportions to form the different substances found in nature. While Empedocles's terminology strikes the modern ear as a somewhat bizarre example of anthropomorphism, it is more important to realize that his introduction of the concept of "proportion" made an enormous contribution to the later development of the theory of matter.

But in what manner did Empedocles explain magnets? An answer to this question comes to us through Alexander of Aphrodisias (fl. c. 200), who is best known as a commentator on the works of Aristotle. In *Quaestiones*, Alexander attributes the following explanation of magnetism to Empedocles:

> The iron is carried towards the magnet by the *effluences* from both and the pores of the magnet which are commensurate with the [effluences] from the iron.... For the effluences from [the magnet] push away the air on the pores of the iron and move [the air] which rests on them like a lid; and when this is removed the iron follows the effluence which flows all together. And the effluences from [the iron] travel to the pores of the magnet, and because they are commensurate with them and fit into them, the iron too follows the effluences and is carried [along].[5]

To the best of our knowledge, this was the first theory of magnetism in history to be based on micromechanics, and actually it went much further than just magnetism. According to Theophrastus (c. 372–c. 288 B.C.), a disciple of Aristotle's, Empedocles described all sensory perception in the same way, theorizing that perception occurred when a substance released some kind of effluence that matched the pores of a particular sensory organ.[6] In other words, every function or action, whether physical or physiological, could be explained equally in terms of mechanics. We must remember, though, that to some extent Empedocles was a product of his times: in the ancient Mediterranean, there was no clear boundary between the physical and the physiological, between the inanimate and the animate. Empedocles thus believed that all objects had tiny, invisible pores and that all actions with respect to any object—including the process of a living thing perceiving the object via its sensory organs—involved stimulation by invisible fluids or solid particles entering and leaving these pores. Until fairly recent times, this theory served as the definitive model used by mechanists and atomists to explain a wide range of actions.

In his commentary, just after the passage quoted above, Alexander points out that, even if we accept the existence of hypothetical effluences, Empedocles's theory of how magnets work does not explain why it is only iron that flows toward the magnet—that is, "why … the magnet does not follow its own proper effluences and is [not] moved towards the iron." Implicit in his critique was the premise that magnets themselves were always stationary.

This premise, in fact, was held during the time of Empedocles as well, almost 650 years earlier. Diogenes of Apollonia was, roughly speaking, a contemporary of Empedocles's. According to biographer Diogenes Laertius, Diogenes of Apollonia (like the earlier Anaximenes) subscribed to monism:

> Air [claimed Diogenes of Apollonia] is the universal element … Air by condensation and rarefaction generates the worlds.[7]

In this respect Diogenes of Apollonia differed from Empedocles, but the two had similar notions about magnetism; the former also hypothesized the existence of something similar to effluences in an attempt to explain why magnets and iron did not exert reciprocal force upon one another. His theory, as described by Alexander, was that

> all the things that are malleable both naturally emit a certain moisture from themselves and attract it from outside, some more and some less, but that bronze and iron emit most, a sign of which is both that something is burnt off and consumed from them in fire, and also that

> when they are smeared with vinegar and olive-oil they rust; they are affected in this way because the vinegar attracts the moisture out of them.... The iron both attracts and emits more moisture; the magnet, being rarer and more earthy than the iron, attracts more moisture from the adjacent air than it emits. Well, it admits into itself that [moisture] which it attracts that is akin to it, but rejects that which is not. The iron is akin to it, and for this reason it attracts [the moisture] from [the iron] and admits it into itself, and through its attracting of this it draws the iron too [towards itself], on account of the continual attraction of the moisture in it. The iron does not also attract the magnet, because the iron is not so rare as to be able to admit the moisture from it all together.[8]

Essentially, the two men's theories are very similar, except for Diogenes's explicit statement that the attractive force between the iron and the magnet is not reciprocal and his use of the word "moisture" instead of "effluence." Whatever his success, or lack thereof, in addressing the nonreciprocity issue, we can be sure that Diogenes's theory was among those that came to serve as a prototype to explain magnetic phenomena in mechanistic terms.

Others, such as Democritus, had theories of their own that were different and yet still mechanistic. First, a little background on Democritus: As mentioned earlier, he, along with Leucippus, founded atomism around 400 B.C. Democritus accepted the existence of the void, a concept that had until then been rejected, and he assumed the indivisible atom to be the building block of all existence and to have the properties of impenetrability and extension. He believed that the world was comprised of the void (empty space) and solid atoms, which moved around in the void.

He further described atoms as being particles made of a single, homogeneous substance and varying only in size and shape. He held the diversity of what we perceive with the senses to be due to differences in the shape, position, and arrangement of the component atoms. He believed that "each of the shapes when arranged in a different compound produces a different condition," and further postulated that atoms in sweet things were round and of moderate size, while those in sour things were large, rough, and angular.[9] Aristotle commented that Democritus "reduces Savours to the atomic figures,"[10] and it was precisely this reductionism—which insists that all characteristics perceived by the senses be explained by the geometrical shape, position, or combination of atoms themselves devoid of such characteristics—that later became the foundation of the atomistic and mechanistic theories that persisted to modern times.

Returning to magnetism, it is said that Democritus produced a work entitled *Concerning the Magnet*[11]. This, unfortunately, has been lost to history, so all we can do is to catch a glimpse of its contents through references made by those who came after him.

According to the commentator Simplicius (c. 490–560), Democritus stated that "by nature like is moved by like and things of the same kind are carried towards one another."[12] Originally, the general belief that like attracts like was most probably a product of hylozoism, with scholars projecting onto the world of inanimate objects their experience of seeing animals of the same kind flock together—as in the observation in Homer's *The Odyssey* that "heaven brings birds of the same feather to one another."[13] But Democritus thought that this phenomenon should be explained in mechanistic terms and not merely attributed to divine will. It is precisely here that we encounter the divide between mythology and science. Sextus Empiricus, who lived in the second century A.D., attributed the following explanation to Democritus:

> For animals ... congregate with animals of the same kind—doves with doves, cranes with cranes, and so with the other irrational animals. Similarly in the case of inanimate things, as we can see from seeds which are being sieved and from pebbles where the waves break. For in the one case the whirling of the sieve separately arranges lentils with lentils, barley with barley, wheat with wheat; and in the other case, by the motion of the waves, oval pebbles are pushed into the same place as oval pebbles, and round pebbles as round pebbles, as though the likeness in things contained some sort of force for collecting things together.[14]

Thus, in Democritus's view, the observation that like attracts like should be understood as a result of the mechanical action of a sieve or waves—inorganic processes dictated by shape and motion. Likewise, his teacher Leucippus explains that similar atoms congregate in the vortex created by many atoms.[15]

The general thesis, however, was later expressed by Plato (c. 428–c. 348 B.C.) in *Timaeus* as "all kindred substances are drawn towards one another"[16] and came to be interpreted as the "sympathy between like things," acts of nature that precluded further explanation, belonging to the realm of anthropomorphism or belief in magic. Such interpretations were to endure throughout Europe for some two thousand years, including the period during the Middle Ages when atomism was in decline.

In A.D. 160, for example, the eminent physician Galen (c. 131–c. 201) attributed magnetism to "affinity of quality."[17] Much later, near the end of the twelfth century, English monk Alexander Neckam (1157–1217) wrote in his book *On the Nature of Things* that

> magnets attract each other with their similar parts and repel each other with their non-similar parts.[18]

Thus he attempted to explain the mysterious dual behavior of magnets, which show both attractive and repulsive forces depending on their relative positions; he remained trapped, however, in the dogma that like attracts like and therefore failed to grasp the fact that magnets have opposite (north and south) poles that attract one another. Similarly, in the mid-thirteenth century Roger Bacon (c. 1214–c. 1294) thought that magnets attract iron "because of the similar nature" of the two.[19] Then, moving into early modern times, in 1537 the German physician Paracelsus (1493–1541) observed that "like substances unite with like substances and do not unite with unlike substances,"[20] and the Italian engineer Vannoccio Biringuccio (1480–1539) stated in *The Pirotechnia* (published in 1540) that

> for the same reasons that Nature wishes that like always hunger for like, it appears that [lodestone] hungers for iron.[21]

*The Pirotechnia* came out just three years before the publication of *On the Revolutions* by Nicolaus Copernicus (1473–1543), from which time scientific debate widened to include gravity as well as magnetism. Johannes Kepler (1571–1630), in his *New Astronomy*, published in 1609, completed Copernicus's heliocentric theory with the concept of elliptical orbits and spoke for the first time of a gravitational force between heavenly bodies. Even at that advanced period in history, Kepler held that

> gravity is a mutual corporeal disposition among kindred bodies to unite or join together [and that] the magnetic faculty is another example of this sort.[22]

But I have gotten ahead of myself. Let us return to Democritus, as we are now ready to consider his beliefs about magnets. Noting that "Democritus, too, postulates that like things are moved towards like," Alexander presents Democritus's theory of magnetism as follows:

> [Democritus] assumes that the magnet and the iron are composed of similar atoms, but that the magnet is composed of finer ones; and he thinks that it is rarer and contains more void, and that for this reason its atoms, being more easily moved, are carried towards the iron more swiftly (for [things are] carried towards what is similar), and, penetrating the pores of the iron, move the bodies within it, spreading out through them because of their fineness. The [bodies] when they have been moved are carried outside, [forming] an effluence, and [are carried] towards the stone, both because of [their] similarity [to it] and because it has more void [spaces]; the iron follows these because they are all at once both separated out [from it] and carried along, [and so] it too is itself carried towards the magnet. The magnet is not also carried towards the iron, because the iron does not have as many void [spaces in itself] as the magnet.[23]

Here, again, we find an explanation in which only the iron is moved toward the magnet and not vice versa.

Leaving aside the differences between the four-element theory and atomism, we find that the theories of Empedocles, Diogenes, and Democritus had much in common: they all explained the mechanism of magnetism in very similar terms. Moreover, these three men were the first among scholars to attempt any rational explanation at all—and it is in itself significant that they made these attempts, however crude the arguments and results may appear to us now. Their position was that magnetism should be attributed neither to a soul nor to the operation of divine will or magic power—views that preclude any further explanation—but rather should be understood and explained on the basis of general principles concerning nature. The clear assertion of such a position for the very first time was nothing short of epoch-making. Their contributions, taken together, represented the culmination of achievement in the initial development of natural philosophy—a course that had begun with the Milesian philosophers and the concept of original substance. With the appearance of Socrates, however, the focus of Greek philosophy shifted from nature to morality, and the study of natural philosophy began to decline.

## Section 2—Plato and *Timaeus*

When we consider the great thinkers whose influence in Europe carried over into the Middle Ages, the Renaissance, and recent times, the name Plato comes quickly to mind. Many of Plato's writings are still with us today, but within the huge corpus of Plato's works, how many times does the subject of magnets arise? Only twice. Obviously, magnetism was not a topic in which he was immensely interested.

In one of his earliest dialogues, *Ion,* Plato crafts a conversation between Socrates and the poet Ion, who is skilled at reciting Homer.

Ion wonders aloud why he is able to perform skillfully when reciting Homer and yet finds he performs less well when reciting any of the other poets. Socrates responds that this is a result of "inspiration" and that "all good poets compose their beautiful poems not by art, but because they are inspired and possessed." He then elaborates:

> There is a divinity moving you, like that contained in the stone which Euripides calls a magnet, but which is commonly known as the stone of Heraclea. This stone not only attracts iron rings, but also imparts to them a similar power of attracting other rings; and sometimes you may see a number of pieces of iron and rings suspended from one another so as to form quite a long chain; and all of them derive their power of suspension from the original stone. In like manner the Muse first of all inspires men herself; and from these inspired persons a chain of other persons is suspended, who take the inspiration.[24]

From the above, we can see that Plato and others of his time were already aware that magnets had the power to magnetize pieces of iron, rather than merely attract them. It appears that people in his day were struck by the mysterious phenomenon of small iron rings hung one from the next so that, without actually being linked together, they formed a sort of chain suspended from a single magnet. This phenomenon was reportedly first witnessed on the mountainous, iron-ore-rich island of Samothrace, whence derive copious references to Samothracian rings, or Samothracian iron, in writings from antiquity to the Middle Ages.*

---

*Related to this are also magnesia stones, whose etymology I will discuss later. The term "stone of Heraclea" was also used in ancient times to refer to natural magnets (magnetized iron ore). Historians are divided on whether Heraclea in this case is a reference to one of the ancient cities called Heraclea or to the Herakles (Hercules) of Greek mythology, who possessed superhuman strength.

But, of course, Plato's purpose here is not to expound on magnetic induction (magnetization). He merely makes reference to the wondrous and mysterious magnet in explaining the inspiration of the Muse—the supernatural power with which the works of the great poets seemed to be endowed. In other words, he believed that the trancelike state of poets possessed by inspiration was comparable to the state of iron magnetized by magnetic stones. Clearly, it did not appear to Plato at this point that magnetism required any explanation as a physical phenomenon.

Plato's second and only other reference to magnets is found in *Timaeus,* a much later work in which the influence of Pythagoras is clear. *Timaeus* was translated into Latin in the fourth century and is among the very few of Plato's writings to have been available to the people of medieval Latin Europe. As such, it had a lasting influence on philosophical and theological thought in Western Europe and deserves to be considered in some detail.

In *Timaeus,* Plato envisions a creator of the universe, the Demiurge, who "desired that all things should be as like himself as they could be." According to this quasi-mythological creation story, "the Deity [intended] to make this world like the fairest and most perfect of intelligible beings" (Tim. 30D). To do this, the deity created the four basic elements (fire, air, water, and earth) in conformity with geometric principles, as these could be grasped by reason. In order to render these building blocks of physical substances "as far as possible the fairest and best," the creator needed to use the most simple and basic of all geometric shapes. Arguing thus, Plato assigns a regular polyhedron to the role of forming each element.

His system is as follows: A plane surface is determined by the establishment of three points. Every object occupies space, and this space is enclosed within a number of plane surfaces. Therefore, the most fundamental unit of any object is the triangle. The two basic triangles are (a) the right triangle formed by dividing an equilateral triangle in half and (b) the right isosceles triangle formed by dividing a square in half. Meanwhile, there are only five regular polyhedrons—the tetrahedron, the hexahedron (cube), the octahedron, the dodecahedron, and the icosahedron. Among these, the dodecahedron, each of whose twelve faces is a regular pentagon, is precluded because a regular pentagon cannot be formed from either of the two basic types of triangle. The right isosceles triangle can be used to form a square, which in turn can be used to form a cube. The cube, according to Plato, is the basis from which the element earth is formed. Two right triangles of the other type, meanwhile, can be used to form an equilateral triangle, from which can be obtained the tetrahedron, the octahedron, and the icosahedron. To these three shapes he assigns the roles of forming, respectively, fire, air, and water. Thus, particles of fire, air, and water all have the same triangular faces and as such can easily insert themselves into one another or transmute themselves into one another. In contrast, because particles of earth alone are formed from cubes, they cannot easily undergo any kind of metamorphosis. For this reason, Plato considers earth to be the most inert of all substances.

To the modern mind, of course, all of this reads like pure fantasy. Actually, throughout *Timaeus* Plato himself often remarks that his arguments are merely probable and denies that they are based on established fact, albeit for reasons very different from those that would be put forward today. He believes that only Ideas, that is, everlasting and unchanging Forms, can be genuinely understood and that therefore only they are subject to academic inquiry. In other words, it is the world of Ideas that people are capable of apprehending through reason, and only in that world can there be the possibility of certain knowledge. Alongside the world of Ideas exists the visible world of objects perceived by the senses. These seen objects, abundantly subject to change, are mere shadows, or imitations, of Ideas, and therefore rigorous theorizing about them is impossible. At most, one can speak only of probability, of best-possible supposition.

Given how fundamental the theory of Ideas is to the totality of Plato's works, one must conclude that *Timaeus*—which lays out detailed theories and arguments on subjects ranging from physics to politics—represents a departure for Plato from his most deeply held beliefs.

Plato's theory of regular polyhedrons, for all its seeming detachment from reality, is essentially not very different from modern theory of elementary particles. Plato was the first to decisively embrace the idea that the ultimate structure of physical substances, although not directly perceivable by the senses, had to be something simple and quantifiable, and that it should therefore be possible to develop a rigorous mathematical understanding of the constituents of matter. In modern physics, a so-called special unitary group, "SU(3)," is used to describe certain features of quarks, and physicists talk about "symmetries with respect to SU(3) transformation" and classify subatomic particles according to the "irreducible representations of SU(3) groups." It goes without saying that Plato's theory does not compare with its modern counterpart in terms of either experimental evidence or mathematical elaboration. Nonetheless, this ancient thinker does deserve some credit for having had an inkling of the theoretical physics that was to emerge more than two thousand years after his time.

## Section 3—Magnetism according to Plato and Plutarch

Plato held that the world thus formed is made up of "being," "space," and "generation." Space, which he defined as "a place that contains something," appears to be somewhat different from the void so emphatically posited by Democritus and the other atomists, and Plato himself made a subtle distinction between his thinking and the theories of those earlier philosophers.

Concerning whether or not the four basic elements could be transformed into one another, Plato's approach also differed from that of Empedocles, father of the four-element theory. But Plato actually borrowed much from both Empedocles and Democritus. Moreover, by theorizing, in *Timaeus*, that the characteristics of the elements were determined solely by geometric shapes, in a fundamental way Plato managed to reconcile the four-element theory with the basic tenets of the atomists. For example, in seeking a basis for the proposition that "fire is hot," Plato took note of the characteristics of the regular tetrahedron, which he held to be the elemental building block of fire, and attempted to find heat in "the fineness of the sides, and the sharpness of the angles, and the smallness of the particles, and the swiftness of the motion" of tetrahedrons. In other words, he said that "all this makes the action of fire violent and sharp, so that it cuts whatever it meets," and that fire's elemental figure (the tetrahedron) "naturally produces that affection which we call heat" (Tim. 61E). This argument, which reduces the sensible qualities of things to the shapes and movements of their constituent particles, is based on logic that can only be called atomistic.

Accordingly, the arguments concerning magnetism in *Timaeus* are thoroughly mechanistic. In that dialog, the subject of magnetism comes up when Plato is theorizing about the nature of breathing.

By likening breathing to the action of a water pump, Empedocles is considered the first thinker to employ fluid mechanics in explaining breathing,[25] but in *Timaeus* Plato, too, understands respiration to be an action involving merely the dynamics of fluid. Denial of the void and adoption of the premise that matter is impenetrable are fundamental to his argument, which is as follows:

> Seeing that there is no such thing as a vacuum into which any of those things which are moved can enter, and the breath is carried from us into the external air, the next point is, as will be clear to every one, that it does not go into a vacant space, but pushes its neighbour out of its place, and that which is thrust out in turn drives out its neighbour; and in this way everything of necessity at last comes round to that place from whence the breath came forth, and enters in there, and following the breath, fills up the vacant space; and this goes on like the rotation of a wheel, because there can be no such thing as a vacuum. (Tim. 79B)

In other words, when something is pushed, that object will in turn push something else. Because voids cannot exist, however, the space previously occupied by the object that has just been dislodged must be quickly filled by something else, and thus occurs a rapid cyclic movement of the medium. This, according to Plato, is the only possible explanation for what is perceived to be an object being drawn from one place to another. In other words, so-called magnetism is actually the result of objects coming into direct contact with one another:

> As to ... the marvels that are observed about the attraction of amber and the Heraclean stones,—in none of these cases is there any attraction. (Tim. 80C)

It is interesting to note, though, that this reference to the amber phenomenon† (subsequently discovered to be the result of attraction via static electricity) is, as far as we know, the first in all of Western literature.[26]

But remember, Plato brings up amber and Heraclean stones (magnets) only in passing as he attempts to explain the mechanism of breathing. And in any case, his argument is extremely obscure. It might be possible to place magnetism and static electricity in the same category of phenomena, but lumping these together with breathing as well goes a little beyond the pale. Perhaps for this reason, this particular section of *Timaeus*—despite fairly wide knowledge of the work as a whole in medieval Europe—was largely ignored by fellow scholars all the way down until mathematician and physician Girolamo Cardano (1501–1576) took notice of it in the sixteenth century.

The one exception, however, was philosopher and biographer Plutarch of Chaeronea (c. 46–c. 125). In his *Moralia*, he cites Plato's argument regarding magnetism in great detail and then goes on to amplify it. Using *Timaeus* as a springboard, Plutarch develops an original theory that deserves attention in its own right. So allow me to jump ahead some 450 years from the time of Plato to describe in detail that theory, which is sometimes called "cyclical propulsion."

Plutarch flatly rejects the existence of an attractive force, asserting that "amber does not attract any of the objects placed near it as the loadstone does not either, nor does any of the things in their

† In science history this is often referred to as the "amber effect," but in fact the attraction that takes place is a phenomenon accompanying friction and is not produced directly by the amber itself; I therefore prefer the term "amber phenomenon" as somewhat less misleading.

neighbourhood spring to them of itself." Instead, he explains the force of amber and magnets as follows:

> The loadstone emits certain effluvia which are heavy and like wind, and the continuous air, forced back by these, pushes the air that is before itself, and that air, moving around in a circle and settling again upon the vacated space, forces the iron back and drags it along with itself....
>
> Amber contains a substance like flame or wind which it ejects when its pores have been opened by friction of its surface; and this substance, when it has escaped, has the same action as that from the loadstone has but because of its tenuousness and weakness drags along the lightest and driest of the things in the neighbourhood, for it is not strong and does not have weight or impetus capable of expelling an amount of air with which to master the larger objects as the loadstone does.[27]

One of the chief shortcomings of Plato's analogy, indeed the one that made his analogy all but incomprehensible, was his failure to posit any substance discharged from either Heraclean stones or amber as a counterpart to the "breath" that was discharged from living bodies during the act of respiration. This is where Plutarch steps in, supplying the missing piece out of the workings of his own mind. He claims that the magnet and amber each discharges a particular gaseous substance—called, respectively, in modern terminology, a "ponderable fluid" and an "imponderable fluid"—and that these substances press against the air, instigating a chain of motion (cyclical propulsion) that ultimately causes iron and other objects to be pressed up against either the magnet or the amber, as the case might be. Presumably, he does a little borrowing from Empedocles and Diogenes of Apollonia, who both proposed the idea of an effluent (a substance discharged from the magnet). With Plutarch, however, for the first time in history we see the power of magnets and the power of amber recognized as two separate phenomena requiring separate explanations. Moreover, Plutarch plants the seeds of what would eventually, much later, come to be called "magnetic effluvium" and "electric effluvium." Presumably, his introduction of two types of gaseous substances—ponderable magnetic fluid and imponderable electric fluid—relates to the fact that, whereas magnets are strong enough to attract and hold something as heavy as iron, amber is comparatively weak and can attract only much lighter objects.

Plutarch is also aware of another difference between amber and magnets, namely, that the latter can attract only iron. Directly after the passage quoted above he speculates as follows:

> How is it then that the air pushes and presses against the loadstone neither stone nor wood but only iron? This, to be sure, is a difficulty that confronts equally those who think that the cohesion of the bodies comes about by the loadstone's attraction and those who think that it comes about by conveyance of the iron, but Plato might provide a solution in the following way. Iron is neither exceedingly loose in texture like wood nor exceedingly close like gold or stone but has pores and passages and corrugations which by reason of their irregularities conform to the air; and the result is for the air, however in its motion to the loadstone it may fall upon the iron, not to slip off but, intercepted by certain lodgements and counterpressures with meshes that conform to it, to force the iron back and push it on before itself.

Moreover, it is noteworthy that Plutarch looks to friction itself, and not to the heat generated through friction, as a factor in causing the amber phenomenon. This is in contrast to later scholars, whose thinking was precisely the reverse. A resolution of this confusion, with confirmation that friction itself was responsible for the tendency for certain objects to adhere to amber, did not come until 1600, with the publication of *On the Magnet* by William Gilbert (1544–1603).

In any event, the explanations for magnetism and electricity found in Plato's *Timaeus* and Plutarch's *Moralia,* which were mechanistic and based on the principle of proximity, were largely ignored until the sixteenth century. In medieval times, *Timaeus* was seen as exclusively a work of interest to Christians. The chief influence of *Timaeus* on people's understanding of magnetism was as a conduit for the transmission of an even older idea: Democritus's notion of like attracting like was mistakenly said to have originated with Plato and was passed on as such from antiquity all the way through the Middle Ages.

## Section 4—Aristotle's *Physics*

The other thinker who, along with Plato, represents Greek philosophy and science is the great master Aristotle, whose impact on European thought has been no less than that of Plato's. Born in Stagirus, a Greek colony and seaport on the coast of Thrace, Aristotle studied at Plato's Academy for twenty years before establishing his own school. Not only did he found the Lyceum, he also originated a vast and complete philosophical system, encompassing everything from logic to natural philosophy, which was to have an especially profound influence on Europe in the late medieval period, from the thirteenth century on.

And yet, if Plato seems indifferent to the phenomenon of magnetism, then Aristotle appears even more so. When we stop to contemplate the almost mind-boggling level of scholarship pursued by Aristotle, who in his system of philosophy attempted to explain and create a place for virtually every object and phenomenon known to exist in the natural world, it is difficult to avoid conjecturing that his seeming indifference to magnetism was actually deliberate.

Aside from attributing to Thales the idea that the magnet has a soul, Aristotle mentions magnets in only one other place, and only in passing. In *Physics*, he argues for the existence of a divine being called a "first mover" and invokes the magnet merely to illustrate what he means by an entity that moves other entities but is not itself acted upon.[28] In order to get an idea, then, of what Aristotle thought of magnetism, we need to know more about how he defined the concept of a first mover. For this purpose, it is worth taking a glance at some of what he had to say about nature in general.

In contrast to Plato's assertion that the Ideas were true reality, Aristotle believed that the world of objects perceived by the senses was fundamental and was within the realm of the knowable. For this reason, although Aristotle carried on the four-element theory as part of his intellectual inheritance, he placed in it a completely different framework from that established by Plato. Whereas the elder philosopher had explained the qualities of elements via their shapes, his one-time disciple believed, in contrast, that qualities were a basic given and that the elements were their manifestation in material form. He believed that anything capable of being known to the senses was capable of being touched, and he argued that the qualities of objects came in pairs of opposites: hard vs. soft, rough vs. smooth, viscous vs. brittle, and so on. Moreover, he claimed that all of these pairs of characteristics could be reduced to two broader binary categories: hot vs. cold and dry vs. wet, with hot and cold regarded as active, and dry and wet regarded as passive qualities.[29] From these two came four subcategories: hot and dry, hot and wet, cold and wet, and cold and dry. Paired with each of these subcategories was a substratum, or carrier, in the form of one of the traditional four elements: fire, air, water, or earth.[30] In this respect, Aristotle's four elements, though adopted from Empedocles, were quite a large step closer to the real world as perceived by the senses.

Aristotle's four-element theory is decisively different from earlier versions in the following two ways. First, his theory allows for the possibility of the fundamental elements themselves going through qualitative change. In other words, an element itself can be transformed if the basic characteristic of that element changes into its opposite. For example, the melting of ice is explained by saying that the ice changes from cold and dry to cold and wet, while the vaporization of water is interpreted as a change from cold and wet to hot and wet. His formulation in general terms is that everything "comes to be or passes away from, or passes into, its contrary or an intermediate state. But the intermediates are derived from the contraries."[31] Thus, as earth turns to water, which turns to air, which turns to fire, one element changes into the element that is qualitatively closest to it, and in this way does natural activity occur. For this reason, the objects in our sublunary world, which are made from the four elements, must necessarily undergo a coming-to-be and a passing away.[32]

The second difference is that Aristotle also assigns to each of the four elements a place in a strict spatial hierarchy. Air and fire, being essentially light, belong in a place far away from the center of the universe, up in the hollow of the celestial sphere believed to hold the moon, while earth (i.e., soil) and water, being essentially heavy, belong down in the very center of the universe. According to Aristotle's geocentric model, the Earth is the center of not only our solar system but the entire universe. This is why, according to Aristotle, when stone or water is forcibly raised to a height and then released, it falls straight to the ground. In other words, he views such phenomenon as spontaneous movement whereby bodies seek to return to the place where they naturally belong; heavy bodies fall to the ground not because they are pulled by an attractive force of the Earth, but rather because their natures impel them to move in the particular direction of the universe's center. Aristotle does not recognize the existence of what we now understand to be the Earth's gravity. He similarly notes that, in the absence of any breeze, fire and smoke rise in a straight line, again as a spontaneous movement of seeking their natural location. Aristotle distinguishes these types of motions, which he refers to as "natural" motion, from what he calls "compulsory," "unnatural," or "violent" motion, which is contrary to nature.[33] The latter type includes, for example, the movement of a stone being tossed either horizontally or straight up in the air, or the movement of fire or smoke fluttering in the wind. This, then, sums up Aristotle's explanation of the kind of movement—in the narrow sense of bodies changing their locations—observed on the Earth. I should add here, however, that Aristotle considers both natural and compulsory motion on the Earth to be nonperpetual; all terrestrial objects, he theorizes, must eventually cease their motion.

In contrast to the transience seen on the Earth, the universe above the moon was a place of eternal sameness. In *On the Heavens*, Aristotle notes that "in the whole range of time past, so far as our inherited records reach, no change appears to have taken place either in the whole scheme of the outermost heaven or in any of its proper parts." The fixed stars seeming never to have changed their positions relative to one another since ancient times, and all of the celestial bodies seeming to revolve around the Earth eternally, Aristotle posits a fifth element as the fundamental building block of all heavenly bodies. This element, called "ether" (sometimes "aether" or "aither"), is fundamentally different from any of the four elements found on the Earth. He defines ether as "some simple body which revolves naturally and in virtue of its own nature with a circular movement." This fifth element "can possess no lightness or heaviness at all" and is "ungenerated and indestructible and exempt from increase and alteration." In this sense, he considers ether to be "perfect" and, accordingly, to be responsible for the natural and eternal movements of the heavenly bodies, which he concludes are "prior to [all the formations we know] and more divine than they."[34] In the final analysis, we can say that the world Aristotle describes is derived directly and logically from everyday experience. After all, his theory did serve, in its time, as a basis for explaining in physical terms why the Earth seemed to stand still while the stars moved across the sky.

Aristotle also holds that "everything that is moved is moved by something,"[35] but he makes a distinction between things that are moved by something else (inanimate things) and things that move by themselves (animate things). In the case of inanimate things, he identifies two types of motion—"the motion of all things that are in motion is either natural or unnatural and violent"—and asserts that, whether natural or compulsory, in inanimate things it is caused by "something other than themselves."[36] The difference between the two types of motion is that "when fire or earth is moved by something the motion is violent when it is unnatural, and natural when it brings to actuality the proper activities that they potentially possess."[37]

It is rather easy to see what Aristotle means by compulsory motion being motion caused by an external agent. He refers to the agent responsible for any so-called compulsory motion as an "efficient cause." A catapult, for example, is the efficient cause of a stone's being hurled. The action of the catapult itself, furthermore, is the result of some other efficient cause, and thus is an instance of an object's being acted upon by some outside agent (e.g., another object, or the air) and then in turn becoming an agent acting upon yet another object.

But what about Aristotle's idea that natural movement is also caused by an external agent? The logic of this assertion is harder to grasp, but can be explained with two concrete examples. First, when water is heated, steam rises. According to Aristotle's theory, through the action of the heat, the water, which is potentially light, becomes steam, which is actually light, and then travels upwards. The second example is the falling of a boulder that has been resting on a platform. When the platform (an obstacle to natural movement) is taken away, the natural tendency of the boulder to fall is unlocked and actualized. In other words, the agent creates an opportunity for natural movement to take place. Thus, not only compulsory motion but also natural motion is caused by some external agent.

When we consider the movement of any given inanimate object and attempt to trace its direct cause, and then the cause of this cause, and so on, it seems eventually we must come to the movement's ultimate cause—in other words, to a first mover. (This is unless, of course, we allow for infinite regress.) In *Physics*, Aristotle maintains that the first mover is in itself immovable and is the cause of eternal celestial movements, namely, the diurnal movement of the stellar sphere and the movements of the planets, the sun, and the moon. He concludes that this ultimate cause of motion is itself without parts and without magnitude, that it is not itself a body in any sense of the word. In *Metaphysics*, he proposes that this "unmoved mover" of the heavenly bodies is none other than God—the ultimate cause of all things, eternal, goodness itself in its highest form.[38] This first mover began by setting in motion the outermost celestial sphere, which held the fixed stars; these in turn set in motion the planets, sun, and moon, which in turn were responsible for changes in the Earth's seasons and weather and for the circulation of air in the Earth's atmosphere.

This, then, is Aristotle's picture of the universe. But where does the magnet fit in? Now here is a problem, because the magnet, despite being terrestrial, is a mover that itself has no mover, and nowhere in Aristotle's system of natural philosophy or his version of the four-element theory can we find a place for it. Moreover, Aristotle holds that "the things moved and the movents [i.e., agents causing movement] must be continuous or in contact with one another."[39] He adopts more or less *a priori* the idea that every action must be local. But the magnet's effect on iron appears to directly contradict this premise, so in this respect as well the magnet has no place in Aristotle's natural philosophy.

Up to this point, however, we have been talking only about Aristotle's views on the nature of inanimate things. What happens when we expand the discussion to include his ideas about living things?

Aristotle's definition of a living entity is "that which is in motion but is moved by itself and not by anything else."[40] In other words, living things have both a body (which is in motion) and a soul (which is the agent of that motion). Aristotle considers the soul to be the "cause or source of the living body"; makes an unambiguous distinction between living things and inanimate objects; and concludes equally unambiguously—though his argument is rather difficult to follow—that the soul "[originates] movement in the body" and "cannot be moved."[41]

From this, it seems reasonable to surmise that Aristotle would consider the magnet to have a soul and be alive. But in terms of ordinary experience, magnets are a form of mineral and therefore not things that can easily be classified as living. In chapter 9 of book 3 of *On the Soul*, Aristotle counts among the powers of the soul the power (shared by both plants and animals) to take nourishment and grow, the powers (of animals alone) to move from place to place and perceive, and the powers (of humans alone) to conceptualize and reason. Again, we find here no place for the magnet. This is what leads me to the conjecture, stated earlier, that Aristotle's neglect of magnets was no less than deliberate.

Much later, Thomas Aquinas (c. 1225–1274), who adopted Aristotle's philosophy, found a place for minerals in his elaborate hierarchy of souls. Only then, for the first time, could magnetism finally be accounted for within an Aristotelian framework, as we shall see later.

## Section 5—Theophrastus and Later Aristotelians

Aristotle was all but silent with regard not only to magnets but to minerals in general. The only monograph we have discussing minerals from the standpoint of Aristotelian physics was written by Theophrastus, Aristotle's most faithful disciple. Theophrastus was born in Eresus on the island of Lesbos in the Aegean Sea. He first studied under Plato and later moved to the Lyceum, where he became Aristotle's protégé. Upon the death of the great master, Theophrastus took over as head of the Lyceum, a post he retained for thirty-five years. In this capacity, he served devotedly as the central figure in the advancement of Aristotelian thought, and yet he was no intellectual sycophant. In particular, he was critical of his mentor's teleology and also advanced a different version of the four-element theory, in which fire was singled out as having unique characteristics.[42]

Theophrastus was, according to biographer Diogenes Laertius, "a man of remarkable intelligence and industry" who referred to himself as a "pedant."[43] This man of letters is supposed to have been a prolific writer, but very few of his works have actually been preserved. One of the most notable of these texts is his *History of Plants*, and another, more relevant to our discussion, is *On Stones*. The latter work, written in or around 321 B.C., has been preserved in its entirety, and a heavily annotated English translation (published with the original Greek text) is available.[44] In their introduction, the translators refer to the work as "an attempt to classify mineral substances on the basis of Aristotelian principles" and "so far as we know, the first attempt to study mineral substances in a systematic way."

*On Stones* begins with a listing of the various visual and tactile characteristics of stones, including color, transparency, brightness, brittleness, smoothness, and so on. To these attributes, the author adds the following:

> Stones, however, have these differences and in addition there are others that depend on their power of acting on other substances, or of being subject or not subject to such action. For some can be melted and others cannot,

> some can be burnt and others cannot, and there are other differences of this kind.... some have the power of attraction and others can test gold and silver, such as the stone called the Heraclean and the one called the Lydian. (OS 4)

Theophrastus's treatment comes across as a hodgepodge of the various characteristics of minerals—sensory properties, such as appearance and texture; chemical and physical properties, such as combustibility and solubility; the presence or absence of magnetism; and so on—with no single common thread. For our purposes, however, at least one of his observations is highly noteworthy: his recognition of attractive force as an attribute of certain minerals. In this respect, he places amber and magnets together in the same category:

> And since amber is also a stone ... the power of attraction would belong to this too. The stone that attracts iron is the most remarkable and conspicuous example. This also is rare and occurs in few places. This stone too should be listed as having a similar power. (OS 29)

What is noteworthy here is the fact that he defines the presence or absence of an attractive force as a whole mineralogical category unto itself, from which we can infer that he acknowledges the possibility of there being other minerals aside from magnets and amber that share this characteristic. In fact, he also discusses something called *lyngourion* (lynx-urine stone), saying that it

> has the power of attraction, just as amber has, and some say that it not only attracts straws and bits of wood, but also copper and iron, if the pieces are thin. (OS 28)

If we conclude that lyngourion is actually a different substance from amber, then it follows that Theophrastus was here the first to record the existence of something other than amber possessing the quality of static electricity. But there are several theories about what lyngourion really is. In the first century, Pliny (23–79) mentions "lyncurium" in his *Natural History* but claims that it is identical to amber and describes Theophrastus's account as "wholly false."[45] It is probably the case that Pliny was right and lyngourion is nothing more than amber. Even so, it is interesting and significant that *On Stones* is the earliest known work to state that amber can attract not only straw, wool, and so on but also small bits of metal—in short, almost all kinds of matter—indicating a power qualitatively different from magnetism.

"As the metals have been discussed in another place," writes Theophrastus at the beginning of *On Stones*, "let us now speak about the stones." This wording echoes Aristotle's *Meteorology*, where he explains that there are "two kinds of bodies that originate in the earth, 'fossiles' and metals"; and classifies "fossiles" as "stones that cannot be melted" and "metals" as bodies that are "either fusible or malleable."[46] Believing that metals could be melted while stones could not be, subsequent scholars further concluded that, within the scheme of the four elements, metals were derived chiefly from water, while stones were derived chiefly from earth. Magnets, then, were classified with stones, given that they did not easily melt. This designation of magnets as a type of stone rather than as a metal persisted until fairly recently, as is indicated by the comment by Robert Boyle (1627–1691) that "considering the greatness of its specific gravity, the malleableness and other properties, wherein iron differs from loadstone, I cannot easily believe, that … a metal should be turned into a stone."[47]

But above all, what we must note about *On Stones* is that it merely mentions the existence of substances possessing the power of attraction: attempts to explain this power, which appear in the writings of Greek philosophers from Empedocles to Plato, have been abandoned. Aristotelian philosophy, at least at that point in history, shortly after Aristotle's death, left the matter of elucidating the mechanism of magnetic attraction utterly untouched.

* * *

The ancient Greeks were the first to attempt to explain magnetism. Some, such as the atomists and Plato, attempted to explain the magnet's action in terms of an invisible substance providing contact as an intermediary agent, while others attributed it to the work of a living soul. Although they diverged onto these two separate roads, they were alike in being the original pathfinders in the discovery of the concept of force.

Let's jump ahead for a moment to Gilbert, who discovered at the end of the sixteenth century that the Earth itself is a magnet. Two thousand years after the time of Aristotle, Gilbert maintained that "magnetick force is animate, or imitates life," and that "Thales, not without cause, held that the loadstone was animate."[48] It is really not surprising that Gilbert thought the Earth, as a magnet, to be a living, active entity rather than an inert lump of soil, because he was essentially an Aristotelian, and as such he hearkened to the ancient Greek master's assertion that "what has soul in it differs from what has not, in that the former displays life." Soul, Aristotle had added, was "'the essential whatness' of a body" that has "the power of setting itself in movement and arresting itself."[49] Ironically, this belief of Gilbert's in the lifelike qualities of our planet was at loggerheads with another of Aristotle's beliefs: that the heavens revolved around the Earth.

I have already discussed Aristotle's conception of four Earthly elements and a fifth element, the ether, and their arrangement within a spatial hierarchy. Aristotle also posited a hierarchy of value corresponding to this spatial order, with the ether, as a divine substance, being ranked most highly and soil, the element forming the Earth, being accorded the lowest status and considered the most distant from any relation to living things. This way of thinking was actually not new: recall that the Milesian monists selected either water, air, or fire—never earth—as their candidates for the single original substance forming all of matter. Plato, too, singled out earth as the only one of the four elements that, because of its shape, could not transform itself. So it seems that from the very beginning of Greek philosophy, only earth is relegated to this, shall we say, base position, possessing no soul and no life, and leading inevitably to a definition of the Earth as something dull and inert. This line of reasoning lent a naturalistic basis to geocentric theory—a model of the universe in which the celestial bodies orbited a motionless Earth—and was ingrained deep in the psyches of generations of scholars and scientists.

But then along came Gilbert, who audaciously suggested that this very same Earth was a large magnet, might actually have a soul, and hence could move. Although historians generally have not dwelled upon this, part of the great significance of Gilbert's discovery is that it established a foundation, in terms of natural philosophy, for the shift from a geocentric to a heliocentric (sun-centered) model of the universe. Indeed, it offered for the first time a physical basis for the Earth's movement—however misguided that basis may seem to present-day sensibilities. In short, the focus of a number of controversies about Copernicus's heliocentric theory shifted so that discussion now centered not on astronomy but on physics. Then came Kepler, who under the influence of Gilbert developed a concept of gravitational force between celestial bodies. The details of how all of this unfolded form a major theme of this book, so I will be returning to this topic in later chapters.

Aristotle, the greatest of the ancient Greek philosophers, died in 322 B.C. His passing marked the end of the first of three periods, each lasting approximately three hundred years, into which the history of science in ancient Greece is generally divided.[50] Politically as well, great change was in the wind. Just a short time earlier, in 338 B.C., the allied troops of Athens and Thebes were defeated by King Philip II of Macedonia (382–336 B.C.), resulting in the absorption of Greece into the Macedonian empire. Philip II died just one year later, leaving his throne to Alexander III (356–323 B.C.). The new king, better known as Alexander the Great, had in his youth studied under Aristotle, and died just one year before the great master. Alexander the Great is famous for his conquests, which extended across Asia Minor, Egypt, and central Asia into India. His eastern expedition widened Greek hegemony in one great sweep, but after his death the empire was divided among Ptolemy I of Egypt, Seleucus I of Asia, and Antigonus I of Macedonia, at the dawning of the Hellenistic age. The Greek city-states lost much of their power and came under the thumb of Macedonia. It was around that time that the first stage of Greek science came to an end.

# Chapter 2

# The Hellenistic Age

## Section 1—Epicurus and Atomism

The Ptolemaic dynasty was the most stable of all the kingdoms of the Hellenistic world, and it is there that we find the roots of state-sponsored scientific research. Ptolemy I (r. 323–285 B.C.) and Ptolemy II (r. 285–246 B.C.) established a research complex, including facilities such as a library and a zoo, in Alexandria. The institute, called the Mouseion (or Museum), employed some one hundred researchers gathered from all over the realm. These scholars received a salary from the throne and were thus able to devote themselves entirely to their research. This arrangement led to the appearance of specialists, notable among them the geographer Eratosthenes; the astronomer Aristarchus; the mathematicians Euclid and Hipparchus; and later the great astronomer and geographer Claudius Ptolemy and the geographer Strabo. The Mouseion and its scholars gave rise to ancient Greece's greatest triumphs in the fields of mathematics, physics, astronomy, and geography. And yet in the area of magnets and magnetism, no new knowledge, nor any new viewpoints, emerged from this intellectually fertile ground.

Nonetheless, during the Hellenistic age the two opposing perspectives on magnetism that had originated earlier in Greek history were carved out in sharper relief. On the one hand were the mechanistic/atomistic theories of the reductionists—those who saw everything in terms of the elements constituting matter; on the other hand were the organic, holistic theories of the hylozoists—those who believed that everything possessed life.

The two great schools of thought of the Hellenistic age were Stoicism and Epicureanism. Epicurus of Athens (c. 342–271 B.C.) was the period's foremost advocate of atomism. Like numerous other outstanding scholars of the ancient world, he is said to have been a prolific writer, but most of his texts have been lost. Nonetheless, we have two reasons to believe that he did in fact give consideration to the subject of magnets and magnetism. First, his teachings, including an explanation of magnetism in atomistic terms, were later immortalized by Lucretius (c. 99–c. 55 B.C.) in the poem *On the Nature of Things.* And second, in the second century the physician Galen harshly criticized Epicurus's theory of magnetism by name. For now, let's examine Epicurus's version of atomism by taking a look at one of the few pieces of his oeuvre that have been preserved, his *Letter to Herodotus.*[1]

The letter begins with a denial of the existence of anything that could be considered a god or godlike. This is followed by the assertion that the universe is made up of matter and the void, with matter being composed of atoms. Epicurus argues in the letter that "if there were no space (which we call also void and place an intangible nature), bodies would have nothing in which to be and through which to move." At the same time, he says, matter cannot be born from nothing or return to a state of nonexistence; therefore, there must exist "indivisible and unchangeable elements"—in other words, atoms (LH 10.40). The atom and the void are thus seen as indispensable in establishing conditions for, respectively, the conservation and the variability of matter.

Up to this point, Epicurus has said nothing different from the atomists who came before him, but now he starts to get original. Whereas Democritus had posited just two attributes of atoms—size and shape—Epicurus adds a third, weight: "The atoms ... possess none of the qualities belonging to things which come under our observation, except shape, weight, and size, and the properties necessarily conjoined with shape" (LH 10.54). Somewhat later, he adds that atoms also exhibit "downward motion due to weight" (LH 10.61). He goes on to theorize that atoms interact with one another whenever their downward motion strays from its normal course, which is vertical.

But there is more to Epicurus's theory than just this. He postulates as follows:

> The atoms are in continual motion through all eternity. Some of them rebound to a considerable distance from each other, while others merely oscillate in one place when they chance to have got entangled or to be enclosed by a mass of other atoms shaped for entangling. (LH 10.43)

To simplify this rather complicated explanation: the atoms that "rebound to a considerable distance from each other" are gases; those that "oscillate in one place" are liquids; and those that "are enclosed by a mass of other atoms" are solids. This is in contrast to the four-element theorists, who thought that air, water, and earth were substantively different. Here, Epicurus becomes the first to assert that air, water, and earth are merely different phases of matter, that they differ from one another only in their movements and in the arrangements of their constituent atoms. And as we will see later, his use of the term "rebound" figures greatly in his theory of magnetism.

Epicurus further reasons that our sensory perceptions are a result of atoms shooting forth from the substances to which they belong and stimulating our sensory organs. Regarding vision and hearing, for example, he says that "it is by the entrance of something coming from external objects that we see their shapes" (LH 10.49), and that "hearing takes place when a current passes from the object, whether person or thing, which emits voice or sound or noise, or produces the sensation of hearing in any way whatever" (LH 10.52). In other words, all sensory perceptions owe their existence to "particles conveyed from the object which are of the proper sort for exciting the organ" (LH 10.53). Here, we are back in familiar territory, with Epicurus employing the notion of atoms emitted from substances just as Empedocles had employed the concept of "effluences" more than a century earlier.

In the letter, Epicurus also refers to another work of his, which he describes as an "epitome of the whole system"—that is, a summary or outline of his entire body of scientific writing—created for the benefit of inexperienced students. It is believed that that document, now lost, included Epicurus's theory of magnetism. But as these ideas were, in all probability, faithfully transmitted and then expanded upon by Lucretius in *On the Nature of Things*, let us leave Epicurus at this point and take a look at what his interpreter, the poet, had to say.

## Section 2—Lucretius and his Legacy

The first time we see a comprehensive explanation of the natural world from an atomistic perspective is in Lucretius's epic poem *On the Nature of Things*, written sometime around 70 B.C. His treatment of magnetism there goes beyond a simple description to include a full theoretical explanation based on general scientific principles.

What little we may know about Lucretius's life has been left to us in the form of a very short and rather odd biography written by Saint Jerome (also known by the Latin name Hieronymus) some four hundred years after Lucretius's time: "The poet Titus Lucretius was born. He was driven mad by a love-potion and, having composed in the intervals of his insanity several books which Cicero afterwards corrected, committed suicide in his forty-fourth year."[2] Lucretius was a Roman in the waning days of the republic, and therefore he wrote in Latin. Nonetheless, intellectually he was a successor to the Greek natural philosophers and is therefore best thought of as belonging to ancient Greek civilization during its period of decline.

*On the Nature of Things* would be an important historical document even if it were merely a record of ancient Greek atomism in general and the atomism of Epicurus in particular. But it is more than just that. First, it is also a poem of great literary merit that is studied even today for its significance as a representative piece of ancient Greco-Roman culture. Second, although its explanation of atomism is nothing but an adaptation of Epicurus's thought, the underlying purpose of the poem was actually to free people from their blind allegiance to religion and their fear of death. The poem deserves attention for all of these things, but for our purposes I will restrict most of my discussion to what it has to say about atomism and magnetic theory.

*On the Nature of Things* is huge, comprising six books. In the first book Lucretius declares, "I shall disclose the first-beginnings of things, from which nature makes all things and increases and nourishes them, and into which the same nature again reduces them when dissolved."[3] He immediately goes on to explain that these "first-beginnings" (*primorda*, or atoms) are things that "we are accustomed to call matter, and bodies that generate things, and seeds of things, and to entitle the same first bodies [*corpora prima*], because from them as first elements all things are" (ONT 1.59–60).

He contrasts the study of nature with the ignorance bred by religion: "Assuredly a dread holds all mortals thus in bond, because they behold many things happening in heaven and earth whose causes they can by no means see, and they think them to be done by divine power" (ONT 1.151–54). Reading between the lines, we see here images of the Roman Republic in its last days, with societal upheaval at every turn as a result of rampant superstition spread by the various religious cults that were adopted by the masses. Hence the need to shatter religious fear, a goal that Lucretius presumably saw as the supreme purpose of the study of nature: "This terror of mind … and this gloom must be dispelled … by the aspect and law of nature" (ONT 1.146–48).

As a "first principle," Lucretius holds up the notion that atoms (first bodies, or *corpora prima*) are neither created nor destroyed, declaring that "no thing is ever … produced from nothing" (ONT 1.150) and that "no single thing returns to nothing, but all by disruption return to the elements of matter" (ONT 1.248–49).

In explaining the basis for these assertions, he refers to the seed cycle. With the changing of the seasons, seeds sprout into young plants, which grow to be large trees, which then bear fruit.

Against this backdrop of constant change in the natural world is the fact that a given seed will always produce the same species of tree, which will always produce the same kind of fruit; this fruit, in turn, will always yield the same kind of seed from which the process started. And so within change there is also sameness. To the modern ear, it sounds a bit odd to hear biological and hereditary phenomena brought into a discussion of atomism, but remember that in Lucretius's time philosophers still made no strict distinction between animate and inanimate objects; this was as true for atomists and mechanists as it was for hylozoists. Accordingly, Lucretius expresses his conclusion by declaring that "there are such things as consist of body solid and everlasting, which we teach to be seeds of things and their first-beginnings, out of which now all the sum of things has been built up" (ONT 1.501–2).

He also holds up several readily observable phenomena as evidence for the existence of the void: the ability of objects to move, the fact that some objects can be permeated by liquids or gases, and the fact that two objects of the same size can have different weights. He states that the universe consists only of matter (*corpora*) and empty space (*inane*):

> The nature of the universe, therefore, as it is in itself, is made up of two things; for there are bodies, and there is void, in which these bodies are and through which they move this way and that.... Besides, there is nothing which you can call wholly distinct from body and separated from void, to be discovered as a kind of third nature. (ONT 1.419–20, 430–31)

So far, what we have here is atomism, plain and simple. But then Lucretius goes on to explain how the characteristics of various substances are determined, by saying that a small number of qualities—"concurrence," "motion," "order," "position," and "shape"

(ONT 1.685, 1.1021–22)—mix and match to form different characteristics, much like the letters of the alphabet are arranged in various combinations to form words with different meanings and pronunciations: "It is often of great importance with what and in what position these same first-beginnings are held together, and what motions they impart and receive mutually" (ONT 1.817–19; see also 1.908–10).

At this point we find that Epicurus and Lucretius have explicitly added "motion" to the three attributes of atoms posited by Democritus (shape, position, and arrangement) as the determinants of what we perceive with the senses. The importance of this enhancement will not be recognized until the seventeenth century.

Finally, Lucretius drives home the point that the universe as we know it is formed by countless atoms each moving this way and that in accordance with natural law. They are subject neither to divine will nor to any purpose outside of themselves:

> Certainly neither did the first-beginnings place themselves by design each in its own order with keen intelligence, nor assuredly did they make agreement what motions each should produce; but because, being many and shifted in many ways, they are harried and set in motion with blows throughout the universe from infinity, thus by trying every kind of motion and combination, at length they fall into such arrangements as this sum of things consists of. (ONT 1.1021–28)

And so ends book 1 of *On the Nature of Things*. The view of the cosmos expressed there stands in stark contrast to the view that heaven and Earth were purposefully created by a supreme being, and it effectively banishes teleology from the study of nature. For this reason, when Robert Boyle and Pierre Gassendi revived atomism in the seventeenth century, they were compelled to build into it an intellectual compromise with the Christian church: the movements of atoms, it was claimed, were conferred upon them by God at the creation of the universe.[4]

In book 2, Lucretius goes on to discuss the movement and shapes of atoms. To begin with, he asserts that "no rest is granted to the first bodies ... [which are] rather driven by incessant and varied motions" (ONT 2.95–97). Even when an object is motionless at the macrosopic level—the level of the object as it appears to the naked eye—the atoms that form it are constantly moving. And yet, Lucretius explains later, "since you cannot get so far as to see the things [i.e., atoms] themselves, they must necessarily steal their motions too from your sight" (ONT 2.312). This theory bears a distinct resemblance to the kinetic theory of modern molecular physics.

Lucretius theorizes that the shapes of atoms impart to objects not only certain physical characteristics but also characteristics that impinge upon our other sensory organs, such as taste and odor. He argues that the reason light can show through a lamp even though water cannot pass through glass is that light atoms are smaller than water atoms. And olive oil, to take another example, pours less easily than wine "because its elements are larger, or because they are more hooked and entangled more closely" (ONT 2.393–95). Similarly,

> those bodies which can touch our senses [of taste] pleasantly [such as honey and milk] are made of smooth and round atoms, but contrariwise all that seem to be bitter and rough are held in connexion by atoms more hooked, and are therefore accustomed to tear open their way into our senses and to break the texture by their intrusion. (ONT 2.400–407)

Lucretius theorizes in like manner with regard to the other senses.

This explanation of the natural world, with its implication that the structure of objects has fundamentally the same character at both the macroscopic and microscopic levels and that the only difference is one of scale, seems from a modern perspective to be naive or even childish. Yet it was this very theory that in the seventeenth century was revived by Rene Descartes (a mechanist) and Pierre Gassendi (an atomist). Gassendi, for example, virtually parroted Lucretius when he proposed that "frigorific atoms" had pointed edges and teeth, whose stabs or bites produced in people the sensation of coldness; and in 1657, French libertarian Cyrano de Bergerac, who was influenced by Gassendi, wrote that glass was transparent because it had openings that were the same shape as the atoms of fire that passed through them.[5] In other words, modern atomism began with a rehashing of knowledge by then already well over a thousand years old.

## Section 3—Magnetism according to Lucretius

Book 3 of *On the Nature of Things* is an exploration of mind (*animus*) and spirit (*anima*). Even there we see the influence of atomism, Lucretius proposing that "the mind … is a part of man, no less than hands and feet and eyes are parts of the whole living being" (ONT 3.95–97). He reasons that, while the mind instructs the flesh and causes its various parts to move, "none of these things can be done without touch, and further that there is no touch without body"; and therefore that "the nature of mind and spirit is bodily" (ONT 3.161–65). Moreover, the action of the mind upon the flesh occurs more rapidly than the action of any object we can see with the eye; for this reason Lucretius concludes that "the mind … must consist of bodies exceedingly small and smooth and round"

(ONT 3.203–5). I will not go into this particular line of reasoning any further, but will say this: to the extent that Lucretius reduces even the mind to physical substance, as he does here, it is clear that he precludes the position taken by both Thales and Aristotle, who had simply dismissed magnetism by regarding it as a soul-like action.

Book 4, on sensation and love, and book 5, on astronomy and civilization, are outside the scope of our discussion, so I will next move on to book 6, which takes up meteorology and geology. Lucretius offers rather nimble and accessible explanations of thunder, lightning, tornadoes, rain, earthquakes, and volcanoes that are all in the atomist tradition. And then finally, he comes to magnetism. Here, one gets the unmistakable impression that Lucretius finds himself in rather rougher waters that he has difficulty navigating. He broaches the topic thus:

> I will begin to discuss by what law of nature it comes about that iron can be attracted by that stone which the Greeks call magnet from the name of its home, because it is found within the national boundaries of the Magnetes. (ONT 6.905–9)

With imagery reminiscent of the so-called Samathracian rings, he then mentions what was by then already common knowledge: the magnet's power to both attract and magnetize iron; or, as he puts it, that the magnet

> makes a chain out of little rings hanging from it. (ONT 6.913)

In order to lay a foundation for explaining this phenomenon, he establishes three premises about substances in general. The first is that sensory perception occurs because objects continually emit certain atoms that are specific to each sense organ (ONT 6.931–34). The second is that all objects have pores, which contain empty space (ONT 6.940–41). And finally, the third, illustrated with the observation that atoms of light are able to pass through glass and atoms of heat are able to pass through metal, is that "since there are many pores to be found in different things, they must be endowed with different natures" (ONT 6.981–83).

Having thus prepared the groundwork, Lucretius then proceeds to build his explanation. Bear with me as I present this in the philosopher's own words, so that you can appreciate the complexity of his construction:

> In the first place, it must be that very many seeds flow out from this stone [i.e., the magnet], or, let us say, a current which by its blows beats away all the air that lies between the stone and the iron. When this space is made empty and a large place becomes vacant between, at once the first-beginnings of the iron gliding forward into the empty space fall in a body together, and the result is that the ring itself follows and passes in this way as a whole. And indeed there is nothing that has its first elements more intertwined, nothing more closely connected together and coherent, than the substance of strong iron with its chilly roughness. For this reason it is less surprising ... if the large number of bodies emanating from the iron cannot move into the void without the ring itself following; this it does, and follows until it has reached that very stone and clung to it by unseen attachments. (ONT 6.1002–16)

In other words, the atoms released by the magnet strike the air, and thus a void is formed there. Atoms from the iron then rush in to fill the void, and the iron itself moves along this same path until it reaches the magnet. There it remains, because it is held by "unseen attachments." These, too, are elaborated upon:

> When the textures of things have fallen into such a relation to each other that the empty places of this answer to the full places of that, the empty places of that to the full of this, here is the best conjunction. It is also possible that some pairs may be held in coupling as if they were linked with sort of [*sic*] rings and hooks, which seems to be rather what happens between this stone and iron. (ONT 6.1084–89)

Here Lucretius first describes a bond similar to that formed between brick and mortar or between wood and glue, then contrasts it with the bond between magnets and iron, which he postulates to be mechanical, taking the form of "rings and hooks."

Lucretius's arguments up to this point do serve as some sort of explanation for magnetism, despite being rather simple and naive. Indeed, the idea of adhesion with hooks and rings was one upon which seventeenth-century mechanists depended for their own theoretical constructions. But the logic of his explanation for the force between magnets and iron makes it rather difficult to answer two crucial questions: How is it that magnets not only attract but also repel iron? And why do they act selectively, attracting only iron and no other metals, not to mention nonmetallic substances?

Lucretius deals with the first question as follows:

> Sometimes also the iron recedes from this stone … I have even seen Samothracian iron dance, and at the same time iron filings go mad in a bronze bowl, when this

> magnet stone was applied underneath: so eager seems the iron to escape from the stone. When the bronze comes between, all this quarrel is caused, because doubtless when the current from the bronze has come first and taken possession of the open channels of the iron, afterwards comes the current from the stone to find all full in the iron, and no way to swim through as it had before: it is therefore compelled to beat and buffet the iron texture with its flood; and in this way it rejects from itself and sets moving through the bronze that which without the bronze it often sucks back. (ONT 6.1042–55)

This passage appears to be the first formal documentation in history of the observations that (a) magnets can repel as well as attract iron and (b) this power is not mitigated when another metal is interposed. But Lucretius posits a cause-and-effect relationship between these two observations, believing that the repulsion of iron by the magnet is the result of bronze being placed in between and filling the pores of the iron with its own particles before the magnet has a chance to act. It appears likely that Lucretius just happened to use a bronze bowl in making his observation and then generalized about the principle involved on the basis of what he saw in that particular instance. Moreover, the logic of his underlying theory does not lend itself to an explanation of why the same substances might be bound by attraction on one occasion and yet separated by repulsion on another. Practically speaking, he leaves himself with little choice other than to conclude that repulsion occurs by coincidence as a result of some third intervening substance rather than because of some essential property of magnets themselves.

Moving on to the second question, regarding the selectivity of magnets, Lucretius offers the following theory:

> In this connetion, you must not allow yourself to feel surprise that the current from this stone has not the power to propel other things also. For some stand firm by their weight, of which kind is gold; while others, because they are of a body so porous that the current flies through unrestricted, cannot be propelled anywhere, in which kind the substance of wood is seen to be. Between the two, then, is the nature of iron, and when it receives certain minute bodies of bronze, the result is that the magnet stones drive it by their flow. (ONT 6.1056–64)

This explanation fails to be in the least convincing. Indeed, Lucretius leaves unsolved the question of why other metals of nearly the same specific weight as iron are not affected by magnets. But then again, it would be absurd to criticize him on these grounds. After all, the human race did not manage to come up with a compelling theory of why magnets are selective, attracting and repelling only iron and a few other metals, until the twentieth century.

In the end, we must remember that however simplistic, limited, or faulty we may find Lucretius's theory of magnetism today, it was the starting point for early modern mechanism and atomism: over a millennium and a half after Lucretius, theories put forward by Descartes and Gassendi were, leaving religion aside, equally naive. At a minimum, we need to credit Lucretius as one among a small group of ancient thinkers (including also Empedocles, Diogenes, and Democritus, as reported by Alexander of Aphrodisias; Plato in his later years; Plutarch; and Epicurus) who not only described magnetism as an observed phenomenon but also made some attempt at explaining it in rational terms. And with Lucretius the reductionist tradition, at least with respect to ancient Greece, reached its end.

## Section 4—Galen's *On the Natural Faculties*

As we will see later, the writings of Galen contain a harsh attack on Epicurus and his brand of atomism. Galen was born in Pergamon, a former kingdom of Asia Minor that had come under the control of the Romans. He studied at various places, including Alexandria, which had long been the world's most important center for medical research. Having become well versed in both natural philosophy and the history of medical thought from ancient times onward, Galen became embroiled in an intellectual controversy, from which he emerged victorious. He garnered overwhelming respect in Rome and was made court physician by the Roman emperor Marcus Aurelius (r. 161–180). One might therefore wonder why I have written about him here, rather than in chapter 3, where I discuss the Roman Empire. I believe that Galen's thoughts needs to be seen in the context of Greek science because his writings embodied the medical knowledge of ancient Greece, including, most notably, that of Hippocrates, the Father of Medicine himself. Moreover, Galen himself wrote in Greek.

In many cases, Galen's works are not simply expositions of his own thoughts on medicine. Instead, they tend to contain excessive praise, indeed adulation, of Hippocrates and also to take issue in rather strident terms with the ideas and practices of other medical professionals of his time. This lack of decorum brought Galen much criticism. Over a millennium later the Italian physician Girolamo Cardano noted that "Galen ... bec[a]me involved in a disgraceful contention"[6]; and in modern times a scholar of Greek philosophy remarked of Galen that "particularly offensive are his childish vanity and his snobbery.... No doubt the best of him is borrowed plumes."[7]

But it was precisely because of Galen's quarrelsome nature that the teachings of earlier physicians and the various opposing schools of his own time have been handed down to our day. The fact that he did not mince words in his attacks on other theories has meant that two distinct viewpoints, namely elemental reductionism and organic holism, have been carved in sharp relief into the historical record.

Because this book is not intended as a work on the history of medicine, I will not be delving deeply into Galen's medical theories and opinions. Rather, my reason for even mentioning Galen, who passed on the torch not only of Hippocrates but also of Aristotle, is that he so vehemently criticized atomism, and with it Epicurus's explanation of magnetism, in the process of his dramatic clashes with the atomists of his day.

Galen's chief work, *On the Natural Faculties*, written in or shortly after the year 160, lays out in detail many of his thoughts on medicine and science as well as his criticism of atomism.[8] The following passage is from book 3:

> *There are two kinds of attraction*, that by which a vacuum becomes refilled and that caused by appropriateness of quality; air is drawn into bellows in one way, and iron by the lodestone in another. (ONF 3.15:206)

For Galen, these two phenomena are analogous to, respectively, respiration and digestion. In the first instance, the lungs absorb air in the mechanical manner of a vacuum being refilled; in the second case, the stomach and intestines absorb nutrients through their membranes in a physiological process involving "appropriateness of quality." Magnetism, according to Galen, is a phenomenon of the latter type. What is most notable about Galen, for our purposes, is his argument that magnetism occurs in the same manner as the physiological activity of a human organ.

Galen maintains that the processes of growth and nutrition are two chief characteristics of living things (both plants and animals) and that these are both products of nature (*phusis*). Moreover, he holds that "feeling" (i.e., sensation in general) and voluntary motion are two further characteristics possessed by animals alone and that these are products of, and are regulated by, the soul. This formulation is clearly influenced by Aristotle, who states that "'nature' means the genesis of growing things" and counts among the powers of living things "the nutritive, the appetitive, the sensory, the locomotive, and the power of thinking."[9]

Another of Galen's claims, that "our bodies are compounded out of the Warm, the Cold, the Dry and the Moist" (ONF 2.8:116; see also 2.8:111) is also based on Aristotelian thought, in this case on Aristotle's theory of four qualities. Further, Galen's pathology posits four bodily fluids corresponding to Aristotle's four elements: blood, which is warm and moist; yellow bile, which is warm and dry; phlegm, which is cold and moist; and black bile, which is cold and dry. He attributes illness to a lack of harmony in the balance or blending of these four fluids. Actually, this so-called humoral pathology is not new; Galen adopts it from the ancient Greeks. (10) But more importantly, it is through him that it later passes on to medieval Europe, where it is embraced as central to medical theory and practice. Moreover, Galen interprets change in any substance to be a transition—a qualitative change—from one to another of the four qualities (hot, cold, dry, or moist). In his words, "there are, in all, four mutually interacting *qualities*, and … to the operation of these is due the genesis and destruction of all things that come into and pass out of being" (ONF 1.2:5; see also 1.3:8). This, of course, echoes Aristotle's *On Generation and Corruption*. Galen's medicine, then, can be thought of as the medicine of Hippocrates bolstered by the natural philosophy of Aristotle.

And now we come to the heart of what, for our purposes, makes Galen so historically important. Galen considers the issue of whether or not qualitative change in substances is possible to be the crux of a decisive split between himself and the atomists. "There have arisen … two sects in medicine and philosophy among those who have made any definite pronouncement regarding Nature," he declares. One consists of those who believe that "all substance which is subject to genesis and destruction is at once *continuous* and susceptible of *alteration*." Galen as a member—or more precisely, as the leader—of this school maintains, for example, that the process whereby food is turned into blood, muscle, and so forth is a process not only of absorption but also of genuine qualitative change. Similarly, growth is a process not only of an increase in height or weight but also, again, of genuine qualitative change. The other sect holds "substance to be unchangeable, unalterable, and subdivided into fine particles, which are separated from one another by empty spaces." These thinkers consider any phenomenon viewed as change to be superficial only and explainable as a result of the mingling and separating of unchangeable particles of substance (atoms) (ONF 1.12:27).

Galen claims that the fundamental point of difference between these two schools of thought is the acknowledgement or rejection of what he calls the "natural faculties" or simply "Nature." The school represented by Galen asserts as follows:

> Nature is not posterior to the corpuscles, but is a long way prior to them and older than they; and … it is Nature which puts together the bodies both of plants and animals; and this she does by virtue of certain faculties which she possesses—these being, on the one hand, attractive and assimilative of what is appropriate, and, on the other, expulsive of what is foreign. (ONF 1:12:28; see also 1.13:38)

For Galen, Nature as personified by the human body entails a unified whole whose functioning takes precedence over the functioning of individual organs. It is only when brought together as part of a single organism that each organ assumes its characteristic role. In short, Galen espouses a form of what we can call organic holism. In contrast with this is the other school, which we can refer to as elemental reductionism. According to Galen, they hold that "there does not exist any substance or faculty peculiar either to Nature or to Soul, but that these result from the way in which the primary corpuscles, which are unaffected by change, come together" (ONF 1.12:28). According to this perspective, Galen notes, "nothing is naturally in sympathy with anything else," and for this reason he calls the atomists "ignorant of Nature's faculties, both that attracting what is appropriate, and that expelling what is foreign" (ONF 1.13:39).

Galen concludes that each part of the body possesses four faculties—attraction, preservation, assimilation, and expulsion—and that the existence of these faculties is a basic fact of "Nature" that can be neither further reduced nor explained. For this, Galen was criticized by one modern philosopher who deems that "much of Galen's talk of 'natural faculties' is of little value, being no more than a restatement of the phenomena to be explained."[11] While we must concede that this criticism is consistent with the reductionist perspective, we must also take note of how limited that perspective was: however clever the reductionists were at crafting ready "explanations" for the physiological functions of the various organs, the tool they were working with—atomism—was still too crude and immature for that task, and as a result their explanations were fanciful at best. In contrast, Galen's theories were based on actual observation; he developed, in his own way, a phenomenological understanding of the metabolism of various animals and used this understanding to describe and classify the functions of the different organs. This, it should be admitted, was an accomplishment of no little value after all.

## Section 5—Galen and the Dispute over Magnetism

The problem was not so much that Galen oversimplified matters by making acceptance or rejection of his so-called natural faculties the defining issue in his dispute with the atomists. More to the point, he made that position dogma and widened it indiscriminately to the point of applying it even to the physical activity observed in inanimate and inorganic substances.

In *On the Natural Faculties,* Galen makes the question of whether or not to admit attraction a major focus of his opposition to atomists. In turning his diatribe to the field of medicine, he chooses as his victim Asclepiades, the first atomist in the discipline of physiology. Asclepiades was born in Greece just slightly before Lucretius, but eventually moved to Rome, where as a physician he seems to have enjoyed some fame. Galen argues that "each drug attracts *that humour which is proper to it,*" and then judges it nonsensical that Asclepiades chooses to disbelieve this (to Galen) obvious fact (ONF 1.13:44). Galen uses the term "attraction" in a broad sense to include absorption, inhalation, suction, pulling, and so forth, and he mentions magnetic attraction as one example of it.

So now we see the term "attraction," which Galen began by designating as one of the natural faculties of living things and organic substances, interpreted far more broadly, as a characteristic of medicines in general. In fact, Galen goes even further, attributing this quality to a mineral, the magnet. To the modern mind at least, this conception of attraction seems incongruous. Somewhat later in his treatise Galen widens the concept's application even more:

> It is not only cathartic drugs which naturally attract their special qualities, but also those which remove thorns and the points of arrows such as sometimes become deeply embedded in the flesh. Those drugs also which draw out animal poisons or poisons applied to arrows all show the same faculty as does the lodestone. (ONF 1.14:53)

As if this were not enough, in book 2 Galen likens absorption of nourishment to the lodestone's attraction of iron (ONF 2.7:106) and moreover declares that

> we shall ascribe to the semen a faculty for attracting blood similar to that possessed by the lodestone for iron. (ONF 2.3:85)

One could explain all this by calling it simply a form of hylozoism, but it is more relevant to note that Galen's argument makes an enormous logical leap. At the dawning of modern times, Gilbert addressed precisely this issue, proclaiming that

> Galen errs . . . when he expresses the view that whatever agents draw out either the venom of serpents or darts also exhibit the same power as the loadstone.... Drugs against poisons or darts have no relation to, no similitude with, the action of magnetical bodies.[12]

But from the time of Thales, magnets had been thought to possess some sort of power akin to life, and even in Galen's day the distinction between living and nonliving objects remained blurred. Considering this history, it is really not so strange after all that Galen would consider magnetism to be a physiological phenomenon.

Galen refers explicitly to Epicurus's theory of magnetism and probably wrote his rebuttal having had access to this ancient atomist's writings as primary sources, so it is worth taking a closer look at this portion of Galen's text. He gives Epicurus credit for having—unlike Asclepiades—at least recognized the existence of attraction as a force of nature, but he calls Epicurus's explanation for magnetism "perfectly unconvincing":

> Epicurus … allows that iron is attracted by the lodestone, and chaff by amber. He even tries to give the cause of the phenomenon. His view is that the atoms which flow from the stone are related in shape to those flowing from the iron, and so they become easily interlocked with one another; thus it is that, after colliding with each of the two compact masses (the stone and iron) they then rebound into the middle and so become entangled with each other, and draw the iron after them. (ONF 1.14:45)

In short, explains Galen, Epicurus's claim was that "all attraction takes place by virtue of the *rebounds* and *entanglements* of atoms" (ONF 1.15:59). Having interpreted Epicurus in this way, Galen goes on to criticize the atomist as follows:

> [Epicurus] always stated the observed fact, although he gives an ineffective explanation of it. For, that these small corpuscles belonging to the lodestone rebound,

> and become entangled with other similar particles of the iron, and that then, by means of this entanglement (which cannot be seen anywhere) such a heavy substance as iron is attracted—I fail to understand how anybody could believe this. Even if we admit this, the same principle will not explain the fact that, when the iron has another piece brought in contact with it, this becomes attached to it. (ONF 1.14:47)

Galen then reports that

> I have seen five writing-stylets of iron attached to one another in a line, only the first one being in contact with the lodestone, and the power being transmitted through it to the others. (ONF 1.14:48)

He then widens his rebuttal further, producing a long-drawn-out harangue, the crux of which is the failure of Epicurus's atomism to account for the phenomenon of magnetization, as seen when a piece of iron comes into contact with a magnet and then attracts another piece of iron.

First, he assails Epicurus for claiming that atoms emitted by a magnet could possibly hold a chain of iron scraps, for these particles would have to be very large in number and therefore extremely small in size. Implicit in his criticism is the idea that these small particles would need to have tiny hooks from which the heavy scraps would be physically suspended, and that this is impossible:

> [Epicurus says,] "but these corpuscles must be looked on as exceedingly small, so that some of them are a ten-thousandth part of the size of the very smallest particles carried in the air." Then do you venture to say that so

> great a weight of iron can be suspended by such small bodies? (ONF 1.14:49)

Second, Galen then argues, because magnetized iron scraps hang from both the bottom and the sides of a magnet, there would have to be multiple "hook-like extremities by which they interlock with each other" protruding from both the sides and the bottom of each atom—an idea that he calls "absurd."

Third, he finds irrational Epicurus's idea that attraction occurs when atoms emitted by a magnet bounce off a nearby piece of iron. In cases where more than one piece of iron scrap becomes suspended from a magnet, the mechanism posited by Epicurus entailed, according to Galen, a ridiculous scenario in which one atom bounces off the nearest iron scrap while another atom passes through this piece of metal and then bounces off the next.

In short, Galen ferrets out and exposes all of the weaknesses in Epicurus's theory and then uses them as a basis for rejecting atomism in general. But he does not then proceed to replace the atomistic model with one superior or more suitable. His bottom line on the subject is that "everything which exists possesses a faculty by which it attracts its proper quality" (ONF 1.14:55), and he considers this a fact of nature. According to Galen, a magnet's power to attract iron is a faculty akin to life and, just like the power of a living creature to take nourishment from food and drink, an irreducible primary that defies further explanation.

## Section 6—Alexander of Aphrodisias

Galen's kind of hylozoism appears also in the thought of Alexander of Aphrodisias, an Aristotelian philosopher active in the late second and early third centuries, just after the time of Galen. Alexander, who probably hailed from the city of Aphrodisias in the Caria region of southwestern Asia Minor, is known for having influenced the philosophers of sixteenth-century northern Italy around the city of Padua, especially Pietro Pomponazzi. Alexander also revived, in Athens, the Peripatetic school (whose name derives from the Greek *peripatetikos*, meaning "to walk about," after either Aristotle's tendency to walk while he lectured or the walkways [*peripatoi*] at the Lyceum). The school that Alexander headed operated from 198 until 211. He also wrote numerous annotated works on Aristotelian philosophy, igniting the final sparks of ancient Aristotelianism.

We have already glimpsed Alexander, whom I introduced as a commentator on the mechanistic and atomistic explanations of magnetism advanced by Empedocles, Diogenes, and Democritus. But what I have not yet mentioned is the reason for his extensive discussion of these earlier thinkers: Alexander's purpose was to lay the groundwork for a critique of that type of reductionistic theory.

We have also seen Empedocles's theory of effluences criticized on the grounds that it does not explain why iron moves toward magnets but magnets do not move toward iron. Alexander goes further, though, criticizing the failure to explain why the force of magnetism is selective, acting on iron alone.

> Why will the iron not sometimes be moved even without the magnet, when the effluences from it travel all together towards some other thing? For why is it only the effluences from the magnet that are able to move the air which rests on the pores of the iron like a lid and holds back the effluences? Moreover, why is nothing else carried towards anything else in this way, although he [i.e., Empedocles] says that many things have pores commensurate to one another with regard to their effluences?[13]

In this respect, Alexander voices similar criticism of the theory put forward by Diogenes of Apollonia, who thought that magnetic attraction involved the release of fluid by metals:

> But if this is so, why is it just the magnet that attracts the iron, while nothing else [is attracted] by what is akin to it—for example bronze and lead? For these too emit and attract moisture, and there are certain things that are akin to these too, as the magnet is to the iron. Moreover, why does the magnet not also attract certain other solid things, which emit more moisture, among which is bronze?[14]

Alexander also takes up Democritus's idea that like attracts like, criticizing its inadequacy as an explanation for the difference between magnetic attraction and the amber phenomenon (static electricity):

> Someone might accept that the magnet and the iron are composed of similar [bodies], but how [could anyone accept this of] amber and chaff? ... there are many things that are attracted by amber; if it were composed of similar [bodies] to all of these, then these too, being composed of similar [bodies] to one another, would attract one another.[15]

The difference between magnetism and static electricity, at least in terms of their visible properties—one attracting only iron and the other attracting a variety of objects—was apparently already well known in Alexander's time. One of Alexander's accomplishments was to notice that this difference pointed to an Achilles' heel in mechanistic and atomistic explanations of magnetism.

From this point, Alexander advances his own conclusion, that magnetism is by nature a nonmechanistic form of action. First, he asserts that magnetism is not an example of action through a medium (local action) in that it does not exert its power by moving against objects in direct contact with it:

> Some things attract by force and contact; and these are moved themselves when they cause movement. But the magnet does not [attract] in this way, for it is unmoved.[16]

Second, he argues that the definitive difference between magnetism and static electricity is that the former acts at a distance without exerting any influence on the intervening air or moisture. The magnet, that is, does not attract iron

> to itself by means of the intervening air or water, drawing [them to itself]. For it would attract the things that are on the surface of [the iron] and lighter by drawing the intervening air towards itself, as amber and the cupping-glass do by the hot within themselves. But [the magnet] does not [attract things] in this way. For in those cases the fire, being moved and passing outwards, draws in the adjacent moisture, and, coming into contact, attracts what is adjacent by what it draws in; but the magnet only [attracts] the iron. For if it drew in the intervening air, all the things in the air that are lighter than the iron would be carried towards the magnet before the iron.[17]

This passage marks the first time in recorded history that any thinker had explicitly pointed out the ability of magnets to transmit their power to iron not only without direct contact but even with the presence of air or water in between, and without using those intervening substances as intermediaries. On the basis of this fact, Alexander understood that magnetic attraction did not occur mechanistically through a medium and therefore was not in that sense a form of local action.

Alexander's criticism of mechanism and atomism—his argument for what magnetism is not—is persuasive. But what about his argument for what magnetism actually is? This, to be frank, fails to convince, as it is rather difficult to fathom:

> Nourishment ... and everything that is an object of desire and appetite attracts a living creature, not by making what is between itself and the object of appetite like itself (for what is between does not become nourishment, nor is this attracted) but rather, what intervenes is set in motion by the object of appetite and transmits the form to what is set in motion, as occurs in the case of seeing. It is in this way that the iron, too, is carried towards the magnet; [the magnet] does not attract [the iron] to itself forcibly, but rather by desire for that which it lacks itself but the magnet possesses.[18]

While the point here is not quite clear, Alexander seems to be arguing that iron is attracted to magnets in the process of seeking nourishment, just as living creatures seek food, smelling the air and so on, and are instinctively drawn to what will nourish them. This is similar to Galen's argument, but with the roles of the iron and the magnet reversed. In other words, Alexander theorizes that not only magnets but also iron itself acts similarly to the way living things act. He sums up his argument by concluding that "it is not only things that possess sensation and soul that have a desire for what is natural to them; this is so with many things that are without soul, too."[19] Here, then, we can sum up our analysis of Alexander's theory of magnetism by concluding that he dodges intellectual difficulties by taking the hylozoistic position that magnetism represents some form of life—end of story.

* * *

Before we leave our discussion of antiquity, let's step back from the trees and take a look at the forest. Broadly speaking, the Greek philosophers fell into two camps concerning the mysterious and wonderful power known as magnetism. First were the atomists (principally Democritus, Epicurus, and Lucretius) and the micro-mechanists (principally Empedocles, Diogenes, Plato in his later years, and Plutarch.) These reductionists rested their theory on the premise of local action. Second were those who saw magnetism as divine and soul-like (principally Thales, Plato in his earlier years, and Aristotle) and those who subscribed to organic and holistic theories, arguing that magnetism was somehow akin to life (principally Galen and Alexander.) All those in the latter camp were alike in seeing magnetism as a form of action at a distance that could be explained no further.

Later in this book, we will see how this intellectual rift repeated itself in early modern times, when thinkers inquiring into the nature of gravity divided themselves into mechanists on the one hand and Newtonians on the other. In the first group were those who revived the reductionist (mechanistic and atomistic) theories of the ancients in fundamentally the same form, with the premise that all action had to take place through some intermediary. In the second group were Newton and his adherents, who championed the idea that gravitational force between the planets and other celestial bodies was simply a fact of nature. Newton considered the mechanistic models constructed by Descartes to be fantastic and dismissed them, criticizing Descartes for "feigning hypotheses." The Cartesians, for their part, criticized the Newtonian theory of gravity whereby bodies (i.e., the planets and so on) exerted force on one another through the void of empty space. They argued that Newton had done no more than to restate the very phenomenon that they demanded be explained.

All of this was where the debates that played out in ancient Greece eventually led, but this revival of those earlier struggles took well over a thousand years to surface. After the criticisms by Galen and Alexander that appeared in the late second and early third centuries, Europe seemed to lose sight of the mechanistic-reductionist view of magnetism almost completely. I say "almost" because in the twelfth century the philosopher William of Conches (c. 1100–1154) made reference to Lucretius's *On the Nature of Things* in his treatise *Philosophia mundi* (Philosophy of the world). But until an almost complete manuscript of Lucretius's poem was discovered in 1417 by the Italian humanist Poggio Bracciolini and then printed in 1473, this perspective was essentially gone from the intellectual landscape. Plato's *Timaeus* was one of the very few works of Greek philosophy that continued to be read in medieval Europe, but we do not see references to its mechanistic theory of magnetism in the scholarship of the Middle Ages. We do see a revival of Platonism in the third century, but according to *On Abstinence from Killing Animals*, written by the neo-Platonist Porphyry (c. 233–c. 304),

> a magnet gives soul to iron placed near it, and the heaviest iron becomes light as it rushes up to the breath of the magnet.[20]

This, clearly, is an extension of the arguments espoused by Thales and Aristotle. And in the medieval period, the influence of mysticism (with its traces of the Orient) overlapped with the organic and holistic theories of Galen and Alexander.

Notice that until this point we have been considering only the attraction between magnets and iron. Interestingly, it seems that the ancient Greeks were not aware of the force that magnets exert on one another.

A further sidenote is that the ancient Greeks apparently did not realize that lodestones and magnetized needles pointed north, and mariners of that time therefore did not enjoy the advantages of navigation by magnetic compass. In his 1590 work *Natural and Moral History of the Indies*, Spanish Jesuit José de Acosta, who had been engaged in missionary activity in the New World, noted that "there is no word in Latin or Greek or Hebrew for such a remarkable object as the compass," and asserted that "the ancients did not know of" the lodestone or its "properties and remarkable virtue … in navigation."[21]

And further, Timoteo Bertelli, a nineteenth-century Italian historian, conducted a study[22] of over seventy related works written in Greek and Latin between the sixth century B.C. and the tenth century A.D. and found not even a single reference to magnets pointing northward, much less to the application of this phenomenon in navigation, astronomy, or geodetic surveying.*

We have noted that ancient attempts at explaining magnetism were neglected by Europeans for about a thousand years, as were efforts to study magnets in a scientific manner. But this does not mean that interest in the subject waned. The mysterious phenomenon of magnetic attraction continued to be a much-talked-about source

---

*Albert the Great states in his *Book of Minerals* that "Aristotle in his *Lapidary* says 'the corner of a certain kind of magnet has the power of attracting iron towards *zoron,* that is, the North; and mariners make use of this. But another corner of this magnet attracts in the opposite direction, towards *aphron,* that is, the South Pole'" (*Book of Minerals*, 148). This attribution of the *Lapidary* to Aristotle, however, is spurious; the work likely originated in Byzantium, Syria, or Persia, and was then added to and revised numerous times before being translated into Latin. The words *zoron* and *aphron* appear to be from Hebrew, and we can do no more than conjecture about when and where this isolated reference to the north-finding property of magnetized iron might have been appended.

of fascination. This interest, however, was pursued from a very different perspective from that which was to emerge later with the birth of modern physics. What sort of perspective this was we shall soon see.

## Chapter 3

# The Days of the Roman Empire

### Section 1—Aelian and Roman Science

After the death of Alexander the Great in 323 B.C., the Greek world was divided into the Hellenistic states under the rule of the Macedonians. By that point, Greek civilization was already in decline. In 146 B.C. Macedonia fell to the Romans, and in 133 B.C. Rome gained more territory when the kingdom of Pergamon bequeathed its land to the nascent empire. When Rome defeated the Ptolemaic dynasty in 30 B.C., the demise of the Hellenistic era was complete. With the provinces of Gaul and Hispania having already been conquered, Roman hegemony over the Mediterranean world was now secured. Meanwhile, after a century of bloody civil war, Emperor Octavian succeeded in transferring Rome from a republican to what was in effect an imperial system. Thus began a generally stable period lasting over two hundred years. The latter part of this period, known as the era of the Pax Romana, saw Rome reach its height of power under the successive leadership of the so-called Five Good Emperors, who reigned from A.D. 96 to 180. In this sense, they took over where Alexander had left off with his program of imperial expansion. However, "for all her avid exploitation of her eastern conquests, Rome had let slip the greatest prize they had to offer through inability to absorb and hand it on."[1] In fact, Rome succeeded in taking firm possession of no more than a few meager scraps of Greek culture when Alexandria fell into its hands.

It is true that the Roman Empire was home in its early years to Lucretius and Galen and later to the neo-Platonists and the philosopher Boethius, and in that sense it did carry on with some of the speculative natural philosophy of the ancient Greeks until roughly the sixth century. For the most part, however, those efforts merely exploited the Greek legacy and were not inherent in the Roman tradition. Bertrand Russell sums up the general consensus on Roman culture thus: "The Romans invented no art forms, constructed no original system of philosophy, and made no scientific discoveries. They made good roads, systematic legal codes, and efficient armies."[2] For this reason it is very rare in science history to find Rome mentioned by itself; instead, one finds references to the "Greco-Roman period," with Rome as a kind of hitchhiker on the Hellenistic bandwagon.

But even if we today are hard-pressed to consider the Roman view of nature as science or are inclined to say that their culture represented a regression in intellectual endeavor, we must concede that the Romans had their own characteristic way of understanding the forces of nature and of viewing nature in general. At least with respect to magnets and magnetism, the writings of ancient Rome reveal a perspective that is notably different from anything bequeathed to the Roman era by the Greeks, and characterizes that era's own distinctive view of natural forces. Moreover, that understanding of nature was to have a huge influence on medieval

European thought, as we shall see later on. In this chapter, therefore, I shall take up the ideas that emerged in Rome as peculiarly Roman rather than as extensions of Greek science.

Of course, from the perspective of science history there is no denying that logic and rationality, the characteristics that gave Greek philosophy and science their great luster, were lost in the transition from the earlier culture to the later. One document making this plain is *Historical Miscellany*, written in Greek by a Roman named Aelian (c. 175–c. 235) around the year 200. This work, a collection of Greek anecdotes and other miscellany, was written for the entertainment of the Roman leisured class and is not especially academic in tone. Aelian himself is not among those whose pictures generally appear in the annals of science; William Harris Stahl's monograph *Roman Science: Origin, Development and Influence to the later Middle Ages*, for example, makes no mention of him whatsoever. But since only well-educated Romans of an intellectual bent read and wrote in Greek in Aelian's time, his work is therefore interesting as a glimpse into what aspects of Greek civilization the educated Roman elite found noteworthy.

The book begins with a quotation from Aristotle's *History of Animals* on the behavior of octopuses and spiders and then goes on to discuss the thoughts and deeds of various Greek philosophers in no particular order and with no coherent purpose to tie it all together. It offers frivolous gossip, such as that "Socrates' favourite boy was Alcibiades, and Plato's was Dion," and that "the poet Sappho … is described as beautiful by Plato"; but amid such Who's Who-style information there are no references whatsoever to what any of those scholars actually had to say about serious subjects.[3] According to at least one source, talking about Greek culture was in vogue among the Romans of the second and third centuries[4], so we can surmise that doing so in the Greek language was, for the educated elite, a way of showing off. But what appears on the pages of *Historical Miscellany* merely scratches the surface of what can properly be thought of as genuine Greek culture. And for precisely this reason, the book reveals the unvarnished truth about how educated Romans appraised and treated the legacy of Greek civilization.

Indeed, *Historical Miscellany* was not the only work of its kind. In the Roman Empire there emerged an entire genre of compilations of nonsensical fluff thrown together without regard for logic or organization. Most of these were edited in a style that can be compared, if one is being generous, with an abridged encyclopedia or perhaps a user's manual. Among these, one stands out as being the most comprehensive and thorough: Pliny's *Natural History*. This voluminous book, written around the year A.D. 70, spawned a great many imitators. One other noteworthy work from that era is *The Greek Herbal* written by Dioscorides around the year A.D. 60. Let us turn our attention now to these two books, taking up *The Greek Herbal* first.

But do not put Aelian out of mind entirely, for we will be returning to him at the end of this chapter for a look at his views on magnetism.

## Section 2—Dioscorides and *The Greek Herbal*

Pedanius Dioscorides (c. 40–c. 90) was born in Cilicia, a region of Asia Minor that had by that time already been incorporated into the Roman Empire. He was educated in Pergamon and Alexandria, and was probably influenced in his youth by what remained of Hellenistic science and medicine. He became a military physician in the army of Emperor Nero (r. 54–68) and collected material for *The Greek Herbal* while traveling with the troops. He selected and identified a huge assortment of medicinal substances, including some six hundred varieties of plants, eighty medicines derived from animals, and fifty varieties of minerals with pharmacological properties. In *The Greek Herbal*, he not only identified and classified these substances but also recorded their indications, methods of preparation, instructions for use, and so forth. In some senses, this endeavor was a carrying forward of Greek tradition within the context of Roman society—especially given that the work itself was originally written in Greek—and so one might wonder why I chose not to take up Dioscorides in the previous chapter. But in fact *The Greek Herbal* was not merely a revisiting or reiteration of the Greek medical literature; rather, it contained original research and insights and was therefore, for its time, a contemporary compilation on medicine and herbalism. Moreover, the work contained no discussion of the abstract, speculative notion of *cause* and was instead merely a listing of medicines and information concerning their prescription and efficacy. It was intended, after all, as a manual for practical application. In these respects *The Greek Herbal* parted from Greek tradition and is therefore best seen as an example of Roman scientific literature.[5]

In his preamble, Dioscorides roundly criticizes the Asclepiads, a cult of self-styled healers who worshipped the Greek god Asclepius:

> [The Asclepiads] have transmitted the powers of Medicines and their examination cursorily, not estimating their efficacy by experience, but by vain prating about the cause.... Thus Niger, who seems to be man of special note amongst them ... in the face of plain evidence ... sets down many ... falsehoods, which are tokens that he acquired his information not by his own observation, but had it only from the false relation of hearsay.

Dioscoride's own book contains very few indiscriminate references to folklore, most of his material being backed up by his own firsthand observations of the efficacy of the medicines he is writing about. As a work of natural science, *The Greek Herbal* is conspicuous for its time in this respect. Because it was so well regarded, the book was translated into Latin and numerous copies were produced, adorned with exquisite illustrations; for over a thousand years it was a wellspring for further medical scholarship throughout Europe. Even today, the influence of this work remains with us in both the common and the scientific names of many medicinal plants.

Galen, who lived not long after the time of Dioscorides, is said to have preferred Dioscorides of all previous herbalists; and in the sixth century, Roman statesman and author Cassiodorus recommended the book when establishing a hospital in one of the monasteries that he founded, commenting that Dioscorides "described and drew the herbs of the field with wonderful exactness."[6] Much later, in the twelfth century, William of Conches commented that Dioscorides had described medicinal herbs clearly and thoroughly; and a century after that, Roger Bacon went so far as to credit Dioscorides with demonstrating the possibility of prolonging ordinary life spans.[7] Regard for *The Greek Herbal* continued into early modern times as well. The Latin text was printed first, in 1478, and the Greek text in 1498. In 1531, Spanish humanist Juan Luis Vives, in *On Education,* included *The Greek Herbal* in a list of books to be read by students, along with Aristotle's *History of Animals* and Theophrastus's *History of Plants.*[8] And in mid-sixteenth-century England, the writings of Dioscorides were every bit as influential as those of Galen and Hippocrates in the fields of medicine and pharmacology.[9] *Commentario di Dioscorides*, first published in Italian in Venice in 1544, was popular enough to be translated into Latin, Bohemian, and French.[10] Near the end of the sixteenth century, Flemish physician Jean Baptiste van Helmont, while undergoing his training, remarked that reading *The Greek Herbal* made him aware that the science of herbalism had made no progress in the time since that work had appeared.[11] Agnes Arber, a researcher of the history of botany, makes the following remarks in her book *Herbals* (1938): "Up to the height of the renaissance period, and later, *De materia medica* [*The Greek Herbal*] was accepted as the almost infallible authority.... In the seventeenth century, the fame of Dioscorides was unabated."[12]

Much of *The Greek Herbal* is given over to medicinal herbs, but the work also mentions a hundred different medications containing minerals. With regard to mercury (Hg) Dioscorides writes: "Putting an iron Spoon having Cinnabaris in an earthen pot, they cover the Cup dawbing it about with clay, then they make a fire under with coals: and ye soot that sticks to ye pot, being scraped off & cooled, becomes Hydrargyrum" (GH 5:110). This is the first known description, of how mercury is in fact extracted. It has been said that "the ancient herbals and lapidaries are full of the 'astrological botany' and 'astrological mineralogy,'"[13] but *The Greek Herbal* is a notable exception, containing very little in the way of folklore or superstition.

Very striking, for this reason, is the fact that Dioscorides's description of the magnet relies almost solely on oral folk tradition. There could perhaps be no plainer indication of how magnets and magnetism were understood at that point in history than the following entry in *The Greek Herbal*, which I quote in its entirety:

> Of ye Load-stone that which draws iron easily is ye best, & of an azure-like colour & thick, but not too heavy. It hath a faculty

> of drawing out gross humors being given with melicrate ye weight of 3 Oboli [& they say that this hath a discerning faculty of a woman that is chaste & of her that plays ye adulteress with another man, for if any do set it secretly upon the beds for ye chaste woman, who loves her husband, she being overborne into sleep, with a certain natural faculty of ye stone both opens her hands towards her husband, and cleaves close to him; but the other being troubled in dreams with foul labours, falls out of ye bed. Also when two men carry this, it freeth them from all strife, & it causeth concord, and being born on ye breast, it assuageth people]. (GH 5:148, brackets in source)

This represents a complete about-face from the types of arguments previously advanced in Greek science that we examined in the first two chapters of this book. Unlike earlier thinkers, who were intensely interested in seeking explanations for magnetism, the Romans were focused rather on finding practical applications for the phenomenon: whereas the Greeks had looked for root causes, asking *why*, the Romans looked for efficacy, asking *how*. And there is a further difference between the two cultures in the nature of their inquiry: among the Romans we find not only a lack of distinction between physical and physiological activity but also a melting away of the boundary between the natural and the supernatural.

It is believed that peoples as far back as the ancient Assyrians and Egyptians spoke of magnets (magnetized iron ore) as having medicinal properties[14], and we can surmise that the magnet-as-medicine must have been talked about and handed down in the ancient Greek world as well. Indeed, it is said that Alexander the Great recommended the use of magnets in treating gout and epilepsy, along with two other means of cure: the blood of a swallow and the urine of a boy.[15] And in *On the Magnet* the sixteenth-century English physician William Gilbert remarked that Galen had noted the supposed medicinal power of the magnet in his book on medicine.[16]

All of this serves to show that Dioscorides's inclusion of the magnet as a substance with such medicinal properties as the ability to purge gross humors (although it is unclear to which fluids he was referring) was nothing new.

Nonetheless, Dioscorides's discussion of the magnet as a tool for detecting female infidelity strikes us today as ridiculous in the extreme, the inconsistency between this and his other descriptions, which are based on empirical evidence, being rather large, to say the least. In fact, of all the minerals discussed in *The Greek Herbal,* there are only six to which supernatural powers are attributed—the magnet and five others. Regarding the other five, he discusses mainly their medical properties, only briefly mentioning nonmedical applications, such as their use as talismans. In other words, the entry on magnets is very conspicuously unique, and it gives a clear indication of how profoundly the people of that time were impressed with magnets as an utterly singular phenomenon.

And yet, it would be missing the point to dismiss this account of magnetic power as mere superstition and prescientific nonsense; that would be a failure to grasp the outlook on nature that people had in that place and at that time. To the people of ancient Rome, the physical power of magnets to attract iron was on the same plane as its supposed physiological power to draw fluids out of the body. And also on the same plane, in their perception, were its purported powers to detect feminine infidelity and to mediate conflicts. Today we would insist that the former two claims be subjected to scientific experimentation to test their validity and that the latter two be rejected as mere superstition. But such modern discrimination is just that: a modern understanding possible only to modern minds. Our role as historical inquirers demands that we understand the way in which people of the era we are investigating viewed the natural world, including elements of their perspective that we deem superstitious.

Moreover, as we will see in the next chapter, Dioscorides's claim that magnets could identify adulteresses—a notion that not only seems preposterously unscientific from a modern perspective but was also seen as indecent and paganistic from the Christian standpoint—in fact took hold and was repeated throughout Europe for the next thousand years, and then some.

## Section 3—Pliny and *Natural History*

One essential landmark in our exploration of ancient Roman science is Pliny the Elder's *Natural History*, a vast work consisting of thirty-seven books that has been published by the Loeb Classical Library in a ten-volume edition containing both the original Latin text and an English translation. This monumental encyclopedia serves as a sort of index to the Roman Empire's intellectual inheritance from earlier antiquity with respect to knowledge about nature; indeed, it is itself an embodiment of that legacy and shows how it was preserved and passed on to the nations of Europe.

Pliny was born the son of an affluent knight around the year A.D. 23 or 24 in what is now northern Italy. He went to Rome at a young age to pursue his studies and seems to have received an education typical for the upper classes of his day. He later held military and administrative posts in Germania and Hispania, during which time he broadened his knowledge and perspective. According to his nephew, Pliny the Younger, "he combined a penetrating intellect with amazing powers of concentration and the capacity to manage with the minimum of sleep."[17] Known for being twice as industrious as his peers and having a voracious appetite for learning, Pliny was in fact true to his nephew's word: in between performing his many duties as a knight, he continued to write relentlessly, producing a number of works, including a history of the German wars in twenty volumes. Unfortunately, all of his writings have been lost except for *Natural History*, which was written in his later years, completed in the year 77, and published posthumously.

The life of this great scholar ended with the famous eruption of the Mount Vesuvius volcano that began on August 24 in the year 79. At the time, Pliny was a naval commodore at Misenum on the Bay of Naples, just across the bay from Vesuvius. Disregarding the dangers, he traveled by boat to the scene of the catastrophe, compelled both by a desire to rescue local residents and by intellectual curiosity. His body was found days later. The details of the tragedy were recorded in vivid detail by his nephew in a letter to the historian Tacitus. According to this account, Pliny's boat made it to the shore near Vesuvius, Pliny dictating his observations of the eruption to a scribe while spurring on his terrified helmsman. The direct cause of Pliny's death is thought to have been sulfur inhalation. He is one of the first people in history known to have martyred himself to the cause of scientific inquiry.[18]

The idea that permeates *Natural History* is that every natural object possesses its own characteristic power and function and serves some particular purpose for human beings. Pliny believed that the focus in examining any object or phenomenon should be on discovering its utility, and that this was completely unrelated to any further ideological or philosophical speculation. *Natural History* is characterized by, first, an uncritical acknowledgment of anything deemed potentially useful; second, an uncommon curiosity about rare natural objects, or what could be called "natural wonders"; and third, a voracious acquisitiveness for knowledge in both of those categories. Pliny himself claimed to have included in the encyclopedia over twenty thousand items and to have used as references some two thousand works by 473 scholars (from both within and outside the Roman Empire). What he included within the scope of the concepts "natural" and "knowledge," however, were not only natural objects that he observed and phenomena that he verified firsthand but also—and these actually make up the bulk of the collection—reports from scholars who preceded him, folklore handed down from ancient times, and information from foreigners and other travelers he was in contact with.[19]

Pliny's attitude toward all of this folklore and hearsay is inconsistent. At certain points, he sizes up the mythical Arabian Phoenix as a "fabrication" (NH 10.2), cites Greek lore concerning werewolves as an example of "Greek credulity" (NH 8.34), and calls Sophocles's claim that amber is formed from tears shed by birds in the lands beyond India "childish and naive" (NH 37.11). Reading these passages alone, one might conclude that Pliny applied vigorous critical standards in his thinking, but in fact he contradicts this impression in numerous other places. He asserts, for example, that "horses burst themselves which, carrying a rider, follow the tracks of wolves" (NH 28.81); that "certain it is that they [glow-worms] are the offspring of [the Pleiads] constellation" (NH 18.67); and that "it acts as a charm for a man to spit on the urine he has voided" (NH 28.7). He also uncritically repeats claims of the existence in India of humans with the heads of dogs and of a so-called umbrella-foot tribe of one-legged men who lie on their backs and use their giant feet as shelter from the sun (NH 7.2).

In Pliny's worldview, then, the lines between myth and reality, hearsay and fact, and imagination and proof were blurry to the point of being almost nonexistent. His standard for judging authenticity and his criteria for what constituted rigorous inquiry were arbitrary, subjective, and inconsistent. As a result, his encyclopedia is, by modern standards, a mixed bag of everything ranging from the scrupulously factual and practical to the absurdly superstitious and fictitious. *Natural History* is a laundry list not only of the useful but also of the rare, the mysterious, and the amusing, gathered from among the rumor and folklore that Pliny happened to come into contact with, and often selected precisely on the basis of rarity, mystery, or amusement value.

Nonetheless, Pliny was read throughout Europe for over a millennium and a half. The simple fact that all thirty-seven books of *Natural History* have survived fully intact is an indication that the encyclopedia was indeed read both widely and continuously. In the eighth century, the Venerable Bede, in *De natura rerum* (On the nature of things), quoted from Pliny's encyclopedia while discussing the movements of the planets and recommended that people read its second book for a better understanding of the subject.[20] In fact, much of that treatise by Bede is based on *Natural History.* Pliny's name comes up again not only in such scholarly treatises as William of Conches's *Philosophia mundi* and John of Salisbury's *Metalogicon*—both written in the twelfth century—but even in *Ruodlieb*, an anonymous epic poem thought to have been written in the previous century that mentions Pliny and describes the effects of some of the medicinal herbs written about in *Natural History.*[21] Also in the twelfth century, an abridged version of *Natural History* was produced for the king of England,[22] and Pliny is referred to once more in Hugh of Saint Victor's *Didascalicon.*[23] He is taken up again in the thirteenth century in both *On Animals* and *Book of Minerals* by Albert the Great, as well as in *The Art of Falconry*, written by Holy Roman Emperor Frederick II of Sicily[24]; and in the fourteenth century in English cleric Richard de Bury's *Philobiblon*, which describes Pliny's encyclopedia as a "massive work on natural history."[25] Also in the fourteenth century, French philosopher Nicole Oresme quoted from Pliny's *Natural History* more than once in his book on cosmology.[26] And these are just a few examples.

The influence of *Natural History* even extended beyond the Middle Ages into early modern times. Because it was read so widely throughout the medieval period, it was one of the few great ancient works that did not have to wait to be rediscovered during the Renaissance. In Italy, it was the first manuscript on natural history to be produced on a printing press. No sooner had the first printing shop opened in Venice in 1469 than *Natural History* was published in a state-of-the-art edition, with further print runs following in 1470, 1473, 1476, and 1479.[27] Christopher Columbus not only read the work but was profoundly influenced by it. *A Description of the Northern Peoples*, written by Olaus Magnus of Sweden in the sixteenth century, quotes Pliny extensively, and around the same time Gonzalo Fernández de Oviedo y Valdés of Spain wrote that he had compiled his *Natural History of the West Indies* "in the manner of Pliny."[28] Even Georgius Agricola, the German physician considered the father of the modern sciences of metallurgy and mineralogy that emerged in the mid-sixteenth century, notes approvingly in *De natura fossilium* (translated as *Textbook of Mineralogy*) that "Pliny ... is the only writer who has discussed minerals at any length"[29] and indeed cites him throughout the book as an authority in that field. Agricola's *De re metallica* (On metals), furthermore, cites *Natural History* incomparably more than it does any other book. Similarly, *Natural and Moral History of the Indies*, published in 1590 by Spanish Jesuit José de Acosta, describes Pliny as "a conscientious historian of natural things" and mentions him far more often than it mentions Aristotle.[30]

Not only was Pliny's *Natural History* widely read; its contents, including many of its absurdities, were accepted as fact. In book 4, for example, Pliny refers to "a genial region with a delightful climate" on the far side of a snow-covered mountain near the North Pole, where the inhabitants are extremely content in their pleasant surroundings (NH 4.12). This fanciful, wholly invented tale was believed by none other than Roger Bacon, one of the greatest scholars of the thirteenth century, who not only accepted that the land in question "was found by certain experience" and that the story of its existence was based on "the experience of men who were there,"[31] but then discoursed with great earnestness on how such a place could be possible. Scholar of medieval history Charles H. Haskins comments that "Pliny appealed powerfully to the medieval love of the marvellous,"[32] and indeed the people of the Middle Ages found such supernatural phenomena to be not only interesting but credible as well.

Precisely for this reason, *Natural History* provides us with valuable evidence of how both ancient Romans and medieval Europeans viewed and interacted with nature.

## Section 4—Pliny's Conception of Magnets as Living Creatures

Pliny's thoughts on the magnet are scattered here and there throughout *Natural History*. They can be found, for example, in 2.98, 10.1, 34.42, 36.25, 36.66, and 37.15. Magnets, interestingly, are mainly dealt with in discussions of the characteristics of stones (bk. 36) and gemstones (bk. 37), but are not mentioned in the discussion of metals (bk. 38). From the time of Theophrastus, magnets had never been considered a form of metal, and as we shall see in a later chapter, it was probably because of Pliny's influence that this classification remained the status quo until early modern times.

Book 36 begins with a discussion of marble and then meanders through some of the rare structures known to the Roman world that caught Pliny's fancy: the Pyramids and Sphinx of Egypt, the Lighthouse of Alexandria, the novel buildings of Rome, and so on. But then all of a sudden in 36.25 we come to the following:

> As we pass from marble to the other remarkable varieties of stone, no one can doubt that it is the magnet that first of all comes to mind. For what is more strange than this stone? In what field has Nature displayed a more perverse wilfulness? ... What is more impassive than the stiffness of stone? And yet we see that she has endowed the magnet with senses and hands. What is more recalcitrant than the hardness of iron? We see that she has bestowed on it feet and instincts. For iron is attracted by the magnet, and the substance that vanquishes all other things rushes into a kind of vacuum and, as it approaches the magnet, it leaps towards it and is held by it and clasped in its embrace. And so the magnet is called by the Greeks by another name, the "iron stone," and by some of them the "stone of Heracles." (NH 36.25)

This is followed by a description of the various magnets found in different geographical locations. In describing the differences between them, Pliny declares that

> the most important distinction is between the male and female varieties, while the next lies in their colour.

In Pliny's terminology, the male, or *mas*, are the ones that hold their attractive power for a long time, while the female, or *foemina*, are those that lose that power more quickly. With regard to color he then notes:

> It has been ascertained that, the bluer a magnet is, the better it is. The palm goes to the Ethiopian variety, which in the market is worth its weight in silver.[‡]

It appears that magnets of all types were rather rare and valuable objects in Pliny's day.

The above excerpt, where Pliny speaks of iron "leaping" toward the magnet, suggests his recognition of the principle of action at a distance, that is, the idea that a force, in this case magnetism, can operate even without coming directly into contact with the object upon which it is acting (in this case, iron). Notice how odd, how surprising, Pliny finds this phenomenon: among stones, which are generally stiff and insensible natural objects, only the magnet has "senses and hands," "feet and instincts." Moreover, it has the power to control, to pull to itself from a distance even something as stubborn and resistant as iron.

To add just one sidenote here: Pliny explains the origin of the term

---

[‡] Ordinary natural magnets (magnetite; $Fe_3O_4$) never appear blue, even when they contain impurities. There is another substance, however, called maghemite (a form of oxidized iron; $\gamma$-$Fe_2O_3$) that has magnetic properties and is a bluish-blackish color. Magnetized iron ore is transformed into maghemite through exposure to the atmosphere. It is thought that in ancient times magnets were generally found and harvested either from the surface of the ground or in the form of relatively shallow deposits of iron ore. We can therefore surmise that when Pliny and Dioscorides described certain magnets as "blue," they were referring to maghemite. See Blackman, "The Lodestone," 328.

"magnetic stone" (*magnes lapis*) by saying that the magnet was first discovered by a shepherd named Magnes when an iron nail in his sandal (or, some say, the tip of his cane) was attracted by a natural magnet (NH 36.25). As we have already seen, Lucretius had a different explanation, attributing the term to the fact that these stones were found in the region of Magnesia in Asia Minor. Both of these etymologies were repeated frequently for a very long time.

Something else striking about the above excerpt is the use of life-ascribing terms, such as "senses," "male," and "female," in reference to magnetic attraction, magnetization of iron, and variations among magnets in their ability to retain magnetic force. Bear in mind that this type of expression is intended literally and is no mere rhetorical flourish. Pliny employs similar language in another passage on magnetization, or magnetic induction:

> Iron is the only substance that catches the infection of [the lodestone] and retains it for a long period, taking hold of other iron, so that we may sometimes see a chain of rings; the ignorant lower classes call this "live iron," and wounds inflicted with it are more severe. (NH 34.42)

Thus, every effect of the magnet is likened to a biological action or physiological process. Pliny understands magnets to be alive, and from there he reasons that magnets can act upon the human body as well, offering medicinal benefits. We will encounter more about the medical properties attributed to magnets later. For now, suffice it to note the specificity of Pliny's prescriptions, as in the following:

> All magnets … are useful for making up eye-salves if each is used in its correct quantity, and are particularly effective in stopping acute watering of the eyes. They also cure burns when ground and calcined. (NH 36.25)

Pliny also ascribes health benefits to amber, another substance known for having an attractive force:

> Even to-day the peasant women of Transpadane Gaul wear pieces of amber as necklaces, chiefly as an adornment, but also because of its medicinal properties. Amber, indeed, is supposed to be a prophylactic against tonsilitis and other affections of the pharynx. (NH 37.11)

The first part of Pliny's claim is somewhat dubious, as a contemporary of his, Tacitus, reports in *Germania* that the people of Gaul did not appreciate the value of amber.[33] Nonetheless, it is a fact that this notion about Gallic jewelry was widely circulated in Rome. Pliny also claims that "amber is found to have some use in pharmacy.... [and] is of benefit to babies when it is attached to them as an amulet" (NH 37.12).

Here is another instance where Pliny places pharmacological power in the same category as supernatural power. It is fair to say that Pliny was a mere hair's breadth away from designating both magnets and amber as objects of nothing less than magic.

## Section 5—Sympathy and Antipathy in the Natural World

What is most noteworthy about Pliny with regard to magnetism is not only that he classifies magnets into several different types but also that he is the first to write about the ability of magnets to attract one another. This he does with reference to one particular variety:

> The test of an Ethiopian magnet is its ability to attract another magnet to itself. (NH 36.25)

Pliny himself may or may not have been aware of the originality of this observation, but it was certainly epoch-making in the sense that all discussion by earlier thinkers had been limited solely to the attraction observed between magnets and iron.

The repulsive force of magnets seems to have interested the people of Pliny's time, although recognition of this phenomenon did not lead to an understanding of the concept of magnetic poles. To continue with Pliny's discussion of the "Ethiopian" magnet:

> Also in Ethiopia and at no great distance is another mountain, (the ore from) which on the contrary repels and rejects all iron. (NH 36.25)

Elsewhere he relates the following legend about India:

> There are two mountains near the river Indus, the nature of one of which is to hold all iron and that of the other to reject it; consequently if a man has nails in his shoes, on one of the mountains at each step he is unable to tear his foot away from the ground and on the other he cannot set it down on the ground. (NH 2.98)

These claims presumably originated with stories brought back from the Middle East as a result of Alexander the Great's conquests. The information itself is of questionable credibility, but that is beside the point. What is interesting here is that, as far as we can tell from these passages, Pliny and others of his time were completely unaware that two identical magnets would sometimes attract and at other times repel one another, depending on how they were placed; individual magnets were seen as being capable of either attraction or repulsion, but not both. Until William Gilbert rejected the idea in 1600, Europeans continued to speak of magnets that drove away iron, calling them "theamedes."[34]

As historians we can make another observation about *Natural History* that is quite remarkable. Recall that with this encyclopedia Pliny is supposed to have scrupulously and enthusiastically gathered together practically every bit of information on the natural world known to both oral and recorded history, including details about the characteristics and capabilities of magnets, selected without regard for their truth or falsehood. It is astonishing, therefore, that he makes not a single reference to any of the theories advanced by the ancient Greeks, from Empedocles and Democritus and all the way to Lucretius, about the cause of magnetism—not a word about effluences, atoms, or anything of the kind. To begin with, Pliny had almost no understanding of the theoretical science of the Greeks.[35] As a result, we see in his work no attempt to explain magnetism in rational terms, no attempt to analyze cause and effect with respect to the strange properties of magnetic stones or the wondrousness of nature to which they bear witness. In the words of one historian of Roman science, "like a child, [Pliny] stood in awe of nature and had little more than a child's sense of connection between cause and effect."[36]

Accordingly, Pliny displays almost no intention of presenting an integrated understanding of nature from a consistent perspective. The closest he comes to such an aim is where he explains and classifies diverse objects and phenomena in terms of sympathy and antipathy. At the beginning of book 20, in which he discusses the properties of various herbs, he writes:

> Herein will be told of Nature at peace or at war with herself, along with the hatreds [*odium*] and friendships [*amicitia*] of

> things deaf and dumb, and even without feeling. Moreover, to increase our wonder, all of them are for the sake of mankind. The Greeks have applied the terms "sympathy" [*sympathia*] and "antipathy" [*antipathia*] to this basic principle of all things. (NH 20.1)

To illustrate, he continues with the following examples:

> ... water putting out fire; the sun absorbing water while the moon gives it birth; each of these heavenly bodies suffering eclipse through the injustice of the other.... the magnetic stone draws iron to itself while another kind of stone repels it.

He returns to the theme in book 37:

> Throughout the whole of this work I have tried to illustrate the agreement and disagreement that exist in Nature, the Greek terms for which are respectively "sympathia," or "natural affinity," and "antipathia," or "natural aversion." (NH 37.15)

In other words, sympathy (friendship) and antipathy (hatred) form a dichotomy that, throughout all of *Natural History*, is the guiding principle by which the natural phenomena that Pliny writes about can be articulated and systematized. Between fire and water, for example, there is antipathy, and so also between the sun and water. Between the moon and water, meanwhile, and also between magnets and iron, there is sympathy. This, then, is Pliny's sole attempt throughout his entire massive encyclopedia to create a framework, a schema, with which to understand the myriad interactions taking place among the objects of the natural world.

Pliny's formulation of this dichotomy was clearly influenced by Empedocles, but we must take care not to equate the concepts of sympathy and antipathy with the more limited concepts of attraction and repulsion as understood in modern physics, for Pliny's scheme goes beyond mere mechanics. In 37.11 of *Natural History* Pliny points out that amber attracts straw and clothing, and then in the following paragraph notes that

> when rubbing with the fingers draws forth the hot exhalation [*caloris anima*], amber attracts straw, dry leaves and linden-bark, just as the magnet attracts iron.

Apart from the fact that he attributes amber's attractive force not to friction itself but rather to the heat produced by friction, and that he fails to clearly grasp the difference between magnetism and static electricity, we must concede that here Pliny has more or less accurately described the phenomenon of static electricity.

Even so, in this passage Pliny does not explain the physical property of amber's attraction in terms of sympathy. This is notable because elsewhere he uses the concept of antipathy in explaining the ability of pigs to protect themselves from the poison of salamanders (NH 29.23). On the one hand, then, he does not employ his own framework regarding amber's attractive force (as modern physics understands it), but on the other he does apply that framework to something as mysterious as a supposed connection between pigs and salamanders.

In other words, we can think of Pliny's sympathy-antipathy dichotomy as a highly ambiguous, or even symbolic, construction for describing the workings of nature in terms of harmony and disharmony, friendship and hatred. He applies the dichotomy not only to what we today consider to be matters of natural science but also to the supernatural. Commenting on Pliny's use of the term *vis* for "force," Max Jammer remarks that "Pliny's uncritical use of *vis* ... for the designation of physical, chemical, medical, and occult forces is partly responsible for the obscurity of his writings, which appear as a peculiar mixture of natural science and superstition."[37] Indeed, Pliny draws no clear distinction between natural science and superstition, nor does he grasp them as belonging to different levels of understanding.

This point is eloquently illustrated in Pliny's description of two rather bizarre relationships. At one point, he says that "the diamond, the rare delight of Wealth, unbreakable and invincible by all other force, is broken by goat's blood" (NH 20.1). And elsewhere, he explains that

> the "adamas" has so strong an aversion to the magnet that when it is placed close to the iron it prevents the iron from being attracted away from itself. Or ... if the magnet is moved towards the iron and seizes it, the "adamas" snatches the iron and takes it away. (NH 37.15)

Smith and Lockwood's *Latin-English Dictionary* defines *adamas* as "the hardest steel," giving its etymology as a derivation of a Greek word, ἀδάμας, meaning "invincible." But the word was used variously to refer to hard stones in general and to diamonds in particular. And in fact, many of the words used today to refer to this most precious of gems are derived from the word *adamant* (meaning "hard stone"): *diamant* (French), *diamante* (Spanish, Italian, Portuguese), *diamond* (English), and *Diamant* (German, Dutch). Judging from the fact that Pliny describes *adamas* as "the rare delight of wealth," I think it is reasonable to guess that he was referring to diamonds, and the following comment by him puts this interpretation almost beyond doubt: "The most highly valued of human possessions, let alone gemstones, is the 'adamas,' which for long was known only to kings, and to very few of them" (NH 37.15).[§]

Assuming that Pliny is referring to diamonds, it follows that he is arguing that diamonds and magnets repel one another, that diamonds interfere with the ability of magnets to attract iron, and that diamonds can be destroyed by the blood of a goat. All of this seems utterly unfounded to us, but not so to people of earlier times: in the third century, Roman writer Julius Solinus declared that "there is a hidden, natural disharmony between *adamas* and the magnet, rendering magnets unable to attract iron when *adamas* is placed nearby."[38] Thus it appears that, at least at that time, the idea that diamonds could render magnets impotent, and presumably also the idea that diamonds could be destroyed by goat's blood, were not particularly held in doubt. In the next chapter, we will see that even Saint Augustine, the greatest thinker of the early Christian world, spoke of a strange connection between magnets, diamonds, and the blood of the goat. Mysterious as these relationships seem, the power of magnets to attract iron was long ago seen as no less so: if the latter were possible, people thought, then the former must be possible as well. According to Pliny, furthermore, these phenomena were to be seen within the schema

[§] Somewhere along the line, however, the word *adamas* as used in Latin came to include the magnet within its meaning. It is believed that this meaning was derived from the term *lapis adamans* (loving stone), which had its roots in the verb *adamare* (to love). Gilbert called the French word *aimant* (magnet) a corruption of *adamant* (*On the Magnet* 1.2:11), but the former, too, is more likely a derivation of *aimer* (to love). And the words in Spanish and Portuguese for magnet, *imán* and *imã*, respectively, appear to be derived from *amar,* also meaning "to love." Thus, from the Middle Ages on, the word *adamas* has been used to mean either steel, or magnet, or diamond, requiring that we infer from the context which meaning is intended in any given case.

of sympathy and antipathy. Similarly, his contemporary, Plutarch, gave the following examples of sympathy and antipathy: "A wild bull is quieted and made gentle if bound to a fig-tree; amber moves and attracts all light things, except basil and whatever is wet with oil; the loadstone does not attract iron rubbed with garlic."[39] Magnets and garlic, diamonds and goat's blood, magnets and diamonds—all of these, it was firmly believed in ancient Rome, were in antipathy to one another.

## Section 6—Claudian and Aelian

Let us turn now to a poem titled *Magnes*, written in the fourth century, before Christianity took hold in Rome. The poem's author is Claudian (c. 370–c. 404), born in Alexandria and known as the last poet of pre-Christian Rome. First, an excerpt:

> There is a stone called the loadstone; black, dull, and common. It does not adorn the braided hair of kings nor the snowy necks of girls, nor yet shine in the jewelled buckles of warriors' belts. But consider the marvellous properties of this dull-looking stone and you will see that it is of more worth than lovely gems and any pearl sought of Indian amid the seaweed on the Red Sea's shores. It lives on iron and feeds on its inflexible nature; iron is its food and nourishment; from iron it recruits its strength. This seemingly inedible food, circulating throughout its body, renews its hidden powers. Without iron the loadstone dies; its bulk wastes away from lack of nourishment and thirst parches its emptied veins.[40]

The last line of this excerpt presents for the first time the idea that magnets lose their power when not in contact with iron. It is not clear whether Claudian bases this claim on firsthand experience or on some bit of folklore he encountered. In any case, he clearly expresses an understanding of magnets as living creatures. His depiction of iron as a food that nourishes the magnet was well noted by others and repeated often.

As the poem continues, Claudian describes the scene at a temple where an iron sculpture of Mars, the god of war, and a lodestone sculpture of Cytherea (Venus), the goddess of love and beauty, are enshrined in symbolic marriage:

> Cytherea, without quitting her station, attracts her husband to her, and recalling the scene of which heaven was once witness, clasps Mars to her bosom with amorous breath.

We do not know whether or not these sculptures ever actually existed, but it has been suggested that, if they did, then fine hidden strings must have been used to hold the figures in their embrace, as it would have been simply impossible for iron and magnet sculptures of any size to maintain their attraction in such a pose.[41] If this still-life performance in fact did take place, then we can assume that the viewers thought they were witnessing a miracle, or else an act of magic. Leaving aside the question of the story's veracity, it is true that there was much talk in those days about the use of magnets and their power to act at a distance to perform miracles or magical feats. In *Natural History* Pliny writes of an attempt made in Alexandria to construct a temple with an arched ceiling made of lodestone from which an iron sculpture could be suspended (NH 34.42). It seems that the project failed to get off the ground, as it were, but Pliny's reference to it provides another example of how the masses of ancient Rome associated the magnet with magic and miracles.

The notion of magnets as magical objects seems to have come from Egypt. Earlier in this chapter I discussed Aelian, who wrote in the third century. Among Aelian's surviving works is *On the Characteristics of Animals*, which was written in Greek and is on a par with his *Historical Miscellany*. Its gist is summed up well in the following comment: "Aelian also accepted conventional wisdom about animals from the nonphilosophical side of Greek life: not only the popular wisdom of an Aristophanes or an Aesop, but also that of myths and legends."[42] *On the Characteristics of Animals* has this to say about magnets:

> If the bone from [a hawk's] tibia is put beside gold it attracts and draws it to itself by some inexplicable fascination, persuading it to follow even as, [the Egyptians] say, the stone of Heraclea [i.e., loadstone] somehow bewitches iron.[43]

This excerpt is especially interesting in our attempt to understand how the Romans understood magnets and magnetism. First, it transmits in a straightforward manner folklore handed down from ancient Egypt; and second, it presents a view of magnets as objects possessing supernatural and unequivocally magical power. Note that both of these elements are absent from the writings of Greek philosophers and scientists. Proclaiming magnets to be objects of magic was tantamount to declaring that no further theories about or analysis of their properties was possible; and having adopted this position, the Romans utterly abandoned the attempt, exemplified so clearly by Greek philosophers, to *explain* magnetism.

Presumably, local superstitions and folk religion existed in ancient Greece as well. These would have been spurred especially by contact with the non-Greek ways of the Orient as Alexander the Great pushed his way eastward, bringing more of the known world under his rule. And without doubt, folklore surrounding the profoundly mysterious powers of the magnet would have been part and parcel of that trend. It is said that Alexander had a high regard for superstition. In *The Curious Lore of Precious Stones,* George Frederick Kunz writes that the emperor had his soldiers carry magnets "as a defence against the wiles of the jinns, or evil spirits"[44] when they were away in battle.

Nonetheless, among the writings of the Greek philosophers there are almost no references to folklore or superstition with regard to magnets. And it is precisely on such grounds that ancient Greece is considered to have been the birthplace of science. I hasten to add, however, that Greece's claim to this title is due to the work of an extremely small number of scholars and that the overwhelming majority of the population was spiritually colored in a decidedly darker hue of superstition. Philosopher Bertrand Russell puts the point well: "I think that Greek literature and art are probably very misleading as regards popular beliefs.... We must not, therefore, judge of popular religion in Greece ... by the works of poets and aristocratic philosophers."[45] Although they may have left no record in the literature of their time, the ancient Greek masses were surely well steeped in primitive religious beliefs and superstitions concerning natural objects. Some two centuries before the birth of Christ, astrology in particular is known to have flourished amid the tumult of the half century leading up to Rome's conquest of Greece, as did the "theory of occult properties or forces immanent in certain animals, plants, and precious stones."[46]

Such theories took even greater hold in Roman society, spreading to the upper classes, including intellectuals and those of cultural refinement—as is evidenced by the excerpt from Aelian quoted above. In short, the decline of philosophy and science, integration with the Orient, and, most especially, contact with Egyptian civilization all combined in such a way that prescientific beliefs in the magical properties of magnets came to be shared by educated and ordinary folk alike.

* * *

And so in ancient Rome we see the formation of the prototype for the way in which the magnet and magnetism, as well as the forces of nature in general, were understood in the Christian medieval period. The view that magnets were in some way alive; the notion that magnets had not only physical but also physiological characteristics and, moreover, supernatural powers; and the idea that all of nature operated through a mesh of interlocking relationships based on either sympathy or antipathy—this entire perspective was formed in ancient Rome.

"The science of the Dark Ages," points out one historian,

> had a spiritual kinship with Roman science from its very beginnings. The symptoms are clearly seen in Pliny: inability to comprehend Greek science or to distinguish between absurd anecdote and sober theory, between ungrounded opinion and brilliant original thinking.[47]

As this suggests, Roman "science," led by Pliny, played a huge part in shaping the Middle Ages. Indeed, it is significant in this regard that Lynn Thorndike, the well-known investigator into the history of medieval science, begins his voluminous *A History of Magic and Experimental Science* with Pliny, not Aristotle. The departure represented by Pliny is likewise my reason for treating Rome separately rather than follow the example of most science historians, who lump Roman and Hellenistic civilization together under the rubric "Greco-Roman era."

The fact is that the Roman view of nature, and in particular the web of relationships the Romans called sympathy and antipathy, had a huge influence on medieval European thought all the way up until the Renaissance. In general, that is, natural objects came to be valued for their supposed supernatural powers affecting human fate and, more specifically, for their purported physiological, psychological, and pharmacological effects on the human body and soul. The power of magnets was a typical case in point: magnetism was considered and debated primarily as something with medical effects and more broadly as a phenomenon that influenced the body, the soul, and, ultimately, even destiny itself.

But with the Middle Ages came Christianity, a new thread woven into the intellectual web, and one that requires a new chapter in our story.

Chapter 4

# Christianity in the Middle Ages

## Section 1—Saint Augustine and *The City of God*

Having emerged around the Jordan River valley in Judea, a province that had been governed by Rome since the early days of the empire, Christianity soon spread throughout the entire Mediterranean region. Initially, its popularity was confined to the lower classes, its adherents suffering many years of persecution at the hands of those in power. As the Roman Empire declined, however, the new faith acquired a strong following among the upper classes as well. Christianity was legitimized by Emperor Constantine I in 313 and declared the state religion of the Roman Empire by Theodosius I in 380. This was just after the Germanic invasions began and a mere fifteen years before the Roman Empire was split into eastern and western halves. The penetration of the originally anti-Roman, anti-establishment Christian faith into the ruling classes of Rome was partly due to the missionary efforts of Saint Paul the Apostle, himself a Roman citizen. His teachings led to the modification of certain Christian tenets, and the pious nature of Roman society also contributed to the religion's adoption.

When Constantine converted to Christianity, in effect he was merely substituting Christ for the supreme deity associated with Roman sun worship. It is therefore hard to imagine that his conversion engendered any major religious conflict among the Roman governing class. Just as the oppressed and despised lower classes, which included slaves and people of the provinces, had initially sought salvation in Christianity, the ruling elite, fearful of imminent extinction, sought in Christianity the means to shore up a social order on the verge of collapse.

And so came the establishment of Christian society and the beginning of the European Middle Ages. The greatest ideologue in Christendom at that time was Saint Augustine (354–430), born in North Africa and eventually ordained bishop of Hippo (located in what is now Algeria). His philosophy determined the course of medieval thought, and his influence on the European mind was to continue, at least superficially, for close to a millennium.

Augustine equated Plato's world of Ideas with the kingdom of God in heaven, in relation to which mundane reality—the world of nature and man—was subordinate and evil. Consequently, he placed the study of nature beneath that of the Bible. *The City of God*, to the writing of which he devoted the later years of his life, was written in response to the sacking of Rome by Alaric, king of the Visigoths. Augustine is thought to have begun writing this work in order to refute the idea that God had forsaken the Romans; he laid down his pen, after twenty-two volumes, some fifty years before the fall of the Western Roman Empire. Essentially, the book declares that even if Rome were to be overrun by barbarians and the Roman Empire were to collapse, this would not signify the absence of God or the victory of paganism, for God does not concern himself with the vicissitudes of earthly states, nor is his kingdom on Earth. Augustine exhorts his readers to believe instead in God's kingdom in heaven.

Augustine refers to the magnet—our main concern here—near the end of the fourth chapter of book 21:

> We know that the loadstone has an astonishing power to attract iron; and when I first saw this phenomenon I was utterly amazed. I saw, I most certainly saw, an iron ring snatched up and held aloft by a stone! And then it seemed as if the stone had given its own power to the iron which it had snatched up, and made it a joint property; for this first ring was applied to another ring, which it lifted aloft; and the second ring clung to the first just as the first ring clung to the stone; in the same way a third ring was added, then a fourth; and in the end there was a kind of chain of rings hanging, the rings not being joined together internally, by the interlinking of their circles, but adhering to each other from outside. Who could fail to be astounded at this property of the stone, which was not merely inherent in it but also passed on through so many objects suspended from it, and bound them together by invisible connections?
>
> But much more astonishing than this is an experiment with the loadstone which I learnt about from my brother and fellow-bishop, Severus of Milevis. He told me that he had seen Bathanarius, sometime count of Africa, when the bishop was at dinner at his house, produce a loadstone and hold it beneath a silver dish on which he placed a piece of iron. Then he moved his hand, with the stone in it, underneath the dish and iron moved about on the dish, following his movements; and there was no effect on the dish in between, while the stone was being drawn by the man backwards and forwards underneath at extremely rapid speed, and the iron was rushing to and fro on top under the influence of the stone.[1]

Augustine was clearly amazed by all of this, but of course his discoveries were not new, Plato, Pliny, and others having recorded similar examples long before. Even the mention of the fact that magnetic force was not blocked by an intervening metallic dish was not without precedent: Lucretius had reported, in *On the Nature of Things*, the effect of a magnet on iron filings placed in a bronze bowl.

A few pages after this observation, however, Augustine writes:

> The loadstone by some insensible power of suction attracts iron, though it will not stir a straw.[2]

Obviously this remark was made to contrast magnetism with the amber phenomenon, the knowledge that amber attracts straw also being by that time nothing new. Augustine's understanding of the facts exceeded Pliny's only in the sense that he recognized somewhat more clearly that what we now know as magnetic force and electrical force were two separate phenomena. The larger difference between the two thinkers concerns the attitude each believed people should take toward such marvels as lodestone and amber.

In the same passage, Augustine cites, along with the power of lodestone, two other examples of marvels observed in daily life. One is the "fact"—which he claims to have confirmed for himself—that the flesh of a dead peacock does not putrefy. The other is the observation that quicklime is contrary to nature: whereas most substances, when burning, will be cooled and their fire extinguished by water while oil will feed their flames, burning quicklime is further heated by water but not by oil.

Leaving aside Augustine's claim about peacocks, we must concede that the observations about quicklime and magnets, at least, can easily be reproduced by experiment. His point is that even if man is incapable of explaining marvelous phenomena, their existence can be confirmed by observation. Furthermore, he reasons, if man is not capable of explaining even such common wonders as those found in everyday life, then the incomprehensible nature of God's miracles, which are far more rare, cannot be considered sufficient reason to refute their existence: "The fact that a rational explanation cannot be given for something," he argues, "does not mean that it could not have happened in the past, or that it could not happen in the future." Despite this, however,

> the unbelievers demand a rational proof from us when we proclaim the miracles of God in the past and his marvellous works that are still to come, which we cannot present to the experience of the unbelievers. And since we cannot supply this rational proof of those matters ... the unbelievers assume that our statements are false.[3]

In presenting this argument, Augustine asserts that the reason we cannot explain miracles is that "they are beyond the powers of the human mind." In other words, miracles and marvels are God's revelation; they are a manifestation of his greatness, and it is not for the frail and limited human mind to explain them. All that man is permitted to do is to discern God's will—salvation—revealed within nature.

In *The City of God,* Augustine shows no intention of searching for a logical and understandable *explanation* for the marvels of magnetism and magnetization. On the contrary, he regards the very desire to seek reasons for such natural wonders—what we now think of as simply "intellectual curiosity"—as something on a par with man's physical appetites, something forbidden, to be repudiated and avoided through self-restraint. In *Confessions*, his spiritual autobiography, Augustine is explicit about this:

> In addition to our bodily appetites … the mind is also subject to a certain propensity to use the sense of body, not for self-indulgence of a physical kind, but for the satisfaction of its own inquisitiveness. This futile curiosity masquerades under the name of science and learning, and since it derives from our thirst for knowledge and sight is the principal sense by which knowledge is acquired, in the Scriptures it is called *gratification of the eye*.… And for the same reason men are led to investigate the secrets of nature, which are irrelevant to our lives, although such knowledge is of no value to them and they wish to gain it merely for the sake of knowing. It is curiosity, too, which causes men to turn to sorcery in the effort to obtain knowledge for the same perverted purpose. And it even invades our religion, for we put God to the test when we demand signs and wonders from him, not in the hope of salvation, but simply for the love of the experience.[4]

This implies not merely that the study of nature as an end in itself is distinct from religious belief but rather that the two are in direct opposition. Even if Augustine's actual view is not quite so categorical, he certainly rejects reason unaccompanied by faith. In *Confessions*, he states bluntly that "the study of astrology does not interest me."[5] Indeed, many Christian intellectuals of that period knew nothing of the elaborate system of Ptolemaic astronomy and passed down a childish cosmology based on the Bible and Plato's *Timaeus*.

## Section 2—Forces Inherent in Natural Objects

Augustine's philosophy came to have a profound effect, particularly upon the European intelligentsia, throughout the Middle Ages.

When Gervase of Tilbury (c. 1155–c. 1234), for example, considers the magnet in his *Recreation for an Emperor*, written from 1209 to 1214, in both content and understanding his treatment of the topic is straight from Augustine.

*Recreation for an Emperor* is a compilation of remarkable things and events that Gervase came upon or read about in his travels around England and southern Europe. As the title implies, it was intended solely for the entertainment of Holy Roman Emperor Otto IV rather than as a scholarly work. I mention it here to illustrate how long Augustine's influence lasted. Gervase devotes only three paragraphs to the magnet, and these are found in the first chapter. The first of those paragraphs relates the example of the iron rings described in *The City of God*, the second deals with the iron in the silver dish, and the third reads as follows:

> We have recorded these things concerning the properties of stones in order that our limited understanding may marvel at what we are unable to explain, due to the ignorance of our weak nature. For, as Augustine says, when we talk about divine miracles, past or future, which we do not have the power to demonstrate to them in practice, unbelievers demand an explanation from us which we are not able to give. Miracles, indeed, are beyond the capacity of the human intellect. But then they reckon that much of what we say is untrue, although they are themselves unable to give an explanation even of things which we see every day.[6]

Here Gervase practically quotes Augustine verbatim, and his attitude of embracing ignorance—holding that marvels should simply be accepted as marvelous and that any further scrutiny is sacrilege—was one that Augustine had not only permitted but actively sanctioned. And because of Augustine's lasting influence in this respect, natural science was effectively halted, at least ostensibly, for close to a thousand years, with the result that rational understanding of magnets and magnetism saw virtually no further progress in Europe until later in the thirteenth century.

This does not mean, however, that all interest in magnetism was completely stifled. To be sure, we find during that period almost no natural science for the purpose of what Augustine referred to as "gratification of the eye," but the impulse to understand the workings of nature persisted. Furthermore, the ancient view of all natural objects as having not just physical and physiological but also supernatural powers not only was handed down throughout the Middle Ages but actually waxed stronger as time went on.

Specifically, not only does Augustine give credence to miracles; he also fails to refute the irrational folklore that was the legacy of Roman society. In fact, to take an example from the previous chapter, Augustine repeats without the slightest trace of criticism Pliny's absurd claim that the diamond prevents magnets from attracting iron yet is itself destroyed by goat's blood. In *The City of God*, he writes that "the diamond is a stone which, so we are told, neither steel nor fire nor any other force can prevail over, except goat's blood," and then the following:

> I will now add something which I have read about this loadstone: that when a diamond is placed next to it, the stone does not draw any iron; and if it has already drawn iron to itself, it lets it go as soon as the diamond approaches.[7]

In this way, many bizarre tales that originated in the Orient, Egypt, or Rome were legitimized and subsequently passed down through the centuries. The claim that diamonds obstruct magnetic force is recorded in *Etymologies*, written in the seventh century by Spanish archbishop Saint Isidore of Seville (c. 560–636) and noted as a comprehensive compilation of all scientific learning from ancient times to the Middle Ages.[8] Much later, the following appears in *Mandeville's Travels*, written around 1360 by English author Sir John Mandeville:

> ... lay the diamond upon the adamant [i.e., magnet] and lay a needle before the adamant. And, if the diamond be good and virtuous, the adamant draws not the needle to him while the diamond is there.[9]

Indeed, as we will see later, this bit of lore was cited uncritically by many thinkers—from well-known philosophers, such as Albert the Great in the thirteenth century and Nicolaus Cusanus and Pietro Pomponazzi in the fifteenth century, to the engineers and natural philosophers of the mid-sixteenth century, such as Vannoccio Biringuccio and Georgius Agricola—until it was finally refuted by Giambattista Della Porta in the second (1589) edition of *Natural Magick*.[10]

The idea that diamonds are destroyed by goat's blood is also cited in Isidore's *Etymologies*,[11] as well as in an eleventh-century work by Marbode, bishop of Rennes[12] and in the twelfth-century *Metalogicon* by John of Salisbury.[13] It is mentioned in German narrative epics, such as the late-twelfth-century *Erec* by Hartmann von Aue and the early-thirteenth-century *Parzival* by Wolfram von Eschenbach,[14] and even Albert the Great makes reference to it.[15] Moreover, despite Roger Bacon's mid-thirteenth-century testimony in *Opus majus* (Great work) that he had proved this premise false through experimentation,[16] it is repeated yet again even in a sixteenth-century book on alchemy written by Paracelsus in which he claims not to recognize the authority of ancient texts.[17]

In any case, it is remarkable, from our modern perspective, that people, including the foremost scholars of their time, could believe for a thousand years in such a strange network of sympathetic and antipathetic relationships: magnets exposing female infidelity, diamonds obstructing magnets, goat's blood destroying diamonds. I must repeat, however, that while the people of the Middle Ages regarded such phenomena with wonder, they did not doubt their existence. In this way, they continued to believe that natural objects had inherent power, without questioning whether the character of that power was physical, physiological, or supernatural.

## Section 3—The Absence of Medical Theory in Christianity

Augustine rejected science for science's sake. His was an attitude of expedience: whenever Christians needed knowledge of the natural or other secular sciences to help them interpret the Bible, they could simply borrow it from the pagans. In *On Christian Doctrine* he states that in his view the purpose of learning is "to seek the will of God" in all of the Scriptures.[18] The second volume of that work is a guide to the study of the Scriptures. There he explains that, because the Bible includes many "figurative expressions" based on natural phenomena, and because "an ignorance of things makes figurative expressions obscure," "things that have been written about the location of places, or the nature of animals, trees, plants, stones, or other objects ... are valuable for the solution of enigmas in the Scriptures."[19] In other words, active adoption of pagan culture is encouraged as long as it contributes to the study of the Bible, which means that all knowledge known to the pagan world is appropriate for inclusion in biblical curricula. Lacking its own theories of natural science, Christianity could not, in practical terms, ignore conventional learning.

The absence of inherent scientific principles in Christianity had an especially practical and direct impact in the area of medicine. The tradition of creative, theoretical Greek medicine, which began with Hippocrates, ended for the most part with Galen's death. Instead of further developing this heritage, Rome diminished it, partly due to the influences of indigenous folk beliefs and Oriental mysticism. Even from a generous viewpoint, the Roman Empire's contributions to medical science and practice were limited to military medicine and the concepts of public hospitals and urban sanitation. Virtually nothing new was added to theory. Medical practice in Roman society thus had a strong magical element, a fact reflected in the Latin verb *medicare*, which means both "to heal" and "to use magic." A comment by Pliny in his *Natural History* makes the point well: "Nobody will doubt that [magic] first arose from medicine"[20].

As for ascetic Christianity, in its infancy it placed medicine and medical treatment below faith. In his *Regulae fusius tractatae* (Greater monastic rules), Archbishop Basil of Caesarea (c. 330–379), who preceded Augustine by half a century, deems medical treatment itself to be against Christian doctrine, remarking that "to place one's hopes for health in the hands of a doctor is behavior befitting cattle." According to Basil, "illnesses for which medical treatment appears effective are not always due to improper diet or other bodily factors," because "illness is often an admonition of our sins intended to lead us toward a change of heart." In other words, suffering from sickness is often a divine punishment for original sin and an opportunity granted by God for atonement. When we become ill, therefore, we should "acknowledge our sins ... and silently endure the pain given us without relying on a physician."[21] Healing can be realized only through salvation by God, which appears in the form of a "miracle." This view rejected medicine as an autonomous branch of learning and most certainly repudiated any reliance on pagan theories. Indeed, as one historian puts it, "Greek medicine was a pagan art for which there was no room in the early church."[22] Although this type of fundamentalism could not persist forever, and compromises would eventually have to be made, Christian society essentially viewed our worldly sufferings as a test from God and therefore relegated medicine, which sought to cure the ills of the body, to a very low status. In fact, the Catholic church continued to affirm the connection between illness and sin until modern times.[23]

Once it had been adopted as the state religion of the Roman Empire and the compromises had begun, Christianity had to look beyond itself on matters pertaining to medical care. Utterly lacking in the fundamentals of both theory and practice, the church was forced to rely in large part on the meager legacy remaining from prior generations of Greek and Roman scientists and practitioners. In his *Etymologies*, for example, Isidore cited an explanation of humoral doctrine inherited from Greece that, although rather humdrum, was then passed down for the remainder of the Middle Ages. And the works of Discorides and Pliny, which one might have expected early Christians to dismiss as pagan documents, were widely used as medical references, particularly in matters of pharmacy.

But the search for medical and pharmacological knowledge was not confined to the legacies of Greece and Rome. Folk remedies derived from magic and spells indigenous to Gaul and areas north of the Rhine are thought, rather, to have dominated the medical practice of the day. Christianity, linked to political authority in the waning Roman Empire, promulgated its teachings within the upper echelons of provincial ruling classes as it expanded from the Mediterranean northward into the interior of the European continent, thereby strengthening its ties with secular powers and consolidating its organizational structure. The official adoption of Christianity by Gallic and Germanic leaders—their alliance with the Church—was a different thing from the permeation of Christianity into the masses, and there was a considerable time lag between the two processes. Accordingly, no matter how Christianity might have proclaimed its catholicity, in reality indigenous religions continued to affect the lifestyles and spiritual lives of the lower classes for much longer.

It is clear from recent research that the European Middle Ages was "a time during which, over a long period and mainly passively, paganism resisted the introduction of the new faith, which had the backing of an ever-expanding church."[24] Even in the Roman Empire, of which Christianity was ostensibly the state religion, "the countryside, in which the greater part of the empire's population lived, went on being pagan, in the sense of preserving local religious beliefs, for a long time."[25] The permeation of Christian ideology into every aspect of society did not occur overnight. Accordingly, medical treatment administered to the masses continued to consist of either Oriental medicine or else practices that were undeniably pagan and magical, such as folk remedies passed down in remote rural areas by midwives and spell-weavers and in cities by barbers (an occupation at that time considered disreputable).

In *On Christian Doctrine*, Augustine refers to healing spells and other forms of magical medicine as "superstition," declaring that "all arts pertaining to this kind of trifling or noxious superstition ... should be completely repudiated and avoided by the

Christian." Bear in mind that when he speaks of "superstition" he contrasts it not with scientific rationality but rather with Christian doctrine: he repudiates it not for being unscientific but rather for being a vestige of paganism. He believes in the existence of supernatural power, but he shuns its use because he shuns magic. He sanctions the administering of medicinal treatments whose effectiveness derives from natural powers, but the boundary between natural and magical power remains unclear; it is often impossible to determine which of the two forces is supposed to be at work when a remedy is said to effect a cure. Concerning popular folk medicine, Augustine goes so far as to say that "the Christian should avoid it the more cautiously the more it seems to be efficacious in doing good."[26] It is as if he admits from the outset that church orthodoxy is impotent where healing is concerned. The fact was that in the Middle Ages "magic and spells were prerequisite to healing from the start," and "the church itself cloaked magic practices in the garb of Christianity."[27]

The Middle Ages was a time of high infant mortality, when reaching adulthood unscathed was almost a matter of luck. The majority of the population were continually worn down by hard labor, nature's violence, and desperate poverty; their lives were consumed with the overwhelming tasks of protecting themselves from illness, defending their crops from nature's whims, guarding their families from the brutality of their rulers, and raising their progeny. On top of this, they lived in constant fear of the next plague, a fear that was shared by people from all classes of society, whether intellectuals or simple folk, farmers or city dwellers. Regardless of how inexcusable it may have appeared from the perspective of official Christian doctrine, it was inevitable that the masses looked to paganism for salvation, perpetuating magical customs from previous generations and using natural objects and symbols as talismans to ward off evil spirits. As historian Richard W. Southern notes, "secular studies had a life of their own, independent of any theory about their place in the general scheme of Christian knowledge."[28]

Consequently, a magical view of nature was passed on to successive generations, and statements concerning magnetism and electricity were understood within that framework. Having established this context, let us turn our attention to *De lapidibus* (On stones), written by Marbode, bishop of Rennes (c. 1035–1123), in the eleventh century. Though I make a great leap in time here from Augustine to Marbode, this is because during that interval the invasions of barbarian hordes resulted in the loss of much that was inherited from Rome—just as the making of that empire had come at the price of so much Greek culture laid to waste. It was not until the time of Charlemagne (Charles the Great) that culture on the European continent began showing signs of recovery.

## Section 4—Marbode's *De lapidibus*

Conventional histories of physics, and especially of electromagnetism, generally hold that early developments in the study of static electricity occurred slowly, beginning with the ancient discovery of the attractive force of amber and then continuing with the gradual discovery of other materials that exhibited the same type of action, until, with the growth of such empirical knowledge, static electricity eventually came to be recognized as a universal force not limited to specific substances. According to this established view, the first example of that process can be seen in an observation made by the Venerable Bede concerning the attractive force exhibited by jet. As leading advisor on ecclesiastical and educational affairs to Charlemagne, the Northumbrian scholar

Alcuin of York (c. 735–804) also made great contributions to the cultural restoration of the Roman Empire, but it was Bede who developed the kingdom of Northumbria into one of the most prominent and independent centers of learning in eighth-century Western Europe. In his *Ecclesiastical History of the English People*, written in 731, Bede describes jet as

> a black jewel which can be set on fire and, when burned, drives away snakes and, like amber, when it is warmed by friction, it holds fast whatever is applied to it.[29]

The next significant reference to static electricity is found in a long-overlooked eleventh-century description by Marbode of chalcedony, a translucent quartz:**

> Warmed by the sun's rays or, by rubbing of the fingers, it attracts chaff to it.[30]

That Marbode's reference was overlooked by later scholars may be because of the influence of William Gilbert, who in 1600 (as we will see in a later chapter) included chalcedony among substances

---

**Marbode's discovery of the attractive property of chalcedony (*calfectus radiis solis vel digitorum fricatu paleas ad se trahit*) has been completely overlooked in histories of electromagnetism. Joseph Priestley's *The History and Present State of Electricity*, first published in 1767, was the first history of electricity as a science. The corrected and enlarged fourth edition, published in 1775, states, "Amber and jet were … the only substances which, before the time of Gilbert, were known to have the property of attracting light bodies when rubbed." See Priestley, 2–3. Likewise, Marbode's discovery was subsequently ignored by both Edmund Hoppe and Paul Fleury Benjamin, who wrote histories of electricity at the close of the nineteenth century. Even in a fairly detailed treatise by Duane Roller and Duane H. D. Roller in the twentieth century, jet and amber are cited as the only substances to have been known to exhibit the so-called amber effect until Girolamo Fracastoro's discovery of the diamond in the sixteenth century. See Roller and Roller, "The Prenatal History of Electrical Science," 351.

that do not exhibit electricity. In any case, merely to muse about history's unfolding in hindsight, imposing on it a logically consistent, modern-day perspective, does not serve the purpose of understanding history as it was made moment by moment. After all, those who lived in the Middle Ages saw the phenomena we know today as electrostatic force and magnetic force from a completely different perspective and understanding of the world.

The above reference to chalcedony is from a passage Marbode wrote concerning the twelve jewels mentioned in the Apocalypse of John (Book of Revelation) in the New Testament. Marbode's best-known work, however, and one that had great historical influence, is *De lapidibus*, a 732-line didactic poem. This work's basic theme is encapsulated in four lines near the beginning:

> And hoar experience surely doth attest
> The native virtue by each stone possessed.
> Though in the herb a potent virtue lurks
> Greatest of all that which in jewels works.[31]

The poem describes sixty types of stone, the virtue (*virtus*) concealed within each one, and the ways in which they are useful to man. By "useful" Marbode implies not only their direct medicinal effects but also a much broader range of meaning, including, for example, the ability of various stones to confer peace of mind or ward off evil. The poem therefore contains information of both a practical, medical nature and a pagan, magical nature. According to science historian George Sarton, Marbode's lapidary "contained no Christian elements."[32] For this reason, some scholars have hypothesized that Marbode wrote the poem before his ordination as bishop of Rennes in 1096, but there is no solid basis for this argument. "In [post-Pliny] Latin literature," notes Max

Jammer, "the concepts of physical force and occult agencies are indiscriminately intermingled. The term *virtus*, in general, denotes them both."[33] Even more accurate would be to say that no clear distinction at all was made between those two types of power.

In the poem, Marbode discusses such phenomena as magnetic force and electrostatic force as typical examples of the spirits that inhabit stones and the magical powers held by jewels. In that sense, *De lapidibus* is a kind of magical inquiry aimed at identifying ways to extract and exploit the supernatural force within stones for the benefit of man. Let us explore this subject by examining the views of Marbode and other thinkers on jet and other stones.

Jet is a dense, granular, pure-black coal. Dioscorides claimed that it could be used to treat gout, diagnose epileptics by suffumigation, and cure hysteria in women, and that its vapors could heal latent gynecological conditions.[34] According to Pliny, not only was jet effective against toothache and "king's evil" (scrofula, a form of tuberculosis), but its smoke could expose someone posing as a virgin. Isidore also mentions this ability to determine virginity.[35]

In the eighteenth stanza of *De lapidibus* Marbode describes jet's virtues in a similar fashion. First, it cures dropsy (edema), dissolved in water it firms up loose teeth, its fumes help to bring on menstruation, it is effective in treating epilepsy and stomach disorders, and it hastens childbirth for women in labor; second, it wards off evil and serves as a test for virginity; and third, "to itself it draws / If warmed by friction near adjacent straws."[36] From a modern-day perspective, the first are physiological and medicinal properties, the second are supernatural and magical powers, and the third is distinguished by clearly signifying a physical force of attraction. This type of categorization, however, is possible only in a modern context; in Marbode's understanding, all three types of action were treated on the same level, as part of the same dimension. In particular, there was no discrimination between physiological and magical effects; both types were simply seen as potentially useful given a good understanding of the secrets of how they worked.

Stepping back to examine the way in which notions about stones changed over time, we can see that belief in their mystical, supernatural powers, far from waning after Dioscorides and Pliny, actually intensified even into the eleventh century, by which time Christianity had spread widely and its authority was pervasive. The idea that a certain stone called hyaenia had supernatural powers is a case in point. Pliny was skeptical of the claim: "'Hyaeniae' … are, it is said, obtained from the eyes of the hyena … When the stones are placed under a man's tongue, they are alleged to foretell the future, if we are foolish enough to believe such a thing."[37] Concerning this same stone, however, Isidore likewise remarks in his *Etymologies* that "hyaenia is a stone found in the eye of the hyena and they say that if it is placed under the tongue of a man he foretells the future";[38] and in the fourteenth stanza of *De lapidibus* Marbode tells us that "The Hyaenia … / On mortals can prophetic gifts bestow / And give the power the future to foreknow."[39] In the latter two cases, not a trace of criticism or even of skepticism remains.

In the nineteenth stanza of *De lapidibus* Marbode describes the purported effects of magnets. These include suppressing dropsy, dispersing pain from burns, and revealing infidelity in a wife, as well as fostering peace between antagonists, nurturing love between newlyweds, and conferring the gift of eloquent persuasion. He also relates that, before burgling a house, thieves would sprinkle powdered magnet onto smoldering embers in the belief that the smoke from the magnet caused the soul of the house's master to leave the building, making it possible for the thieves to do as they pleased. Similarly, he claims that the gemstone "Hebites" lessens the pain of childbirth, prevents miscarriages, causes wealth to accumulate, and makes people happy.[40] Thus Marbode was interested in gemstones not only for their medicinal virtues but even more so for their magical properties.

## Section 5—Hildegard von Bingen

The notion that stones have supernatural powers can be found scattered throughout European literature of the Middle Ages. One example is the epic *Parzival*, written by the German knight Wolfram von Eschenbach between 1200 and 1210. This work mentions various gems, including garnets and diamonds, and claims that "many of these stones make people feel better … and are also useful as medicine. Anyone who has a knowledge of these jewels and is capable of recognizing their value will acknowledge these effects."[41] Another example is *La quête du graal* (The quest for the grail), a work of unknown authorship written in the French language in the 1220s that refers to "the virtue of jewels, the force of herbs"[42] In *King Horn*, an English verse romance written around the same time, we find reference to a ring set with jewels of miraculous efficacy; [43] and in *The Romance of the Rose*, a popular French poem written in the latter half of the thirteenth century, we find these words: "The buckle was made of a stone that had great power and virtue, for he who wore it on himself feared nothing from any poison; no one could poison him."[44] Around the same time, Archbishop of Genoa Jacobus de Voragine (c. 1230–1299) wrote in *The Golden Legend* of "a precious stone … which gives sight to the blind, restores hearing to the deaf, enables the dumb to speak, and instils wisdom into the foolish."[45] *Reinke de Vos*

(Reynard the fox), a German poem written sometime between the mid-thirteenth and the mid-fourteenth century, tells of a Jew "with a profound knowledge of medicinal herbs and the efficacies of stones" and of a jewel that can cure myriad illnesses, remove all suffering, and save one from any disaster.[46] In the much later *Pandosto*, written by the English author Robert Greene in 1588, we find it observed that "the stone Hebites is not so much liked for the colour, as for vertue."[47] Still later, Christopher Marlowe, in his famous play *The Tragical History of Doctor Faustus*, penned around 1592, has a character declare, "He that is grounded in astrology / Enrich'd with tongues, well seen in minerals / Hath all the principles magic doth require."[48] It appears that knowledge of "minerals" (gemstones) was considered indispensable to magic. And in *The Faerie Queene*, by Marlowe's countryman and contemporary Edmund Spenser, we find mention of a wounded knight's ring that "did restore / His weakned powers, and dulled spirits whet, / Through working of the stone therein yset."[49]

The appearance of such references to the mysterious powers of gemstones in so many literary works suggests that these powers were also talked about widely among the masses. One thing we must keep in mind, however, is that medieval European literature represents the world as it was viewed by only a handful of Christian intellectuals. The overwhelming majority of the illiterate masses did not see nature on the same level and did not share the views expressed by ecclesiastics such as Augustine, Bede, Isidore, and Marbode, who were fluent in Latin and represented a very limited elite. My point here is that, since even the elite accepted the works of such pagans as Dioscorides and Pliny, we can assume that the magical view of nature remained much stronger still among the masses, who were more susceptible to the influence of folk beliefs. It is next to impossible to verify most oral tradition, which tends to vanish into the dark shadows of history without ever being set to paper. Fortunately, however, Hildegard von Bingen (1098–1179), a unique woman who served as abbess of a Benedictine convent in the Rhineland, has bequeathed to us in her writings a glimpse of the indigenous perspective on nature.

From a young age Hildegard had visions. She entered a convent at the age of eight but did not begin writing about her visions until the age of forty-three. Although she lived her life as a nun and rose to the position of abbess, in the male-dominated world of the medieval Christian church she was not necessarily ranked among the intellectual elite. She received a rudimentary education in the Bible and theological works at the convent, but unlike her male counterparts at monasteries, she was not offered a systematic curriculum of Scholastic learning or scriptural interpretation, and her Latin was reportedly unpolished. From 1150 to 1160 she wrote *Physica*, a treatise on natural science and medicine. In this, under such headings as plants, elements, trees, stones, fish, birds, animals, reptiles, and metals, she records over five hundred items, describing each object's medical and pharmacological properties as well as noting related information in the manner of a natural history. In her consistent classification of substances into the categories of hot, cold, wet, and dry, and also her ideas on healing "like with like," we can clearly discern the influence of the four-humor theory found in Greek philosophy and medicine. In the volume on stones she records twenty-six items, and her inclusion of magnets under stones rather than metals also suggests the influence of Greek and Roman natural history.

Quite unlike the other books of her time, however, there is no evidence in the individual items recorded in *Physica* that Hildegard based her work on the medical texts or encyclopedias written by

Dioscorides, Pliny, or Isidore. Rather, her work seems to have been based on the empirical knowledge of magical medicine accumulated by folk practitioners and passed down to succeeding generations. The book's content is, indeed, highly peculiar. One recent study maintains that Hildegard's *Physica* was derived from "oral transmission, based in Germanic tradition and supplemented by [women's] own experience and observation," and moreover that "no other extant medieval text gives such a detailed picture of popular herb-lore."[50]

Near the beginning of the volume on stones, Hildegard writes:

> Precious stones are born from fire and water; whence they have fire and moisture in them. They contain many powers and are effective for many needs. Many things can be done with them—but only good, honest actions, which are beneficial to human beings; not activities of seduction, fornication, adultery, enmity, homicide, and the like, which tend toward vice and which are injurious to people. The nature of these precious stones seeks honest and useful effects and rejects people's depraved and evil uses.[51]

Although she views stones through the moral lens of Christianity, we can infer from her statements that belief in powers abiding in stones was popular and widespread.

The eighteenth item in the same volume is the magnet. This entry is so exceedingly peculiar that I must quote it here in its entirety:

> Magnesian stone is hot [*magnes calidus est*]. It is born from the spume of certain poisonous worms [*venenosus vermis*] that inhabit particular sand and water, but are in sand more than water. One kind of poisonous, sluglike worm lives around particular water and dwells in it. It sometimes ejects its spume

> onto land in a certain area where iron is customarily forged. When another venomous worm, living around and in that water or feeding on the earth where the iron is made, sees it, he runs eagerly to that spume. He pours out his poison, which is black, on the other worm's spume. The poison forcefully penetrates it, hardening it into stone. And so, Magnesian stone has the color of iron and naturally pulls iron after it, because it is coagulated by poison nourished by earth which produces iron. The water next to which the stone lies attenuates and diminishes most of the poison by washing over it in frequent inundation.
>
> If a person is mad or in any way tortured by phantasms, smear the Magnesian stone with his saliva, then rub the top of his head and his forehead with the stone, and say, "You, raging evil, cede to that virtue by which God changed the strength of the devil who fell from heaven to human goodness." He will recover his senses. The fire of this stone is both useful and harmful: the fire it has from the iron-bearing earth is useful; the fire it has from the poison of the worms is harmful. When stirred up by warm, salubrious human saliva, it overthrows the noxious humors that disturb human understanding.[52]

Whether examined in terms of the formation of magnets, their strange powers, or their uses, this bizarre explanation is completely different from any of the Greek or Roman accounts that we have encountered so far. Perhaps such tales were passed down in the subcultures of Germanic societies enshrouded in dense forests far from the civilization of the bright, sunny Mediterranean. The chanting of spells to draw out the magnet's power, a pagan act of witchcraft clearly violating Christian teachings, was a superstition also probably derived from indigenous folk beliefs. Nor are magnets the only such example. Concerning agate, Hildegard advises: "Every night before a person goes to bed, he should carry a clearly visible agate through the length and then the width of the house, in the pattern of a cross. Thieves are then less able to exercise their wills and so profit less in thievery."[53] This unusual scholar thus did more than combine the sign of the cross with pagan folklore; by asserting that it was God's will that we use the power of earthly stones, she actively incorporated folklore and indigenous religion into the framework of Christianity.

## Section 6—Albert the Great's *Book of Minerals*

Strange notions about minerals, such as that adulteresses can be identified by magnets and virgins by jet, were passed down from one generation to the next throughout Christian-dominated medieval Europe. Concerning magnets in particular, it has been pointed out that from the second to the twelfth centuries

> the lodestone is known, principally through medical uses, uses in magic, and as an example of occult powers, in the West.[54]

German philosopher Albert the Great (c. 1200–1280), who served as a lector in the Dominican priories in Paris and Cologne in the mid-thirteenth century, summarized what was known about magnets at that time in his *Book of Minerals*. Appointed prior provincial of the German province of Teutonia in 1254 and special envoy to Pope Urban IV from 1263 to 1264, Albert was an elite of the highest order in the Christian church, and he was also renowned as the most erudite man of his time. The *Book of Minerals* was written in either the 1250s or the 1260s, around the same time as Hildegard's *Physica*. Despite its considerable length, I quote here in full his passage concerning magnets:

> Magnes or *magnetes* ... is a stone of an iron colour, which is mostly found in the Indian Ocean, [where] it is said to be so abundant that it is dangerous to sail there in ships that have the

> nails outside. It is also found in the country of the [Troglodites]. I myself have seen one found in the part of Teutonia called the province of Franconia, which was of large size and very powerful; and it was extremely black, as if it were iron rusted and burnt with pitch. [Magnet] has a wonderful power of attracting iron, so that its power is transferred to the iron and then that, too, attracts: and sometimes many needles are seen, thus suspended from one another. But if the stone is rubbed with garlic it does not attract. And if an *adamas* is placed on it, again it does not attract, so that a small *adamas* in this way [can] restrain a large magnet. In our own time a magnet has been found that attracted iron from one corner and repelled it from another. And Aristotle says that this is another kind of magnet. One of our Order, a careful observer, has told me that he had seen a magnet belonging to the Emperor Frederick, which did not attract iron, but on the contrary, the iron attracted the stone. Aristotle says that there is still another kind of magnet that attracts human flesh. In magic it is reported that [magnet] is marvellous for calling up phantoms, principally or especially if incantations and magic signs are used, according to the teachings of magic. And taken in honey-water, it is reported to cure dropsy. They say, too, that if the stone is placed under the head of a sleeping woman, it makes her turn at once to her husband's arms, if she is chaste. But if she is adulterous, she is so alarmed by nightmares that she falls out of bed. They say also that thieves entering a house place burning coals in four corners of the house and sprinkle upon them the powder of this stone; and then those who are sleeping in the house are so harassed by nightmares that they rush out and leave the building. And then the thieves steal whatever they want.[55]

In this book, Albert, like others before him, distinguishes between stones and metals and classifies magnets as stones. He also mentions two types of magnets: those that attract iron and those that are attracted by it. With this, he dispels the misconception that magnetic attraction is one-sided, an error that had been passed down from ancient Greece. Simultaneously, however, he replaces it with yet another misunderstanding—the notion that there are two different types of magnet. Della Porta, as we will see, was the first, centuries later, to point out the latter mistake.

As will be discussed in more detail in the next chapter, it was around Albert's time that Western Europe rediscovered Aristotle. Albert eagerly embraced Aristotelian natural philosophy and was one of the first to value the study of nature as an end itself rather than as a revelation from God. In fact, as Lynn Thorndike points out, "in his works on natural science Albert does not merely repeat past ideas whether of Aristotle or others, but adds chapters of his own drawn in large measure from his own observation, experience, and classification."[56] Yet this same thinker accepts all the stories about magnets handed down from generation to generation. He also treats all the magnet's supposed properties—the physical force of magnetic attraction, the spiritual ability to expose adultery, the medicinal effect of curing edema—on the same footing, as all belonging to the same category of "the powers of stones." Likewise, he accepts another baseless ancient relic, the notion that garlic renders a magnet incapable of attraction, believing it to be just as solid a fact as the power of magnets to attract iron.

These views remain unchanged even in the thinking of Albert's successor Thomas Aquinas, who perfected Scholasticism and is considered to have been one of the most learned men of the entire European Middle Ages. Thomas not only claims that

> a magnet attracts iron, and … a sapphire cures an abscess[57]

but also unhesitatingly affirms that

> if a magnet is greased with [garlic],[††] it cannot attract iron.[58]

As both phenomena are, in his eyes, of equal mysteriousness, he views them as equally real.

In the *Book of Minerals*, Albert, like Marbode, painstakingly describes peculiar powers of a wide range of stones in addition to the magnet. Concerning jet, for example, he asserts that "if rubbed it attracts straws."[59] Unlike in earlier theories, here heat is no longer given as a prerequisite for such attraction, and in this respect Albert's description removes a misunderstanding that had persisted from the time of Alexander of Aphrodisias and was repeated by both Pliny and Marbode. From this remark alone, we might be tempted to conclude that by Albert's time people had accumulated more experience of static electricity, that their observations had become more accurate, and that their understanding had deepened. But the truth is not quite so simple: in the same passage about jet Albert reports that "experience shows that if water in which it has been washed is strained and given with some scrapings [of the stone] to a virgin, after drinking it she retains it and does not urinate; but if she is not a virgin, she urinates at once" (brackets in source). Clearly, how one interprets "experience" varies according to one's perspective on it.

[††]The original Latin reads: "Si magnes aliis perungatur, ferrum attrahere non potest" [*Expositio in libros physicorum Aristotelis* (*Commentary on Aristotle's Physics*) 7.3.903]. The standard English translation by R. J. Blackwell et al. is: "If a magnet is greased with other things, it cannot attract iron" (p. 461). However, *aliis* should be regarded as the ablative plural not of the indefinite pronoun *alius* (others), as Blackwell et al. have it, but of the neuter noun *alium* (garlic).

Albert also makes the following general remarks:

> Many indeed seem to doubt whether there are in stones any of the powers which are regarded as belonging to them, such as curing abscesses, expelling poison, reconciling the hearts of men, bringing victory, and the like; and they assert that there is nothing in a composite substance except [what is due to] its constituents and the way they are combined.... But the opposite is proved most convincingly by experience: since we see that the magnet attracts iron and the *adamas* restricts that power in the magnet. Furthermore, it is proved by experience that some *saphirus* cures abscesses, and we have seen one of these with our own eyes.[60]

Even allowing his claim to have "seen one of these with our own eyes," we see nonetheless that Albert takes just a few limited examples of actually observed mineral phenomena—the medicinal effects of certain minerals, the effect of magnetism on iron—and uses them to justify and expand upon the idea that all stones possess particular powers of one kind or another.

Regarding magnets in particular, many magical and spiritual powers were presumably attributed to them. The use of incantations and other magic was believed to impart to the magnet even more marvelous effects, a "fact" plainly acknowledged by Albert himself and also attested by Hildegard. Another who wrote about magnets around the same time was the thirteenth-century Franciscan monk Bartholomew of England. His writings, too, contain legends about magnets' magical effects, but we also find among them the comment that "the magnet sometimes attracts iron from one angle and repels it from itself in another" angle.[61] In the long passage I quoted near the beginning of this section, Albert also refers to a magnet "that attracted iron from one corner and repelled it from another." The complexity and peculiarity of magnetic force were thus gradually being noticed, but proper comprehension and working out of the facts required a still-missing factor: the discovery of the directionality of magnets and magnetized needles.

* * *

According to Kunz's *The Curious Lore of Precious Stones*, the beliefs that spirits reside in stones and that gems are endowed with magical powers are found in folklore from every part of the world. In ancient Rome such ideas had probably been transmitted from Egypt or Asia, and they were certainly not compatible with Christian doctrine. It is known that the penchant for gemstones that pervaded the Roman upper classes was condemned by the early Christian church. This was not simply because it represented extravagance but precisely because gemstones were associated with magic. Far from dying out with the spread of Christianity, however, magical and pagan beliefs concerning the powers of stones actually became more firmly rooted. Zircon (jacinth) and emerald (smaragd), for example, were reportedly worn as protection against infection during outbreaks of the bubonic plague in the fourteenth century.[62]

Nor was such lore restricted to the Middle Ages. Oral traditions concerning the inherent powers of gemstones continued into the seventeenth century in Europe. Anselmus Boetius de Boodt, regarded as the originator of modern geology, asserted in 1609 that "it is very necessary to observe that many virtues not possessed by gems are falsely ascribed to them."[63] Nonetheless, half a century later, in a long treatise written from the then new mechanist perspective and titled *An Essay about the Origin and Virtues of Gems*, even Robert Boyle, the English natural philosopher who, along with Galileo in Italy and Descartes in France, was at the vanguard of the seventeenth-century scientific revolution, wrote: "I will not indiscriminately reject all the medicinal virtues that traditions and the writers about precious stones have ascribed to those noble minerals."[64]

Thus, the perception that gemstones and magnets were magical persisted unabated in Europe. In fact, Christianity never completely supplanted indigenous folk religion. Instead, it chose for the most part to quietly tolerate and coexist with pagan elements and folklore by cloaking them in Christian doctrine, though occasionally branding one or another for rejection as heretical or magical. Although medieval Europe was ostensibly dominated by Christian ideology, at a deeper level pre-Christian, pagan, indigenous, and folk spirituality lingered on and, in its own way, continued to arouse people's interest in nature.

This trend was particularly conspicuous in the field of medicine and healing. In the Middle Ages, medical treatment had a strong indigenous, magical, and pagan character. Interest in magnetism during this period was predominantly focused on its supposed medicinal properties, understood in a broad sense that included mental healing and psychic or spiritual effects. For this reason, it was intimately tied to the study of pagan and magical arts, despite the fact that European society was dominated by Christianity. Even leading scholars of the medieval Christian world—Marbode in eleventh-century France; Bartholomew in thirteenth-century England, and the illustrious thirteenth-century philosopher Albert the Great of Germany—openly proclaimed the magnet's supernatural, magical powers. This alone is a striking testament to magnetism's unique hold on the imaginations of the people of that time.

Chapter 5

# The Discovery of Magnetic Directionality

## Section 1—Medieval Society in Transition

The medieval European understanding of magnetism was for a long time deeply entwined with magic and witchcraft. In the thirteenth century, however, this began to change. The transition was represented by three men whose approaches to the matter were quite different from one another: Thomas Aquinas, a Dominican monk from southern Italy who perfected medieval Scholasticism; Petrus Peregrinus, an engineer from Picardy who penned the *Letter concerning the Magnet*, the world's first treatise on experimental physics; and Roger Bacon, a Franciscan friar from England who is regarded as the originator of empirical science. All three men were active during the same period, the latter half of the thirteenth century, particularly around the late 1260s: Bacon produced his main works, *Opus majus* (Great work), *Opus minus* (Lesser work), and *Opus tertium* (Third work), in England from 1266 to 1268; Thomas his *Commentary on Aristotle's Physics*, *The Soul*, and other works in Paris from 1269 to 1272; and Peregrinus his *Letter concerning the Magnet* in 1269 in Lucera, southern Italy.

That three such thinkers appeared at that time can be attributed to the fact that it was a transitional period in European society in general. In the case of magnetism, the introduction of the mariner's compass (magnetic compass) highlighted a new discovery: magnetic directionality. It was initially believed to be the North Star, or "Heaven's pole," that pulled the tips of magnetic needles northward, and directionality was seen as validating the astrological concept of heavenly bodies or heaven itself acting on terrestrial objects from a distance, a concept that would later have tremendous impact. Before delving further into this subject, however, let us take a brief look at the social changes that occurred in Europe at that time.

The Iberian Peninsula had been under Muslim rule since the eighth century, but during the Reconquista Christian Spain retook Cordoba and Seville in 1236 and 1248, respectively. In 1266 French nobleman Charles d'Anjou, acting under papal authority, conquered the kingdoms of Sicily and Naples, which at the time shared a unique dual culture incorporating elements of both Islam and Christianity. Thus, by the mid-thirteenth century, a large portion of the Iberian Peninsula, the entire Italian peninsula, and the island of Sicily had been secured under Christian rule, and Europe as we know it today began to take shape. This period also marked the beginning of European contact and trade with the previously little-known East. John of Plano Carpini returned from his mission as papal legate to the Mongol Empire in 1245, followed in 1255 by William of Rubruck, an emissary of France's King Louis IX to the same region, and the Venetian merchant Marco Polo and his father set out for China in 1271.

The European world was clearly undergoing a metamorphosis. At the institutional level, the first signs of the modern nation began to appear mid-century, first with the establishment of the Parlement of Paris around 1250 and then with the election of the first English parliament in 1265. By around the end of the century, "the nation gradually replaced the church community as the institution to which the people felt political allegiance"[1] and Europe "was divided into a grouping of regions which prefigure the later division into national states."[2]

These political and social changes did not happen in a vacuum. Rather, they were preceded by the discovery of the science and philosophy of ancient Greece through contact with the Islamic and Byzantine civilizations, as well as by almost two centuries of what can fairly be called the industrial and agricultural revolution of the Middle Ages. Noteworthy technological innovations included increased use and diversified application of the waterwheel as a source of power and, above all, marked technical advances in agriculture due to equipment improvements, notably the development of iron farming implements. In particular, wheeled heavy ploughs suitable for cultivating the heavy soils in the wet climate north of the Alps were introduced, along with the use of iron-shod horses to draw them. Cultivation methods also changed, with a shift in the eleventh century from a two-field to a three-field system of crop rotation. Thanks in part to a comparatively stable, mild climate, these changes resulted in a significant increase in agricultural productivity.[3] In addition, land clearing and reclamation by drainage expanded the total area of cultivated land, thereby contributing to a dramatic increase in the European population from the middle of the eleventh century. Population growth was particularly marked from the thirteenth century and continued into the early fourteenth century.[4]

Though occurring quite gradually by the timescale standards of today, this industrial and agricultural revolution had far-reaching effects, one of which was to encourage the formation and development of towns and cities in a wave of unprecedented urbanization. The rapid improvement in agricultural productivity eroded the traditional manorial system and shifted the economic base from autarky to trade of surplus products. This, coupled with the relaxation of feudalistic restrictions, bode well for increased stratification of the peasant class and also led to the development of towns as commercial and transportation centers that played an increasingly important role in society.

This trend became especially pronounced from the beginning of the thirteenth century. In France during the reign of Louis IX (r. 1226–1270) the population of freemen residing in towns increased and the scope of their activities expanded. Moreover, in order to strengthen central authority, the king appointed members of the urban elite to positions within the structure of the government itself, resulting in the emergence of an intellectual bureaucracy and a governmental framework befitting a modern nation. To secure sufficient wealth for the national treasury, Louis IX also strengthened his ties with the increasingly powerful merchant class (burghers) and in return granted towns more local autonomy and privileges. In Germany, the first self-governed city had emerged in 1190, and this wave of urbanization continued until the mid-thirteenth century, with city dwellers acquiring ever-greater political and economic power. From the twelfth to thirteenth centuries, groups that did not fit into the traditional three-class hierarchy of religious leaders (clergy), fighters (nobility and warriors), and laborers (peasants) emerged and began to make their presence felt in such new roles as bureaucrats, merchants, and skilled tradesmen and other manufacturers—in short, the future bourgeoisie. By furthermore acquiring literacy for commercial purposes, this new class laid the foundation for an entirely new culture and brought an end to the clergy's monopoly on written culture.

Parallel with the development of cities and towns, from the late twelfth century came the emergence of the university as an educational institution. Several universities—Paris, Bologna, Padua, Montpellier, and Oxford—grew out of master-student associations or guilds, and by the mid-thirteenth century these were firmly organized and established. Thus, by that time the transition from the monastery and cathedral school to the university as the arena of higher education was almost complete[5], ending the monastery's role as the custodian and transmitter of knowledge, a role it had carried out, after a fashion, since the sixth century. One further change worth noting from this period is the appearance of the mendicant orders, including the Franciscan, established in 1209, and the Dominican, established in 1216. Unlike previous orders, which sought to distance themselves from the secular world in favor of cloistered religious activity, the mendicant orders were based in cities and actively involved in society. Because they viewed scholarship as a central component of their religious precepts and emphasized the importance of thorough research, they were able to supply the emerging universities with highly capable human resources. Indeed, "most of the great thirteenth- and fourteenth-century theologians were members of the mendicant orders."[6]

Having thus briefly reviewed the social environment that heralded the emergence of Thomas Aquinas, Petrus Peregrinus, and Roger Bacon, we are now ready to consider the intellectual and philosophical changes that took place in Europe during that period.

## Section 2—The Rediscovery and Translation of Ancient Philosophy

Contact with Muslim civilization, which was far more advanced economically and culturally than the agrarian Christian kingdoms of the day, was the decisive factor in the transformation of the medieval European perspective on nature and approach to its study. Through exposure to that civilization, Europeans discovered not only the learning developed by the Muslim world but also the science and philosophy of the ancient Greeks, particularly Aristotle, that Muslim scholars had preserved and studied.

People tend to associate medieval European-Muslim contact with the Crusades, which raged from 1096 until in effect coming to an end in 1270. In fact, however, that protracted conflict, almost two centuries of barbaric military campaigns, brought Europe little in the form of cultural benefit. Rather, as we will soon see, European access to the wisdom and learning of the more advanced Muslim civilization was obtained far from the clamor of the Crusades.

Perhaps the earliest thinker in Latin Europe to recognize the superiority of Islamic science[§§] and to devote himself to acquiring and transplanting that knowledge was Gerbert d'Aurillac (c. 950–1003). Gerbert is an intriguing figure well worth a brief digression from our main story to note his contribution to its context.

---

[§§] "Islamic science," or "Arabic science" as it is sometimes called, refers here to the science conducted by people who drove cultural development using the Arabic language from the mid-eighth century into the fifteenth century in lands conquered by Islam. Such scholars included not only Muslims but also Jews and Nestorian Christians, and not only Arabs but also Persians, Turks, and Hebrews. For convenience, I refer to the works of all of these as "Islamic science."

Gerbert was born in the middle of the tenth century, was probably of poor peasant stock, and was educated as a monk in a Benedictine monastery. For a child of peasants, monastery life was at that time virtually the only means of escape from that class. After receiving what appears to have been a very thorough education, in 967 Gerbert went to Catalonia in northeastern Spain, where he spent three years studying mathematics, astronomy, and music. As the majority of the Iberian Peninsula was under Muslim rule, it is certain that he came into contact with Islamic science while studying there. Being blessed with exceptional talent and good fortune, he was selected by the pope to serve as master of a cathedral school in Rheims, where he gained renown as an outstanding educator and brilliant scholar. In 999 he became the first French pope, in which capacity he retained the title Sylvester II until his death in 1003. In light of the rigidity of medieval society, Gerbert's rise from lowly origins to such an illustrious station solely on the merits of his ability is extraordinary, and in that sense he led a very modern life. He was also very modern in his thinking, a fact borne out by his reported proclivity for "seeking consolation in philosophy rather than in prayer." While on the one hand a follower of orthodox beliefs, at the same time he sought to combine faith with reason, claiming that "the Divinity … made a great gift to men in giving them faith while not denying them knowledge."[7]

Gerbert introduced Islamic astronomy and mathematics to Europe. He studied Ptolemaic astronomy, which had been passed down in Islamic society, and used this knowledge to make a sophisticated armillary sphere. He is said to have introduced arabic numerals into European mathematics, making calculations far easier than they had been with roman numerals, and to have revived the use in Europe of the abacus, which had been known to the ancients.[8]

Peter the Venerable (c. 1092–1156) was another pioneer who was convinced that European culture lagged far behind that of the Orient. At a very young age, he was appointed abbot of Cluny, the greatest abbey in Europe and mother-house to many other abbeys over which it maintained autonomous control. Acutely aware of the West's marked inferiority in terms of both culture and access to information, Peter realized the limitations of trying to conquer the Islamic world by force. He therefore spent vast sums of money purchasing Greek and Islamic literature, assembled a team of translators, and strove to introduce those works to the Western world. Although motivated not by religious tolerance but, on the contrary, by a desire to expose the errors in Islamic doctrine and thereby triumph over heretics through reason and ideology, he nonetheless deserves credit for his attempts to find solutions to that great cultural conflict through dialogue. The first Latin translation of the Koran was produced under his direction.

Obviously, the emergence of such men as Gerbert and Peter the Venerable could not have occurred without European contact and communication with Islamic society, and much of that contact occurred on the Iberian Peninsula and the island of Sicily.

At the beginning of the eighth century, Muslims overthrew the Visigothic Kingdom on the Iberian Peninsula and established an Islamic society in Spain that continued for more than seven centuries until the fall of Muslim-ruled Granada in 1492. Likewise, Sicily, once home to such luminaries as Empedocles and Archimedes, was conquered by Arabs in 902 after a brief post-Roman period of Ostrogothic rule and three centuries under the Byzantine Empire. Sicily remained under Muslim rule until it was retaken by the Normans in the second half of the eleventh century. In both Sicily and the Iberian Peninsula, the result of Muslim rule was the establishment of a culture and economy that far surpassed the rest of Europe. The Arabs, who possessed highly advanced irrigation technology, constructed irrigation works that transformed the two regions into fertile farmland and introduced to European soil such new crops as cotton, mulberry, sugar cane, palm, and oranges, and these efforts dramatically improved both the quality and quantity of agricultural production. The Muslims also developed mines; took up beekeeping, horse breeding, and silk production; and promoted commerce. By the ninth and tenth centuries, they completely controlled all shipping on the Mediterranean, with a firm hold on the ports at Palermo and other key cities. Both Palermo and Cordoba flourished, expanding into thriving cities with populations of over three hundred thousand in the tenth century,[9] unrivalled even by Latin Europe's largest cities, Paris and Rome.

Nor was prosperity the only remarkable feature of these Muslim societies. Although merciless in their response to those who took up arms against them, the Muslims accepted those who submitted peacefully, and never forced Christians or Jews to convert to Islam. According to one expert,

> even when conquered by Islam, Christian churches retained their civil right to be the spiritual leaders of their congregations. Moreover, they were permitted to keep and acquire property and to receive donations. Muslims were forbidden to interfere in the lives of Christians, be it concerning their doctrine, their faith, or their church regulations. With the exception of slaves, all Christians, whether clergy or laymen, were given complete freedom to travel anywhere in the Muslim or even the non-Muslim world.[10]

Of course there were restrictions, and undoubtedly there was social discrimination as well. Christians were strictly forbidden to proselytize their faith to Muslims or to express contempt of Islam. But on the whole, Christians and also Jews were, as "peoples of the Book," allowed to practice their faiths and coexist peacefully with their Muslim neighbors as long as they paid a special tax.

From the seventh century, after the death of the Prophet Muhammad, the formerly nomadic Arabs had embarked on a large-scale campaign of conquest. This not only served to expand their territory and economic activity but also contributed to rapid cultural growth, because the Muslims not only were tolerant of other religions but moreover enthusiastically embraced the culture and technology of the peoples they conquered. Islamic society, including Turks and Persians, absorbed the philosophy, medicine, and other sciences of the Greeks via the Byzantine Empire as well as the mathematics and astronomy of India. Translation of works into Arabic was enthusiastically promoted, an undertaking that owed much to papermaking technology transmitted from China around the mid-eighth century. The House of Wisdom in Baghdad, which was founded in the ninth century by the Abbasid caliph and became the center of Islamic scholarship, was originally established for the express purpose of translating Greek texts into Arabic. The philosophy and science of the ancient Greeks, long lost by the Latin world, was thus carefully preserved and eagerly studied in the Islamic world. Moreover, Muslim scholars made many contributions of their own in such fields as mathematics, optics, medicine, and astronomy.

European contact with Islamic civilization, and through it the rediscovery of ancient culture, accelerated with the reconquest of the Iberian Peninsula and Sicily. Although revolts by Christian armies against Islamic rule had occurred sporadically from almost as soon as the Muslims conquered the peninsula, it was not until the fall of the Umayyad Caliphate in 1031 that Christian Spain gained military superiority and the Reconquista gathered momentum. With the capture of Toledo by Alfonso VI in 1085, most of the northern half of the peninsula surrendered to Christendom. At first, the conquering Christians did not expel the Jews or Muslims who remained in Spanish territory under their rule. While this was only a temporary policy owing to the Christians' lack of the necessary manpower to reoccupy the territory themselves, the result was the creation of cosmopolitan centers like Toledo and later Cordoba (recaptured by the Christians in 1236) where the highly advanced Arab-Islamic culture intermingled with Latin-Christian and Jewish cultures.

In Sicily, it was a force of Norman mercenaries that overthrew Muslim rule. After consolidating their foothold in southern Italy, then under the control of the Byzantine Empire, the Normans launched their invasion in 1061, just several years before the Norman Conquest of England, and by 1091, five years before the Crusades began, they had the entire island of Sicily in their possession. When they took Palermo in 1072, the Norman mercenary leaders, Robert Guiscard and his brother Roger I (who later became Count of Sicily), wisely rejected the idea of stamping out Islam and instead sought conciliation with people of different religious and ethnic backgrounds. This appeasement policy continued without change after the coronation of Roger II in 1130, which marked the establishment of the Norman Kingdom of Sicily, and throughout the subsequent reigns of William I and William II.

Under Norman rule, Latin, Greek, and Arabic were all used as official languages; both the Julian and the Islamic calendars were used; and Roman law, the Koran, and Norman common law were all respected. Muslims, Byzantines, and Jews were appointed to important posts, and it is said that Arabs and Greeks made up the core of government administration while Muslim troops comprised the main force of the kingdom's army. Some observers conclude from this that the conquering Normans practiced religious "tolerance," but their realistic self-interest probably played a greater role. The majority of the island's population was either Arab or Greek, and moreover the Arabs were at the hub of economic activity. The rustic Norman knights, skilled only in warfare, were forced to depend on the Arabs to successfully run their large commercial center, and they needed to keep Muslim military forces in their service because they did not completely trust the pope. At the same time, being country folk they were awed by the dazzling Palermo culture that Muslim rule had produced and overwhelmed by the city's tremendous economic vitality. In Sicily, as on the Iberian Peninsula, "the defeated culturally enslaved their victors."[11] Thus, in that corner of medieval Europe, too, there emerged a different world—a culture within a culture.

Raised in Palermo, the crown jewel of Arab and Byzantine culture, Roger II, the first ruler of Sicily's Norman Kingdom, stabilized the kingdom through skillful diplomacy and thereby made it the most prosperous in Western Europe. A cosmopolitan fluent in French, Latin, Greek, and Arabic and an educated man who loved learning, Roger gathered scholars from Europe and the Islamic world to his palace. This special regard for scholars continued under subsequent rulers. In the late 1150s Henricus Aristippus, a top adviser to William I, brought back copies of many works from the Byzantine imperial library and translated Plato's *Meno* and *Phaedo* as well as some of Aristotle's *Meteorology* from the Greek. Similarly, Emir Eugenius, another high-ranking official in the Sicilian government, translated Ptolemy's *Optics*.[12] For almost a century until it was seized by Holy Roman Emperor Henry VI in 1194, Palermo was the most cosmopolitan city in Europe, a fusion of Latin-Christian, Greek-Byzantine, and Arab-Muslim cultures that provided a breath of fresh air in a medieval Europe stifled by Christian dogma. It was the strong-willed Frederick II, son of Henry VI, who subsequently pushed Palermo once again to the height of culture, a development to which we will return later in this chapter.

As mentioned earlier, when Europe came into contact with the Muslim world on the Iberian Peninsula and Sicily from the end of the eleventh century and into the twelfth century, she gained access not only to a sophisticated Muslim culture but also to an impressive legacy of ancient Greek philosophy and science, especially medicine and physics. Progressive Europeans who recognized the high scholastic standards of the Greeks and of Muslim culture and technology—or, perhaps more accurately, who were shocked by the inferiority of their own culture—vied with one another to translate such works into Latin. This translation movement, which began in the early twelfth century and continued until the mid-thirteenth century, was accelerated by the occupation of Constantinople in 1204 during the Fourth Crusade and the subsequent influx of many handwritten copies of such works into Europe. In the century and a half leading up to 1260, large portions of the oeuvres of such Greek philosophers and scholars as Aristotle and Archimedes were translated into Latin from Arabic or the original Greek with the help of polyglot Jews in the former Muslim centers of Toledo, Cordoba, and Palermo, as well as in such Byzantine trade centers as Venice and Pisa.

The quantity of translations completed during this period is mind-boggling, as evidenced by the list of these works included in the first volume of Alistair Cameron Crombie's *Augustine to Galileo*: a cursory glance is enough to convey the energy and passion that the leading intellectuals of Europe at that time devoted to absorbing the new learning they encountered. This deluge of sophisticated Muslim and Greek scholarship and philosophy swept across Europe as if a dam had burst, and it formed the basis for later scientific development in Western Europe. The establishment of universities in the twelfth century is intimately linked with this movement: "Indeed," notes one historian,

> the university was the institutional means by which western Europe would organize, absorb, and expand the great volume of new knowledge; the instrument through which it would mold and disseminate a common intellectual heritage for generations to come.[13]

The fact that medieval universities studied science not through observation and experimentation but rather through books would subsequently become a hindrance to further progress in science. This is a topic I will address later; suffice it for now to say that it was from this period that Western Europe began to take off intellectually.

## Section 3—The Introduction of the Mariner's Compass

With the introduction of the mariner's compass, medieval Europeans discovered the phenomenon of directionality both in magnetized iron and in magnets themselves. Precisely how and when the use of compasses began is a mystery and somewhat outside the scope of this book, the chief purpose of which is to examine how human understanding of magnetism affected the development of the concept of force in physics. Nonetheless, because awareness of directionality entailed a profound shift in European understanding of the magnet, it is worth taking a brief look at the historical record for the few hints it offers about the specifics of how this discovery took place.

Could the new knowledge have come from China? At the end of the eleventh century Chinese scientist Shen Kua (c. 1031–1095) recorded in his work *Brush Talks from a Dream Brook* (also known in English as *Dream Pool Essays*) the fact that iron needles rubbed against a magnet point north-south. This is the oldest confirmed report of the phenomenon of directionality.[14]

At the end of the sixteenth century William Gilbert claimed that Marco Polo had brought back knowledge of the mariner's compass from China,[15] but in fact European mariners were already using the compass a century before the explorer returned from the Orient in 1295. Another frequently encountered theory is that the Islamic world acted as a conduit, passing this knowledge from China to Europe, and the eminent historian George Sarton suggests that "Muslim sailors were probably the first to apply the directive property of a magnetic needle to navigational purposes."[16] But there is no concrete evidence supporting any of these theories. It is more plausible to suppose that Europeans discovered the compass on their own, as references to this device are seen at earlier dates in Western European literature than in Islamic literature.[17]

But where in Europe might the compass have originated, and when? In the *Dictionary of the Middle Ages*, published in 1983, the entry "compass, magnetic" declares that the device "is believed to have been invented in the Mediterranean region, probably at the Italian port of Amalfi, which was engaged in shipping magnetic ores from the mines of Elba."[18] This story originated in the fifteenth century with the Italian poet Antonio Beccadelli and the historian Flavio Biondo, was recorded in works by Della Porta and Gilbert in the sixteenth century, and continues to be mentioned to this day.[19] Judging from the account that, around the time the Norman Kingdom of Sicily was established and the Crusades began, "ships from Italian cities became the undisputed rulers of the entire [Mediterranean] sea region,"[20] it certainly seems likely that at that juncture in history some form of magnetic compass was either developed in the Mediterranean or else became known to sailors there through contact with their Islamic or Byzantine counterparts. There is, however, no proof.[¶¶]

Another theory suggests that the Vikings used magnetic needles and passed the knowledge on to their descendants, the Normans. As evidence, proponents of this theory claim that the extensive sea voyages involved in the Vikings' discovery of Iceland in the ninth century and of Greenland in the tenth century would have been impossible without compasses.[21] This opinion is supported by the following statement in Peregrinus's *Letter concerning the Magnet*:

---

¶¶ Weighing in favor of the Amalfi-origin theory is the existence of fifteenth- and sixteenth-century mariners' compasses equipped with compass cards and inscribed with directional names, such as Greco (northeast), Libeccio (southwest), and Scirocco (southeast), known to have been used by Mediterranean sailors to denote directions. The magnetic compass said to have been invented at Amalfi may have been just this type of advanced compass, equipped with a compass card. One source provides a report of a compass said to have been manufactured in Amalfi between 1295 and 1302; that compass (called a *bussola*) is said to have had an almost perfect compass card showing thirty-two directions and to have been housed in its own box (*bussolo*). See Hazard, "Early History of the Mariner's Compass," 180; Aczel, *Riddle of the Compass*, 61; and E.G. R.Taylor, *Mathematical Practitioners*, 7, 97, 106.

> Such a stone, in general, is found in northern regions, and is reported by sailors in all ports of the northern seas, as for example, of Normandy, Picardy, and Flanders.[22]

Once again, however, this is only conjecture.

But when did Europeans themselves first make reference to the use of magnetic needles for navigation on the open ocean? As far as we know now, the earliest such mentions are by Alexander Neckam (1157–1217), an English monk-scholar of Saint Albans; by the French poet and monk Guyot de Provins (fl. 1184–1210); and by the bishop of Acre in the Kingdom of Jerusalem, Jacques de Vitry (1165–1240).

Neckam, in his *De naturis rerum* (On the nature of things), appears to have been the first. (There is some discrepancy among scholars as to when this work was actually written, but many agree that it appeared sometime around the end of the twelfth century.) Neckam reports that

> sailors crossing the sea, when because of overcast skies they lose the sun's light, or when the world is wrapped in the darkness of night, and they do not know what cardinal point the ship is headed toward, put a needle above the lodestone; and the needle revolves until, after its motion has stopped, its point faces due north.[23]

Neckam had traveled to Paris and Italy, and it is thought that he came upon this notion while somewhere in Italy. Note, however, that the description is rather vague, which suggests that he was recording inaccurate hearsay, not personal experience or firsthand observation. It is also possible that some of Neckam's words were inadvertently omitted from the relevant copy of *De naturis* *rerum*, which was handwritten. Perhaps, for example, after "put a needle above the lodestone" he had originally included some such elaboration as "and allow the needle to revolve freely [by some means or other]."***

In *Cosmos*, Alexander von Humboldt, the foremost naturalist and geographer of the early nineteenth century, claims that the first recorded mention of the compass in Europe was by French poet Guyot de Provins[24] in *La bible*, a poem over two thousand lines long. A prose translation of the relevant part is as follows:

> There is an art which the sailors have, which cannot deceive. They take an ugly brown stone, the magnet, to which iron willingly attaches itself, and touching a needle with it, they fix the needle in a straw, and float it on the surface of water, whereupon it turns infallibly to the Pole-star. It makes no

***In his *De nominibus utensilium* (On the names of useful goods), Neckam writes that "he who wants to have a well-equipped ship [should] have a needle under a dart; for the needle will rotate and revolve until the point of the needle looks towards the East, and thus sailors perceive in which direction they ought to go, when the Little Bear is hidden in disturbed weather." (Quoted in May, "Alexander Neckam and the Pivoted Compass Needle," 283 [Latin and English]. This passage can also be found in Hellman, "Die Anfange der magnetischen Beobachtungen," 127 [Latin]; Smith, "Precursors to Peregrinus," 34 [Latin and English]; and Benjamin, *Intellectual Rise in Electricity*, 129 [Latin and English].) This statement has long puzzled scholars. In 1858, d'Avezac proposed that this description resulted from a mistake in transcription of the original and that "under a dart" (*jaculo suppositam*) should be "upon a dart" (*jaculo superpositam*) and "toward the East" (*orientem*) should be "toward the North" (*septentrionem*) (see Benjamin, *Intellectual Rise*, 129). The intended meaning would thus have been "a needle supported on a pivot" and "the point of the needle points toward the North." This remains the subject of some controversy, but as it is not the intent of this book to explore such arguments, I refer interested readers to the studies cited in this note. In any case, we can surmise that the compass to which Neckam referred was a so-called dry compass, that is, one having a needle that pivoted in air rather than moving on the surface of water.

> mistake about this, and no mariner doubts it. And when the sea is all dark, and no stars nor moon can be seen, they light a lamp with which to see the needle, and thus they go not astray from their course.[25]

Although the time when this was written cannot be identified with any precision, it was around the end of the twelfth or the beginning of the thirteenth century, about the same time or perhaps a little after Neckam's *De naturis rerum*.

Then there is the following excerpt from Jacques de Vitry's *Historia Hierosolimitana*, a history of the Holy Land and the Crusades up to the year 1218:

> There are … in eastern regions precious stones of admirable virtue and incredible rarity. The adamant is found in farthest India, and is of a clear and reddish colour; its size does not exceed that of the kernel of a hazelnut. It is so hard that it resists all other metals, but it can be broken down by the fresh and warm blood of a goat. Fire does not cause it to grow warm; it attracts iron unto itself through some hidden power. After an iron needle touches the adamant it always turns towards the North Star, which, as though it were the axis of the firmament, is not moved, but the other bodies revolve around it.[26]

He goes on to claim that

> when placed next to a magnet, [the adamant] does not allow the latter to attract iron.

This also appears to have been written around the beginning of the thirteenth century, but the confusion of the properties of both diamonds and magnets in the single word *adamant* (*adamas*), indicates that this report was also probably based on hearsay.

Note that the above statements refer to the compass not as a newly invented device but rather as an object of common knowledge, suggesting that magnetic needles had been in use for some time before those pieces were written. Immediately following the abovequoted passage from *De naturis rerum*, Neckam remarks, "So should a prelate guide his subjects in the sea of life," using the magnet as a metaphor for the attitude that should be adopted by high-ranking church figures. Similarly, the quoted lines from Guyot's poem are preceded by the words "Our apostolic Father ought to be the Pole-Star, which never moves" and followed by "This art is unfailing … so also ought our Father to be"—a criticism of the then incumbent pope's irresolution, which Guyot contrasts with the constancy of both the star and the north-pointing magnetic needle.

In all probability, sailors began using the magnetic needle much earlier than was first recorded by the monks who heard tale of this technological innovation. For one thing, we can assume that seamen were not in the habit of recording in writing the nature or content of their work. Historian Lynn Thorndike notes that the thirteenth-century theologian Thomas of Cantimpré touched upon the mariner's compass but also that "those in possession of the secret of the mariner's compass were long afraid to reveal it for fear of being suspected of magic."[27] A. Crichton Mitchell makes the same observation in his painstakingly detailed research on the history of the mariner's compass as recorded in ancient and medieval literature.[28] Furthermore, it is highly conceivable that the use of the magnetic needle was considered a commercial secret among traders.[29] Yet while such details remain shrouded in mystery, the impact of the compass on European understanding of magnets and magnetism cannot be disputed.

## Section 4—The Discovery of Magnetic Directionality

By the turn of the thirteenth century, then, it was known at least in southern Europe that an iron needle rubbed against a magnet would align itself north-south, and this knowledge was exploited in navigation. A clear distinction should be made, however, between that discovery and the discovery of the north-south polarity of magnets (natural lodestones) themselves, which is another matter. Nonetheless, many historians have failed to make that distinction, viewing the two discoveries as one and the same.

In a book written at the end of the nineteenth century, for example, historian Park Benjamin, having just discussed in detail a reference Neckam made to the use of magnetic-needle compasses, notes as follows:

> This operation must have required considerable thought for its invention. It was necessary to discover, *first*, that a lodestone bar would, when free to turn, place itself longitudinally in a north and south direction: *second*, that an artificial lodestone could be made by rubbing a needle with the natural lodestone: *third*, that such a needle would place itself north and south in the same way as a lodestone ...[30]

But at the beginning of the thirteenth century Europeans knew only that an iron needle rubbed with a magnet pointed north, and that was all they needed to know to make or use a compass. They were as yet unaware that magnets—for them this meant natural lodestones—aligned themselves north-south. Moreover, the fact that iron rubbed with a magnet becomes a magnet as well was something not understood until the modern age. In medieval times lodestones and magnetized needles were seen as completely different entities; the concept of manmade magnets did not yet exist.

Even in a work of the twentieth century, by science historian Joseph Needham, we find the following:

> Among the Europeans the directive property of the lodestone … is mentioned by Alexander Neckam in +1190—and then by many others, such as Guyot de Provins in +1205 and Jacques de Vitry in +1218.[31]

In reality, however, neither Neckam nor any other writer of that time ever stated the observation that lodestones point north-south. In fact, as far as we know, the first European to record the directive property of lodestones themselves was Michael Scot (c. 1175–c. 1235), a scholar who served King Frederick II of Sicily from 1227. In Scot's *Liber particularis* (Book of particulars), written at the request of the king, we find the following:

> … the stone whose virtue attracts iron to it which is the lodestone [*calamita*] and it shows the place of the north star. And it is another stone of the lodestone type which repels iron from itself and shows the place of the southern star.[32]

*Calamita*, translated here as "lodestone," is defined in *The Cambridge Italian Dictionary* (1962) as "magnet, lodestone." Scot describes calamita as "the stone whose virtue attracts iron to it" (*lapis qui sua virtute trahit ferrum ad se*) and elsewhere in *Liber particularis* reports that calamita "reconciles wives to their husbands" (*calamita reconciliat uxorem ad maritum*). Clearly Scot uses the word *calamita* to mean lodestone and not magnetic needle. Guido Delle Colonne, a Sicilian poet and judge of Messina, echoes Scot's usage when, in a poem written in Italian around the year 1250, he remarks that "the calamite is a stone (*che calamita perta sia*)."[33]

One may wonder why I have gone into such detail here regarding a passage from the pages of science history that not only has been largely ignored by modern researchers but also expresses a misconception about lodestones: the idea that certain types point north and attract iron while others point south and repel iron. But the excerpt from *Liber particularis* quoted above is nonetheless highly significant in two respects: it is the earliest known indication that people were aware of north-south directionality in magnets (natural lodestones) themselves, and whereas Neckam and Guyot merely reported hearsay, Scot spoke from firsthand knowledge.

Concerning the former point, note that, elsewhere in the same work, Scot comments that

> through the lodestone with a needle it is known where north star is [*per calamitam scitur ubi est tramonta cum acu*].[34]

From this it is clear that Scot not only knew of the directionality of magnetized needles but also recognized that magnets themselves shared this property, and made a distinction between these two observations.

Concerning the latter point, in a 1922 study historian Charles H. Haskins notes that "Scot has plainly gone beyond the books and conducted his own experiments, leading at times to new results."[35] Likewise, Thorndike, who studied Scot's handwritten manuscripts, asserts that they "reflect Michael's own observation and experience."[36] Here, at least, are two scholarly voices that support my interpretation of Scot's place in our chronicle. Given the likelihood that his descriptions of the magnet's directionality were based on his own experience and experiment, his observations mark an important turning point both in how Europeans understood magnetism and in how they studied it.

## Section 5—Michael Scot and Frederick II

Michael Scot, along with his patron Frederick II, represented a watershed in the European understanding not only of magnetism but of nature in general. Let us take a moment, then, to consider more closely the intellectual environment in which they lived.

Scot was invited to Frederick's court as an astrologer. Pope Honorius III, pontiff from 1216 to 1227, referred to Scot as one "who flourishes among other men of learning with a singular gift of science,"[37] from which we can infer that Scot was quite well known in the Latin-Christian world in the first half of the thirteenth century. After his death, he was portrayed as a magician in Dante Aligheri's *Divine Comedy* (*Inferno* 10.115–17) and in Giovanni Boccaccio's *The Decameron* (Eighth Day, Ninth Story)—further clues that his intellectual excellence was widely recognized. Scot is known to have translated a work by Nur al-Din al-Bitruji, a leading Arab astronomer of the twelfth century (known also by the Latinized name Alpetragius) while he was in Toledo in 1217. Scot also compiled Latin translations of a number of Aristotle's works on animals under the title *De animalibus* (On animals), a book used by Albert the Great as well as by Robert Grosseteste.[38] Roger Bacon later attested in his *Opus majus* that "from the time of Michael Scotus, whose translations with authentic expositions of certain parts of Aristotle's works on nature and metaphysics appeared in the year of our Lord 1230, the philosophy of Aristotle has grown in importance among the Latins."[39]

Michael Scot was thus among the thinkers who played leading roles in the dramatic shift that occurred in the post-Augustine Latin-European view of nature. But his role went far beyond merely translating some of Aristotle's works. The rediscovery of Aristotle in Europe was accompanied by commentaries written by Muslim philosopher Averroes (1126–1198; the Arabic original name Ibn Rushd) from Cordoba. As we will see in more detail in the next chapter, in Aristotelianism the natural world is eternal and nature operates in accordance with inherent principles, a view that precludes the creation of heaven and Earth by a transcendental being (God) existing outside the world. This view is of course directly opposed to the teachings of Christianity, Judaism, and Islam, and therefore could not be countenanced from a religious perspective. Averroes, however, explained that religious truth (faith) and intellectual truth (philosophy) are separate, and that therefore a proposition can be theologically false yet philosophically true. In Western Europe this thinking was later dubbed "Latin Averroism," branded the "doctrine of double-truth," and condemned by the Catholic church. Nonetheless, Averroes's recognition that philosophy could lead to conclusions different from those of faith, even though faith was never wrong, had a powerful impact. The penetration of Averroism into the faculty of arts of the University of Paris in the thirteenth century led to decisive changes in the European Christian view of nature, and it was none other than Michael Scot who went down in the history of medieval philosophy as "the most energetic translator of Averroes."[40]

Given that Averroes wrote his commentaries in the mid-twelfth century, we must acknowledge that Michael Scot was startlingly astute in perceiving Averroes's greatness as early as the beginning of the thirteenth century. "Thanks to Michael Scot," observes Sarton, "Averroism reached and influenced western philosophy at a time when the majority of Muslims were still unaware of it," and "for the first time the Latin world was allowed to know of a Muslim achievement while it was really fresh and alive."[41] From the mid-thirteenth century, many Scholastics undertook to study Aristotle, basing their research on Averroes's commentaries.

One final point worth mentioning here is Scot's choice of which works of Aristotle's to translate. He did not choose, for example, *Prior Analytics* or *Posterior Analytics*, which regard deductive proof based on first principles as the fundamental form of knowledge and learning. Instead of that kind of impatient quest for universal concepts or the "essence of things," he opted to translate works like *History of Animals*, an immense compilation of individual facts and descriptions of natural history, and thereby emphasized and reinforced the importance of inductive, empirical natural science concerned with individual phenomena.

Frederick II (c. 1194–1250), referred to as *stupor mundi* ("wonder of the world") by his contemporary, the English historian Mathew Paris, brought Michael Scot to Palermo and favored him with his patronage. Frederick was born to Henry VI, of the House of Hohenstaufen, and Constance, daughter of Roger II, the first ruler of the Norman Kingdom of Sicily. After being raised in Palermo—a city that was then in some ways a world apart from Europe—and instructed in kingcraft by Pope Innocent III, Frederick became the most powerful ruler in Europe, eventually serving simultaneously as Holy Roman Emperor, King of Sicily, and King of Jerusalem. The appellation *stupor mundi*, however, was based not only on the extent of his power but moreover on his approach in exercising that power and his philosophy concerning it; in both of these respects he was far ahead of his time.

Endowed with exceptional political ability, Frederick II restored order to the Kingdom of Sicily, which had been thrown into confusion after the death of William II; stripped the feudal lords of their privileges; and became the first ruler in Europe to establish an absolute monarchy. In particular, his Constitutions of Melfi, enacted in 1231, represented the first codification of administrative law in Western Europe. They concentrated military power and jurisdiction in the sovereign and prescribed, among other things, state management of industry, methods of state control including direct and indirect taxation, state administration of commercial activity in cities and towns, and a system of rank for government officials based not on land grants but on the principle of a monetary salary.[42] Henry the VII of England (r. 1485–1509) and Henry IV of France (r. 1589–1610) are said to have laid the foundations for absolute monarchies north of the Alps, but Frederick preceded them by some two and a half and three and a half centuries, respectively. In order to train state officials, in 1224 Frederick also established the University of Naples, the first university in Europe not connected to the church. This university would in time welcome among its students Thomas Aquinas, who was born in either the year of the university's establishment or the year after and was the son of Frederick's vassal, Count Aquino.

Unlike most European rulers, who were caught up in the frenzy of the Crusades, and despite having been raised as a favorite of the pope, Frederick II took a very objective view of his country's relationship with Islamic civilization. He was excommunicated for defying a papal command to dispatch troops. Thus chastised and isolated from the church, he promptly changed tack, organized his own army of crusaders, and conquered Jerusalem through negotiation rather than by force. He transcended the mold of contemporary Catholic monarchs and prevailing modes of thought in the medieval Christian world. At home, he attempted to maintain the religious tolerance traditional to the Kingdom of Sicily, though in reality religious persecution against Muslims by the Roman Catholic clergy and feudal lords increased as Muslim representation in the population fell. The resultant uprisings eventually forced Frederick to relocate the Muslim population to the southern Italian city of Lucera. Thus the policy of conciliation that had prevailed since the kingdom was founded essentially came to an end. After Frederick's death, Charles d'Anjou, who with the pope's backing overthrew the Kingdom of Sicily, attacked the Muslim colony of Lucera. It was while stationed with Charles's troops on the outskirts of Lucera during that battle of encirclement that Petrus Peregrinus wrote his *Letter concerning the Magnet*, the earliest modern monograph on magnetism.

Frederick II, who inherited the cosmopolitan nature of his grandfather, Roger II, was fluent in a number of languages. He was also a patron of the arts and sciences and gathered many of the greatest minds of his time to his palace—not only such Christians as Michael Scot and algebraist Leonardo Fibonacci, one of the first to introduce arabic numerals to Christian Europe, but also Jewish, Arabic, and Greek scholars. In having his personal poets compose in the vernacular he anticipated Boccaccio and Dante, who wrote their works in Tuscan and thus set the stage for that dialect to eventually become established as the standard language of Italy. At the same time, Frederick had scholars translate works by ancient philosophers such as Aristotle. In these two respects alone, "the reign of the Emperor Frederick II holds an important place in the transition from medieval to modern culture," and "it is under Frederick ... rather than in the days of Petrarch, that the real beginning of the Italian Renaissance is to be sought."[43]

More than just a supporter of the arts and sciences, Frederick II was a scholar in his own right, as demonstrated by his treatise *The Art of Falconry*. This is by no means an amateurish work scribbled in his leisure time; completed after thirty years of painstaking study and planning, it represents an imposing scholarly treatise concerning the ecology and anatomy of falcons. The following passage from the foreword is particularly worthy of note in our present account:[†††]

> We discovered by hard-won experience that the deductions of Aristotle, whom we followed when they appealed to our reason, were not entirely to be relied upon, more particularly in his descriptions of the characters of certain birds.
>
> There is another reason why we do not follow implicitly the Prince of Philosophers: he was ignorant of the practice of falconry—an art which to us has ever been a pleasing occupation, and with the details of which we are well acquainted. In his work, the *Liber Animalium*, we find many quotations from other authors whose statements he did not verify and who, in their turn, were not speaking from experience. Entire conviction of the truth never follows mere hearsay.[44]

Although Frederick II acknowledges Aristotle as the "Prince of Philosophers," at the same time he does not hesitate to point out inaccurate information based solely on hearsay, and he shows himself to be aware of Aristotle's inadequacies. Consider, for example, the following statement by Frederick from the same work:

> Birds have two kidneys, one on the right, one on the left side. They lie close to the vertebrae under the iliac bones and extend

[†††] This passage is preceded by these words: "As the ruler of a large kingdom and an extensive empire we were very often hampered by arduous and intricate governmental duties, but despite these handicaps we did not lay aside our self-imposed task and were successful in committing to writing at the proper time the elements of the art." I agree with Haskins's opinion on the work's authorship: "That Frederick himself was the author can no longer be doubted.... If he did not actually write the book with his own hand, he at least directed its composition and dictated the greater part of its substance." Haskins, "The *De Arte Venandi cum Avibus*," 341.

> toward the anus. The urine is excreted from the kidneys by way of the anus through the uriniferous tubules, which pass below and in close contact with them. As the urine is passed with the feces, the bird does not require and does not, indeed, possess a urinary bladder.[45]

Unlike Aristotle's observation in chapter 16, book 2 of *History of Animals* that birds have no kidneys, it is clear that Frederick's account is based on actual dissection. The superiority of this book on falconry is also apparent in comparison with superficial descriptions by Frederick's contemporary Albert the Great in *On Animals*, such as that "no bird urinates."[46]

In the chapters that follow I will discuss in some detail the changes and developments that occurred in the European view of nature from that time forward, but before that let us pause and glance ahead at the overall winding path traced by that intellectual evolution. The medieval Christian philosophy of revelation that had prevailed since the time of Augustine underwent major revisions with the discovery of Aristotle in the thirteenth century. These changes were followed by a revival of intellectual interest in magic during the Renaissance until, in the sixteenth and seventeenth centuries, renewed criticism of Aristotle spurred the ascent toward modern science. In that sense, Aristotle represents "a sort of tragic hero striding through medieval science"[47] who was criticized on both sides, first by early medieval Christians and again later by modern empirical scientists. That Frederick II took issue with Aristotle's reliance on hearsay clearly falls into the latter category, demonstrating that the monarch was two to three centuries ahead of his time not only in politics but in philosophy as well.

In general, the intellectuals of Europe, from Dioscorides and Pliny to Isidore, Marbode, and Albert the Great, appear to have accepted ancient folklore uncritically. In particular, anything passed down in written form tended to be unconditionally trusted. In contrast, Frederick II was the first to declare his own experience to be superior to the authority of existing documents, and in that sense he was a pioneer of empirical science.

* * *

Let us review the shifts covered in this chapter. Two key developments occurring mainly in the twelfth and thirteenth centuries revolutionized Latin Europe's understanding of nature and precipitated a philosophical and social transformation. These included, first, the discovery of Aristotelian philosophy—the greatest legacy of the ancient Greek world—and of the works of then-contemporary Islamic philosopher Averroes; and second, the development of towns and cities and the emergence of universities.

In terms of magnetism in particular, the discovery of directionality in magnetic needles and magnets set the stage for the subsequent discovery of magnetic poles. Moreover, the fact that the phenomenon of directionality was also perceived as a straightforward example of the influence of the heavens on terrestrial bodies precipitated a momentous change in the concept of force. The various phases of this philosophical and social transformation were expressed at roughly the same time, beginning in the late 1260s, in the philosophies of Thomas Aquinas, Roger Bacon, and Petrus Peregrinus. And here is where we begin to glimpse the embryo of modern physics.

Chapter 6

# Thomas Aquinas and his Understanding of Magnetism

## Section 1—The Structure of Knowledge in Christian Society

The rediscovery of the science and philosophy of ancient Greece, most notably the works of Aristotle, by Christian society in Western Europe in the twelfth century threatened to introduce a crack in the structure of knowledge previously upheld by the church and to shake its spiritual unity.

Until that time, Christian society had allowed for the study of nature within the context of religious faith. The purpose in doing so was not to satisfy human curiosity, nor was it to improve people's lives in the material sense. Rather it was to seek divine revelation, and implicit in that approach was the idea that anything pointing the way to God's love was worthy of study. Origen of Alexandria, one of the fathers of the early Christian church, said early in the third century that "everything which exists or happens in this world is arranged by the providence of God."[1]

Accordingly, the study of nature was used as a vehicle for presenting to the masses, in the form of allegories and metaphors, the ideology that God created everything in heaven and on Earth for the purpose of teaching humankind moral righteousness and faith in the divine. Consider *Physiologus*, a collection of allegories of uncertain authorship that appeared in Greek in the second century and was translated into numerous languages, including Latin. This book, described by some as one of the best-selling works of the Middle Ages, alongside the Bible, was read widely and over a long period of time. Despite its title, which means "the naturalist," *Physiologus* was in fact a typical example of the use of storytelling with the aim of providing moral instruction, presenting religious symbolism, and illustrating God's power, which human beings could not possess.[2]

The situation in higher education was much the same, with a similar approach being visible in the universities founded during the twelfth century. Fundamental to all university curricula were the liberal arts, consisting of the so-called trivium of rhetoric, logic, and grammar and the so-called quadrivium of arithmetic, geometry, astronomy, and music. These core curricula prepared students to then pursue specialized courses of study in theology, law, and medicine. Faculties of theology offered courses based on texts that interpreted and gave commentary on the Bible and the writings of the church fathers; from the perspective of theology, the liberal arts, which involved the study of language and worldly facts, were no more than basic preparation for true higher learning: such subjects were considered mere "handmaidens" to theology, their only purpose being to aid in the study of God's word.

An example of this thinking comes from Saint Bonaventure (c. 1220–1274), who lectured at the University of Paris from 1248 to 1255 and was elected minister general of the Franciscan order in 1257. He wrote a work titled *The Soul's Journey into God* in

1259, in which he describes the status of natural science as a branch of learning thus:

> When the soul considers its Triune Principle through the trinity of its powers, by which it is an image of God, it is aided by the lights of the sciences which perfect and inform it and represent the most blessed Trinity in a threefold way. For all philosophy is either natural or rational or moral. The first deals with the cause of being and therefore leads to the power of the Father; the second deals with the basis of understanding and therefore leads to the wisdom of the Word; the third deals with the order of living and therefore leads to the goodness of the Holy Spirit.
>
> Again, the first, natural philosophy, is divided into metaphysics, mathematics and physics. The first deals with the essences of things; the second with numbers and figures; and the third with natures, powers and diffusive operations. Therefore the first leads to the First Principle, the Father; the second to his Image, the Son; and the third to the gift of the Holy Spirit.[3]

Bonaventure's assertion here is that all sciences, including physics, mathematics, and the other subjects that make up natural philosophy need their existence justified by a system, based on religious faith, that is of a higher order than they are. This system, he declares, is the Trinity, and the only legitimate function of science is to guide people toward greater understanding of the Trinity as the internal architecture of the Supreme Being.

Until the works of Aristotle were rediscovered beginning around the second half of the twelfth century, medieval Europeans depended on Christian doctrine and Platonism for their understanding of the natural world. Or, to be more specific, they depended on the Bible (and especially on the Book of Genesis) and on Plato's *Timaeus*. The mythology of *Timaeus* presents the

Demiurge as the transcendental being who created the physical world out of pre-existing elements, imposing an order modeled on the world of Forms, or Ideas. This construct of Plato's was rather easy for medieval Christians to accept, as it paralleled their search for divine revelation in the natural world and their desire to explain the creation of heaven and Earth in terms of God's will. Saint Augustine believed that the only thing missing from Platonic philosophy was a doctrine of the Incarnation.[4] This view enabled Augustine and the other church fathers to simply replace the Demiurge with the Christian notion of God and adopt Plato's theory of Ideas, modifying it to form a theory called exemplarism. Platonism, then, was not seen as a threat to Christian thought. As we shall see in the next section, however, Aristotelianism was another matter entirely.

## Section 2—Aristotle and the Discovery of Nature

Aristotle's observations about nature were incompatible with Christian doctrine, as the two systems of thought had fundamentally different theories of knowledge. As we have already seen, Augustine subordinated the study of nature to the study of the Bible and preached that any natural scientific inquiry motivated by intellectual curiosity was detestable. Saint Bernard, in the twelfth century, was among those who stated the case unequivocally: "There are some who wish to learn only in order to know, and this curiosity is unworthy of a man."[5] In contrast, Aristotle had begun *Metaphysics* with the statement that "All men by nature desire to know," thus giving his blessing, so to speak, to intellectual curiosity.[6]

Moreover, while Christianity spoke of the creation of heaven and Earth, of final judgment, and of miracles, Aristotle painted a different picture of existence. He argued that "the heaven as a whole neither came into being nor admits of destruction … but is one and eternal, with no end or beginning"; and that "nothing can happen contrary to Nature considered as eternal and necessary."[7] In other words, the world according to Aristotle was not created purposefully by some transcendental being exernal to it, nor did it owe its nature and continued existence to the arbitrary will of such a being. At bottom, the difference between the Christian and the Aristotelian worldviews entailed, first, seeing the world as having a beginning and an end versus seeing it as something eternal, and second, seeing nature as something created versus seeing it as something that came into being of itself. One implication of the Aristotelian view is that the world can and should be considered an object of study, that truths about nature are discernible and verifiable by means of rational inquiry. He sums up his position eloquently: "since nature is for the sake of something, we must know this cause. We must explain the 'why' in all the senses of the term."[8] In other words, nature has its own causes and purposes, which are knowable.

Aristotle's huge corpus of writings provided people of the Middle Ages with a wealth of observations concerning the natural world that had previously been hidden from their view, which was dimmed by the tinted glasses of Christian teachings. Moreover, it gave them a conceptual apparatus and a logical framework with which to integrate these observed facts—in other words, a set of principles with which to comprehend nature. But the full impact of his legacy was even greater: in addition to facts and principles, the work of Aristotle, when rediscovered in the medieval period, gave to the people of that time a *new orientation* vis-à-vis nature and disposed them to examine nature's secrets in new ways.

Gradually, the seeds of change began to sprout among European intellectuals. They came to realize that order and change in the natural world were governed by laws of nature itself—by intrinsic powers and purposes—and that nature and its laws were meant to be understood by human beings through a process of rational inquiry. It was because of this new orientation that the twelfth century, when translation of Aristotle's works began in earnest, has come to be thought of as the age of the discovery of nature.

French philosopher William of Conches, in his *Philosophia mundi*, illustrates well the process of transformation from theology to science as the generally recognized foundation for intellectual inquiry. He holds that when the Bible tells us that the world was "created," we show no disobedience to the Holy Book when we attempt to explain *how* creation took place. Instead, he claims, we have "an obligation to seek a basis for all things." Thus, although he accepted the Bible as a source of truth, he did not believe in the kind of piousness that led people to swallow its teachings whole and without question. William began with the premise that God had initially created the various elements of matter, but that the process by which the world subsequently came into being was impelled by the natural workings of the elements themselves. And so we see that by the twelfth century, even with biblical truth still held supreme and with creationism still taken for granted, times had in fact changed: the autonomous function of nature and the existence of intrinsic natural law could no longer be ignored.[9]

While William and other intellectual pioneers of the twelfth century were advancing only cautiously, and while only some of the works of Aristotle had been translated and circulated, it was still possible for natural philosophy to be kept subordinate to theology.

But as time went on, the contradictions became obvious and could no longer be smoothed over. The Aristotelian and the Christian theories of nature belonged to two completely different universes; sooner or later, the world according to the pagans and the world according to the church were destined to clash.

Early in the thirteenth century, the church saw danger looming. At that time, the greatest centers for Aristotelian studies were in Paris and Oxford, particularly the faculty of arts at the University of Paris. In 1210 the provincial synod of Sens banned the works of Aristotle—both original texts and commentaries—that pertained to natural philosophy. And in 1215 the pope sent a special delegation to the University of Paris to repeat this ban. Thus, to outward appearances at least, research and teaching of Aristotle from 1210 to 1240 were limited to his works on logic and ethics.

In 1231, Pope Gregory IX found it necessary to renew the ban initiated in 1210, from which we can infer that there were at least a few known violations. Two years earlier in Toulouse, a new university had opened and attempted to lure students away from Paris by advertising that it offered courses on Aristotle, suggesting that the restrictions set in place by the church had, if anything, fueled rather than dampened enthusiasm for the Greek philosopher among young scholars. Across the English Channel, at Oxford, research and teaching of Aristotle proceeded throughout this period without obstacle. As a practical matter, the dam could not hold, and by 1240 the ban at Paris was widely disregarded. When Englishman Roger Bacon taught liberal arts at the University of Paris from 1240 to 1247, included in his courses were a wide range of works by Aristotle, among them *Metaphysics*, *Physics*, *On Generation and Corruption*, and *On the Soul*. Thus, though the road to success was not always smooth, Aristotle's vast, powerful, and rational system of natural philosophy eventually won the hearts and minds of professional scholars and students alike.

Then in 1255 came a decisive turnaround: on March 19 of that year, the arts faculty at the University of Paris officially decided to incorporate almost all of Aristotle's writings into its courses.[10] Even until that point, the faculty, far more secular than the faculty of theology, had tended to study and discuss the ancient philosopher's works somewhat independently, without reference to the creed of Christianity. But the 1255 decision was an open declaration that Aristotelian philosophy was now a part of its core curriculum. The significance of this development was profound: until that point, the liberal arts had been considered supplementary and subordinate to *theology*; from then on, however, they were to be considered supplementary and subordinate to *philosophy*. In function, if not in actual name, the faculty of arts was now the faculty of philosophy, and its truths had a voice of their own that was independent of theological doctrine and dogma.

In time, this shift would lead to a movement, begun by Aristotelians Siger of Brabant, an arts master at Paris, and Boethius of Dacia, one of the university's leading arts masters in the late 1260s and early 1270s, toward accepting as fact the inferences arrived at by deductive reasoning based on rational philosophy.

From 1255 to 1260, Siger studied liberal arts at the University of Paris, and upon being graduated he began to teach there. From the beginning of his career as a professor, he espoused the theories of Averroes concerning the relationship between religion and philosophy. He believed that the universe was eternal; that there existed a single intellect shared by all individuals; and that the soul of the individual could be destroyed. From the standpoint of Christianity, these were dangerous ideas.

And so there was a split. Throughout the decade beginning in 1260, those who called themselves Christian scholars divided into two camps. On one side were Bonaventura and others who proclaimed the supremacy of theology over philosophy, holding belief in the truth of Christian teachings above all other knowledge. Opposing them were Siger and his ilk, who advocated what has been criticized as a "theory of twofold truth," whereby they accepted the content of divine revelation as fact but also persistently clung to rational philosophy and sought to pursue its intellectual consequences to the fullest extent. Theologians, understandably, continued to eye philosophers with suspicion and animosity.

## Section 3—Saint Thomas Aquinas

With the spread of Aristotelian philosophy, Christian theology thus faced a crisis. One man above all has been credited with averting this crisis by successfully integrating Aristotelian thought into biblical teachings. He began his career as a professional scholar in 1256 with an appointment to the faculty of theology at the University of Paris, the undisputed leader in its field at that time, and his name was Thomas Aquinas.

Thomas Aquinas (c. 1225–1274) was born to an aristocratic family in Naples, which was at the time, along with Sicily, a meeting place for the Islamic and Byzantium cultures—a conduit between East and West. Being the youngest child, Thomas was not to inherit title to any territory and instead, in accordance with custom, was sent at an early age into monastic life. At age five he entered the

Benedictine abbey at Monte Cassino, not far from his home, and beginning in 1238 he spent six years studying liberal arts at the University of Naples.

Holy Roman Emperor Frederick II not only was the founder of the University of Naples but also had control over the Monte Cassino abbey. Largely for this reason, Thomas's family strongly opposed his 1244 transfer from the Benedictines to the Dominican order, a move that suggested disloyalty to the emperor in favor of the pope. Another reason for their displeasure involved status. Entering the Benedictine order, which had a powerful network of monasteries spanning a wide geographic region, put a young man in line for eventual promotion up through the ranks of the clergy. The Dominicans, in contrast, were among the newer mendicant orders, and it was not considered fitting for a young man of noble birth to become one of a class of people who depended on charity for their livelihood. But for Thomas, the Dominican order meant much more, for it was here that he met Albert the Great, who became his teacher and set him on the intellectual path that he would journey for the rest of his life. The illustrious elder scholar had realized early on that Greek philosophy and science could be of benefit to the church, and it was with him that Thomas spent the period from 1248 to 1252 at the Dominican abbey in Cologne studying not only theology but Aristotelian philosophy as well. It is known that most of the works of Aristotle had been translated into Latin by the mid-thirteenth century,[11] so his system of thought was more or less fully accessible to medieval scholars by the time Thomas began his studies in earnest.

At Albert's recommendation, in 1252 Thomas moved on to the University of Paris, at that time the center for philosophical and theological research, and in 1256 he was appointed to the university's theology faculty, which was without equal in the Christian world. By that point in time, the faculty of arts had placed Aristotelian philosophy at the center of its curriculum and spoke freely of philosophy as a self-contained discipline independent of theology. Thomas spent the remainder of his career, until around 1273, tenaciously pursuing the ambitious goal of integrating Aristotelian philosophy into Christianity so that the two could coexist peacefully. He died in 1274, at about the age of fifty, while his greatest work, *Summa theologiae* (Summary of theology), was still in progress.

Earlier, in 1259, Thomas left the University of Paris for several years (until 1264) and returned to Italy, where he wrote *On the Truth of the Catholic Faith*. His aim in this work was to refute and persuade Muslims and Jews, who did not accept the premises or veracity of Christianity, so he had to present a rational argument on the strength of philosophy, the same type of intellectual arsenal that his intended audience used. The book altogether avoided Christian assumptions and instead developed arguments leading to the conclusion that there were no contradictions between the truth of divine revelation (*fides*) and the truth of science (*scientia*) as demonstrated by so-called natural reason. Thomas conceded that certain parts of *fides* were beyond human reason and could be grasped only by relying on the power of revelation. Or, to state his position more bluntly, human reason was frail, too frail to ascend to the height occupied by *fides*. On the other hand, he acknowledged that certain things could be comprehended through reason; reason, he believed, did not contradict faith but rather had its own limited realm of applicability.

By 1268 Thomas had returned to his teaching post at the University of Paris, where he was to spend three more years. This was his most prolific period, as well as the stage in his life where his enthusiasm for Aristotle's naturalistic theories reached its peak. In the period from 1269 to 1272 alone, Thomas penned three commentaries on the works of Aristotle: *Commentary on Aristotle's Physics*, *Commentary on the Metaphysics of Aristotle*, and *The Soul.*

With these in hand, Thomas pursued the creation of his *Summa theologiae*, through which he aimed to integrate Aristotelian philosophy with Christian theology. The result was the establishment of a new philosophy—rigorous Scholasticism—that was essentially a reformulation of Christian theology utilizing Aristotle's rational philosophical system. In practice religious dogma took precedence over philosophical principle, as can be seen in this passage from *Summa theologiae*: "The divine will can be manifested by revelation, on which faith rests. Hence that the world began to exist is an object of faith, but not of demonstration or science."[12] For this reason, it might perhaps be more accurate to say that what Thomas achieved was not so much an integration of religion and philosophy but rather a very clever defense of Aristotelianism crafted in such a way as to remain consistent with Christian teachings.

The divide between Aristotelian philosophy and the other intellectual factions at the University of Paris can be described as follows. On the one hand were the conservatives, including Saint Bonaventure and most of the professors attached to the theology faculty, who staunchly insisted on the supremacy of biblical teachings. On the other hand were the Averroists—in other words, the philosophical rationalists and radical Aristotelians—of whom Siger topped the list. The latter intellectuals, who were probably few in number, were professors attached to the arts faculty. Thomas clearly highlighted the differences between himself as a philosophical theologist and the philosophers on the arts faculty when he advanced a criticism of the Averroists in his book *On the Unity of the Intellect Against the Averroists*). Nonetheless, many clergymen and academics, quite understandably, associated him with Siger and others on the Averroist side, simply because he approved—however he may have tempered his views—an Aristotelian philosophy. The religious conservatives believed that Thomas went too far in compromising with pagan ideas, or, worse, that he was a sympathizer of Siger's, or perhaps an accomplice to this thinker, who was considered dangerous. In 1270 Bishop of Paris Étienne Tempier condemned thirteen propositions drawn from the writings and lectures of Siger and his followers; in 1277—Thomas had passed away by then—Pope John XXI called upon Tempier to investigate and report on the crisis brewing at the University of Paris, whereupon the bishop issued a new condemnation branding 219 propositions heresy. Thomas, along with Siger, was found guilty of having espoused heretical ideas.

Some fifty years after his death, Thomas's reputation was fully rehabilitated. The scholar was canonized in 1323, and in 1325 the bishop of Paris reversed the 1277 condemnation of those propositions among the 219 that had originated with Thomas. Thomas's theology having thus gained official approval in Christian Europe, Scholasticism took shape and later came to dominate the spiritual scene in Europe for the remainder of the medieval period.

It would be interesting to examine the specific logic that Thomas used in order to bring the pagan theories of Aristotle into harmonious coexistence with the Gospels, which resulted in Scholasticism's taking root with the approval of the church. This, however, would require a major digression that would be too far outside the scope of this book. Suffice it to understand one extremely important result of this process: by giving his blessing, so to speak, to the idea that there was no contradiction between faith and philosophical truth when arrived at by means of natural reason, and that therefore religion could safely embrace philosophy practiced in this manner, Thomas guaranteed that reason would thereafter be entitled to a sphere of independent operation. In other words, he granted reason the privilege of being exercised even outside the context of seeking a deeper understanding of divine revelation. This paved the way for scholars to pursue rational inquiry into the workings of nature as an end in itself, without considering questions related to revelation and even without being motivated by any theological interest.

It is true that, in the end, Scholasticism became a hindrance to the progress of natural science, erecting, if anything, even higher barriers than had previously existed. But the responsibility for this outcome lay primarily with the methods of Scholasticism itself, in particular with the form of its proofs. Paradoxically, although Thomas's goal had been to integrate philosophy and religion, in the thirteenth century his theories actually had the reverse effect of hastening the separation of natural science from theology.

## Section 4—Aristotle and his System of Causality

Thomas Aquinas's philosophy and, above all, his theories concerning natural philosophy were basically adopted straight from Aristotle. His explanations of the four original elements and of the cosmos, for example, are virtually identical with those of the Greek master. Nowhere in his writings were magnets or magnetism taken up as the focus of discussion. Judging from the occasional references he made to the subject, however, it is clear that he attempted to understand the characteristics of the magnet, too, in terms of Aristotle's framework of logic. In fact, the various intellectuals who revived Aristotelianism in the thirteenth century—including Roger Bacon, the central figure of chapter 7—were in almost perfect agreement in their understanding of magnetism, and particularly of magnetization. Before we pick up our discussion of magnetism again, therefore, it will be useful to stop and take a look at Aristotle's schema of causality.

In his *Physics*, Aristotle defines "motion" very broadly. He considers the concept to include, in addition to locomotion (i.e., change in location) changes in characteristics and amount or size as well, asserting therefore that "there are three kinds of motion, local, qualitative, and quantitative."[13] Included in the idea of motion as understood by Aristotle would be the growth or metamorphosis of living organisms (whether animal or plant) as well as any and all chemical changes or changes in the physical state of substances, such as the melting of ice, the vaporization of water, and the expansion of gases when heated.

Aristotle perceives things that undergo change within a twofold framework of *form* and *matter.* He uses the term "form" in association with the term "matter": "We ... describe in both ways what brazen circles are; we describe both the matter by saying it is brass, and the form by saying that it is such and such a figure."[14] For example, to Aristotle the "form of a desk" is the concept of a particular shape and structure upon which one can read or write. Accordingly, "form" is that which defines the essence of a thing and that which distinguishes it from other things; in certain cases it may be the purpose for which it was made (its design), or may prescribe the state of its matter. In any of these senses, form is required if matter is to be some definite thing or state of some kind. Every material substance, therefore, results from the union of two elements, one permanent (matter) and one variable (form).

Aristotle also maintains that "being is twofold."[15] Any entity X, while existing actually as X, simultaneously holds latently within itself the possibility of becoming something else, some entity Y. To state this another way: X, which is capable of becoming the actual being of Y by acquiring the form of Y, can be considered the *potential being* of Y even while having not yet taken on Y's form. In this sense, all change (in other words, motion broadly defined) exhibited by bodies can be understood as *change from potential being to actual being*. When ice melts, for example, its actual being changes to water, which in its frozen state had been potential water. Similarly, when a craftsman builds a wooden desk, the wood (which can be thought of as the desk's matter and whose potential being had been a desk) attains the form of a desk and thereby undergoes a change in its actual being. Aristotle's unique schema for understanding change as movement from potential being to actual being was a carefully considered solution to problems concerning the nature of existence and change that had bedeviled philosophers ever since the time of Parmenides.

Thomas adopts this framework for the most part without revision, but there is one area that he elaborates upon and renders more precise in his *The Principles of Nature*. Relying on Aristotle's *Metaphysics*, Thomas identifies two kinds of existence: when we say a person exists, for example, that denotes a *substantial* existence, but when we say a person is white, the person's whiteness is a contingent or *accidental* kind of existence. Thomas likewise identifies two corresponding types of forms, substantial and accidental: "What makes something exist substantially is called *substantial form*, and what makes something exist accidentally is called *accidental form*."[16]

In *On Being and Essence*, Thomas further posits that "in substances, something that is one through itself is brought about from substantial form and matter, with one certain nature resulting from their union, which properly is placed in the predicament *substance*"; and that "a certain essence results from [that] union."[17] In other words, substantial form, which is distinct from accidental form, gives a thing the essence that makes it truly that thing. Thus, in Scholasticism, the causes of things are explained by defining their substantial forms.

## Section 5—Thomas Aquinas and Magnetism

Thomas Aquinas, Roger Bacon, and others made frequent use of Aristotle's unique schema of causality, according to which potential being is transformed into actual being. As we saw in our discussion of the ancient Greeks, Aristotle himself avoided making specific claims about the action of magnets, with the exception of his suggestion that magnets were soul-like, lifelike beings that happened to have the power to move iron. Despite this reticence, Thomas and other like-minded scholars attempted to develop an understanding of magnetism within the framework of Aristotelian theory. In a typical example of these attempts, Jean of Saint-Amand, a cleric and a contemporary of Thomas's, proposed that the magnetization of iron occurred by "exciting the active power which exists incomplete in the iron, which is born to be completed by the form of the magnet."[18] As nineteenth-century physicist and philosopher Pierre Duhem observed, this was an apt description of the phenomenon.[19]

Aristotle declares in *Physics* that "it is a fundamental assumption with [the physicist] that motion is ultimately referable to nature herself."[20] For *nature* here he uses the word *phusis*, signifying the natural forces that lead bodies to behave in accordance with their intrinsic properties. In his *Commentary on Aristotle's Physics*, Thomas states that "the principle of a natural motion is in the natural body which is moved," and that the key concept required in any discussion of movement and change is *the nature of the thing*:

> Natural things differ from the non-natural insofar as they have a nature. But they differ from the non-natural only insofar as they have in themselves a principle of motion. Therefore, nature is nothing other than a principle of motion and rest in that in which it is primarily and per se and not per accidens.[21]

If we are restricting ourselves to a discussion of natural philosophy as it attempts to understand movement and change, then we can understand the *form* of a natural thing to be equivalent to *nature* in the sense used by Thomas. In *Commentary on the Metaphysics of Aristotle*, Thomas states that "nature is a principle of action and motion in the thing in which it is present" and also that "the nature of a thing is that in which the process of natural generation is terminated, i.e., the form."[22] It follows from this thinking that the human soul is equivalent to human nature and, at the same time, to the human form. Similarly, whatever principle it is that causes magnets to behave in the singular way that they do is the magnet's nature and, at the same time, the magnet's form.

Thomas recognized, however, the existence of a hierarchy among the forms of the various natural things. He classified the magnet as a mineral and placed its form above those of the four terrestrial elements (which possessed the active qualities of being hot or cold and the passive qualities of being dry or wet), but below the forms of the plants and the animals. Thomas's most prolific period, and the time during which he consolidated his own system of thought, was the period from 1269 to 1273. In *On Spiritual Creatures*, which he wrote sometime in or around 1269, Thomas discusses the order of forms as follows:

> Now it must be borne in mind that the more perfect a form is, the more does it surpass corporeal matter. This is clear from induction in regard to the various orders of forms. For the form of an element does not have any activity but the one which takes place through active and passive qualities, which are the dispositions of corporeal matter. But the form "mineral body" has an activity that goes beyond active and passive qualities, and is a consequence of its species by reason of the influence of a heavenly body; for instance, that a magnet attracts iron, and that a sapphire cures an abscess.[23]

He discourses in a very similar vein, but in much greater detail, in *The Soul*, which was written around the same time:

> We find among the forms of lower bodies that the higher a form is, the more it resembles and approaches higher principles. This can be seen from the proper operation of forms. For the forms of the elements, being lowest [in the order of forms] and nearest to matter, possess no operation surpassing their active and passive qualities, such as rarefaction and condensation, and the like, which appear to be material dispositions. Over and above these forms are those of the mixed bodies, and these forms have (in addition to the above mentioned operations) a certain activity, consequent upon their species, which they receive from the celestial bodies. The magnet, for instance, attracts iron not because of its heat or its cold or anything of this sort, but because it shares in the powers of the heavens.[24]

Thomas does not state explicitly that the form of the magnet, or of minerals in general, is a soul, but actually it corresponds to the soul of animals and plants. In *On Spiritual Creatures*, just above the form of mineral comes the "vegetative soul," which takes in nourishment and is capable of growth, and above that comes the "animal soul," which can move from one place to another and has the power to see, hear, and desire. And above these comes the "human soul," which has the power of understanding and is the most perfect of all the forms that exist. In the chain of existence, then, Thomas places the iron-pulling magnet right in between nonliving things and plant life. Although he makes no reference to Aristotle in this context, it is clear that Thomas's hierarchy is firmly founded on this passage from the Greek master's *History of Animals*: "Nature proceeds little by little from things lifeless to animal life in such a way that it is impossible to determine the exact line of demarcation, nor on which side thereof an intermediate form should lie."[25]

But at the same time, Thomas's discussion modifies and in a sense surpasses Aristotle's theories concerning the soul. "Under the heavenly bodies according to Aristotle," notes Thomas, "the only animated bodies are those of animals and plants."[26] As we will see in more detail later, Thomas's intention was to increase the number of souls. Bearing this aim in mind, we can interpret the medieval theologian's densely populated hierarchy of souls thus: he took Thales' understanding of magnetic force as soul-like, elaborated upon it in Aristotelian terms by increasing the total number of souls, and thus brought it into conformity with Aristotle's system of natural science.

Thomas's understanding of magnetism can be seen most clearly in the passage below from his *Commentary on Aristotle's Physics*. The passage is Thomas's response to Aristotle's reduction, in *Physics*, of "the motion of things that are moved by something else" in two ways, namely to mere "pulling and pushing."[27] Thomas first observes that "to move something toward one's self with respect to place occurs in three ways." The second of these "three ways" is magnetic attraction, which he describes as follows:

> In another way it can be said that a thing pulls because it moves something toward itself by altering it in some manner, from which alteration it happens that the thing altered is moved with respect to place. In this way a magnet is said to pull iron. For just as a generator moves heavy and light things insofar as he gives them a form through which they are moved to a place, so also a magnet imparts some quality to iron through which the iron is moved toward the magnet.

Thomas then enumerates three characteristics that distinguish magnetism from the phenomenon—known in modern times as gravity—of bodies moving by the force of their own weight:

> First a magnet does not pull iron from any distance, but only from nearby. But if iron were moved to a magnet only as to an end, as a heavy thing is moved to its own place, it would tend toward it from any distance.
>
> Secondly, if a magnet is greased with [garlic], it cannot attract iron. It is as if [the garlic] either impede[s] the alterative force of the iron or else change[s] it to its contrary.[‡‡‡]
>
> Thirdly, in order for a magnet to attract iron, the iron must first be rubbed with the magnet, especially if the magnet is small. It is as if the iron receives some power from the magnet in order to be moved to it. Thus a magnet attracts iron not only as an end, but also as a mover and an alterer.[28]

In other words, a heavy body will move toward its natural place—the center of the universe—from any distance, no matter how far;

‡‡‡ Regarding my interpolations in this quotation, see chapter 4, footnote *.

and weight is a quality that can neither be granted nor taken away through the action of another body. Because magnetism differs

from gravity (i.e., weight) in these three respects, Thomas regards magnetism as having a special set of properties and a special kind of form. As things possessing these properties, magnets confer upon iron this same form (the nature of the magnet). This conferring of form is the phenomenon witnessed when a magnet attracts iron. Part of the basis for this line of reasoning is Thomas's premise that "any power extends to certain things insofar as they share in one nature."[29]

Essentially, this was the first attempt at a theoretical understanding of the force exerted by magnets on iron that made use of the logic of Aristotelian natural philosophy. Folklore concerning such matters as the medical properties of gemstones and the power of garlic to interfere with magnetism continued to linger. But Thomas freed himself of the conception, passed down until the time of Albert the Great, that magnetism was something magical and mysterious. Instead he made what we can consider the first attempt to explain magnetism and magnetization in rational terms in a way that was not a type of mechanistic reductionism.

Another point worth noting relates to the first of the three characteristics of magnetism posited by Thomas and cited above. Historically, this is the first time we see reference to the idea that the range of magnetic force is finite. From this subtle piece of awareness came, eventually, the concept of the "sphere of virtue."

## Section 6—The Heavens and their Effect on Magnets

In the writings of Thomas cited above we find the peculiar ideas that magnetism occurs "by reason of the influence of a heavenly body" and that the magnet "shares in the powers of the heavens." This distinctive line of thinking emerges even more clearly in the following excerpt from *On the Unity of the Intellect Against the Averroists*:

> We see in many instances that a form is indeed the act of a body made of a mixture of elements, and yet it has some power which is not the power of any element, but which belongs to such a form by reason of a higher principle, for example, a celestial body; just as a magnet has the power of attracting iron, and jasper of checking the blood flow.[30]

Thomas states the case in more general terms in *Summa theologiae*: "The natural forces of natural bodies result from their substantial forms which they acquire through the influence of heavenly bodies; wherefore through this same influence they acquire certain active forces."[31]

Such a conception is not found in the writings of Aristotle. Thomas, as we have just seen, believed that magnets had a form conferred upon them by the action of heavenly bodies. It is really no surprise that Aristotle did not share this notion, for the ancient Greeks knew nothing of compass needles and magnetic poles. In *Meteorology*, however, Aristotle does maintain that "the whole world surrounding the earth … has a certain continuity with the upper motions: consequently all its power and order is derived from them."[32] While it is true that Aristotle is here explicitly referring to celestial influence on terrestrial bodies, we also know that he considered these types of effects solely from a physical standpoint, specifically, as action occurring through the physical medium of air. Aristotle estimated the distance from the heavens to the surface of the Earth to be much shorter than we now know it to be, and he also believed all of this space to be filled with air. Thomas, on the other hand, considered the influence of heavenly bodies on earthly matters from a standpoint that was both metaphysical and theological.

Let's take a deeper look at the difference between the two perspectives.

As we saw much earlier, Aristotle theorized the existence of a "first mover," that is, some entity analogous to God, that was responsible for the activity of the celestial sphere that housed the fixed stars. Below this were the spheres that held the outer planets, whose activity was directed by some nonmaterial entity that itself was subject to the dictates of the first mover. In his *Commentary on the Metaphysics of Aristotle*, Thomas discusses this set of relationships and the basis for Aristotle's theory as follows:

> Since among bodies one which contains is more formal, and is thereby nobler and more perfect … and since the sphere of a superior planet contains that of an inferior planet, therefore a superior planet … must have a higher and more universal power than an inferior planet, and must produce more lasting effects … This is also the reason why the effects of the planets appear in lower bodies in accordance with the order among the planets.[33]

In Aristotelian natural philosophy, the power that moves the planets comes from angel-like beings construed as nonmaterial entities of a separate kind of substance. Thomas adopts this idea and integrates it into Christian theology. In one of his last works, *Treatise on Separate Substances*, he writes that if "one should wish to examine diligently the words of Sacred Scripture, he will be able to gather from them that the angels are immaterial, for Sacred Scripture calls them certain powers"; and that "the spiritual substances whom we call angels, because they are closer to the First Cause, carry out the divine providence in a more universal way."[34]

But Thomas argues that Aristotle was mistaken in his theory that the number of these nonmaterial beings was limited to the number of movements seen in the heavens:

> But the immaterial substances surpass corporeal substances much more than the heavenly bodies surpass elementary ones. In view of this, the number, power, and disposition of immaterial substances cannot be adequately grasped from the number of heavenly movements.[35]

In other words, according to Aristotle, the angels—the nonmaterial beings controlled by the first mover—play no role in the universe other than to move the planetary spheres, whereas Thomas gives them a wider role. Thomas believes that these nonmaterial beings exist in greater number than the celestial spheres and that they exist in numerous locations below the heavens and perform a variety of tasks.

Or we can put it this way: According to Thomas, both Aristotle and Plato had held the view that "the highest souls of the heavens exercise a providence over all the lower souls and over the whole generation of lower bodies; and in the same way, that the higher souls exercise a providence over the lower ones." But Thomas also remarks that "Aristotle did not posit any souls intermediate between the souls of the heavens and the souls of men, as did Plato." Therefore, "there are many things which are evident according to senses, for which an explanation cannot be given on the basis of what Aristotle teaches,"[36] because Aristotle's beliefs concerning the characteristics of the four elements do not lend themselves readily to an explanation of many of the movements and changes seen on Earth. For these reasons, Thomas believes it is necessary to look to the stars as an additional source of influence upon terrestrial bodies.

This theme is evident again in *On the Truth of the Catholic Faith*, which Thomas wrote in the early 1260s:

> Since a man has both the ability to choose and to carry out what he chooses, he may at times be assisted by higher causes in regard to both or he may also be hindered. In regard to choice, of course, as we said, man is either disposed by the celestial bodies to choose something, or he is enlightened by the guardianship of the angels, or even he is inclined by divine operation. But in regard to the carrying out of the choice man may obtain from a higher cause the strength and efficacy needed to accomplish what he has chosen. Now, this can come not only from God and the angels, but also from the celestial bodies, to the extent that such efficacy is located in his body. For it is obvious that inanimate bodies also obtain certain powers and abilities from the celestial bodies, even beyond those which go along with the active and passive qualities of the elements, which, doubtless, are also subject to the celestial bodies. Thus, the fact that a magnet attracts iron is due to the power of a celestial body, and so have certain stones and herbs other hidden powers. So, nothing prevents a man, too, from getting, as a result of the influence of a celestial body, a certain special efficiency in doing some bodily actions, which another man does not possess: for instance, a physician in regard to healing, a farmer in regard to planting, and a soldier in regard to fighting.[37]

Thomas's perspective on the power of the heavens was remarkably consistent from early in his academic career all the way through to the end of his life. Consider this statement from *The Disputed Questions on Truth*, which he wrote shortly after he began teaching at the University of Paris: "There are in those bodies some effects of the heavenly bodies which are not caused by heat or cold, such as the attraction of iron by a magnet"[38]; and then this one from *Summa theologiae*, his final work: "natural bodies acquire certain occult forces resulting from their species through the influence of the heavenly bodies."[39] Another case in point comes in his commentary on Aristotle's theory, described in *Meteorology*, that metals are formed from a compound of moist vapors and smoky vapors, Thomas noting that "celestial virtue" is needed to make this composition of gases possible.[40]

This idea of heavenly bodies causing the phenomenon of magnetism was entirely new: it is not found in any of the theories propounded at any time from ancient Greece—not even by Aristotle—to the early Middle Ages. The reason for this is unmistakable: leaving aside whatever theoretical considerations Thomas himself may have posited as the basis for his thinking, the appearance of such an idea at that time was an immediate result of the discovery by Europeans of the directionality of magnets and magnetized needles. The influence of this discovery continued to be felt well into the Middle Ages by such thinkers as Jean Buridan (c. 1295–c. 1358), a Scholastic philosopher, and Marsilio Ficino, a fifteenth-century neo-Platonist whom we will discuss in detail later. Here is an excerpt from Buridan:

> We can say this of the magnet: It is doubtless that the stars and heaven have a number of diverse qualities and powers and produce a number of diverse effects here on Earth. It is

> therefore plausible that magnets have one part that is especially susceptible to the influence of stars near the North Pole and another part more susceptible to influence from stars near the South Pole.... Clearly this distinction within magnets could only have come from heaven.[41]

In any event, the remarkable discovery that iron needles that had been rubbed against magnets continued thereafter to point northward—as though pulled by the North Star—clearly convinced thinkers of the late Middle Ages that some heavenly body (the North Star itself) was acting directly upon magnets and influencing their properties. This is a point that bears emphasis, for the impact of the discovery of magnetic directionality on the way Europeans understood nature from the latter part of the medieval period all the way to the beginning of modern times was tremendous. In particular, the discovery, far from being limited to merely an enhanced understanding of magnets, provided an empirical foundation not only for astrology, with its belief in the power of the stars and the planets, but magic as well.

* * *

In his writings, Thomas was constantly discussing one or another issue of natural science. At bottom, these wide-ranging references were united by a single viewpoint: the idea that the attributes and behaviors of any given thing could be deduced through a process of logical reasoning as long as one began with a correct understanding of the true natures of things. In other words, a fact could be considered proven if it were deduced from a general principle. This idea, which formed the basis for the rational ideologies of Scholasticism, first came into being in ancient Greece and was then transported into the Europe of the thirteenth century as part and parcel of the works of Aristotle. In *On Being and Essence*, written sometime between 1250 and 1260, Thomas asserts that "a thing is intelligible only through its definition and essence" and that "essence is that which is signified by the definition of a thing."[42] For him, truth was to be found in the world of words: "When I desire to grasp the essence of a stone, I must arrive at it through a process of reasoning."[43] Accordingly, he placed little emphasis on observation and experimentation, even within the realm of natural science. In fact, we cannot find anywhere in Thomas's writings a single observation about the natural world that goes beyond what was left to history by Aristotle.

To begin with, Thomas—unlike his mentor Albert the Great—had never had much interest in natural philosophy per se. Instead, his interest lay in the principles of theology and metaphysics, and he discussed natural philosophy, which he considered to be one of the "lower" branches of learning, only to the extent that it contributed to an understanding of those principles. In his *Commentary on the Metaphysics of Aristotle*, for example, he includes a very long, drawn-out treatment of Aristotle's model of the cosmos whose sole purpose is to offer a proof of the existence of God: "Hence it follows that the whole universe is like one principality and one kingdom, and must therefore be governed by one ruler. Aristotle's conclusion is that there is one ruler of the whole universe, the first mover … whom above he called God, who is blessed for ever and ever."[44] And in Thomas's view, this "one ruler," God, does more than to merely keep the celestial spheres in motion: "The intention of the First Mover and Orderer, namely God, must extend not only to certain beings but even to the very last ones. Therefore all things are subject to His providence." The "executors" of this divine providence are the"lower separate intellects … that we call angels."[45] It is not only that these beings move the heavenly bodies, but also that these bodies—for example, the fixed stars, the planets, and the moon—have souls of varying levels and impart the power of the angels to the minerals, plants, animals, and humans of the Earth.

In all probability, however, Thomas's theory of magnetism as a power imparted to lodestones by heavenly bodies was not founded on this metaphysical basis alone. It seems quite clear that his beliefs were also based on empirical knowledge of magnetic polarity and of the fact that magnetized needles pointed northward. In this sense, Thomas embodied Scholasticism and all the significance and limitations that that philosophy was to have for the formation of modern natural science.

## Chapter 7

# Roger Bacon and the Propagation of Magnetic Force

### Section 1—Roger Bacon and his New Perspective

Like his contemporary Thomas Aquinas, philosopher Roger Bacon (c. 1214–c. 1294) was greatly influenced by Aristotle. Unlike Thomas, however, he recognized the sterility of Scholasticism's preoccupation with analysis of ancient texts. Moreover, he was more modern than other intellectuals of his day in emphasizing the role of both mathematics and experience in natural philosophy and in focusing on a hands-on and practical approach to scholarship.

Bacon was born in England and began his advanced studies at Oxford University. As mentioned in chapter 6, in the 1240s he lectured on the works of Aristotle at the faculty of arts of the University of Paris. This was the decisive period during which the university's ban on most of Aristotle's works was widely ignored, leading to their official approval in 1255, at which time the faculty essentially became the faculty of philosophy, independent from theology. Bacon's intellectual conversion appears to have occurred in just that period, namely, the late 1240s, and by the 1250s he had become a Franciscan friar. In 1266, Pope Clement IV (r. 1265–1268) expressed interest in Bacon's views, and it was at his urging that Bacon hurriedly wrote *Opus majus*. This he completed in 1267, and he followed it immediately with *Opus minus* and *Opus tertium*. These three volumes were essentially intended to convey to the pope Bacon's convictions concerning the intellectual approach to missionary work that should be adopted by the Christian world. Described as a "persuasion," the series constituted an earnest and heartfelt petition rooted in Bacon's perception that, as he puts it in *Opus majus*, "Christians are few, and the whole broad world is occupied by unbelievers; and there is no one to show them the truth."[1]

By the mid-thirteenth century, Christendom had witnessed the abject failure of the Crusades and come to the painful recognition that it was outnumbered and surrounded by nonbelievers. The superiority of the Islamic world in both technology and economic power had become all too apparent. While some people were enlightened by the sophisticated knowledge of ancient times, Scholasticism, which still adhered obstinately to the Bible and the writings of the church fathers, was impotent. Similarly to Thomas's *On the Truth of the Catholic Faith*, Bacon insisted that Christians could not hope to persuade anyone to convert through self-righteous arguments based exclusively on the Bible and an uncritical dependence on Christian assumptions. "Wherefore," he asserts in *Opus majus*, "we must seek for reasons in another way which is common to us and to unbelievers, namely, philosophy" (OMj, 793). This statement established, from a broad purview that objectified and relativized Christendom, the justification and necessity for Christians to study philosophy.

But this did not mean that Bacon placed philosophy above religion.

On the contrary, as a friar he naturally shared the medieval clergy's belief in the supremacy of theology. In *Opus majus* he seeks to show "that there is one wisdom that is perfect and that this is contained in the Scriptures," and declares that "one science is the mistress of the others, namely, theology, to which the remaining sciences are vitally necessary" (OMj, 36). Thomas also recognized the value of philosophy, but he believed that an insurmountable gap existed between human knowledge and divine knowledge, considered theology to surpass the bounds of philosophy, and thought philosophical knowledge (*scientia*) was no match for the wisdom of divine revelation (*sapientia*). Bacon, on the other hand, rejected such a separation between the two disciplines. Rather than seeing philosophical study as supplementary, he believed that it was indispensable to understanding the Scriptures and central to the study of theology: "there is one perfect wisdom which is contained in the Scriptures ... to be unfolded, however, by philosophy" (OMj, 65).

Bacon includes the knowledge of nonbelievers within his concept of philosophy. In fact, he says, "philosophy belongs in an especial sense to unbelievers, since we have derived all of our philosophy from them." Despite this, philosophy is not in conflict with the "divine truths," that is, the teachings of the Scriptures, because "the power of philosophy is a vestige of the divine wisdom given by God to man" and therefore is "in perfect accord with the divine wisdom" (OMj, 793). Bacon adds that, "since Philosophy is the forerunner of religion and disposes men to it, the Christian religion is the only one that should be adhered to" (OMj, 808). In other words, the truth given to us by God is one, and therefore, when correctly understood, philosophy—even pre-Christian philosophy developed by pagans—must inevitably lead man to Christianity.

It is also worth noting that Bacon's concept of philosophy included the secular branches of learning, namely, mathematics, physics, astrology, and alchemy. Basically, he believed that making use of secular knowledge and pagan knowledge was essential if Christianity was to become a truly universal faith with a strong and productive theology capable of convincing and converting nonbelievers. After all, Europeans in Sicily and on the Iberian Peninsula were confronted with the astounding reality that "wherever the influence of the Muslims, those sons of Satan, is exerted, everything prospers, springs flow, and the earth is covered in flowers."[2] Christians could not hope to conquer by complacently preaching a doctrine of salvation in the afterlife.

The essence of Bacon's ideas concerning learning and science is to be found in his advocacy of "experimental science" or "experimental art."[3] Experimental science, he argued, possessed three characteristics—referred to as "prerogatives"—that set it apart from all other sciences. This argument is laid out in part 6 of *Opus majus*, his principle work.

According to Bacon, the first so-called prerogative of experimental science is that "it investigates by experiment the notable conclusions of all [other] sciences" (OMj, 587). In contrast to the conventional science of his day, in which individual phenomena were logically deduced from general principles, Bacon insisted that, even if those principles could be induced from experience, this in itself was not enough to demonstrate the veracity of what was deduced from them; the results of that process must be confirmed by experiment. For Bacon, however, experience consists of two types, "exterior sense" and "interior illumination," the latter being experience that is conferred by divine grace and directly verifies truth. Therefore, when he talks about proof through experience (sometimes translated as "experiment") he does not necessarily mean the modern scientific procedure of confirming a hypothesis empirically. In any case, he deserves credit for being the first to demonstrate through actual experiment the absurdity of the notion, unquestioningly passed on by European intellectuals for over a millennium from Pliny forward, that "the diamond cannot be broken except by goat's blood" (OMj, 584).

The second prerogative of experimental science is that it is "able to give us important truths within the confines of the other sciences, which those sciences can learn in no other way" (OMj, 615). In other words, some things can be known and proved only through experience, whose value is this ability to investigate unexplored territory within existing sciences and to reveal undiscovered knowledge. As one example Bacon points out that "the experimental art supplies the defect of medicine" (OMj, 618).

Bacon's third prerogative of experimental science "arises from those properties through which it has no connection with the other sciences, but by its own power investigates the secrets of nature." Specifically, it consists in "the knowledge of the future, the past, and the present" (i.e., astrology) and in "wonderful works by which it excels in the power of forming judgments the ordinary astronomy dealing with judgments" (OMj, 627). He refers here to those sciences that disclose powers of nature previously concealed from man (although according to Bacon they were known to the ancient sages) and that allow man to manipulate those powers—to engage in what, in modern terms, might be referred to as the technological exploitation of natural forces based on experience.

It is true that Bacon includes within "experimental art" the magical and the fantastic, such as alchemy to discover "the medicine which would remove all the impurity of a baser metal, so that it should become silver and purest gold" (OMj, 627); as well as "ever-burning lamps" and, "against foes of the state," "important arts [by which] without a sword or any weapon requiring physical contact [one] could destroy all who offer resistance" (OMj, 629). But while we may question this indulgence in fantasy, we must recognize that his arguments were a refreshing challenge to the barren views of conventional learning. Bacon deserves credit for his forthright criticism of Scholastic views, which were extremely narrow and impoverished when compared with the reality presented by nature itself, and of the arrogance of Scholasticism's claim that everything in the universe could be deduced from first principles and the nature of things (by which was meant merely their definitions), if only accurate logic were used.

*Opus majus* ardently proposed that the best and most urgently needed strategy for consolidating Christian society and gaining supremacy over nonbelievers was the exploration of nature's secrets and the technological application of natural forces. Bacon concludes part 6 as follows:

> Although other sciences do many wonders … yet all things of such wonderful utility in the state belong chiefly to this science [experimental art] … For this science teaches how wonderful instruments may be made, and uses them when made, and also considers all secret things owing to the advantages they may possess for the state and for individuals; and it directs other sciences as its handmaids, and therefore the whole power of speculative science is attributed especially to this science. And now the wonderful advantage derived from these three sciences [the three prerogatives of experimental art] in this world on behalf of the Church of God against the enemies of the faith is manifest, who should be destroyed rather by the discoveries of

> science than by warlike arms of combatants ... Moreover, the Church should consider the employment of these investigations against unbelievers and rebels, in order that it may spare Christian blood, and especially should it do so because of future perils in the times of Antichrist, which with the grace of God it would be easy to meet, if prelates and princes promoted study and investigated the secrets of nature and of art. (OMj, 633–34)

The gist of Bacon's passionate appeal to Pope Clement IV was that Christendom, surrounded as it was by nonbelievers, should rise above the fruitless arguments of Scholasticism, with its preoccupation with words and its refusal to learn from experience, and forge ahead along the path of experimental science, which was more practical. Bacon thus identified the purpose of science to be domination over nature and its exploitation for the benefit humanity. It is this above all that defines his perspective and sets him apart from previous European thinkers.

## Section 2—Bacon on Mathematics and Experience

Bacon was influenced heavily by the works of Aristotle, for by the time he was studying at Oxford University during the 1230s, many of the Greek master's works, including *Posterior Analytics*, were already available in Latin; unlike at the University of Paris, moreover, at Oxford they were not banned. Aristotle thus made a direct and very major impact on Bacon. Bacon assimilated Aristotle's philosophy and then surpassed it by placing greater emphasis on the importance of experience and of mathematics.

Before we go into how the ideas of Bacon and Aristotle were alike and different, let's first get an understanding of what Aristotle saw as the value in comprehending nature and as the proper means of doing so, particularly as he expressed these views in *Posterior*

*Analytics.* And to help gain this understanding, let's first take a look at Plato and then compare the perspectives of these two ancient philosophers.

Plato believed that certain knowledge in the true sense of the word was only possible in the world of so-called Forms (also referred to as Ideas). Furthermore, Plato thought that geometry conferred ideal knowledge. For example, one proposition of geometry is that a triangle inscribed in a circle will always be a right triangle if one of its sides is a diameter of the circle. According to Plato, it is possible to prove this proposition by argument based on several self-evident and irrefutable truths and on the definitions of a triangle and of a circle. The triangle and the circle, in his system, are universal Forms that transcend the actual, individual shapes drawn on paper, the latter being necessarily inexact to some extent as a result of their having been made concrete and specific.

Thus, Plato argued that the proposition is not obtainable by measuring actual triangles or circles, nor does it need to be verified by measurement in order to be deemed correct, as long as there has been no error in reasoning. For the world of visible and tangible things, in contrast, "we are not able to give notions which are altogether and in every respect exact and consistent with one another."[4] It is impossible, thought Plato, to have truly exact and objective knowledge concerning the world perceived by the senses; subjective guesses were the best that could be expected. Rather than aiding understanding, reliance on the senses carried the danger of actually steering one wrong. Thus, Plato considered induction based on experience to be inferior to mathematical proof as a method of obtaining knowledge, and moreover he considered these two methods to be opposed to one another.

Returning to Aristotle, let's first note one point of similarity between his view and Plato's. In *Posterior Analytics*, Aristotle observes concerning true knowledge of things that "we know the cause on which the fact depends, as the cause of the fact and of no other, and ... that the fact could not be other than it is"; and he calls this knowledge by "demonstration."[5] In other words, accurate knowledge of a thing can be obtained only when its various characteristics are implied by precise reasoning from the definition of the nature of that thing.

In contrast to Plato, Aristotle gives empirical science, which Plato had relegated to a much lower rank, a legitimate position. Aristotle recognizes the necessity and the value of empirical and inductive knowledge: "it is clear that we must get to know the primary premisses [sic] by induction."[6] In his view, in order to achieve knowledge by demonstration we must know in advance the premises from which reasoning (deduction) can begin, but at the same time induction based on the senses is essential if we are to know those first premises.

Accordingly, Aristotle differed from Plato in believing that sense perception assisted understanding and was in fact indispensable to it. In his natural philosophy, knowledge begins with sensory perception and experience of specific objects and phenomena. Human beings have been endowed with "a congenital discriminative capacity which is called sense-perception"; "out of sense-perception comes to be what we call memory, and out of frequently repeated memories of the same thing develops experience"; and when the same thing is experienced repeatedly, knowledge—that is, knowledge of facts—is born. Thus, Aristotle asserts that while "it is impossible to come to grasp universals except through induction," "induction is impossible for those who have not sense-perception."[7] To be sure, universals are "that which is prior for knowledge," yet "things which are more knowable and obvious to us" should be "prior in relation to perception."[8]

Concerning mathematics, although Aristotle said that "it is the business … of the mathematicians to know the reasoned fact,"[9] and ideologically recognized the importance of mathematical knowledge in astronomy and music, in general he believed that in the terrestrial world only qualitative knowledge was possible, and he himself did not treat any phenomenon mathematically except for the phenomenon of motion, defined in the very narrow sense of changes in the locations of bodies.

Aristotle classified perceivable bodies into two pairs of opposite characteristics: cold versus hot and wet versus dry. As long as he did so there was no need to quantify those qualities logically. In other words, the fact that these characteristics were opposing qualities meant that they were heterogeneous and therefore could not be expressed either in terms of relative degree or in terms of numerical units. This closed the door to the modern, quantitative concepts of temperature and humidity. Even the terms heavy and light when used by Aristotle were qualitative and therefore did not lead to the quantitative concept of weight. Thus his terrestrial physics remained qualitative.

And now we come to Bacon and his natural philosophy. Bacon is emphatic about the importance not only of empirical methods, but also of mathematics. In his perspective, in fact, knowledge has two pillars: mathematical reasoning and empirical proof. On the one hand, he believes that "without experience nothing can be sufficiently known" (OMj, 583), but on the other hand he also holds that "the things of this world cannot be made known without a knowledge of mathematics" (OMj, 128). This does not mean, however, that there are two types of knowledge, but rather that in order for knowledge to be exact and complete both mathematical (demonstrable) and empirical (sensory) knowledge are necessary and that the two must be used to complement one another.

In fact, although Bacon accepts Aristotle's distinction between the celestial and terrestrial realms, he also, like Thomas, deems that the former exerts influence on the latter. And in positing that influence he demonstrates that it is both possible and necessary to bring mathematical knowledge to bear on the study of nature in all parts of the universe. As he puts it: "The heavenly bodies are known only through quantity, as is clear from astronomy. Therefore all the categories depend on a knowledge of quantity of which mathematics treats" (OMj, 120). "But," he maintains,

> celestial things are the causes of terrestrial. Therefore these terrestrial things will not be known without a knowledge of the celestial, and the latter cannot be known without mathematics. Therefore a knowledge of these terrestrial things must depend on the same science. (OMj, 129)

On this basis he concludes that "these terrestrial things cannot be learned without mathematics" (OMj, 129). This view, while inheriting much of Aristotle's natural philosophy, transcends both the narrow view of Plato's Forms (which devalued sensory perception) and the limitations of Aristotle's restriction of his inquiries to an understanding of qualitative properties alone. What Bacon advocates, then, is a mathematical natural philosophy.

Thus, part 4 of *Opus majus* is devoted to extolling mathematics as the foundation for the other sciences. Here is a striking example:

> In mathematics we are able to arrive at the full truth without error, and at a certainty of all points involved without doubt; since in this subject demonstration by means of a proper and necessary cause can be given.... In mathematics alone are there demonstrations of the most convincing kind through a necessary cause. And therefore here alone can a man arrive at the truth from the nature of this science.... Wherefore it is evident that if in other sciences we should arrive at certainty without doubt and truth without error, it behooves us to place the foundations of knowledge in mathematics ... Mathematics alone ... remains fixed and verified for us with the utmost certainty and verification. Therefore by means of this science all other sciences must be known and verified. (OMj, 123–25)

Note, however, that the relationship between mathematics and experience presented in Bacon's philosophy is actually quite peculiar. While on the one hand he bases the accuracy of knowledge upon mathematical reasoning, on the other hand he says that it is by sense perception that man verifies the truth of mathematics:

> In this subject [i.e., mathematics] it is possible to have for all things an example that may be perceived by the senses, and a test perceptible to the senses in drawing figures and in counting, so that all may be clear to the sense. For this reason there can be no doubt in this science. (OMj, 124)

Accordingly, the only conclusions reached through mathematical reasoning about which man can have any confidence are those that he has also directly verified through perceptual experience:

> There are two modes of acquiring knowledge, namely, by reasoning and experience. Reasoning draws a conclusion

> and makes us grant the conclusion, but does not make the conclusion certain, nor does it remove doubt so that the mind may rest on the intuition of truth, unless the mind discovers it by the path of experience ... This is evident in mathematics, where proof is most convincing. But the mind of one who has the most convincing proof in regard to the equilateral triangle will never cleave to the conclusion without experience ... Aristotle's statement ... that proof is reasoning that causes us to know is to be understood with the proviso that the proof is accompanied by its appropriate experience. (OMj, 583)

Certain knowledge cannot be attained without mathematical reasoning, yet only when the conclusion is verified by experience can it be accepted without reservation. This is a pioneering expression of a physics that is mathematical and experimental at the same time.

## Section 3—Robert Grosseteste

Bacon proposed a highly significant new concept concerning magnetic force: the spatial propagation of magnetic action. He derived his inspiration for this in the theory of optics put forward by English philosopher and theologian Robert Grosseteste (c. 1175–1253). Grosseteste, who came from Suffolk, served at one time as chancellor of Oxford University, lectured on theology at a Franciscan friary in Oxford, and for eighteen years beginning in 1235 was bishop of Lincoln, the largest diocese in medieval England. He was one of the first to advocate the concept of experimental and mathematical natural science, and he developed an intricate theory of optics.

Let's take a closer look at Grosseteste and his contributions to science, which were important both in themselves and in the impact that they had on Bacon's research. The name Grosseteste was apparently a nickname meaning "great head," and true to this appellation Grosseteste was one of England's outstanding intellects at that time. In *Opus majus*, Bacon remarks that in order to translate scientific treatises one must be well versed not only in the two languages concerned but also in the scientific content, and in this respect he singled out Grosseteste for praise: "Master Robert, called Grosse-Teste … alone knew the sciences" (OMj, 76).

Indeed, Grosseteste not only translated Aristotle's *Nicomachean Ethics* into Latin but also wrote commentaries on Aristotle's *Physics* and *Posterior Analytics*. Of particular importance are his adoption of the distinction between the two types of knowledge (inductive and deductive) described in *Posterior Analytics*; his belief that the study of nature began with a knowledge of facts based on experience but then needed to continue with an exploration of fundamental causes; and his accurate assessment of the value of empirical knowledge based on sense perception. But Grosseteste's philosophy of science is worthy of mention not merely because he emphasized the importance of empirical knowledge, but also because he so highly evaluated the role of mathematics and particularly of geometry in understanding the natural world. And his focus on geometry was primarily derived from his theory of light—his metaphysics of light.

This unique theory is laid out in *De motu corporali et luce* (On corporeal motion and light) and in *On Light*, which Grosseteste is believed to have written while he was at Oxford. *On Light* opens with these words:

> The first corporeal form which some call corporeity is in my opinion light [*lux*]. For light of its very nature diffuses itself in

> every direction in such a way that a point of light will produce instantaneously a sphere of light of any size whatsoever, unless some opaque object stands in the way.[10]

In short, as Grosseteste posits a few lines later, light has the nature of "multiplying itself ... instantaneously in all directions."

To fully appreciate the above passage, we need to be aware that wherever the word "light" appears in it, in the original Latin Grosseteste used the word *lux*, referring not to physical light that was perceptible by the senses, but rather to a metaphysical entity that exists prior to all material bodies. In other words, *lux*, which expands of itself instantaneously and three-dimensionally, can invest primary substance without dimension—"first matter"—with three-dimensional extension. While this argument is based on Aristotle's schema of causality, which held that primary entities in potentiality become physical entities in actuality, Grossesteste is unique in identifying light (*lux*) as the agency of that change. To him, *lux* actualizes all matter, giving it tangible embodiment, and as such it occupies center stage in metaphysics and natural science as "corporeity itself" or the "first corporeal form."

In the second half of *On Light*, Grosseteste proposes a bold theory concerning the beginning of the universe, where *lux* features in a sort of Big Bang. First, the firmament, the celestial sphere, is formed as the maximum possible extension of *lux*. Although this metaphysical light expands infinitely, because it begins as an infinitesimal point, the firmament created by its infinite multiplication is a finite sphere that houses the fixed stars and establishes the outer limits of the material universe. Then, as a result of the expansion in all directions, and concentration within the celestial sphere, of *lumen* (physical light) reflected by the firmament, the spheres of the planets, the sun, the moon, and the Earth itself formed one after the other.

This theory describes the genesis of an Aristotelian hierarchical universe. Of course, Grosseteste's argument was likely based on the special role that light plays in the biblical account of Genesis, which begins with God's command, "Let there be light." But the most remarkable thing about Grosseteste's theory is that the creation of the universe is presented not as God's will or plan but rather as the spontaneous unfolding of the natural world in accordance with the laws of nature. As discussed in the previous chapter, this idea—that nature was governed by laws that were meant to be understood—was at the heart of a major change in the direction of thought sparked by the introduction of Aristotle's theories into medieval European thought.

In *Concerning Lines, Angles, and Figures*, which Grosseteste is believed to have written near the end of his time at Oxford, he develops his ideas on the propagation of all actions in nature based on this metaphysics of *lux*.[11] All actions in nature, he argues, are the result of the expansion of *lux*, and therefore their propagation occurs in the same way as the diffusion of *lux*, in a manner he describes as a "multiplication of species" or "multiplication of power":

> A natural agent multiplies its power from itself to the recipient, whether it acts on sense or on matter. This power is sometimes called species, sometimes a likeness, and it is the same thing whatever it may be called ... It acts in a single manner whatever it encounters ... But the effects are diversified by the diversity of the recipient, for when this power is received by the senses, it produces an effect that is somehow spiritual and noble; on the

> other hand, when it is received by matter, it produces a material effect.[12]

Several paragraphs after that he notes that

> the sphere is required by the multiplication of a power; for every agent multiplies its power spherically, since it does so in every direction and along all diameters—upward, downward, before, behind, to the right, and to the left.[13]

Those willing to indulge in drawing anachronistic parallels might claim that this one sentence foreshadowed the divergence theorem (also known as Gauss's theorem), a mathematical law that was discovered and proved centuries later. Without going quite this far, we should still note that Grosseteste's concept of "multiplication of species" or "multiplication of power" occupies a very important position in our discussion because it was adopted directly by Roger Bacon as the model for his theory of the spatial propagation of magnetic action.

The word "species" is derived from the Latin verb *specere* (to see). It originally referred to "appearance, shape, or aspect" and then came to have such meanings as "figure," "vision," "image," "idea," and "species." As a term used in taxonomy it relates directly to "genus" and means a group subordinate to a genus and comprising individuals that share certain common attributes. Along with words denoting "figure," "image," and the like, it refers to the factors involved in distinguishing a certain thing as separate from others that are in certain respects similar to it. The term "idea," on the other hand, is based on Plato's *idea* and Aristotle's *eidos* (form). But Grosseteste's "species" was also called "power" (*virtus*) and at times "similitude" (likeness), and in *Concerning Lines, Angles, and*

*Figures* he even uses it in the following way: "when reflection occurs from rough bodies, the species is dissipated and the action is weak."[14] Taking these things into consideration, we can say that Grosseteste's "species" is a more physical concept than Aristotle's *eidos* (form) and means specifically the action that an agent exerts on something else.

Let's return to Grosseteste's model for the propagation of all physical actions, which followed the three-dimensional, isotropic radiation of *lux*. In *Concerning Lines, Angles, and Figures* he presents the proposition that "nature acts in the briefest possible manner," and from this concludes that three laws—the basic laws of geometric optics—govern the scheme of the propagation of action.[15] First, in a homogenous medium the propagation of an action is rectilinear. Second, when an action is projected into an opaque medium, it is reflected in such a way that the angle of reflection and the angle of incidence are equal. Third, when projected into a transparent medium, if the second medium is denser, the action is refracted toward the medium and if it is less dense it is refracted away from the medium. Grosseteste was, in fact, the first to point out on the basis of these three laws that a rainbow is the result of sunlight refracted by drops of water, thereby opening the way to subsequent developments in the theory of the rainbow. But it was his claim that the rules governing *lux* govern all physical and natural phenomena, and not merely optical phenomena, that characterized both the basic thesis of his metaphysics of "light" and its peculiarity. Assuming this claim were true, it would follow inevitably that any model of the propagation of actions would have to rely heavily on the laws of geometry, and for this reason geometric concepts would gain that much more importance in the study of natural philosophy.

At the beginning of *Concerning Lines, Angles, and Figures*, Grosseteste has the following to say about the importance of geometric concepts:

> The usefulness of considering lines, angles, and figures is very great, since it is impossible to understand natural philosophy without them. They are useful in relation to the universe as a whole and its individual parts. They are useful also in connection with related properties, such as rectilinear and circular motion ... Now, all causes of natural effects must be expressed by means of lines, angles, and figures, for otherwise it is impossible to grasp their explanation.[16]

This predates by as much as four centuries a similar statement by Galileo:

> Philosophy is written in this grand book, the universe, which stands continually open to our gaze ... It is written in the language of mathematics, and its characters are triangles, circles, and other geometric figures without which it is humanly impossible to understand a single word of it.[17]

It was Roger Bacon who adapted Grosseteste's metaphysical theory of the multiplication of species and recast it in a form more in tune with natural philosophy. Indeed, he used this model as the basis for his theory of the propagation of action through a medium.

## Section 4—Multiplication of Species according to Bacon

Bacon believed that all action, including not only corporeal action but also spiritual action, was propagated by multiplication of species. Here are his words, in which Grosseteste's influence is clearly evident:

> Every efficient cause acts by its own force which it produces on the matter subject to it, as the light [*lux*] of the sun produces its own force in the air, and this force is light [*lumen*] diffused through the whole world from the solar light [*lux*]. This force is called likeness, image, species, and by many other names, and it is produced … by spiritual substance as well as corporeal. (OMj, 130)

Bacon presents this same theory independently and in more detail in *On the Multiplication of Species*[18], which was written around the same period. In it he explains that the word "species" is

> meant to designate the first effect of any naturally-acting thing. And to explain this meaning of "species" with an example, we say that the *lumen* of the sun in the air is the species of the solar *lux* in the body of the sun … [and] *lumen* is that which is multiplied and generated from that *lux* and which is produced in air and other rare bodies, which are called media because species are multiplied by their mediation. (OMS, 5)

While it is clear that Bacon adopts this theory and phraseology from Grosseteste, his argument differs in several fundamental ways. For one thing, although Bacon, like Grosseteste, argues that the propagation of *lux* causes multiplication of species, to Bacon *lux* is only one example of many types of action, whereas *lux* as defined by Grosseteste has a special role as the primary agent that causes all other actions. Bacon sees all actions on the same level, and he does not postulate a special, meta-level primary agent as the initiator of those actions. Rather, for him, it is the dynamism of propagation, that is, multi-plication of species, that is the common form of all actions.

The fact that Bacon repeatedly uses the term "first effect" in *On the Multiplication of Species*, as in the above quotation and in other passages on multiplication of species—his comment, for example, that "a species is the first effect of an agent" (OMS, 7)—indicates that he perceives the propagation of action to be action through a medium. This is his second and decisive point of departure from Grosseteste's line of reasoning, and it represents a shift in direction from metaphysics to physics. Bacon explains the idea as follows:

> The active substance of the agent, touching the substance of the recipient without intermediary, can alter, by its active virtue and power, the first part of the recipient that it touches. And this action flows into the interior of that part. (OMS, 53)

In other words, "proximity is required as a necessary condition of action" (OMS, 63–65), and therefore the agent can act only on things with which it is directly in contact. Accordingly, if the agent and the recipient are spatially separated, there must necessarily be a medium that fills the space between, and only the part of the original agent that touches the medium—the part that is directly in physical contact with it—can generate that species. This is what Bacon means by "the first effect." This species, "the first effect," draws out the second species, "the second effect," from the part of the medium with which it is in contact. "The first part of the recipient, having been transmuted and possessing species in actuality, transmutes the second part, the second part transmutes the third, and so on" (OMS, 91).

In this way, the species multiplies as a sort of chain reaction within the medium and, as a result, an action is propagated. In his first example, solar *lux* from the sun, the source of light, generates and multiplies the species *lumen* in the medium of the atmosphere, and this is further propagated through the air. (At that time, it was believed that the atmosphere extended throughout the entire universe, that is, within the sphere of the fixed stars.) This point is explained in more detail in *Opus majus* as follows:

> The species … which is produced in the first part of the air is not separated from that part … but the species forms a likeness to itself in the second position of the air, and so on. Therefore it is not a motion as regards place, but is a propagation multiplied through the different parts of the medium. (OMj, 489–90)

In terms of light, this is an explanation of what in modern language we would call wave theory rather than particle theory.

As a consequence of this reasoning, Bacon naturally concluded that the propagation of all actions, including light, required a finite amount of time:

> All authors, whether sacred or others, who state that light is multiplied instantaneously, are to be understood in regard to a divisible instant, which is imperceptible time, and not in regard to a real instant, which is the indivisible terminus of time, just as the point is of the line. (OMj, 491)

Or, in different words, "light is multiplied in time, and likewise all species of a visible thing and of vision" (OMj, 489). In Grosseteste's view, *lux* itself expands instantaneously in the shape of a sphere, and therefore multiplication of species, the primary agent of which is light, takes place without a medium. With Bacon, for the first time, a medium is seen as necessary for the propagation of all actions, including light, and therefore the speed of propagation has to be finite.

Concerning the mode of the propagation of action, Bacon maintains that the way in which actions are propagated in nonliving mediums are the same as that for "rays": rectilinear in homogenous mediums, refracted at the interface between mediums with varied density, and reflected on the surface in opaque mediums. The path of propagation for all actions is to be expressed as a straight line or a refraction of this line. In addition, he also mentions propagation of actions in animated mediums, and the path of propagation in this case is winding. These four types of propagation set off directly from the agent and are called "principal multiplication," and the paths of all four are expressed as lines. In addition to these, there is also "accidental multiplication," described as follows:

> The fifth line along which a species proceeds differs from all the aforementioned, since it does not come from the agent [itself], but from any of the four aforementioned lines; and therefore it comes not from the object that produces the species, but from the species. Thus the species travelling along this line is the species of a species, just as the light in the corner of a house comes [not directly from the sun, but] from the solar ray entering through the window. For the latter ray comes from the sun along a rectilinear, refracted, or reflected path; and therefore it is the principal multiplication. But the light that comes from the [principal] ray to other parts of the house is an accidental multiplication. (OMS, 103; brackets in source)

In other words, secondary multiplication of the action occurs from each point along the ray of light, and this Bacon calls accidental multiplication; he uses this concept to explain the diffraction of light which causes the spreading of light beams into shadow space.

This astounding argument concerning the mechanism of propagation seems almost to anticipate Huygens' principle on the propagation of waves, formulated in the seventeenth century.

Of course, in the history of science we must refrain from reading too much into past events or adding excessive meaning on the basis of modern hindsight, and to claim that Bacon's theory of multiplication of species pioneered field theory in modern physics would be an exaggeration. After all, Bacon's multiplication of species was still based on Aristotle's schema of causality even where it concerned action through a medium, and it was certainly not a mechanistic explanation like that of Huygens. It is true, however, that Bacon for the most part eliminated the metaphysical connotation from Grosseteste's theory of species multiplication, and by doing so rewrote it as a theory more in keeping with the discipline of physics.

## Section 5—The Propagation of Magnetic Force as Action through a Medium

Roger Bacon is often spoken of as a pioneer of modern natural science because of his emphasis on mathematics and experience. Yet the schema by which he viewed nature and the theories he expounded as interpretations of individual natural phenomena were in fact straight from Aristotle.

In *Opus minus*, published in 1267, Bacon presents the following observations and interpretation of magnets:

> Iron follows the part of a magnet that touches it, and flies from the other part of the same magnet. And the iron turns itself after moving to the part of the heavens conformed to the part of the magnet which it touched. Now four parts of the world are surely distinguished in the magnet, namely, east, west, north and south, and they can be recognized by an experiment in which it is well shown to what part of the heavens a given part

> turns. And then if the iron is touched by the northern part of the magnet, it will follow that part however it is moved about, namely, upwards or downwards, to the right or left, and in all different positions. And so much is it captivated that if the iron is put [to float] in a vessel full of water and the hand [with the magnet] is placed under the vessel, the part touched dives into the water in the direction of the magnet. And if the magnet is brought beyond us anywhere above the iron, the part touched flies up towards whatever place the magnet is brought. And if the other end of the magnet is presented to the same part of the iron, this part flies from it as from an enemy, as a lamb from a wolf. And when the magnet is removed the touched part turns itself towards the place of the heavens similar to the part of the magnet.
>
> Ordinary philosophers do not know the cause of the common observations concerning this part, and believe that the Nautical Star [Pole Star] brings it about. But it is not the star that effects it but the part of the heavens; and the three other parts of the heavens, namely, south, east, and west, operate as well as the north. Similarly they do not take note of the fact that these four parts of the world may be distinguished in the magnet. But many attribute [the effects observed] to one part, which agrees with the Nautical Star in natural property.[19]

Bacon understood the magnet to be a microcosm of the universe divided into east, west, north, and south, and he believed that the direction toward which a magnetized piece of iron pointed was determined by the part of the magnet with which the iron had been in contact. As will be seen in the next chapter, Petrus Peregrinus was to inherit this point of view, but on the basis of his own experiments and observations, he did not perceive the east-west and north-south poles to be equal, but rather attached greater importance to the north and south poles.

It is interesting that Bacon, in the above excerpt, asserts that the "Nautical Star" is not responsible for conferring on magnets or magnetized iron their northward directionality. It has been boldly speculated that this assertion shows Bacon knew about magnetic declination (the phenomenon whereby magnetized needles deviate from true north),[20] but this is far-fetched. Rather, here Bacon wishes to emphasize that the source of magnetic attraction was not the Pole Star but instead "a part [or place] of the heavens" (*locus coeli*). This is a quintessential example of Bacon as an Aristotelian.

According to Aristotle, when a heavy body falls, it heads not toward the Earth but toward the center of the universe as its proper *place*. Since, in his model, the center of the Earth and the center of the universe are one, no difference is apparent, but he asserts that "if one were to remove the earth to where the moon now is, the various fragments of earth would each move not towards it but to the place in which it now is."[21] In this understanding, what acts on a heavy body is not the object Earth but the place the Earth occupies. Similarly, in the fourteenth century Jean Buridan notes that

> some say … that place is the cause moving the heavy body by a method of attraction, just as a magnet attracts iron.[22]

Thus, to Aristotelians, "place … exerts a certain influence."[23] Bacon most likely considered the power of place to be the greatest validation of his claim that the power exerted on magnets comes not from the celestial body of the Pole Star but from the various *parts* or *places* of the heavens—north, south, east, and west.

Moreover, Bacon perceived changes in matter that result from action in terms of Aristotle's schema of causality, which entailed a transition from potentiality to actuality.

According to Bacon's *On the Multiplication of Species*, "at first, before the action, the recipient is of itself dissimilar to the agent, and through the action it becomes similar," the reason for this being that "the virtue or species produced by an agent is similar to the agent in nature, definition, specific essence, and operation" (OMS, 7). This is probably why the term "likeness" is also used for species. According to Bacon, however, when an action occurs, the propagation of species is not simply one-way, from agent to recipient. Rather, the formation of the species in the recipient is the result of the agent's realizing or manifesting something that already existed potentially or latently in the recipient:

> The agent sends forth a species into the matter of the recipient, so that, through the species first produced, it can bring forth out of the potentiality of the matter [of the recipient] the complete effect that it intends ... The agent directs its efforts to making the recipient similar to itself, because the recipient, as Aristotle intends in *De generatione*, is always potentially what the agent is in actuality, as he there asserts. (OMS, 7; brackets in source)

Bacon also clearly states in *Opus majus* that light in the air (*lumen*), as the species of light (*lux*) given off by the light source, "is not produced by a flow from a luminous body, but by a renewing from the potency of the matter of the air" (OMj, 490). In this sense, Bacon's scheme of propagation differs fundamentally from Huygens's mechanistic portrait of light propagated through a series of vibrations of a medium (ether) set in motion by pulsations of a luminous body. Instead, Bacon postulates that it is impossible for an agent, regardless of its type, to propagate the species in the recipient if the species does not already exist in potential form within the recipient. For example, fire seeks to propagate its species to matter around it, causing it to burn, but this does not mean that anything will burn. Only those things that already possess the nature of fire, that is, flammable substances, will burn, meaning that they will become fire in actuality as the species of fire is drawn out from them as recipients. The species of fire cannot be propagated in matter that does not have this potentiality, which is to say, in nonflammable substances.

Bacon cites colored glass and magnets as good examples of this dynamism. Concerning the fact that sunlight shining through colored glass imparts the same hue to whatever it falls upon, he concludes that "this color is more apparent than real; and it is merely a species [of the colored glass], and therefore it can be produced out of the potentiality of matter, especially that of a mixed body." His reasoning here is as follows:

> The species of the glass is produced in air before it is produced in an opaque mixed body, but it is produced far more weakly in air because of the simplicity of the body of air; and when it reaches a mixed body more suited to color, the species existing in air can call forth a stronger species out of the potentiality of the matter [of the mixed body], just as magnetic virtue is conveyed to iron by air, but is stronger in the iron than in the air because the iron is better suited [to receive it]. (OMS, 55; brackets in source)

As we saw earlier, Bacon's theory of multiplication of species is based on the principle of action through a medium, with both light and magnetic force as examples of propagation through the medium of air. This notion that magnetic force is transmitted through air was voiced by others of that era as well. The twelfth-century Jewish philosopher Maimonides, for example, held that

> the magnet attracts iron from a distance through a certain force communicated to the air which is in contact with the iron.[24]

Likewise, around 1250, Sicilian poet and judge Guido delle Colonne wrote:

> The learned relate that the lodestone
> Could not attract
> Iron by its power,
> *Were it not that the air between them aids.*[25]

Bacon, however, does not stop there but goes on to explain this within Aristotle's schema. He postulates that the medium of the air itself does not possess the potentiality for magnetism, and therefore it is neither magnetized nor attracted to the magnet. This is just the same as the phenomenon of a ray of light passing through colored glass, which, while it colors the object upon which it falls, leaves the intervening air itself almost colorless. From this he concludes that

> the power of a lodestone [*adamas*] or a magnet [*magnes*] is stronger in distant iron than in adjacent air, and yet it passes through the air to the iron. (OMS, 209)

In other words, the magnetic force emitted by the magnet is propagated through the medium of the air, and it is only when it reaches the iron that magnetism, which the iron potentially possesses, is actualized. This is how the magnet makes iron like itself—magnetic—and pulls it toward itself.

Aristotle, an adherent of the theory of action through a medium, was unable to successfully fit into his physics the phenomenon of magnetic force, which appeared to him to be action operating at a distance. It is fair to say that Bacon's theory of multiplication of species overcame, from within the Aristotelian theoretical framework, the limitations Aristotle himself had in this regard. Moreover, it was the first attempt to explain the seemingly remote and magical effect of magnetic force as an example of action through a medium from a standpoint other than that of reductionism, which relied on theories involving atoms or effluences.

In the first half of the fourteenth century, Oxford theologian and philosopher William Crathorn maintained that

> a magnet affects the medium before it affects the iron, even though the form impressed on the intermediary air is not as perfect as the form impressed on the magnet.[26]

With this indication that he understood magnetic force to be action through a medium, the medium in this case being air, Crathorn provided just one example of the unmistakable impact Bacon had on the subsequent development of medieval thought.

* * *

As we have seen, Bacon proposed that magnetic force, like light, was propagated through the medium of air within a finite amount of time. Moreover, he explained this phenomenon within Aristotle's schema of causality, using the framework of potentiality and actuality. Here, let us recall that the term "species" was also called "likeness" and consider together with this Bacon's statement in *Opus majus* that the magnet attracts the iron "because of the similar nature" of the two (OMj, 631). It seems clear that Bacon used Aristotle's theories to reinterpret and demonstrate a new basis for the idea that like attracts like—a proposition that had been around since the time of Democritus and Plato. But to the extent that Bacon took this approach, magnetic force remained mysteriously cloaked in the shadows cast by the limitations of Aristotelian natural philosophy. Bacon advocated experiential science, stressed the importance of experimentation and mathematics in natural philosophy, and proposed the technological application of natural forces as the ideal purpose of intellectual endeavor. In the case of magnetism, however, he did not practice what he preached. As we will see in the next chapter, this was left up to Petrus Peregrinus.

Chapter 8

# Petrus Peregrinus and his *Letter concerning the Magnet*

## Section 1—The Discovery of Magnetic Polarity

Pierre de Maricourt, better known by his Latin name Petrus Peregrinus (Peter the Pilgrim; fl. 1269), probably hailed from Picardy. Little is known about his life; even the years of his birth and death are highly uncertain, although we do know that he was a contemporary of Thomas Aquinas and Roger Bacon. Nonetheless, Peregrinus is a hugely important figure in the history of experimental science in general and of the physics of magnetism in particular.

As far as we know, he was the first person to make the phenomenon of magnetic north-south alignment—a new discovery in thirteenth-century Europe—the subject of research in experimental physics; indeed, he was almost the first to engage in experimental physics in any subject whatsoever. His intellectual legacy, a work entitled *Letter of Peter Peregrinus de Maricourt to Sygerus de Foucaucourt, Soldier, concerning the Magnet* (hereafter *Letter concerning the Magnet*), was the earliest known description of the magnet that was based on purposeful observation and controlled experimentation and then presented as an organized, rational study. While simple and rudimentary, the *Letter concerning the Magnet* in fact meets some of the modern criteria for research in the natural sciences, and because of its modern qualities it stands out among scientific works of the Middle Ages. The work was first printed in Augsburg in 1558, but had circulated prior to that in the form of handwritten copies.[1]

Peregrinus wrote the *Letter concerning the Magnet* as a personal communication to a friend in Picardy on August 8, 1269, while he himself was encamped with the troops of Charles d'Anjou during the siege of Lucera, an Islamic quarter of southern Italy that had been established by King Frederick II. The name "Peregrinus" was often given to soldiers and others returning from a crusade. Given that he bore this appellation at the time of Charles's campaign, it is possible that Peregrinus had also participated in some capacity in the Sixth Crusade (1248–1254), in which Louis IX was roundly defeated. It may also be, however, that the name was given to him based on his involvement in Charles's campaign alone. In either case, what the name does tell us is that Peregrinus most certainly had some form of contact with Islamic society.

The *Letter concerning the Magnet* is divided into two parts. Part 1 contains ten chapters and covers Peregrinus's observations, experiments, and theories concerning the magnet. Part 2 discusses, in three chapters, its practical applications. All in all, the work is best described as a pamphlet, with each chapter being quite short. Part 1 begins as follows:

> Most intimate of friends, at your request I shall explain in familiar language a certain hidden virtue [*occulta virtus*] of

> the lodestone … Nevertheless, we shall communicate in this letter only information regarding the manifest properties of the lodestone, because this information will form part of a treatise in which we shall show how to construct physical instruments. (LCM 1.1)

We can infer from this that Peregrinus's aim was to discuss the characteristics of magnets as observed directly through experimentation and not to delve into the philosophical issues surrounding the root cause and fundamental nature of magnetism. The *Letter concerning the Magnet*, in other words, was not chiefly intended as a metaphysical treatise on magnetism, but rather as an aid in the manufacture of practical devices incorporating magnets.

Chapter 2 of part 1 includes a fascinating description of the characteristics required of people engaged in experimental research, which we will take up later. Chapter 3 relates that the best-quality magnets are those that are tinged with light indigo or other shades of blue, homogenous, heavy, and unblemished, and that, according to sailors in the ports of Normandy, Flanders, and Picardy, they are found in northern lands.

The fourth chapter of part 1 goes into methods of finding the magnetic poles of lodestones:

> You must know that this stone [lodestone] bears in itself a likeness to the heavens … And just as there are two points in the heavens more noteworthy than all the others because the celestial sphere turns about them as upon axes, one of which is called the arctic or north pole and the other the antarctic or south pole, so also in this stone, you should clearly understand that there are two points, one north and the other south. You may arrive at the general determination of these two points in

> various ways. One way is to round the stone with an instrument just as crystals and other stones are rounded, and then let a needle or elongated piece of iron, slender like a needle, be placed on the stone, and a line be drawn along the length of the iron dividing the stone in the middle. Then let the needle or iron be placed in another position on the stone and mark the stone with a line in the same manner according to that position. And if you wish you may repeat this in many places or positions, and there is no doubt that all the lines of this (stone) will converge in two points just as all the meridian-circles of the globe meet in the two opposite poles of the world. Know then that one is north and the other south …
>
> There is another better way of finding these points, namely, that you note the place on the rounded stone, as has been described, where the end of the needle or of the iron clings more frequently or with greater force. For this place will be one of the points determined by the method already described.
>
> In order, therefore, that you may determine one point on the stone exactly, break from the needle or iron a little piece which shall be oblong and about as long as two finger-nails and place it on the spot at which, as already stated, the point has been found, and if it stands perpendicular to the stone, there is no doubt that the place sought is there; if not move it about until it does stand perpendicular. When this has been done mark the point there; and in like manner you will find the opposite point on the opposite side of the stone. If you do this rightly and the stone is homogeneous and select, the points will be diametrically opposite each other just as are the poles of a sphere [*tanquam poli in spera*]. (LCM 1.4)

A similar explanation would be given by William Gilbert some three hundred years later. In another section, I will discuss the significance of this method for identifying magnetic poles, but for the moment I simply wish to draw attention to the fact that the spherical magnet Peregrinus used in his experiments was deliberately shaped in emulation of the celestial sphere.

Chapter 5 deals with the north-south directionality of the magnet and the determination of its north and south poles. According to Peregrinus, the magnet was to be placed on a wooden plate that was then floated on water in a large bowl; the plate would then spin until the magnet had aligned itself in a north-south direction:

> This stone, thus placed, will turn its small vessel about until its north pole will stand in the direction of the northern point of the heavens and the southern in the direction of the southern point. (LCM 1.5)

Peregrinus furthermore found that the resultant alignment was always the same, no matter how he rotated the plate when repeating the experiment.

One could easily overlook the significance of this experiment, but in fact it was the first one ever performed in Europe to verify the directionality of natural lodestone itself (rather than that of a magnetized needle)—a natural phenomenon first hinted at by Michael Scot. In that they contain history's first known remarks on the polarity and directionality of natural lodestone, the fourth and fifth chapters of the *Letter concerning the Magnet* are the most important parts of the entire work. No one before had demonstrated a firm grasp of these phenomena, and in fact one of Peregrinus's claims to fame is as the discoverer of the fact that natural magnets have poles. Indeed, Peregrinus was the first person, as far as we know, to have used the word *polus* (pole) with regard to the lodestone.[2] And it was he who coined the terms he used to describe, respectively, the north and south poles of the magnets he experimented with—*polus septemtrionalis lapidis* and *polus meridionalis lapidis*—although he did also use the word *pars* (part) synonymously with the word *polus* in this context.

Let's jump ahead to chapter 9 of part 1 for a closer look at magnetic polarity. Here, Peregrinus describes the following experiment:

> Take one stone which you may call by *AD*, in which *A* is the north and *D* the south point. Divide it into two parts so that two stones are made from it. After this, place the stone which contains *A* on water so that it may float; you will see that *A* turns towards the north as before. For breaking does not take away the properties of the parts of the stone, if it is homogeneous. Hence the part of this stone at the point of fracture which is *B*, must be the south. Let, then, this stone regarding which we have just been speaking be represented by *AB*; as to the other stone, which contains *D*, if it is placed on water, you will see that *D* is south as at first, because it turns towards the south, if placed on water. But the other part near the fracture, which may be designated by *C*, will be the northern; this stone will therefore be *CD* … and thus you see that the two parts of the two stones which, before the separation, were continuous in one stone, after the separation, were found to be, one the northern and one the southern part. (LCM 1.9)

Thus Peregrinus discovered that natural lodestones not only had north and south poles, but also that these poles could not be separated. Rendering north as plus (+) and south as minus (–), we can restate his observation thus: dividing a magnet having two opposite poles ([+          –]) in two results not in two magnets, one with only a north pole ([+     ]) and the other with only a south pole ([     –]), but instead in two magnets each with opposite poles

($\boxed{+\quad -}$ and $\boxed{+\quad -}$). In modern terms, what he found was that all magnets are dipoles, that it was impossible to create a magnetic monopole. Peregrinus's claim that cutting a magnet did not change its nature, that *B*, at the point of fracture, points south and is therefore the south pole, is not as self-evident as one might think. This principle of the indivisibility of the two poles would become quite important to the development of theoretical electromagnetism from the nineteenth century on, and yet as far as I know the significance of its discovery has been largely overlooked by historians. The first person to recognize Peregrinus's observation as an important and independent discovery, standing apart from the discovery of polarity itself, was nineteenth-century English electrical engineer Sir John Ambrose Fleming.[3]

## Section 2—Inquiries into the Nature of Magnetism

In chapter 6 of part 1, Peregrinus describes another experiment from which he discerned what he considered to be a general rule about magnetism, namely, that

> the northern part in a stone attracts the southern part in another stone, and the southern part the northern. (LCM 1.6)

In this experiment he again made use of a magnet placed on a plate and floated on water in a bowl. He held another magnet with its south pole just at the rim of the bowl near the north pole of the floating magnet. Moving the magnet in his hand along the rim of the bowl, he observed that the magnet in the bowl was pulled along with the one in his hand, causing the plate to turn. He then repeated the experiment with the north pole of the magnet in his hand near the south pole of the magnet in the bowl and observed the same phenomenon. Attempting to place south near south and north near north, he also noted that this caused the magnet in the bowl to move away, and he concluded that

> this is because the northern part seeks the southern part; wherefore it will be seen to flee the northern. (LCM 1.6)

By that time Albert the Great, Bartholomew, and Roger Bacon appear to have been aware that magnets sometimes attracted and sometimes repelled one another, but no one until Peregrinus had integrated this understanding into a general principle, much less established a relationship between this phenomenon and the existence of magnetic poles. In other words, Peregrinus deserves credit for having discovered the fundamental rule of magnetism that there is both mutual attraction between opposite poles and mutual repulsion between like poles. One might say that there is nothing surprising in this, since Peregrinus was also the discoverer of magnetic polarity itself, but in fact the rule is far from self-evident.

In part 1, chapter 7, Peregrinus discusses the magnetization and resultant directionality of iron needles:

> It is known to all who have tried it that when an oblong piece of iron has touched the magnet and has been attached to a piece of light wood or to a straw and is placed on water, one end will turn towards the star which they call the nautical star, because it is near the pole; the fact being that it does not turn toward the aforesaid star but towards the pole [*non movetur ad stellam dictum, sed ad polum*], the proof of which we will present in its own chapter; but the other end will turn towards the opposite part of the heavens. As to which end of the iron turns to which region of the sky, know that that end of the iron which shall have touched the southern part of the stone will turn towards the northern quarter of the sky. (LCM 1.7)

Peregrinus did not recognize that iron needles that had been rubbed with magnets were themselves magnets, and so he does not use the terms "north pole" when referring to such needles. In effect, the ends of the needles rubbed with the north poles of magnets became the "south ends," while the ends rubbed by the south poles became the "north ends."

In chapter 8, Peregrinus presents the conclusions that (a) the north poles of magnets and the south ends of the needles they have rubbed are attracted to one another; (b) the same is true for the south poles and north ends; (c) south poles and south ends exhibit mutual repulsion; and (d) this is also true for north poles and north ends. The phenomenon of magnetization, or magnetic induction, as well as the existence of a force of some kind between magnets and iron, were known to Peregrinus as to others before him, but Peregrinus was the first to explicitly identify the polarity of magnetized objects (albeit without referring to "poles" as such with respect to these objects) and to describe the attractive and repulsive forces that existed vis-à-vis this polarity.

On the basis of these observations Peregrinus enters into his inquiry, in chapters 9 and 10 of part 1, concerning the nature and causes of magnetism. In chapter 9, to the question of why the northern part of the stone attracts the southern of the other and vice versa, he explains that it is because

> the active agent strives not only to join its patient to itself but to unite with it, so that out of the agent and the patient there may be made one. (LCM 1.9)

In other words, he sees it as natural that two magnets, both having two opposite poles, attract each other as they seek to become a single magnet with two opposite poles. As a demonstration of this principle, Peregrinus presents in chapter 9 yet another experiment, again using a magnet *AD* that has been divided into two, *AB* and *CD*. If, he says,

> the two parts [*AB* and *CD*] ... which, before the separation, were continuous ... are again brought together, one will attract the other until they are joined together at the point *BC*, where the break took place. Thus by the natural appetite, they will form one body [*AD*] as at first. An indication of this is that if they are cemented at that point, they will exhibit the same operations as at first. (LCM 1.9)

Peregrinus goes on to interpret this phenomenon as follows:

> The active agent ... strives to unite its patient to itself, but this is done because of the similitude between them. (LCM 1.9)

It would be a misunderstanding of Peregrinus's intention to interpret this passage as indicating that he regards the force between the poles of separate magnets in the same light as the received view that like attracts like (*inclinatio ad simile*). For one thing, this view would not explain the phenomenon, in the case of magnets, whereby opposites attract and the similar repel.

Instead, we must realize that Peregrinus is speaking rather of the behavior of magnetic dipoles. In chapter 9 he observes furthermore that the two magnets formed by cutting a larger one in half could be reunited either at *B* and *C* or at *D* and *A* but could not be reunited if one of the magnets were flipped around, placing adjacent to one another either *A* and *C* or *B* and *D*. He concludes that

> nature which tends towards being and acts in the better way, chooses the first order of action in which the identity is better preserved than in the second. (LCM 1.9)

His true intention is thus to explain the natural tendency of magnets to behave in a manner that preserves their bipolar character. That is, magnets always exist in the form either of [+ −] or [− +] and can combine when aligned [+ −][+ −] (forming a new, larger unit [+ −]) or [− +][− +] (forming a new, larger unit [− +]), but when aligned either [+ −][− +] or [− +][+ −] the units cannot combine but instead repel one another. Historically speaking, this was an entirely new perspective on both the nature and the cause of magnetism and magnetic force.

Chapter 10, the last in part 1 of the *Letter concerning the Magnet*, is entitled "On inquiry whence the magnet receives the natural virtue which it possesses." In Peregrinus's time, one theory in circulation was that magnets (magnetized iron ore) got their properties from the mineral deposits in which they had originally been found, and that the fact that they pointed north was simply because there were many iron ore mines to be found in the northern part of the world. Peregrinus rejects this theory, noting that iron mines exist in various parts of the world and that magnets point not only north but also south. He likewise rejects the theory—also commonly espoused at the time, including by Thomas Aquinas—that the North Star exerts an influence on magnets, evidence for which was seen in the fact that magnets pointed toward this star. In refutation of this belief, Peregrinus notes that the North Star was not in strict alignment with the position of the celestial sphere's central axis of rotation. In fact, in chapter 7 (as quoted above), he states that compass needles point toward not the North Star but rather the celestial pole.

To modern minds it seems baffling that Peregrinus saw magnets as being attracted not by the Earth's pole but rather by the pole of the celestial sphere. Also bewildering is the fact that Peregrinus did not realize that the Earth itself was a magnet, and that this discovery had to wait another three hundred years, until William Gilbert arrived on the scene. This is because we look through the filter of our modern common sense and see Peregrinus's spherical magnet as resembling the Earth—a connection that in fact was not made until well after Peregrinus's time. In the Middle Ages, it was generally believed (again, including by Thomas Aquinas) that magnetism came from the heavens whereas the Earth was an inert and inactive mass. Accordingly, it was only natural for people to assume that magnets were attracted by the celestial poles, and that is why Peregrinus referred to the pole of a magnet that pointed toward the celestial north pole as the magnet's "north" pole.[⁋⁋⁋]

---

[⁋⁋⁋] Gilbert, the first to state that the Earth itself was a giant magnet, considered the Earth's magnetic north pole to be located at its geographical North Pole and its magnetic south pole to be located at its geographical South Pole. But this created a contradiction with the knowledge that had been left to posterity by Peregrinus, who had said that the north pole of a needle rubbed with a lodestone was the end of the needle that pointed to the Earth's geographical North Pole. If this were the case, and if lodestones and, by extension, magnetized needles, were being pulled by the Earth-as-magnet, then this would mean that a north-north attraction was taking place. This contradiction was possibly one reason why Peregrinus did not think the Earth was a magnet. On the other hand, to avoid this contradiction Gilbert said that the end of a magnetized needle pointing to geographical north was in fact the needle's south pole, and that the end pointing to geographical south was its north pole. In modern times, however, we refer to the Earth's geographical North Pole as its magnetic south pole and its geographical South Pole as its magnetic north pole. This means that compass needles point to (geographical) north with their (magnetic) north poles and to (geographical) south with their (magnetic) south poles. Modern usage, then, is actually closer to Peregrinus's terminology than to Gilbert's.

A more crucial point to note is that Peregrinus created his spherical magnet as a model of the *celestial* sphere. In keeping with this notion, he concludes (in chapter 10) that

> it is from the poles of the heavens [*a polis mundi*] that the poles of the magnet receive their virtue

with *mundi* here being used synonymously with *coelum*, another Latin word for the heavens. He immediately goes on to conclude that

> the other parts of the stone receive their influence from other parts of the heavens; so that you may suppose not only that the poles of the stone thus receive influence and virtue from the poles of the world [*a polis mundi*], but also the whole stone does so from the whole heavens [*a toto cele*]. (LCM 1.10)

Roger Bacon as well as Peregrinus considered magnets to be microcosms corresponding to the universe as a whole, and they also believed that the heavens exerted influence not only on the north and south poles but also on the east and west parts of these microcosms. This view presumably formed the basis for Peregrinus's decision to create a spherical magnet as a model of the celestial sphere.

In Ptolemaic (geocentric) theory, the celestial poles were special points in the mechanics of the universe, the centers around which the sphere holding the fixed stars revolved, and so it was not as strange as it may seem to us today to suppose that the celestial poles exerted some physical power on terrestrial bodies. Nicolaus Copernicus theorized that the diurnal rotation of the fixed stars was actually an optical illusion resulting from the rotation of the Earth itself. Until this idea was proposed, the two celestial poles (north and south) were believed to be two special points. In order for scientists to arrive at the notion that the magnetic poles of lodestones were attracted not to the celestial poles but to the poles of the Earth, there first had to be a shift away from geocentrism to heliocentricism, or in other words, from the Ptolemaic to the Copernican world system. In 1600, William Gilbert made use of a spherical magnet that superficially can be seen as an imitation of Peregrinus's magnet but that in fact Gilbert called a "terrella," meaning "little Earth." As we will see later, Gilbert had, before fashioning his terrella, already embraced Copernican theory and accepted the premise that the Earth rotated once each day on its own axis while the celestial sphere stood still.

In chapter 10 of the *Letter concerning the Magnet*, Peregrinus describes a perpetual motion machine consisting of a spherical magnet that has been mounted on pivots placed at the northern and southern tips of its axis so that the magnet is maintained horizontally in a position parallel to the Earth's meridian and can rotate freely on its axis. He claims that a device so constructed should be expected to make one full rotation each day, as a result of being pulled not only at its north and south poles by the north and south poles of the heavens, but also at its east and west tips by the east and west parts of the heavens. In other words, the magnet would rotate in perpetuity, powered by the celestial sphere's diurnal revolution. Peregrinus asserted that any reader who failed in this experiment could attribute this failure to a "lack of skill" in constructing the device properly. We will have an opportunity to revisit this idea later, when we look at how it was viewed by William Gilbert.

## Section 3—Peregrinus's Methods and Aims

Peregrinus's discoveries of magnetic polarity and of the attractive and repulsive forces exhibited by magnetic poles were themselves indisputably epoch-making, but also far ahead of his time were the methods by which he arrived at these discoveries and the aims that led him in this direction of inquiry in the first place.

Most remarkable about the way in which Peregrinus conducted his research was the level of initiative and planning that are apparent from records of his experiments and observations. His methods for determining the location of a magnet's poles (described in chapter 3 of part 1 of the *Letter concerning the Magnet*), and his experiments with dividing and rejoining bar magnets (described in chapter 9), for example, are the earliest examples in the history of magnetic research of controlled experiments designed for a clear and specific purpose.

Especially notable is Peregrinus's manufacture of a spherical magnet (by grinding and polishing natural lodestone) and pairing of this device with a magnetized needle in order to determine the location of the magnet's poles. This was remarkable, first of all, for being the earliest known account of an artificially shaped magnet (a fact that itself has not always been sufficiently appreciated by science historians). To shape the magnet spherically was, moreover, a brilliant idea considering that polarity, or axial symmetry, can be brought to the fore only by contrasting it with isotropy, or spherical symmetry; for if a thing whose shape is not geometrically spherical were to exhibit behavior that is not isotropic, one would be at a loss to decide whether this outcome were a result of the thing's physical properties or merely a result of its shape. Only by making the object of study something geometrically isotropic (a sphere) did it become possible to connect its anisotropic behavior with its physical properties.

In the eighteenth century, philosopher Immanuel Kant, in his

*Critique of Pure Reason*, had this to say about the methodology of modern natural science:

> [Reason] must take the lead with principles for its judgments according to constant laws and compel nature to answer its questions, rather than letting nature guide its movements by keeping reason, as it were, in leading-strings ... Reason, in order to be taught by nature, must approach nature with its principles in one hand ... and, in the other hand, the experiments thought out in accordance with these principles—yet in order to be instructed by nature not like a pupil, who has recited to him whatever the teacher wants to say, but like an appointed judge who compels witnesses to answer the questions he puts to them.[4]

What Kant had in mind when he wrote this is evident in the way he began the paragraph in which the above excerpt appears: "When Galileo rolled balls of a weight chosen by himself down an inclined plane ... a light dawned on all those who study nature." Kant refers here to a famous experiment that Galileo performed in order to verify his law of falling bodies. What Galileo did was to roll balls down smooth, gentle slopes so that their speed could be measured more easily and with greater accuracy than was possible with balls observed in free fall. Slowing the speed at which the balls moved by placing them on slopes not only aided in measuring velocity but also reduced air resistance. The significance of what Galileo did in terms of method was, first, to use ingenuity in order to create conditions more congenial to achieving a desired experimental goal, and, second, to minimize predicted extraneous disturbances. In Kant's view, then, Galileo had provided a model for purposeful, well-planned experimental design that enabled the scientist to achieve something close to ideal conditions in which to pursue his topic of inquiry.

In fact, however, this experimental procedure had already been put into practice long before Galileo. Three and a half centuries earlier, with the studies that led to his discovery of magnetic poles, Peregrinus had established an approach to experimental research that was every bit as modern as Galileo's. The superb originality of Peregrinus's spherical magnet paved the way for Gilbert to discover, some three hundred years later, that the Earth itself was a single, giant magnet. Some scholars have singled out Peregrinus and his work for strong praise, and in light of what we have just been discussing, we must conclude that their words are no exaggeration. Science historian Edgar Zilsel called Peregrinus "the best experimentalist of the Middle Ages,"[5] and the *Letter concerning the Magnet* has been referred to variously as "the first work following an inductive, empirical method in natural science" (Erhard Schlund), "the earliest known work of experimental science" (Paul Fleury Mottelay), and "the finest example of the experimental method in the Middle Ages" (W. P. D. Wightman).[6]

While Peregrinus's experiments have been recognized for their modernity even when measured by today's standards, the cutting-edge quality of his approach is even more apparent when seen in the context of the intellectual and cultural climate of the era in which he lived.

As mentioned earlier, little is known about Peregrinus's background. But he does appear to have enjoyed something of a reputation during his lifetime, and his illustrious contemporary Roger Bacon, in his *Opus tertium* (written about a year before Peregrinus wrote the *Letter concerning the Magnet*), speaks of "Magister Petrus" in admiring terms in a passage on empirical science:

> One man I know, and only one, who can be praised for his achievements in this science. Of discourses and battles of words he takes no heed: he follows the works of wisdom, and in these finds rest. What others strive to see dimly and blindly, like bats in twilight, he gazes at in the full light of day, because he is a master of experiment. Through experiment he gains knowledge of natural things, medical, chemical, and indeed of everything in the heavens or earth. He is ashamed that things should be known to laymen, old women, soldiers, ploughmen, of which he is ignorant. Therefore he has looked closely into the doings of those who work in metals and minerals of all kinds; he knows everything relating to the art of war, the making of weapons, and the chase; he has looked closely into agriculture, mensuration, and farming work; he has even taken note of the remedies, lot-casting, and charms used by old women and by wizards and magicians, and of the deceptions and devices of conjurors, so that nothing which deserves inquiry should escape him, and that he may be able to expose the falsehoods of magicians. If philosophy is to be carried to its perfection and is to be handled with utility and certainty, his aid is indispensable.[7]

Everything substantial that we know about Peregrinus comes to us either from the *Letter concerning the Magnet* or from this passage by Bacon. Calling Peregrinus a "master of experiment" (*dominus experimentorum*), Bacon painted Peregrinus as a man who had become well-versed in practical technical skills and knowledge through direct experiment, and contrasts this with those having a penchant for "discourses and battles of words."

But Peregrinus and Bacon lived in an intellectual and cultural climate where the experimental attitude was discouraged. Earlier we saw that the liberal arts (*artes liberales*) became the foundation of higher education, and it is important here to note that "liberal" in that context meant "befitting free men." This was distinguished from the servile arts (*artes sellulariae*), which were considered fitting only those employed in the service of the upper classes, and essentially comprised the mechanical arts (*artes mechanicae*). These enterprises, which chiefly entailed working with the hands rather than with the mind, were scorned by intellectuals in Europe all throughout the Middle Ages and even to the beginning of modern times. This thinking was a vestige of much older times, of days as long ago as ancient Rome, whose culture had been bolstered by slave labor. As late as the fifteenth century, Flemish anatomist Andreas Vesalius had lamented about "fashionable doctors … in imitation of the old Romans, despising the work of the hand."[8]

And here was Peregrinus, who, according to Bacon, was a master at precisely this type of "despicable" industry. Indeed, in the *Letter concerning the Magnet* Peregrinus himself makes no secret whatsoever of his views on the importance of manual work. In chapter 2 of part 1, he comments thus on the qualities needed by those who engage in experimental research:

> Know then, dearest friend, that an investigator of this subject must have an understanding of nature and not be ignorant of the celestial motions. He must also be clever in the use of his hands [*industriosus in opere manuum*] in order that, by means of this stone, he may produce wonderful effects. For by his carefulness he will be able in a short time to correct an error which in an age he could never do by his knowledge of natural sciences and mathematics, if skill were lacking in the use of his hands. For in occult matters we investigate many things by manual industry, and in general without it we are unable to bring anything to completion. (LCM 1.2)

In an age when the educated classes deplored manual labor, this passage was almost tantamount to the declaration of a cultural revolution. But only "almost," for neither Bacon nor Peregrinus was ahead of his time to the extent that they could have been considered true men of the future. In their time the seeds of change had already been sown, as contact with Islamic society and advancements in technology had begun to foster interest in mechanical technology. One tenth-century pioneer in the study of astronomy, Gerbert d'Aurillac (later Pope Sylvester II), who had studied with Muslims, is said to have created quite a sensation when he crafted, with his own hand, an exquisite armillary sphere. As one scholar puts it, "the twelfth century stands midway between the contempt for the *artes mechanicae* in antiquity and their full acceptance in the Renaissance."[9] As early as the 1120, cleric and philosopher Hugh of Saint-Victor had written in his *Didascalicon* of the importance of practical education in areas including mechanics and agriculture.[10] Albeit at a snail's pace, a transition was under way.

We can surmise from reading both Bacon and Peregrinus himself that Peregrinus, a military engineer, was well educated in both the liberal arts and the mechanical arts. His modernity is nowhere made clearer than in part 2 of the *Letter concerning the Magnet*, an exposition on the practical applications of the experimental results described in part 2 that accordingly begins as follows:

> Having reviewed the natural phenomena of the magnet, let us now proceed to the inventions which depend on knowledge of its natural workings. (LCM 2.1)

In the first chapter of part 2, Peregrinus describes the design and manufacture of a device that makes use of the magnet to determine the azimuth (angle with respect to the meridian) of the sun, the moon, or a star. He includes a diagram of the device (see Figure 8-1), which basically consisted of a magnet (*magnes*) affixed to a float and set upon water in a round container. This wet compass differed from others known in Europe at that time by being the first to make use of a lodestone itself rather than a magnetized needle. It was also unique in having a compass card: a paper disk with graduations designating degrees from 0 to 360 that was fixed to and rotated with the lodestone. (Figure 8-1 shows that east was designated as 0 degrees and north as 270 degrees.) Affixed to the disk was a ruler (*regula*) that rotated independently of the lodestone, and a needle was attached vertically to each end of the ruler. Making use of this device, in daytime one could determine the sun's azimuth from the point at which the needles cast their shadows on the ruler. At night, the azimuth of a celestial body, whether the moon or one of the stars, could be found by bringing the two needles into a straight line with that body.

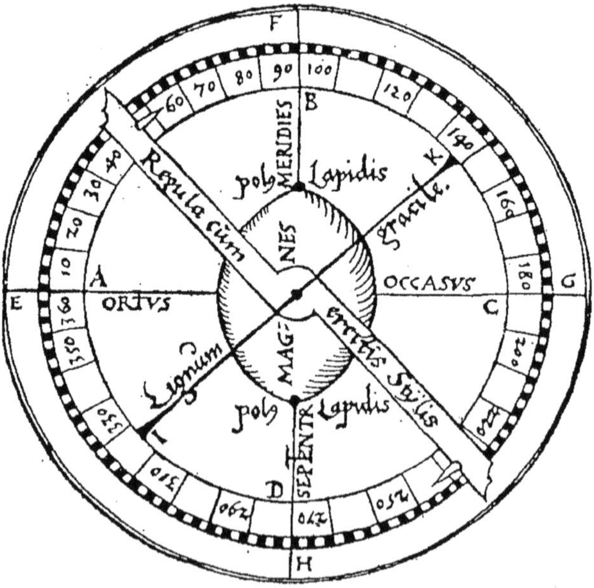


Figure 8.1: Compass for finding the azimuth of the sun, the moon, or any star on the horizon. *Source*: Harradon, "Some Early Contributions to the History of Geomagnetism—I."

Chapter 2 of part 2 discusses an improved model of this device—a dry compass wherein a magnetized needle could rotate freely without benefit of a watery medium. This model consisted of a round container and a clear glass cover, with the needle placed on a pivot inserted at the very center of the device and attached to the inner surfaces of the container and cover.

Chapter 3 begins with the following attention-getting declaration:

> In this chapter I shall reveal to you the way to construct a continually moving wheel, of wonderful ingenuity, in the invention of which I have seen many engaged in vain attempts and wearied with much labor. For they did not perceive that they could effect the accomplishment of this by the virtue or power of the lodestone. (LCM 2.3)

Here Peregrinus describes nothing less than a perpetual motion machine, a device making use of the magnet as an inexhaustible supply of motive power. I will refrain from laying out in detail Peregrinus's description of the device's mechanism, which, quite frankly, is rather beyond comprehension. An attempt to explain the mechanism here would be pointless in any case: scientists now acknowledge that creating a perpetual motion machine is simply not possible, so Peregrinus's device could never have worked as he intended it to no matter how clever its concept or design.

More important for our purposes is to examine Peregrinus's motive in conducting the research behind part 2 of the *Letter concerning the Magnet*. The manuscript was first printed in Augsburg in 1558 under the title, *De magnete, seu rota perpetui motus* (On the magnet or perpetually rotating ring), and historian Park Benjamin has asserted that the true aim of Peregrinus's research was precisely to create a self-moving magnetic motor and thus realize perpetual motion.[11] I will not consider here whether or not Benjamin's claim goes too far, but there can be no doubt that at least one of the chief goals of Peregrinus's research was the harnessing of magnetism as a natural source of motive power. Moreover, Peregrinus was the first person in history to attempt, one might say, to convert electromagnetic energy to mechanical energy. As pointed out by Lynn White, Jr., a scholar of the history of medieval technology, regardless of whether or not Peregrinus succeeded in that attempt, the fact that he conceived of the idea at all was key.[12] This was precisely a practical application of what Roger Bacon intended by his "three prerogatives" of experimental science, which we discussed in the previous chapter.

## Section 4—The Social Context of the *Letter concerning the Magnet*

In chapter 7 we discussed how Roger Bacon's awareness of Islamic technological superiority helped inspire the formulation of his concept of experimental science. We can suppose that in Peregrinus's case the influence of Islamic culture was even more direct. Peregrinus had spent time in southern Italy, which, from the time of the Norman dynasty in Sicily, beginning in the early eleventh century, until the time of Frederick II, who ruled during the first half of the thirteenth century, was Latin Europe's leading point of contact with Islamic culture. While in that region, Peregrinus did not merely come into general contact with Islamic culture; in Lucera (established by Frederick II as a Muslim quarter), he was actually involved in fighting against Muslim troops in an attack launched, with papal sanction, after Frederick's death.

There is nothing implausible, then, about science historian A. C. Crombie's claim that "it was certainly the example of Arab work that stimulated some of the experiments made by Christian writers, for example Roger Bacon and … possibly Petrus Peregrinus."[13] Indeed, a strong case can be made for the idea that both the motivation for Peregrinus's research into the magnet and his opportunity to explore this subject came from his contact with Islamic culture. I do not, however, mean to imply that Peregrinus learned about the magnet and compass directly from Muslims, and there is no evidence that such might be the case. If he absorbed anything from Islamic culture—and this is no more than conjecture—it would have been something more abstract, something along the lines of a new way of approaching the study of natural phenomena, a perspective that had not existed in Europe until that time. Crombie describes such a perspective as a

> special approach to the problems of the world of nature that characterized a strong tradition of Arabic thought … The primary question in this tradition was not what aspects of nature most vividly illustrated the moral purposes of God nor what were the natural causes which would provide a rational explanation of the facts described in the Bible or observed in the world of everyday experience, but what knowledge would give power over nature.[14]

Similarly, historian Philippe Wolff observes that in the Islamic world "the search for a rational explanation of natural phenomena was no longer as fundamental as in Greek science and the desire to acquire power over nature by means of science played a much larger part."[15] In other words, Islamic learning can be characterized by its utilitarian attitude and set apart as something different from both Christian thought and Greek philosophy. And Peregrinus's deepest motivation in carrying out the research that culminated in the *Letter concerning the Magnet* was clearly akin to that attitude; he was driven by a desire that was novel to Europeans of his time and that distinctly bore an emblem of Muslim influence.

In this sense, it is true that Peregrinus's treatise on the magnet was extremely modern and that it transcended the spiritual milieu of medieval Christian society. But there must also have existed some sort of circumstance in that society out of which the *Letter concerning the Magnet* could be born. Peregrinus's accomplishment can be properly understood only within the full context of his time. For example, according to Lynne White, Jr., one challenge that fascinated engineers in thirteenth-century Europe was that of devising a weight-driven clock.[16] The problem of obtaining good sources of motive power was one in which all engineers took an interest, and Peregrinus himself was aware that he was not the only inventor in Europe exploring the possibility of developing a perpetual motion machine.

The physical landscape that Peregrinus and his contemporaries knew was dotted with a great many cathedrals of huge proportions, and there was also at that time a boom in the construction of fortified castles. Construction engineers, of course, had to have both knowledge and skill in keeping with endeavors of this scale. One such man was Villard de Honnecourt, who was also from Picardy and was more or less of Peregrinus's generation. Villard left a portfolio of drawings and other materials, including sketches not only of buildings but also of human figures and of lions and other animals. In addition to sketches, the collection contains diagrams of mechanical toys and various other contrivances. One diagram is of a water-driven saw. We do not know whether any of these designs were actually built and put to use, or whether they remained merely ideas on parchment, but in any case they are extremely interesting. One of the diagrams in the collection is of a perpetual motion wheel powered by gravity, and its caption reads:

> For many days, Masters have discussed how to make a wheel that will turn by itself. Here is one that can be made from an odd number of hammers or with quicksilver.[17]

Peregrinus lived at a time when men of power could no longer depend on an abundance of slave labor, a fact that heightened interest—at the very least among anyone involved in a technical craft—in the development of new sources of power and the study of new types of mechanical devices. One example can be found in *Roger Bacon's Letter Concerning the Marvelous Power of Art and of Nature and Concerning the Nullity of Magic*, attributed to Roger Bacon and written around the same time that Peregrinus was active. "Even if Nature is powerful and marvelous," reads the opening paragraph of this work, "yet Art using Nature for an instrument is more powerful." It goes on to offer numerous predictions about the shape of technology to come, including the following:

> It is possible that great ships and sea-going vessels shall be made which can be guided by one man and will move with greater swiftness than if they were full of oarsmen.
>
> It is possible that a car shall be made which will move with inestimable speed, and the motion will be without the help of any living creature.[18]

Despite the uncertainty over whether or not Bacon himself actually wrote this work, its existence gives us reason to believe that not only Bacon but also numerous other technically minded individuals, anonymous though they may be, rode the same bandwagon of inspiration that anticipated a future fueled by alternative power sources. As Lynn White, Jr., puts it, Bacon (or whoever the true author was) "was speaking not as a lonely visionary but rather for the technicians of his age."[19]

And in fact, during the time of Bacon and Peregrinus, energy consumption in Europe rapidly increased, as is attested by the proliferation of new waterwheels beginning in the eleventh century. Table 8-1 shows the number of waterwheels in operation in various districts of France from the tenth to the thirteenth century. Notice in particular the extraordinary jump that took place in the quantity of waterwheels supplying power in Picardy, the birthplace of Peregrinus and Villard, in the thirteenth century.[20]

Table 8-1: Number of Waterwheels in Operation in France in the Middle Ages

| Century | Aube | Aubette | Forez | Picardy (Year) | Robec | Rouen |
|---|---|---|---|---|---|---|
| 10th | | | | | 2 | |
| 11th | 14 | 1 | | 40 (1080) | | 1 |
| 12th | 60 | 3 | 1 | 80 (1125) | 5 | 5 |
| 13th | 200 | 6 | 80 | 245 (1280) | 10 | 6 |

Along with increased construction of waterwheels came an expansion of the range of their applications. We have evidence of the use of waterwheels as early as the eleventh century not only in flour milling and sawmilling but also in fulling (a finishing process in the manufacture of cloth) and steel forging.[21] Use of windmills also developed in western Europe in the twelfth century.[22] Literature on the history of iron and steel tells us that ironmaking waterwheels came on the scene near the end of the twelfth century, the uses of these machines then expanding from the thirteenth century on to include ore crushing and other smelting-related processes, and resulting in a marked upsurge in iron production output.[23] This in turn led to more widespread use of iron in products, eventually including not only weapons but also farming equipment, horseshoes, and stirrups. With this wave of technological innovation washing over industry after industry, Europe witnessed the burgeoning of a class of keenly intellectual engineers and craftsmen as the driving force of the age.

From the end of the Migration Period (roughly 300–700) until around the twelfth century, Europe remained fundamentally an agricultural society. Aside from those cloistered in the monasteries, people everywhere were either farmers living in small villages and working dawn to dusk in a struggle for mere subsistence, or else members of the feudal aristocratic and chivalrous classes living what was mostly a military life in castles and fortresses. From illiterate peasant to boorish knight, most people led existences untouched by either letters or culture; these, and generally all matters related to the spirit, were the realm of the clergy, whose numbers were small. But from the twelfth century on, with the rise of towns and the establishment of universities, the role of the monasteries in intellectual endeavor declined, and clerics saw their monopoly on knowledge crumble.

Thus, in the thirteenth century a new class emerged, consisting of townspeople who had studied liberal arts at university, learned crafts that enabled them to meet the demands of a more technologically oriented society, and, in short, prepared themselves to earn livelihoods with knowledge as their most important tool.[24] Peregrinus was, as far as we can guess, one among these worldly masters of trades. Therefore, although it is fair to say that he was at the forefront of science and technology in having an approach to research that was not only empirical but also pragmatic, and indeed that he was ahead of most of his contemporaries in this sense, we must not overstate the case by claiming that he was completely ahead of his time. Rather, Peregrinus, a man we know mostly through a single outstanding work, his *Letter concerning the Magnet*, was very much a man *of* his time: a personification of the bourgeois spirit, still fledgling, but slowly rising to prominence.

## Section 5—Jean of Saint-Amand

Peregrinus, for all his accomplishments, left at least one mystery concerning the magnet unresolved. Earlier in this chapter, we discussed his chief contributions to science: through ingeniously designed and well-executed experiments, he discovered that magnets have poles and established the relationship between magnetic polarity and magnetic force; he showed that magnets always exist as dipoles; and he explained an important physical principle behind the fact that like poles repel each other while opposite poles are mutually attracted—the principle that this phenomenon enables magnets to combine in such a way as to preserve their bipolar character. But nowhere did Peregrinus provide a satisfactory explanation for how and why magnetization occurs—why the south pole of a magnet, when it touches iron, confers upon that portion of the iron a north pole, and vice versa. But then it is hardly surprising that the *Letter concerning the Magnet* fails to offer a rational interpretation or convincing explanation for the specific phenomenon of magnetization, since the work's chief purpose was to present practical applications of magnetism rather than to discuss theoretical issues.

Theory of magnetization came from another individual, a physician named Jean of Saint-Amand who practiced in Paris sometime in the late thirteenth century. In one passage of his *Expositio super antidotarium Nicolai* (Commentary on the antidote of Nicolaus), Jean argues that magnetization works because magnets are active agents that actualize the latent magnetic quality of iron, causing the iron, which is passive, to adapt itself to the magnet. This theory was based on Aristotle's schema of causality, which we have already discussed.[25]

Jean shares with both Bacon and Peregrinus the belief that magnets have not only two "poles," north and south, but also two other parts, east and west:

> I say that in the magnet is a trace of the world, wherefore there is in it one part having in itself the property of the west, another of the east, another of the south, another of the north. And I say that in the direction north and south it attracts most strongly, little in the direction east and west. Wherefore the virtues of the poles [*virtutes polorum*] are stronger in it, which is recognized by sailors.

The English phrase "trace of the world" here is, in Jean's original Latin, *vestigium orbis*, with *orbis* signifying either the celestial sphere or the Earth. If Jean is here referring to the Earth, then his words foreshadow the work of William Gilbert, which came some three hundred years later. But judging from the context in which this paragraph appears, it seems more likely that Jean is talking about the celestial sphere and that he considers the magnet to be a microcosm of the universe.

It is not clear whether Jean's belief that magnets have parts corresponding to the four cardinal points reflects Bacon's direct influence or was simply a commonly accepted notion at that time. In either case, by arguing that the north and south ends of the magnet are dominant and exert a stronger force than the east and west ends, Jean—who also used the word *polus* in his work—implies that he is substantially in agreement with Peregrinus's concept of magnetic poles. Also conforming to Peregrinus's ideas is Jean's belief that the north and south poles of a magnet can be found by allowing the magnet to spin freely; the part that then points to geographical north, he says, is the magnet's north pole, and the part that points south is the south pole. It seems highly likely that Jean had read the *Letter concerning the Magnet* and was convinced by much of what he found there.****

How, then, does Jean explain the observation that the point where an iron needle has been magnetized by the north pole of a magnet takes on the property of a magnetic south pole? He speaks, interestingly, of "virtue" and "current":

> Take a magnet of which one part has the property of the south and the other of the north and put a needle above that stone so that it lies on the stone. Then one end of needle touches one end of the magnet and the other the other, and there flows into the end of the needle the virtue of that part

**** Historians are not entirely in agreement about this. Thorndike maintains that Jean flourished in the period from 1261 to 1298, Crombie notes merely that Jean and Peregrinus were contemporaries, and Lynn White, Jr., goes so far as to say that Jean knew of Peregrinus's experiments at the time he wrote *Expositio super antidotarium Nicolai*. Meanwhile, Julian A. Smith, who asserts about the *Letter concerning the Magnet* that "there is little in it that is entirely new" and that Peregrinus had merely gathered together knowledge that had already been discovered by others, claims that Jean's work had been written "possibly earlier" than Peregrinus's. Smith, however, provides no evidence for these claims. I believe it is reasonable to suppose that Jean was influenced not only by Bacon but by Peregrinus as well. See Thorndike, "John of St. Amand on the Magnet," 156–57; Crombie, *Styles of Scientific Thinking*, 424; idem, *Science, Optics and Music*, 69; White, *Medieval Technology and Social Change*, 133; and J. A. Smith, "Precursors to Peregrinus," 73 n. 272, 71.

> which it touches, so that, if it touches the southern, the southern virtue flows in. Then let the needle be raised higher. Then, since there is a current from the magnet through the entire needle placed above it, that part of the needle in which at first there was southern virtue will become northern, as the current from the magnet flows through the whole needle.

Crombie likens this explanation to a nineteenth-century electromagnetic theory, involving something called "tubes of force," that was inspired by Michael Faraday and later fleshed out by James Maxwell Clerk,[26] but I believe that this is reading too much into Jean's words. Thorndike likewise offers an interpretation that misrepresents and far overestimates the scope of Jean's theory, stating that "it seems to approach the conception of positive and negative poles and of an electrical current." Rather than reach for improbable parallels such as these, historians would do better to examine whether or not Jean's argument actually deserves to be called any sort of explanation in the first place and, if so, whether or not the explanation is at all persuasive. Even concerning such basic questions as what Jean actually accomplished, there is room for a diversity of opinion. At the very least, it is clear that Jean discerns in magnetization a subtle issue requiring explanation, and this insight is in itself historically noteworthy.

Unlike Peregrinus, Jean rejects the idea that like poles repel one another. A magnet's south pole, he argues, *appears* to repel another magnet's south pole (or the south end of a magnetized needle), but only because it attracts the north pole: "in attracting the northern part to itself it seems to repel the southern part," he admits, but he is adamant that "it may seem to repel yet does not repel." He thus posits that in fact only an attractive force exists. Indeed, as we will see below, Jean's theory is capable of explaining only the attractive force and not the phenomenon of repulsion.

Jean theorizes that magnetism—again, possessing only the property of attraction—occurs "by multiplying its like and ... exciting the active power which exists incomplete in iron, which is born to be completed by the form of the magnet." Hence,

> the southern part attracts that which has the property and nature of the north, albeit they have the same specific form, and this is not except by some property existing more complete in the southern part which the northern part has potentially and thereby its potentiality is completed.

This notion that an agent actualizes a quality that already exists potentially within the recipient of course comes from Aristotle. And we see still more of the ancient Greek master's influence when Jean proposes that

> in this sort of phenomena like is not actually attracted by like but because it is possible that parts possessing the same form and species should have diverse properties, one being complete, the other incomplete, from which attraction results.

We can conclude from all this that Jean, like Bacon, is employing the Aristotelian causal schema to establish a justification for the age-old belief that "like attracts like."

But even while his use of Aristotle's ideas appears to establish some sort of a basis for attraction, Jean's theory shows itself utterly powerless to shed any light on the phenomenon of repulsion. Jean's understanding of magnetism must be criticized, at a minimum, for lagging behind that of Peregrinus, who recognizes both attractive and repulsive force.

To put this in a wider historical context, recall that Thomas

Aquinas's *Summa theologiae* was written shortly after 1270. This means that at precisely the time when Thomas was establishing the philosophy of Scholasticism, thereby revitalizing Aristotelianism, it was already possible to glimpse the clash that would occur between Aristotelian physics and the experimental study of nature.

* * *

In the history of experimental research concerning the magnet, Peregrinus was a pioneer, the first to demonstrate that magnets were north-south dipoles and that like poles repelled each other while opposite poles were mutually attracted. The late thirteenth century was a time of transition, a period when Europeans seemed ready to shed the old, medieval ideas about magnetism, ideas fraught with mystery and superstition, that had prevailed since the time of Dioscorides and Pliny. This shift started with Thomas Aquinas, who restored the influence of Aristotelian philosophy and gave his sanction to the idea that the rational study of nature did not necessarily violate the teachings of the Bible; it continued with Roger Bacon and his advocacy of experimental science and emphasis on the practical application of natural philosophy; and it was taken still further by Petrus Peregrinus and his trailblazing work in experimental research. And so, as one scholar points out, the thirteenth century came to a close with signs of reform to come.[27] Nonetheless, in the fourteenth and fifteenth centuries at least, there was no further expansion in the study of magnetism stemming directly from these developments—with the sole exception of the work of Nicolaus Cusanus, as we will see in the next chapter.

Despite having theorized about the north-pointing property of magnets, Peregrinus was completely unaware of the role that the

Earth played with respect to this phenomenon. But his ignorance here is quite understandable. After all, in order to see the effect that the Earth had on magnets, people first had to discover the Earth itself. This had to wait for the age of discovery and exploration and for a time when people were ready to dispose of the long-unquestioned assumption that the Earth was an inert, motionless lump of soil. It had to wait for the long, slow passing of the entire fourteenth century and the first half of the fifteenth, during which period Europe wallowed in poverty and stagnation due to waves of famine brought on by adverse climate as well as to repeated epidemics of plague. Eventually, however, much of this hardship passed, and the Middle Ages gave way to the Renaissance.

# Part 2

# Renaissance

Chapter 9

# Nicolaus Cusanus and the Quantification of Magnetic Force

## Section 1—Nicolaus Cusanus and *On Learned Ignorance*

In *The Individual and the Cosmos in Renaissance Philosophy*, German philosopher Ernst Cassirer asserts that "any study that seeks to view the philosophy of the Renaissance as a *systematic* unity must take as its point of departure the doctrines of Nicholas Cusanus."[1] And one of Cassirer's greatest works, *Das Erkenntnisproblem in der Philosophie und Wissenschaft der neueren Zeit* (The problem of knowledge in philosophy and science in modern times), begins with a discussion of Cusanus. Similarly, German scholar Paul Oskar Kristeller considers Cusanus's work to mark the starting point of "Renaissance Platonism," with an influence extending to the fifteenth-century humanist philosophers Marsilio Ficino and Giovanni Pico della Mirandola.[2]

French historian Jacques Le Goff, on the other hand, calls Cusanus "the author of the last great scholastic summa of the Middle Ages."[3] Similarly, a work entitled *La philosophie médiévale*, by French scholar Alain de Libera, ends with a discussion of Cusanus. In other words, these two authors place Cusanus not at the beginning of the modern era but rather at the end of the medieval period.

Then there is the rather cool treatment by American historian Lynn Thorndike in his tome *Science and Thought in the Fifteenth Century* where he asserts that "the name of Nicholas was pushed forward by German investigators in order to antedate the claims made by England for Francis Bacon and by France for Descartes as the founder of modern philosophy."[4] At the risk of seeming to read too much into this, if one were looking for more cynicism of this kind, one could also point to the fact that English philosopher Bertrand Russell does not mention Cusanus's name even once in his voluminous *History of Western Philosophy.*

But regardless of the place in history that Cusanus has been given by philosophers and other intellectuals, for our purpose—the chronicling of developments in the concept of force—it is enough to understand that he was pivotal.

Cusanus was born in 1401 in the tiny German village of Cues on the Moselle. The son of a boatman, he was—under circumstances that are not clearly known—taken into the home of a feudal lord and educated at a Dutch school that ascribed to Christian mysticism. He later studied at both the University of Heidelberg and the University of Padua, obtaining a degree in canon law. After graduating, he continued his research in theology. In 1430 he became a member of the clergy and in 1432 was asked to join the Council of Basel. In the usual scheme of medieval Catholic society, it was rare for the son of a boatman from frontier Germany, culturally a world apart from the ancient city of Rome, to rise so far.

We can attribute Cusanus's professional advancement to unexpected good fortune, to extraordinary talent, to good character, to scholarship, and, last but not least, to the fact that the Catholic church was then in turmoil and in urgent need of all the capable help it could find. Because of repeated famine and pestilence in the fourteenth century, the population of Europe had become severely depleted, and society was in tatters. The Catholic church faced threats externally from the Turkish military and internally from the Avignonese Captivity (1309–1377), the Great Schism (1378–1417), and the Hussite Revolution (1419–1436); and the loss of its prestige and weakening of its structure in the face of all these onslaughts were plain for all to see. As a result, the destiny of Christian society as a whole hung in the balance. More fundamentally, feudalism had reached saturation point, and the established order of medieval life was in the process of dissolving. Thus, the Council of Basel had begun its business in a time of crisis. Cusanus fought hard to help the church regain its position of leadership, and his eloquence earned him the nickname "Hercules of the Eugenians."

Cusanus served under a number of successive popes. In 1437 he was sent to Constantinople—then under military threat from Turkey—to invite the Eastern Church to attend the Council of Ferrara, which had been established to work toward the integration of Eastern and Western Christendom. Upon his return, Cusanus brought with him a large number of valuable manuscripts that had been preserved in the Byzantine Empire. During the journey, Cusanus became versed in the works of Plato and in neo-Platonism under the tutelage of two fellow travelers, Plethon, a Platonist, and Bessarion, a distinguished figure of the Eastern Roman Empire. Cusanus was then sent, as a special papal emissary, to various parts of Germany, including Mainz, Frankfurt, and Nuremberg. He was made a cardinal and in 1450 was also ordained bishop of

Brixen, in the region of Tyrol. For almost two years from late 1450, Cusanus traveled constantly, visiting churches and monasteries in and around Germany on a mission of reform. Cusanus passed away in 1464.[5]

As we can see, Cusanus spent his career not as an academic but rather as an administrator. He gave his life to the cause of upholding the Catholic church as an institution, serving as a right-hand man to multiple popes. This alone kept him busy in the extreme: in those days, a trip from Rome to Germany and its environs was formidable, almost unimaginably so by today's standards. And yet in the midst of carrying out these daunting missions, Cusanus continued his research and his meditations, doing most of his writing in his spare time. Reading Cusanus's works in this context, one is impressed time and again with the fact that he not only freed himself from the complicated knot of Scholasticism but also introduced here and there in discussions of mysticism the sparkle of a more modern and more rational spirit.

It goes without saying that Christianity was firmly at the center of Cusanus's thoughts. But with him we see the assertion of a modern kind of religious tolerance; at a minimum, we find in his work none of the narrow-minded exclusivity characteristic of medieval Catholicism, nor any of its prejudice-filled hostility toward nonbelievers. For example, in *On Peaceful Unity of Faith*, which he wrote amid growing pressure by the Turkish military against the Eastern Roman Empire, Cusanus tells a story in which God receives reports that "for the sake of religion very many men were in armed conflict with one another, and that by physical force men were either compelling their fellow-men to renounce their long-adhered-to religious sect or were inflicting upon their fellow-men death." The story's protagonist is nonetheless convinced that "the few wise men who are rich in the experiential knowledge of all such differences as are observed throughout the world in the different religions can find a single, readily-available harmony; and through this harmony there can be constituted, by a suitable and true means, perpetual peace within the domain of religion."[6]

Cusanus wrote his chief work, *On Learned Ignorance*, in which the bulk of his thoughts on theology are set out, while aboard a ship on his way home from Constantinople.[7] In this work, he refers to God as "the simple, absolute maximum" (OLI 1.4:12), and he takes up the issue not of God's existence but rather of man's understanding of God's existence. His view of understanding can be summed up by his statement that "in every enquiry men judge of the uncertain by comparing it with an object presupposed certain" (OLI 1.1:7). In other words, human intelligence is capable of acquiring knowledge only by the repetition of a limited process of thought in which new phenomena are compared with known facts. It follows from this that human intelligence can never be capable of arriving at the absolute truth of the existence of God, who is without limits. In short, this is because "there is no gradation from infinite to finite" (OLI 1.3:11). Therefore, says Cusanus, "in a way we cannot comprehend, absolute truth enlightens the darkness of our ignorance" (OLI 1.26:61). It is the individual's very recognition of this fact—the realization that he can draw near to God only by acknowledging his own ignorance—that Cusanus terms "learned ignorance."

## Section 2—Cusanus's Conception of the Universe

How this line of thinking was received by other theologians in Europe during Cusanus's time and what sort of role his work played in shaping the theology of his day are questions into which we will not delve deeply here. Suffice it to say that Cusanus's claim that absolute truth was unattainable by human intelligence would lead directly to a relativization of understanding.

In *On Learned Ignorance*, Cusanus makes two chief assertions concerning man's picture of the world and view of nature. The first is that the universe, following from God's infiniteness, is also without limits of any kind. Cusanus's thinking here anticipates the development of the post-Copernican heliocentric theories and theories of the infinity of the universe espoused by Giordano Bruno and Thomas Digges in the late sixteenth century. The second major assertion is that numerical concepts are crucial to the understanding of nature. Here, too, Cusanus foreshadows later thinkers, namely Johannes Kepler, Galileo Galilei, and others who followed them in developing the mathematical sciences.

Regarding the first point, Cusanus argues as follows: The universe "is a creature and necessarily owes its existence to the Absolute Being, God"; and the universe must be limitless, because "nothing that would be a limit to the universe, by being greater than it, is able actually to exist" (OLI 2.1:71). Religious implications aside, this assertion has great significance from the perspective of physics: something that has no limits—no boundaries—cannot be said to have a center. Cusanus's claim therefore clashed head-on with the most important tenet of the universal model espoused by both Aristotle and Ptolemy, who both believed that the universe did have a center, where sat the Earth in perennial stillness. Cusanus asserts instead that

> it is impossible for the motor of the world to have the material earth, air, fire or anything else for a fixed, immovable centre … The earth, which cannot be the centre, must in some way be in

> motion … Just as the earth is not the centre of the world, so the circumference of the world is not the sphere of the fixed stars … It is evident from the foregoing that the earth is in movement. (OLI 2.11:107–109)

This argument, of course, was no more than Cusanus's speculation; he was not led to it by any particular experience, nor by any observations of the physical world. Moreover, he did not advance this argument as a deliberate attempt to replace Aristotelian or Ptolemaic models of the universe with a different model of his own. This is clear first of all from the fact that he merely proposed that the Earth was in motion without expounding in any specific terms about what sort of motion that might be. In this sense, it would be too far a stretch to suggest that Cusanus's argument here was a direct precursor to Copernican theory. But we must acknowledge that the small step Cusanus took by suggesting the possibility of the Earth's motion—or more precisely, by rejecting the preconceived idea of the Earth's absolute rest—was immensely important in the move toward a modern model of the physical world.

Cusanus had an answer, also, to the strongest, most commonsensical argument against the theory that the Earth might be in motion. The fact that people did not perceive the Earth to be moving meant, it was argued, that the theory contradicted everyday experience. Cusanus's rebuttal to this was a very clear assertion that motion in general was something always experienced only relative to something else:

> It is now evident that this earth really moves though to us it seems stationary. In fact, it is only by reference to something fixed that we detect the movement of anything. How would a person know that a ship was in movement, if, from the ship in

> the middle of the river, the banks were invisible to him and he was ignorant of the fact that water flows? (OLI 2.12:111)

*On Learned Ignorance* was written in 1440, so Cusanus's argument that "No movement … is absolute" (OLI 2.10:106) preceded the work of Bruno by one and a half centuries and that of Galileo by almost two centuries. In 1584, in *On the Infinite Universe and the Worlds*, Bruno wrote that

> we cannot apprehend motion except by a certain comparison and relation with some fixed body. Wherefore if we suppose a person within a moving ship in the midst of waters, who knoweth not that the water is in motion, nor seeth the shores, he would be unaware of the motion of the ship. For this reason I might fall into doubt and hesitation as to this quiet and fixity [of our earth].[8]

We can assume that Bruno had read Cusanus and adopted the same argument almost word for word.

In fact, the idea that the Earth and the celestial sphere turned in relation to one another had been discussed a century before Cusanus by French Scholastic Nicole Oresme. But Oresme's conclusion was to reject the possibility of terrestrial motion. He argued that natural philosophy supported equally the possibility of circular motion by the Earth and that by the celestial sphere, but he said that on theological grounds it was impossible to conclude other than that the Earth was completely still.[9] Cusanus argues conversely that, since for theological reasons the universe must be seen as infinite, the possibility of terrestrial motion cannot be denied—or, in other words, that the Earth's possession of this capability is a logical extension of the absence of limits to the universe as a whole.

Cusanus's claim that the universe is infinite also leads to rejection of Aristotle's belief in so-called natural motion—whereby heavy bodies on Earth were said to fall toward the absolute center of the universe—and of the hierarchical universal order underlying this belief. This is because in Cusanus's theory the Earth is no longer a special planet situated at the center of the universe. This shift in the Earth's status allows Cusanus to suppose that, just as heavy terrestrial bodies fall toward the center of the Earth, so too do fragments and other bodies belonging to the sun or to other celestial bodies fall toward the surfaces of their respective homes. As Cusanus puts it, "there is in each a movement of parts that secures the whole" (OLI 2.13:120).

This theory raised the Earth up to the same status as all other celestial bodies, with Cusanus concluding that "it is untrue to say that this earth is the basest and lowest planet" (OLI 2.12:112), a statement that Bruno later quoted verbatim.[10] The Aristotelian model of the universe that had long held sway entailed a spatial hierarchy, with the world situated beneath the moon and composed of four elements—earth, water, air, and fire—distributed from bottom to top and a fifth element forming the heavens above. Along with this spatial hierarchy there was a corresponding hierarchy of value, with the Earth considered the lowest and most humble and bodies closer to heaven perceived as more precious. In one of his philosophical dialogues, Bruno has the voice of Aristotelianism assert that celestial bodies are divine.[11] Such a hierarchy clearly reflected a societal order in which each individual occupied a fixed rank. It was this absolute, hierarchical universe, with all of these implications, that Cusanus rejected. As we will discuss later, in 1600 William Gilbert asserted that "the earth is not so vile and abject a body as it is generally considered."[12] We can see the influence of Cusanus not only in this sentiment but also in the words themselves, although it is unclear whether this influence was direct or came indirectly through Bruno.

Cusanus relativized religions, motions, positions, and celestial bodies. Through a thoroughgoing mysticism in the milieu of late medieval Christianity, he thus anticipated modern intellectual thought.

Considering the relativistic aspect of the celestial bodies, Cusanus regards the Earth as an active entity possessing a power of its own and sees celestial influence as therefore not a one-way street: just as the other planets, the sun, and so forth exert power on the Earth, so too does the Earth exert power on other celestial bodies. He believes it possible not only that the Earth "is a brilliant star having a light, heat and influence distinctively its own and different from those of all other stars" (OLI 2.12:113) but also that "it, as a star, has a similar influence on the sun and its region, as already stated" (OLI 2.12:114).

With such propositions, Cusanus expressed a viewpoint concerning the influence of the Earth on the sun that did not exist in the astrology of the Middle Ages and that, to my knowledge, had never been expressed before in the history of thought. In time, Kepler, influenced by Gilbert, would speak of a gravitational force operating among the celestial bodies, thus planting the seed that would eventually sprout into the theory of universal gravitation. In so doing, Kepler implied that this force was reciprocal—that is, that the Earth pulled on the moon even as the moon pulled on the Earth. In other words, he hinted at what would later become known as the law of action and reaction. It is not an improbable leap of interpretation to say that Kepler took this monumental step forward under the influence of Cusanus.

## Section 3—The Importance of Number in the Understanding of Nature

As we have seen, Cusanus claims that all inquiry involves a process of comparison, and he holds that to compare is to introduce the concept of proportion. This belief in turn points toward the importance of quantity: "Now, while proportion expresses an agreement in some one thing, it expresses at the same time a distinction, so that it cannot be understood without number. Number, in consequence, includes all things that are capable of comparison" (OLI 1.1:8). Cusanus advances both ontological and epistemological arguments as a basis for this assertion of the importance of number in understanding the nature of things.

Near the beginning of *On Learned Ignorance*, Cusanus states the following: "The fact that all things have existence in the best possible way makes it impossible to have plurality of beings without number; for if number is denied then the distinction, hierarchy, relationship, harmony and even plurality of beings must be denied" (OLI 1.5:14). He seeks a basis for this line of reasoning in God's plan, saying that God "created all things in number, weight and measure" (OLI 2.13:119) and that for this reason "each may repose in its own number, weight and measure" (OLI 3.1:127). These words have their origin in *Sapientia Salomonis* (Wisdom of Solomon), one of the books of the Apocrypha. The idea, therefore, was not original and had already received frequent mention before Cusanus. Saint Augustine, for example, makes mention of it in *The Literal Meaning of Genesis*. Augustine, however, says: "The man who knows the words 'measure, number and weight' only in their material sense is like a captive in his limited knowledge"; and he considers these quantitative concepts to have divine character.[13] In the sixth century, Cassiodorus amplified this notion in *Institutiones*, asserting that "God the creator has arranged his dispensations by the use of number, weight, and measure," whereas "the evil works of the devil are not ordered by weight or measure or number."[14] Thus, for centuries up to Cusanus's time, in the Christian world quantitative order was something belonging to the divine realm, and its presence or absence was intimately bound up with a religious understanding of good versus evil.

In contrast, Cusanus asserts the following in *On Learned Ignorance*: "When we measure the size and analyse the elements and study the behavior of things, we make use of the sciences of arithmetic and geometry and even of music and astronomy. Now these same sciences God employed when He made the world" (OLI 2.13:118). What is new, historically, with Cusanus is his observation that numerical order gives human beings a foundation for comprehending terrestrial reality. In other words, our knowledge of the physical world depends upon our ability to create categories with which to consider and organize objects and phenomena; and for Cusanus, these categories are numerical.

In 1450, Cusanus wrote three treatises collectively known as the *Idiota—Idiota de staticis experimentis* (The layman on experiments with weights), *Idiota de sapientia* (The layman on wisdom), and *Idiota de mente* (The layman on wisdom and on mind)—all of which are of interest in our discussion here. First, consider the following passage from *Idiota de mente*, in which Cusanus, in the voice of his layman character, states plainly what he sees as the significance of numbers in human understanding:

> I assert that number is the exemplar of our mind's concepts. Without number, the mind cannot do anything. Neither assimilation, conception, discretion, nor measurement would occur if number did not exist. In the absence of number, things

> cannot be understood as separate and discrete ... Consequently, nothing can be understood without number because it is the mode of understanding.[15]

Numbers, in other words, provide the only certain model that we have for knowing and understanding anything. In *On Actualized-possibility*, a 1460 work, Cusanus defines the act of understanding (*intelligere*) as a process of making things similar (*assimilare*), and says that things are knowable when they can be measured by our understanding (*intellectualiter mensurare*).[16]

For Cusanus, to know with the mind means, precisely, to measure. He holds that cognition can be accomplished only by reducing bodies and phenomena to uniform units of quantity (length, volume, and so forth) and then measuring the particular quantity present. Here we see a conscious shift in the understanding of nature from ancient and medieval interpretations, which were qualitative, to a modern interpretation, which was quantitative.

In this sense, the *Idiota* treatises are of special interest to our discussion not as profound theological or philosophical theory about knowledge of God but rather for revealing Cusanus's ideas about knowledge of reality. *Idiota de sapientia* is presented as a conversation between a poor, uneducated man (the layman/citizen) and a wealthy, learned man (the orator) who have met in a marketplace in Rome. It begins with the laymans's assertion that knowledge is found not in canonical texts written by sages of antiquity but rather amid the tumult of the marketplace and the throng of the street corner. Even by itself, this declaration that inquiry via direct experience takes precedence over academic study provides a striking glimpse of the Renaissance ethos in its infancy. But then the dialogue gets even more interesting:

> LAYMAN: In the first place I want you to describe what you see happening in the forum.
> ORATOR: In one place I see money being counted; in another, goods being weighed. Directly across from us, oil and other things are being measured.
> LAYMAN: These are the works of reason, by means of which men surpass the beasts. Brutes cannot number, weigh, and measure.[17]

Reading this, we can see that Cusanus's statement to the effect that numbers are knowledge was not informed merely by Plato and Pythagoras; rather it was inspired in large part by the development of commerce and the spread of a commodity economy. In this sense, Cusanus was a man of unequivocally modern spirit.

Written in December of 1450, *Idiota de staticis experimentis* contains the following observation in its opening lines: "Although nothing in the world is absolutely precise yet experience demonstrates that the judgment of the balance-scale is one of the truest measurements we have; and so it is in demand everywhere." This serves as a declaration of Cusanus's intent to set forth a means for rendering more accurate the sensory experience that Plato had claimed was incapable of providing true knowledge. Focusing especially on knowledge of the natural world, Cusanus goes on to say (through his layman/citizen character): "I think that by knowing the differences in the weights of things we can more surely understand them and that many more probable conjectures can be made about things." He thus posits a uniform, quantitative basis for knowledge, with weight and specific gravity as the relevant units of measure.[18]

Later in the dialogue, Cusanus describes a botanical experiment that is quite interesting. The layman/citizen suggests that if a man were to put soil in a pot, plant some grass in the soil, and then compare the weight of the soil before and after the grass had grown, he would find only a very small drop in the soil's weight. "From this," the layman/citizen argues, "he may gather that all the herbs received their weight from water. Therefore, the waters being impregnated in the earth attracted a terrestreity, and by the operation of the Sun was condensed into the herb." While it would be an exaggeration to claim that this assertion foreshadowed the discovery of photosynthesis, it did hint at the potential for some sort of chemical transformation due to the heat and light of the sun and did constitute the pioneering proposal of a quantitative experiment in the area of plant physiology.[19] In the drawing of such implications from the mere act of weighing, this approach clearly marked an important shift in thinking about the world.

Cusanus also writes that the composition of metal alloys can be tested by comparing their weights as measured in and out of water. This claim can be stated in more modern terms as follows: Let $W$ be the weight of a certain metal when measured out of water, $W'$ the metal's weight when measured in water, $V$ the volume of metal, and $w$ the water's weight density. Since the buoyancy of the alloy is equal to $Vw$, it follows that $W' = W - Vw$. Accordingly, we can find the specific gravity of the alloy by solving the equation $W/(W - W') = W/Vw$ and then use this value to ascertain the composition of metals contained in the alloy. This all goes back to the discovery by the ancient thinker Archimedes that the purity of the king's gold crown could be determined by comparing its weight in and out of water. Cusanus (again through the layman/citizen) concludes that this method shows "how far short of the truth the alchemist comes in his sophistic science."[20] Alchemists until that time had thought that weight was an accidental quality of substances and that the weight of a substance could increase in the absence of an increase in volume if the substance were to undergo a transmutation. And so they believed that lead could be transformed by alchemy into the heavier substance of gold. Cusanus pointed out that, the specific gravity of each metal being fixed, the composition of a metal substance could be determined by measuring specific gravity; and he may well have been the first thinker to debunk alchemy on those rational grounds.

But Cusanus takes this emphasis on weight to an extreme degree, going so far as to suggest that even the qualitative differences among different substances—that is, their general characteristics—can all be defined in terms of a quantitative difference in specific gravity. In *Idiota de staticis experimentis*, for example, his layman/citizen states that "different things of the same volume never have the same weight,"[21] and he applies this principle in a manner that is undeniably oversimplified, full of logical jumps, and almost dogmatic. At one point, he describes a procedure for measuring pulse and respiration rates using a water clock that operates by releasing a fixed volume of water in a steady stream. This procedure was logical enough, entailing counting a uniform number of heartbeats or breaths and weighing how much water is passed by the clock during that interval. But Cusanus's ventures into the field of medicine go much further. He suggests, for instance, that physicians could learn about the characteristics of medicinal plants merely by finding the weights particular to each plant's roots, stems, fruit, sap, and so forth, and could then use this understanding in prescribing for their patients; and he claims, furthermore, that illness can be diagnosed just by weighing the patient's blood, urine, and so on.[22]

It would appear that Cusanus merely proposed various weight measurements of this kind without actually conducting them.

Nonetheless, we must consider him a pioneer, because in fact the gathering of quantitative data in plant physiology and zoology was attempted by no one until the seventeenth century, when Jean Baptiste van Helmont measured weight gain in maturing plants, and then again in the eighteenth century, when Stephen Hales took measurements of sap and blood. Moreover, Cusanus was not merely a pioneer of specific techniques in a few particular areas; rather, he introduced an entirely new idea with relevance to all fields of scientific inquiry.

In *Idiota de staticis experimentis*, Cusanus repeats the idea that God "created all things in number, weight and measure."[23] The influence of this can be seen quite clearly 120 years later in the words of English mathematician John Dee, who, in a 1570 preface to an English translation of Euclid's *Elements*, a treatise on geometry and number theory, likewise writes that God "hast created all thinges in Number, Waight [sic] and Measure."[24] This remark comes in a discussion of what Dee calls "*Statike*, that is, *the Experimentes of the Balance*." The spirit of this idea thus lingered and ultimately became a guiding principle for the advancement of new sciences by the Royal Society of London, founded in the seventeenth century. William Petty, a pioneer in the field of economic statistics and a founding member of the society, states in his 1690 work *Political Arithmetick* that "instead of using only comparative and superlative Words, and intellectual Arguments, I have taken the course … to express my self in Terms of *Numb*er, *Weight*, or *Measure*."[25] Close to a half century later, in 1727, Hales declares in *Vegetable Staticks* that his method is "to number, weigh and measure," and he calls this "the statical way."[26] Here again we find the direct influence of the ideas of the mid-fifteenth-century thinker Cusanus, ideas that pointed the way for the emergence of modern science in the seventeenth and eighteenth centuries. Cassirer's praise of Cusanus, noted at the beginning of this chapter, was thus based on much more than the narrow-minded patriotism to which Thorndike attributes it.

## Section 4—Cusanus's View of Magnetism

And now, finally, we are ready to examine Cusanus's thoughts on a topic of central interest to us here: magnetic force. As we have seen, Cusanus reduces the qualitative distinctions among all bodies and phenomena to measurable differences in weight. And as is clear from the following procedure, suggested by the layman/citizen in *Idiota de staticis experimentis*, Cusanus's understanding of magnetism rests on the same basis:

> I think the virtue of the lodestone might be weighed, if putting some iron on one scale, and a lodestone on the other, until the balance become even, then taking away the lodestone and putting something else of the same weight on that scale, then by holding the lodestone over the iron so that the scale would begin to rise; by reason of the lodestone's attraction to the iron. Now by taking weight off of the other scale until equilibrium is again achieved, I believe that the weight of what was taken out of the contrary scale would be proportional to the weight of the virtue of power of the lodestone.[27]

This cumbersome method is no doubt intended as a way to derive measurements of magnetic force expressed in a unit of weight. Underlying it, presumably, is the preconception that a magnet's power is proportional to its weight and mass. In other words, Cusanus is suggesting that weight, as a unit of measure, can be used to quantify magnetic force because of a supposed relationship between the two properties. In any case, we can surmise from the simple description above that he had not actually performed the experiment. Nonetheless, we must acknowledge that he brought something totally new to the study of magnets and magnetism simply by recognizing magnetic strength as something to be measured and quantified.* In *Das Erkenntnisproblem in der Philosophie und Wissenschaft der neueren Zeit*, Cassirer describes the quantification and measurement of force as the watershed at which modern science separated itself from the medieval understanding of nature.[28] In *Concepts of Force*, Max Jammer similarly asserts that the difference between the astrological or Hermetic idea of force and the notion of force in modern physics comes down to the absence or presence of quantitative and mathematical determination.[29] It was Cusanus who took the first step in introducing this line of thought. Giambattista della Porta also touched upon Cusanus's method for measuring magnetic force in the sixteenth century; and the first to actually attempt the procedure, albeit with slight variations, was Robert Hooke in the seventeenth century, followed by Pieter van Musschenbroek in the eighteenth.

Of course, we must avoid any overly simplistic portrayal of Cusanus as a pioneer of modern physics on the basis of this contribution

---

*Cusanus's proposed method was mentioned by Gilbert in his 1600 work *On the Magnet* (2:108), but for some reason historians have never accorded it proper attention. Historical overviews by Friedrich Dannemann and J. D. Bernal, as well as Johann Christian Poggendorff's *Geschichte der Physik* (History of physics) and Jammer's *Concepts of Force*, ignore the proposal altogether. Edmund Whittaker refers to it in *A History of the Theories of Aether and Electricity*, stating that "a suggestion of the inverse-square law of attraction between magnetic poles had … been made by the Cardinal Nicholas of Cusa … in 1450" (p. 56, note 1). As we have seen, however, Cusanus never suggested anything so specific, so Whittaker is clearly reading something into Cusanus's work that simply is not there. As far as I know, the only historical work in which Cusanus's procedure for measuring magnetic force has been correctly recorded is Ferdinand Rosenberger's *Die Geschichte der Physik* (History of physics 1:107).

alone. Indeed, it seems he also blindly accepted the ancient myth that diamonds destroy magnetic force, because after his description of the above procedure he adds that

> in a similar manner, the virtue of a diamond may be found since they say it hinders the attraction of the lodestone to the iron.

Cusanus thus conjectures that this supposed power of diamond—its ability to disrupt magnetic force—could be measured and quantified by the same method suggested for the measurement of magnetic force itself, and he came to this notion in part because of his naive acceptance of the idea of natural "antipathy" between diamonds and magnets.[30] In this sense, we must concede that Cusanus's insistence on quantification was an anomaly in his otherwise medieval understanding of nature.

* * *

And so Nicolaus Cusanus, who asserted that quantitative measurement was the foundation for understanding nature, was the earliest known thinker to point out the importance of measuring force—for our purposes, gravity and magnetism. If this line of inquiry had been diligently pursued at the time, there would have ensued a shift from ontological questions concerning why magnetism existed to functional questions concerning how it operated, and the prevailing, qualitative natural philosophy would have at that point begun its metamorphosis into the modern, quantitative science of physics. As it happened, however, that evolution was not to occur for quite some time.

Instead, as the Renaissance unfolded, ideas about magnetism shifted to a Hermetic view of magnetic force as a type of magic. Then, in

1600, Gilbert revived the Aristotelian idea of magnets having a kind of "soul." Meanwhile, Della Porta took up Cusanus's novel proposal of measuring magnetic force as part of the practice of "natural magic." It was from that point on that thinkers and scholars gradually moved toward a modern understanding of magnetic force.

The road to modern science was, however, neither straight nor simple. The modern scientific concepts and methods familiar to us today had to be incubated within a climate of diverse premodern ideas that now seem distinctly unscientific.

Chapter 10

# The Rediscovery of Things Ancient: Magic in the Early Renaissance Period

## Section 1—The Revival of Magic during the Renaissance

The Renaissance, the period lasting roughly from the fourteenth through the sixteenth century, had a wide-ranging impact on European culture and also paved the way for the development of modern science that took place in the seventeenth century. This road, however, was by no means straight: a key contributor to the understanding of magnetism and gravity was, curiously, a renewed interest in magic, particularly so-called natural magic. While at a glance this revival might seem like an intellectual regression, it was in fact an advance, as it fostered development of the concept of force and, in particular, acceptance of the idea of action at a distance. During this period, magnetism was studied solely within the context of natural magic and in general did not come up for discussion except as a typical example of occult power and as evidence for magical and astrological causality.

Magic was suppressed as heresy in medieval Christian society and survived only as an underground movement. Its emergence into the open in the fifteenth century was a sign that the church's ideological control over the people had waned. In particular, Italy, the seat of the Vatican, had split into numerous petty states, and the secular struggles both among those states and between them and the papal government over territory and other assets gradually eroded Italian society, throwing it into confusion. Seizing this opportunity, a newly emerging urban class composed of merchants, artisans, and bureaucrats steadily gained power. The corruption and degradation of the Vatican and the weakening of its structure would lead to the Reformation in northern Europe in the sixteenth century, but a century prior to this, in Italy, these factors had already driven people toward the pursuit of temporal profit, providing the impetus for the Renaissance. To members of the Italian bourgeoisie, who witnessed and were in fact embroiled in this rapid social transformation, the Catholic notions of salvation and the afterlife appeared highly suspect.

In this turbulent period, it was humanism that offered new possibilities for man. Humanism was dedicated mainly to the study of the classics and the unearthing of ancient texts, and characteristic of the movement was the fact that these activities took place outside of academia. The study of the classics in Florence, the center of the Italian Renaissance, began at the end of the fourteenth century with the invitation of Chrysoloras of Constantinople to Florence to teach Greek. The fact that such prominent early humanists as Coluccio Salutati, the man primarily responsible for inviting Chrysoloras, and Leonardo Bruni, who studied Greek under Chrysoloras and translated several of Plato's works, including *Phaedo* and the *Apology of Socrates*, into Latin, both served as chancellor of Florence demonstrates the character of the early Italian humanist movement. The early humanists regarded the ancient Roman Republic and the Greek city-state as ideal forms of social organization and looked to the ordinary lifestyles of the citizens of these democratic societies as models. Their primary concern was the practical business of politics, and they cultivated the democratic system of Florence.[1] For this reason they were contemptuous of Scholasticism, which they saw as being sterile and overly contemplative, yet they did not attempt to replace it with an alternative philosophy or body of scholarship of their own.

It is said that humanism did not evolve into a genuine philosophical movement until later, in a transition first sparked in part by the Council at Ferrara-Florence held in 1438 and 1439 and by the sudden influx to Italy of numerous scholars from the Byzantine Empire following the fall of Constantinople in 1453, which entailed the destruction of the Eastern Roman Empire. Scholars from the Byzantine Empire, where classical culture had been relatively well preserved, brought many handwritten copies of Greek classics to Italy, a practice that gave rise to an ideological and scholarly movement that differed from the early humanist movement with its heavy bias toward rhetoric. In particular, Hermetism and neo-Platonism, which had acquired a decidedly mystical air under the influence of the Byzantine Empire, began to be widely discussed, and magical thought came out of the closet into the light of day.

In the latter half of the fifteenth century, individuals at the Accademia Platonica (Platonic Academy of Florence)—particularly its leader, Marsilio Ficino (1433–1499), and the short-lived Pico della Mirandola (1463–1494)—were at the center of the discovery and introduction of neo-Platonism and Hermetism and the rehabilitation of magical thought. Although the name sounds imposing, the academy was actually a private group of scholars formed in the mid-fifteenth century under the patronage of Cosimo de' Medici, a member of the new bourgeoisie class and founder of the Medici dynasty. Some commentators shrewdly view Cosimo's patronage of the academy as a political strategy to distract the gaze of local citizens away from politics. It is true that as his power grew Florence's claim to democratic politics became increasingly superficial, while the interests of local intellectuals shifted decisively from the practical to the meditative. At the same time, however, political ambition alone cannot have been the only motivation behind his scholarly pursuits. Another impetus cited for Cosimo's founding of the academy was his strong interest in Hermetism owing to the inspiration of Plethon, a mystical pantheistic Platonist from the Byzantine Empire. It is said that Cosimo was particularly drawn to the philosophy because of its supposed power to open up hitherto unknown human potential.

The system of ideas that comprises Hermetism was set down in Egypt during either the second or the third century by Hellenic intellectuals. These scholars were influenced by neo-Platonism, Gnosticism, and Kabbalah (a doctrine of Jewish mysticism), and many Hermetic texts were later compiled as the *Corpus Hermeticum*. Hermetism also took some of its influence from Christianity, though Christians in general thought of it as a pagan doctrine. Although some fragmentary knowledge of its literature was retained during the Middle Ages, it was only in the mid-fifteenth century that the main body of Hermetic writings, which covered a wide range of topics, including theology, philosophy, astrology, alchemy, and magic, became widely known throughout Europe. This dissemination of Hermetic learning resulted from the publication in 1471 of Ficino's Latin translation, commissioned by Cosimo de' Medici, of a selection of handwritten Greek manuscripts that had been brought to Florence.

It was not until the seventeenth century that classical scholar Isaac Casaubon's analysis revealed that the *Corpus Hermeticum* likely originated after Christ. People in the early Renaissance period, including Ficino, believed that it dated back long before Plato to the time of Moses, or even earlier, to an Egyptian priest, Hermes Trismegistus, who was thought to have authored it. They were convinced that it was the forerunner of Platonism and that it also foretold the coming of Christ, because its teachings included monotheism and the creation of a heaven and an Earth, notions that were at least superficially akin to the Christian worldview. In one work, Ficino refers to Hermes Trismegistus—an appellation meaning "Hermes thrice greatest"—as a "divine and wise philosopher" and says: "Inspired by the god Hammon, Hermes discovered arithmetic, geometry, astronomy and moreover introduced the use of letters."[2] In early Renaissance thinking, this, if true, meant that the magical thought running through Hermetism must have been born of ancient, sacred wisdom and therefore could not possibly have been erroneous or evil. Cosimo is known to have ordered Ficino to put aside his translation of Plato in order to work on the *Corpus*, a fact that demonstrates the high esteem in which the latter was held.

This reverence is very hard to fathom without some understanding of the historical background. People in Europe from the late Middle Ages to the Renaissance considered the period of classical antiquity to have far surpassed their own era. And looking even further back, they genuinely believed that the prophets who lived before the Great Flood told of in the book of Genesis were much closer to God than other mortals, and that they were superior human beings who had been entrusted by God with truth. As William of Conches, a twelfth-century pioneer in the introduction of Aristotelian thought and Greek science to Europe, bluntly put it, "the ancients were much superior to our contemporaries."[3] It was this belief that fueled the race by early Renaissance humanists to find ancient texts in the monasteries of Europe. Roger Bacon similarly declared in *Opus majus* that "the full measure of philosophy was given to the same men to whom also the law of God was given, namely, the holy patriarchs and prophets from the beginning of the world"; and that "Aristotle restored the knowledge of the ancients and brought it to light."[4] According to Bacon, the knowledge of the sages of ancient Egypt and Chaldea was the truth that sprang from God, and Aristotle was to be revered for his scholarship because he had discovered the original wisdom of the human race.

It was not until the mid-sixteenth century that Europeans first began to recognize the fallibility of the ancients. Until the voyages of Vasco da Gama, Christopher Columbus, and others to the New World and the southern hemisphere revealed errors in time-honored geographical knowledge, the content of ancient writings generally went unquestioned. No matter how vague or strange certain literature seemed, it was rationalized by such arguments as that by Ficino when he says that "it was the custom of ancient theologians to conceal their sacred, holy secrets in the shadowy guise of allegories lest they be desecrated by the profane and impure" and that therefore "we must consider that the divine mysteries are veiled."[5]

Even Nicolaus Copernicus, who in 1543 established the foundation for modern cosmology with *On the Revolutions of the Heavenly Spheres*, revealed himself to be a true Renaissance man, demonstrating loudly and clearly the climate of the period. Educated in Bologna, Padua, and Ferrara, Copernicus spoke of heliocentric theory not as a new discovery but as a revival of a theory first proposed by the ancient Pythagorean philosopher

Philolaos. He also looked to Hermes Trismegistus to justify placing the sun at the center of the solar system.[6] Thus for Copernicus, too, truth was something to be rediscovered.

Ficino was second to none in his worship of the ancients. In the latter half of the 1450s he became interested in the philosophy of Plato and subsequently spent more than a decade translating into Latin not only *Poimandres*, part of the *Corpus Hermeticum*, but also the complete works of Plato and works by Plotinus, the founder of neo-Platonism. The revival of magical thought and the dissemination of Hermetism during the Renaissance owe much to this series of translations.

The core tenet of Hermetism—that "The Kosmos is first among all living creatures; man, as a living creature, ranks next after the Kosmos"—is elaborated on in the following passage from the *Corpus Hermeticum*:

> The Kosmos then is subject to God; man is subject to the Kosmos; the creatures without reason are subject to man; and God is above all, and watches over all. The divine forces are, so to speak, radiations emitted by God; the forces that work birth and growth are radiations emitted by the Kosmos; the arts and crafts are radiations emitted by man. The divine forces operate by means of the Kosmos, and their operation reaches man by means of the cosmic radiations to which birth and growth are due; and the forces that work birth and growth operate by means of the material elements.[7]

In summary, the whole of the universe is under the divine protection of God; it is a single, living organism filled with various forces. Human beings are organisms next to God and live by virtue of power received from the heavens; their nature is to act upon objects using knowledge and technical skill. This organic worldview formed the foundation of ancient magic, a basic tenet of which was that the mysteries of the cosmos would be laid bare to the chosen investigator who would be able to wield its power with ease. Therefore, people during the Renaissance believed that they could acquire outstanding abilities that approached those of the early sages by seeking out and mastering the knowledge secretly passed down since ancient times through a chosen few. According to the *Corpus Hermeticum*:

> For man is a being of divine nature; he is comparable, not to the other living creatures upon earth, but to the gods in heaven. Nay, if we are to speak the truth without fear, he who is indeed a man is even above the gods of heaven, or at any rate he equals them in power.[8]

As God bequeathed wisdom to humankind alone, humankind possessed greatness. This idea, that men could raise themselves up to the level of the gods, was to definitively change the perception of people's role in the universe, and it appealed strongly to the ethos of the Renaissance.

## Section 2—Foundations for the Spread of Magical Thought

Historians commonly credit Renaissance humanism with the so-called discovery of man, and they consider this to be the movement's greatest achievement. The essence of early humanism is captured succinctly in the title of a 1452 work, *On the Dignity and Excellence of Man*, written by Giannozzo Manetti, a humanist and government official born into a wealthy Florentine merchant family. But certain men, like Pico and Ficino, were not mere humanists; by giving man special rank in the universe, they attempted to provide a philosophical basis for the dignity of man. *Oration on the Dignity of Man*, which Pico wrote impetuously in 1486 at the age of only 23, is said to be "the most widely known document of early Renaissance thought."[9] Near the beginning of the oration, we find God announcing to man that "In conformity with thy free judgment, in whose hands I have placed thee, thou art confined by no bounds; and thou wilt fix limits of nature for thyself." Thus Pico considered man to be not a passive being who must submit to fate but an active being that made autonomous decisions and choices and was able to act upon the world independently. His declaration that "It is given him [man] to have that which he chooses and to be that which he wills" was a sonorous anthem to man, an expression of exalted self-awareness that would have been unimaginable in actual medieval society, bound as it was by the ideology of social rank and status.[10]

Ficino, similarly, believed in the existence of choice. He distinguished five levels of reality—matter (quantity), quality, soul (spirit), the angel, and God, and he saw the rational soul, which was peculiar to the human race, as belonging in the center of the world between matter and God. Humans were privileged in being endowed with the characteristics of both the lower and higher levels of reality, and were ubiquitous in the sense that they could choose to live at either the highest or the lowest levels. And because they existed between the immaterial and the material, individual people represented microcosms of the whole world, containing within themselves that which corresponded to all the constituents of the macrocosm. As such, human beings had the potential to understand and control anything and everything.

The notion that people, if they but desired to do so, could understand and master nature in its entirety, reigning supreme over all creation, was to fundamentally change the medieval concept of the relationship between God and human beings. Essentially, this idea meant that the miracles performed by God could also be performed by mortals, which was precisely what was meant by magic. Thus, surprisingly, magic underwent a revival because of a theory that was actually anthropocentric. Magic was thought to provide the logical evidence for man's activity and independence in relation to nature and to be the power that lay concealed within the wisdom of the ancients. In fact, Pico rejected astrology despite his zealous support of magic precisely because he thought that astrology, which held fate to be dictated by the stars, denied man's free will.[11]

Magical thought rapidly came to exert a major influence on intellectuals throughout Europe after its revival in the latter half of the fifteenth century. There were several reasons for this. As we will see in the next section, the concept of magic was elevated to so-called natural magic and clothed in intellectual garb, and it was thus clearly distinguished from conventional folk practices and magic charms. At the same time, a marked increase in the numbers and power of the bourgeoisie, who did not fit into the three-tiered medieval class structure of nobles, clergy, and peasants, had prepared the groundwork for magic's social acceptance. The emergence of this new class undermined the control that had until then been exerted by feudal and Christian ideology. Being primarily concerned with bettering their current lifestyle, the energetic bourgeoisie found human-centered magical thought, which promised to increase human capacity, highly appealing. Cosimo de' Medici's interest in mastering magic likewise stemmed more from the worldly desire to acquire exceptional powers and to control nature and society than from mere scholarly interest.

The influence of magical thought in Europe was also greatly enhanced by the fact that Ficino's introduction of neo-Platonic and Hermetic works coincided with the emergence of the printing industry. His translation of the *Corpus Hermeticum* was completed in 1463 and published in 1471, with the Tuscan dialect version appearing in 1473. By 1484, he had finished translating all the major works of Plato. Ficino also wrote a commentary on Plato's *Symposium* that was completed in 1469, and he published his main philosophical treatise, *Platonic Theology*, in 1482. This was followed by the widely read *Three Books on Life* in 1489.

Numerous printing shops appeared around the same time in Italy, where book printing and publishing were quite advanced. The first shop was established in Rome in 1464 or 1465, followed by others in Venice in 1469 and Foligno in 1470. More were opened from 1471 to 1472 in cities including Ferrara, Milan, Bologna, Naples, Pavia, Savigliano, Treviso, Florence, Iesi, Parma, and Mantova. In 1480, printing shops were operating in over 110 cities throughout Western Europe, and about fifty of these were located in Italy, with Venice serving as the capital of the printing industry. The number of publications produced in Italy far surpassed that produced in any other country.[12]

This sudden flowering of Italy's printing and publishing industry coincided perfectly with the publication of a series of Ficino's translations and original writings. Because his works, including his translations of Plato, were among the earliest printed books, they came to have far-reaching influence. It is estimated that a single edition at that time consisted of anywhere from three hundred to three thousand copies, with the average being one thousand. A typical print run, to give one example, was the set of 1,025 copies issued of one of Ficino's Latin translations, produced in 1483, of a work by Plato.[13] This far surpassed the number of copies of a work that could be produced by transcription in the monasteries, or even by the more efficient *pecia* system—a method, introduced by universities in the thirteenth century, whereby a manuscript was divided into *peciae* (pieces) and rented out to scribes for simultaneous copying. By the year 1500, as many as fifteen editions of Ficino's translation of the *Corpus Hermeticum* had been printed and published.[14]

And so in Italy from the second half of the fifteenth century, with the mass publication of books as merchandise and the emergence of bourgeoisie readership, the logistical and social foundations were in place for the flourishing of magical thought.

## Section 3—The Magical Thought of Pico and Ficino

At first glance, it seems very odd that printing technology and an urban middle class, unquestionably two symbols of modernization, should be in any way related to such a premodern practice as magic. In fact, however, the magic revived by Renaissance intellectuals had been transformed into a scholarly, refined, and philosophical pursuit far removed from the sinister sorcery or witchcraft of the Middle Ages.[15] According to Keith Thomas's *Religion and the Decline of Magic*, while the "intellectual magic" revived by Pico and Ficino shared some common elements with the "popular magic" that had been passed down from the Middle Ages, they were "essentially two different activities."[16]

In his *Oration on the Dignity of Man*, Pico divides magic into two types: one is cursed and evil, consisting solely of the work and powers of demons (*daemons*); the other, natural magic, seeks to acquire "the knowledge of all nature" and is "the absolute consummation of the philosophy of nature." He describes the effects of the latter as follows:

> Having carefully investigated the harmony of the universe, which the Greeks very expressively call συμπαθειαν (*sympatheia*), and having looked closely into the knowledge that natures have of each other, this second magic, applying to each thing its innate charms, which are called by magicians ιυγγεσ (*iugges*), as if it were itself the maker, discloses in public the wonders lying hidden in the recesses of the world, in the bosom of nature, in the storerooms and secrets of God. And as the farmer marries elm to vine, so the magician marries earth to heaven, that is, lower things to the qualities and virtues of higher things.[17]

To Pico, natural magic was "divine and salutary," and therefore it was permissible.

Ficino, too, spoke of "two kinds of magic":

> The first is practiced by those who unite themselves to daemons by a specific religious rite, and, relying on their help, often contrive portents ... But the other kind of magic is practiced by those who seasonably subject natural materials to natural causes to be formed in a wondrous way.[18]

Ficino, like Pico, called the latter type "natural magic." The above excerpt is from the *Apology* he wrote in the autumn of 1489, near the end of his life, and appended at the back of his *Three Books on Life*, which he had completed in July of that year. The addition was made out of fear that the third volume of the book would be viewed as heretical. There are some who claim that Ficino was a closet believer in demonic magic, but it is not my intention in the present work to delve into that subject.[19] Rather, our primary task here is to define just what natural magic, as opposed to demonic magic, was, what it contributed to the development of man's understanding of nature, and how it affected the subsequent development of natural science.

The dichotomy was not new. The theologian William of Auvergne, for example, mentioned such a division in the first half of the thirteenth century,[20] and in the mid-fourteenth century Nicole Oresme clearly stated that "there are two parts to the magical art: one which operates by means of demons and the other which does not so operate." Oresme claimed that, whereas the former was impossible for ordinary men to perform, to the latter "some natural reason can be assigned." Nonetheless, he judged the performance of magic itself to be criminal and an act incited by the devil.[21] Therefore, the novelty of Renaissance thought on this subject was not the division of magic into two categories but rather the neutralization and sanctioning of "natural magic," meaning magic that was not dependent on demons.

Ficino is thought to have either studied medicine or learned the trade from his father, who was a physician, and his *Three Books on Life* is essentially a medical treatise. But, unlike Pico, he did not renounce astrology, and his medical science was an astrological medicine that claimed to directly make use of the powers of heaven or things imbued with these powers. The main thread woven throughout the book is the idea that the sun and stars affect the human body both directly and indirectly. While Ficino distinguishes between two types of magic in his *Apology*, he also clearly states that natural magic "by natural things, seeks to obtain the services of the celestials for the prosperous health of our bodies."[22] He develops his theory of magic in *On Obtaining Life from the Heavens*, the third volume of *Three Books on Life*,[23] where he claims that through magic, "by means of lower things which are in accord with higher ones, people can in due season somehow draw to themselves celestial things" (TBL 3.15:319). Some of his ideas are difficult to reconcile with modern logic, and we must not attempt to force such a reconciliation artificially; at the same time, however, it is possible to make some sense of his theory of magic by examining it in the context of his cosmology.

According to Ficino, "the world's body is living in every part" and it "lives ... through a soul" (TBL 3.3:255). In other words, the human body and intellect are joined by the soul and, similarly, heaven—a transcendent world comprising the sun, moon, and stars—is a huge living being, specifically, a union of what he calls a *World Body* and a *World Soul.* And therefore the universe, which is comprised of body and soul, contains not just the four elements of earth, water, air, and fire but in addition *spirit*, also known as the *quintessence*, meaning *fifth element* (TBL 3.1). This spirit "is an intermediary between the gross body of the world and its soul; and the stars and daemons exist in it and by means of it" (TBL 3.4:259). This spirit of the cosmos is responsible for animating the world. In Ficino's words,

> just as the power of our soul is brought to bear on our members through the spirit, so the force of the World-soul is spread under the World-soul through all things through the quintessence, which is active everywhere, as the spirit inside the World's Body. (TBL 3.1:247)

In this way, the universal spirit acts as a medium between the soul and the body and is also the medium through which the heavens exert their influence on the things of this world. The things of heaven, namely, the sun, moon, and stars, thus transmit their energy to the Earth through this spirit, activating objects in this world and affecting the human body.

In Ficino's view, these occult qualities or occult powers, which are unusual characteristics or faculties not derived from the four elements, are transmitted from the heavens through the spirit of the cosmos and exist in all things. In chapter 12 of the third book of *Three Books on Life*, titled "Natural and Even Artificial Things Have Occult Powers from the Stars, through Which They Expose Our Spirit to the Same Stars," Ficino writes that

> we do not say that our spirit is prepared for the celestials only through qualities of things known to the senses, but also and much more through certain properties engrafted in things from the heavens and hidden from our senses, and hence only with difficulty known to our reason. For since properties of this kind and their effects could not come into being by elemental power, it follows that they proceed from the life and spirit of the cosmos, particularly through those very rays of the stars. (TBL 3.12:299–301)

Ficino claims that "the occult virtues of things, which are called 'specific virtues' by natural philosophers, are made not by the elemental nature, but by the celestial" (TBL 3.16:323), which is the same point Thomas Aquinas was making when he wrote of the operations of things surpassing active and passive qualities of elements, which they have received from the celestial bodies. Thomas's influence is also clearly evident in Ficino's argument that "The cross ... is a figure which is made by the strength of the stars and serves as a receptacle of their strength; it therefore possesses the greatest power among images and receives the forces and spirits of the planets" (TBL 3.18:335).[24] But unlike Thomas, Ficino sees in these heaven-sent powers a basis for the existence of natural magic that can be stated in terms of natural philosophy.

As we have already seen, the idea that gemstones housed spirits and were endowed with magic powers had been passed down through the Middle Ages from ancient times. Ficino seeks the origins of this specific instance of so-called occult power in the heavens and celestial bodies. For example, emerald, jacinth, sapphire, topaz, ruby, unicorn's horn, and other gemstones believed to possess particular occult properties do so, he asserts, precisely because they are the recipients of celestial powers. "And therefore, not only if they are taken internally, but even if they touch the flesh, and, warmed thereby, put forth their power, they introduce celestial force into the spirits by which the spirits preserve themselves from plague and poison" (TBL 3.12:301). This same mechanism is also evident when he writes that "a somewhat weak power to snatch up straws, given to amber by the heavens, after it has been strengthened by friction and heating suddenly snatches them up" (TBL 3.16:327).

To Ficino, in other words, nondemonic magic—that is, natural magic—is the study of the celestial powers transmitted to terrestrial things through the medium of the universal spirit; and the purpose of such study is to guide and control the flow of this spirit and manipulate the powers and energy of heaven to enhance health and perpetuate life. Ficino maintains that man, uniquely positioned between heaven and Earth, existing simultaneously in both the elemental (material) world and the celestial (spiritual) world, and capable of perceiving both the visible and the invisible realms, can make this power his own through diligence and training. This conviction comes through clearly, for example, in the final chapter of *Three Books on Life* (book 3, chapter 26), where we find the following:

> Agriculture prepares the field and the seed for celestial gifts and by grafting prolongs the life of the shoot and refashions it into another and better species. The doctor, the natural philosopher, and the surgeon achieve similar effects in our bodies in order both to strengthen our own nature and to obtain more productively the nature of the universe. The philosopher who knows about natural objects and stars, whom we rightly are accustomed to call a Magus, does the very same things: he seasonably introduces the celestial into the earthly by particular lures just as the farmer interested in grafting brings the fresh graft into the old stock … The Magus subjects earthly things to celestial, lower things everywhere to higher, just as particular females everywhere are subjected to males appropriate to them for impregnation, as iron to a magnet to get magnetized, as camphor to hot air for absorption, as crystal to the Sun for illumination, as sulfur and sublimed liquor to a flame for kindling, an egg-shell, empty and full of dew, to the Sun for elevation, or rather the egg itself to the hen for hatching. (TBL 3.26:387)

Natural magic emerged during the Renaissance as an alternative to both ritualistic, spell-driven magic on the one hand and demonic magic on the other; it offered something more empirical, technological, and practical despite the peculiar organic worldview that formed its theoretical basis. To the modern mind, natural magic may appear erroneous or even theoretically groundless and, in that sense, merely a form of pseudo-technology. But consider this: although compasses were at first mistakenly believed to be attracted to the North Star or Heaven's Pole, this did not mean that they were a form of pseudo-technology. In fact, the very distinction between truth and falsehood based on the legitimacy of the theory underlying a given technology is a modern one; such distinctions were simply not a part of Renaissance thought.

## Section 4—Magnetic Force as Magical Power

Thus, in the magical philosophy of the Renaissance it was believed that man as a microcosm could absorb and make use of the animating force of the macrocosm—that is, the universe at large—if he mastered philosophy and became well versed in the mysteries of nature. But as we can see in the writings of Ficino and Pico, their concept of force was quite different from ours today, and they did not necessarily think of force as acting in a way that was dynamic or causal.

In his 1469 commentary on Plato's *Symposium*, Ficino states that "in love there is all the power of enchantment." Pico termed this same concept the "harmony of the universe," and in his later years Ficino came to refer to this "power of love" (*vis erotica*) as the "power of Heaven" (*vis coelestis*) and to believe that it was mediated by the spirit. But in his commentary on the *Symposium*, Ficino continues as follows:

> The work of enchantment is the attraction of one thing by another because of a certain similarity of their nature. The parts of this world, like the parts of a single animal, all hanging from one author, are joined to each other by the mutuality of one nature. So, just as in our bodies the brain, lungs, heart, liver, and the rest of the parts all require something from each other, help each other and all suffer when one of them suffers, so in the same way the parts of this huge animal, that is, all the bodies of the world, joined together, exchange natures with each other and are mutually exchanged. From their common relation a common love is born, and from that love a common attraction, and this is true enchantment ... Hence also the magnet draws iron, amber draws the chaff, sulphur the fire, the sun turns many flowers and leaves toward itself, the moon attracts the seas, and Mars always rouses the winds.[25]

We do not see here the cosmological background later developed in *Three Books on Life*, nor the strong belief in human ability to harness the powers of nature. But what does stand out conspicuously is the idea that magnetism and static electricity, and particularly the former, are representative examples of the magical forces that fill the natural world. In this commentary, Ficino explains magnetism as follows:

> The lodestone confers upon iron a certain quality of its own by which the iron, made like the stone, is drawn to it. This attraction, inasmuch as it rises in the stone and attracts toward the stone, is said to be certainly the stone's attraction.[26]

Here he speaks of the magnet's chief property only in terms of its relation to iron: the magnet gives iron power, making it similar to itself and, in the process, attracting it. Of even more interest is Ficino's explanation of the northward directionality of magnets and magnetic needles described in *Three Books on Life*. In that discussion, he deals with the origin of the magnet's chief property, claiming that magnetism is a concrete example of the power conferred by the celestial world on things of this world. In chapter 15 of book 3, the origin of magnetism is clearly placed in the heavens, and specifically in the Pole Star (αof Ursa Minor, the constellation also known as the Little Bear).

> In the lookouts of sailors, to tell them where the Pole is, we see that a balanced needle influenced on its end by a lodestone points towards the Bear; this happens because the lodestone draws it in that direction, since the power of the Bear prevails in this stone; and from there it is transferred into the iron and pulls both toward the Bear. Moreover this sort of power both was infused from the beginning and also grows continually by the rays of the Bear. Perhaps amber behaves in this way towards the other Pole even as it does towards straw. (TBL 3.15:315–317)

In other words, magnetism is implanted in the magnet by the North Star, and this special quality, this occult power, is then shared with iron through physical contact, producing iron needles that point north. Ficino's belief that the magnet's characteristic power has a celestial origin is, again, clearly derived from Thomas Aquinas. Ficino continues the discussion in *Three Books on Life* by wondering

> why does the lodestone everywhere draw iron?—not because they are similar, otherwise lodestones would draw lodestones much more readily and iron, iron … Why then? Both are comprised in the order depending on the Bear, but the lodestone holds the superior rank in the very property of the Bear; iron, however, the inferior. The superior draws what is inferior in the same chain of beings and turns it towards itself, or else otherwise agitates it in some way or other, or influences it by a power infused beforehand. The inferior in turn by the same infusion is turned towards the superior or otherwise agitated or deeply influenced. (TBL 3.15:317)

This is a concrete example of the concept of "higher" and "lower" that we saw earlier where Ficino says of magic that "by means of lower things which are in accord with higher ones, people can in due season somehow draw to themselves celestial things."

We also saw earlier Ficino's claim (in *Three Books on Life*, book 3, chapter 12) that the powers of the different celestial bodies were implanted in gemstones as hidden, or occult, properties and that when introduced into the body they could facilitate medical treatment. This belief was based on the idea that the influential power of the celestial bodies on terrestrial bodies could be transferred to other things, or to people, and then put to use. And in magnetism Ficino saw straightforward evidence for this basic principle of natural magic—whereby people could indeed manipulate the occult properties invested by heaven in terrestrial bodies—because the north-pointing property of iron needles rubbed against lodestones (empowered, it was thought, by the Pole Star) was a fact. Moreover, it was a fact that served as virtually the only concrete example of, and therefore the prototype for, Ficino's natural magic.

And this single example of the magnet was extended to everything else. As we saw earlier, Ficino went so far as to suppose that amber might behave in the same way as the lodestone toward what he called "the other Pole." From the premise that the directionality of the magnet was conferred upon it by the Pole Star, he surmised that the electrostatic force demonstrated by amber must likewise be derived from the heavens and that therefore amber must point toward a different pole, namely the east-west pole. To the modern mind, this very idea is extraordinary, for the magnet is one thing and amber is quite another, and we know today that there are no phenomena to even suggest the possibility that the power of amber comes from the heavens. Yet to Ficino, not only lodestone and amber but all gemstones possessed special powers—occult powers—all of which came from out of the sky. Just as the action of the lodestone was the prototype, that of amber was nothing more than another example of essentially the same phenomenon.

## Section 5—Agrippa's Magic: Nature as Symbol

In the sixteenth century, Cornelius Agrippa von Nettesheim (c. 1486–1535) compiled all the concepts of magical thought that had been revived a century earlier. A German and a contemporary of Martin Luther, Agrippa wandered about Europe and died far from home, defamed, and reduced to poverty.[27] Late in the sixteenth century, Thomas Nashe referred to him in *The Unfortunate Traveller* as "the greatest conjurer in christendome," and Christopher Marlowe, in *The Tragical History of Doctor Faustus*, has Faustus say that he "Will be as cunning as Agrippa was."[28] Thus Agrippa appears to have been notorious throughout Europe as a heretical magician. Yet he is also known to have been a humane lawyer, an enlightened doctor, and a man who fought church authority on behalf of the oppressed during the Inquisition. In one particular incident, he defended a mentally ill woman who had been branded a witch and accused of having had her soul stolen by the devil.[29]

Agrippa sought out ancient and contemporary philosophical, religious, and occult literature, and in 1510 he completed his *Three Books of Occult Philosophy.*[30] In the opening lines, he declares that

> there is a Three-fold World—Elementary, Celestial and Intellectual—and every inferior is governed by its superior, and receiveth the influence of the virtues thereof, so that the very Original and Chief Worker of all doth by angels, the heavens, stars, elements, animals, plants, metals and stones convey from Himself the virtues of His Omnipotency upon us, for whose service He made and created all these things: Wise men conceive it no way irrational that it should be possible for us to ascend by the same degrees through each World, to the same very original World itself, the Maker of all things and First Cause, from whence all things are and proceed; and also to enjoy not only these virtues, which are already in the more excellent kind of things, but also besides these, to draw new virtues from above. (TBOP 1.1:33–34)

This understanding of the world was the essence of Agrippa's system of thought, and he clearly adopted the concept that "every inferior is governed by its superior" from Ficino.

His "Three-fold World" consists of the terrestrial world of material bodies; the world of the sun, moon, and stars; and, beyond that, the world of Ideas. Agrippa further states that, in accordance with these three realms, "all regulative Philosophy is divided into Natural, Mathematical and Theological" branches (TBOP 1.2:35), to each of which he devotes a separate volume of the *Three Books of Occult Philosophy.* "These three principal faculties," he says, "Magic comprehends, unites and actuates; deservedly, therefore, was it by the Ancients esteemed as the highest and most sacred Philosophy" (TBOP 1.2:37). Although Agrippa considers mathematical magic and spiritual magic (which he deals with in the second and third volumes of the trilogy) to be superior, our main interest is his discussion of natural magic (taken up in the first volume), because the purpose of our study is to ascertain how the Western understanding of magnetism developed over time.

"There are," asserts Agrippa, "four Elements, and original grounds of all corporeal things—Fire, Earth, Water, Air—of which all elemented inferior bodies are compounded" (TBOP 1.3:38), and he considers these four elements to be the root of all qualities perceivable by the senses. But of greater importance is that Agrippa believes there is more to the qualities possessed by things in the natural world than just these four elements. He maintains that things are also endowed with their own peculiar and marvelous capabilities, a view evident in the following:

> It is well known to all that there is a certain virtue in the Loadstone by which it attracts iron, and that the Diamond doth by its presence take away that virtue of the Loadstone. So also Amber and Jet, rubbed and warmed, draw a straw to them; and the stone Asbestos, being once fired, is never or scarce extinguished. (TBOP 1.13:65)

This conviction is also evident when he claims that "a Carbuncle shines in the dark ... the Jasper stauncheth blood; the little fish Echeneis stops the ships; Rhubarb expels choler; the liver of the Chameleon, burnt, raiseth showers and thunders." According to Agrippa, these types of unique qualities and actions are derived not from combinations of hot, cold, wet, and dry, which are supposed to be embodied in the four elements, but rather from occult qualities. As he puts it:

> There are also other virtues in things, which are not from any Element, as to expel poison, to drive away the noxious vapors of minerals, to attract iron or anything else; and these virtues are a sequel of the species and form of this or that thing; whence also they being little in quantity, are of great efficacy; which is not granted to any Elementary quality. For these virtues, having much form and little matter, can do very much; but an Elementary virtue, because it hath more materiality, requires much matter for its acting. And they are called *Occult Qualities*, because their causes lie hid, and man's intellect cannot in any way reach and find them out. (TBOP 1.10:59–60)

For Agrippa, furthermore, "the occult properties in things are not from the nature of the Elements, but infused from above" (TBOP 1.15:71). These properties are the powers of the "Celestial Souls" invested in earthly things through the medium of the "Spirit of the World," which Agrippa too refers to as the quintessence. He concludes:

> By this Spirit, therefore, every occult property is conveyed into herbs, stones, metals, and animals, *through the Sun*, Moon, Planets, and through Stars higher than the Planets. (TBOP 1.14:70)

Note, however, how Agrippa's reasoning follows Ficino's argument to the letter. Not only is the latter's influence striking; rather, Agrippa reproduces Ficino's ideas almost verbatim. The only point of originality is his statement that the workings of these occult properties "being little in quantity, are of great efficacy." Later, Francis Bacon suggests a debt to Agrippa in this respect when he cites "instances in which the matter or the efficient cause is slight or small in comparison with the effect or result which follows" as an example of "magical instances." This phenomenon of small causes having large effects—or, in modern scientific terms, this nonlinearity—was the reason why the workings of so-called occult properties were seen as acts of magic. As the definitive concrete example, Bacon cites "a magnet which excites a large number of needles without losing or lessening any of its own virtues."[31]

The most remarkable feature of Agrippa's *Three Books of Occult Philosophy*, however, was not this kind of abstract or theoretical reasoning but rather the author's exhaustive record of all the inherent qualities reported to exist in a vast array of things according to almost two thousand years of ancient and medieval literature. Because Agrippa sought out this information and set it all down in such methodical and minute detail, his trilogy is a sort of "summa of magic" that serves decisively as a final punctuation mark on the literature of the Middle Ages.

The work includes this general statement concerning the effects of the superior on the inferior:

> It is manifest that all things inferior are subject to the superior, and after a manner ... they are one in the other, viz., in inferiors are superior and in superiors are inferior: So in the Heaven are things terrestrial, but as in their cause, and in a celestial manner; and in the Earth are things celestial, but after a terrestrial manner, as in an effect. (TBOP 1.22:87–88)

But the discussion is by no means limited to generalities, as Agrippa crafts an extensive description of things considered inferior to the sun, the moon and each planet, respectively. Things affected by the planet Mars, for example, include fire under the category of elements; iron and red brass under metals; diamond, lodestone, bloodstone, jasper, and amethyst under stones; garlic, hellebore, euphorbia, radish, laurel, wolfsbane and scammony under plants; and horse, mule, goat, wolf, leopard, serpent, dragon, gnat, fly, eagle, falcon, hawk, vulture, pike, and sturgeon under animals (TBOP 1.27:101–102). To us, this is all very puzzling, but to Agrippa—and in the scheme of Renaissance magical thought—nature is a huge accumulation of symbols and metaphors, and all things in nature have a hidden meaning recorded in a codebook invisible to the human eye.

Similarly, Agrippa presents both a general explanation of the sympathy and antipathy between earthly things as discussed since the time of Pliny and an overwhelming number of concrete examples of such relationships: It did not nearly suffice to explain that

> all things have a friendliness and enmity amongst themselves, and every thing hath something that it fears and dreads, that is an enemy and destructive to it; and, on the contrary, something that it rejoiceth and delighteth in and is strengthened by. (TBOP 1.17:75)

Agrippa further claims, for example, that the tendency toward amity, like the pull of the magnet on iron, is to be found in all plants and animals, and he enumerates examples of affinitive relationships—vine and elm, olive and fig, peacock and pigeon, crow and heron, turtle and parrot … The list goes on and on.

In the case of magnets, Agrippa's treatment of the subject of affinity appears to have been based on empirical evidence; however, this is quite obviously not true for the majority of his examples. Consider, for instance, his remark that

> the Soul of one thing went out and went into another thing, altering it, and hindering the operations of it; as the diamond hinders the operation of the loadstone, that it cannot attract iron. (TBOP 1.14:69)

Here, he is concerned not with explaining the mechanism or principle involved but merely with recording one of the myriad examples of sympathy and antipathy that had been handed down from times past. Also based on longstanding folklore are such assertions as that "a diamond doth disagree with the loadstone, that being set by it, it will not suffer iron to be drawn to it" (TBOP 1.18:79) and that "Amber draws all things to it besides garden basil" (TBOP 1.18:81–82). And here again Agrippa produces an exhaustive list of things said to be in antipathy to one another, including scorpion and crocodile, rhubarb and choler, cucumber and oil, mouse and weasel, and emerald and lust. And this information represents but a small portion of the so-called codebook of relationships that the ceaseless efforts of sages from ancient times forward are said to have uncovered.

Agrippa also applies the notion of antipathy to so-called elements, such as water and fire. Taken on its own, this particular example seems fairly reasonable. Obviously, water in sufficient quantities can quench a fire, while a raging inferno can evaporate water. But certain of his other statements make one wonder where in the world such ideas could possibly have come from. Consider, for example, these claims:

> Mercury, Jupiter, the Sun and Moon are friends to Saturn; Mars and Venus enemies to him. All the planets besides Mars are friends to Jupiter, also all besides Venus hate Mars; Jupiter and Venus love the Sun; Mars, Mercury and the Moon are enemies to him. (TBOP 1.17:75)

These assertions are based neither on Scholasticism, with its foundation of definitions and syllogisms, nor on the modern law of cause and effect, but rather on a wholly symbolic view of the universe and the laws that govern it. We in the modern age tend to look at nature objectively, as though we stood wholly outside of it, and so in our world, viewpoints such as Agrippa's essentially do not exist. One could say that his perspective is that of someone who not only is part of nature but has put aside the self and become one with nature, and that it therefore precludes consistent understanding in terms of modern logic.

To Ficino and Agrippa, through the spirit of the cosmos, things of the celestial world and things of the terrestrial world mutually affected each other and acted in sympathy with one another. Within the terrestrial realm too, all manner of things acted upon one another through relationships of sympathy (amity) and antipathy (enmity) that were complex and intricate yet also richly symbolic, and through this web of interactions they formed an enormous organic whole that was eternally in motion. No matter how incomprehensible it may appear to us today, this worldview was shared by the Hermetic philosophers of that time. And not only by Hermetic philosophers: in *Ten Books on Architecture*, Leon Battista Alberti, a prominent fifteenth-century humanist and intellectual, reports the ancient belief that "certain animals and certain things have such a natural aversion to one another, that one will ruin and destroy the other,"[32] and he discusses the use of this relationship to rid houses and fields of harmful insects and noxious beasts.

It was this organic worldview that formed the theoretical basis upon which the possibility of magic could be given credence. Agrippa states:

> Magic is a faculty of wonderful virtue [that] produceth its wonderful effects, by uniting the virtues of things through the application of them one to the other, and to their inferior suitable subjects, joining and knitting them together thoroughly by the powers and virtues of the superior Bodies. This is the most perfect and chief Science, that sacred and sublimer kind of Philosophy, and lastly the most absolute perfection of all most excellent Philosophy. (TBOP 1.2:34–35)

In other words, man is capable not only of appropriately combining and exploiting the power of the sympathetic and antipathetic relationships between terrestrial bodies; he can, in addition, skillfully bring to bear the influences of the celestial world and channel them in a way that is profitable to man. This is the magic of which both Ficino and Agrippa speak.

But whatever the other implications of this worldview, it is clear from the quotations and examples cited above how important the existence of lodestones and man's observations of magnetism and static electricity were to magical thought during this period. While historical and social factors certainly contributed to the rejuvenation of magic during the Renaissance, alongside of these the peculiar behavior of magnets clearly also lent support to the belief in magic as something real. Through direct experience, people saw magnets exert a strong force on iron from a distance, as though using an invisible arm; they saw iron needles transform themselves into objects with magnetic power; and they saw these magnetized needles then pointing in one direction, at a certain star, as if they had been invested with power from the heavens themselves. Even near the turn of the next century, by which time people's understanding of magnetism had further advanced, Giordano Bruno (1548–1600) referred in his *On Magic* (circa 1590) to

> what happens as a result of the powers of attraction and repulsion between things, for example, the pushes, motions and attractions due to magnets and such things

as "'natural magic' in the proper sense,"[33] indicating that the lodestone was still considered proof that natural magic really worked. According to historian Herbert Butterfield, to the Renaissance naturalists

> the whole universe seemed sometimes to be a universe of symbols. The action of the magnet seems to have influenced these thinkers very much. It appears even to have been taken as a typical example of the way things work in nature, so that there was a search for secret magical sympathies between objects.[34]

And indeed, magnetism and the amber phenomenon (static electricity) provided virtually the only realistic evidence for the view that nature was a network of power, an idea that formed the backbone of magical thought.

* * *

For people of the Renaissance, nature was an aggregate of symbols and metaphors, the universe was a gigantic network of power, and magic was nothing less than a profound philosophy and sacred technology that gave man the ability to sense and decipher the meaning hidden within all natural things by becoming one with the universe and to manipulate this network of power, which extended to all things. Most importantly, the magical thought of Ficino and Agrippa became vehicles for open discussion of the conviction that man could, by studying nature, harness the power of the universe and nature's energy. In this way, Renaissance magical thought was to become one of the driving forces behind the subsequent development of scientific technology.

But modern science did not sprint straight out of the starting block from that point in time. Ficino and Agrippa ended their work one step before natural magic approached empirical science. They, and especially Agrippa, remained medieval thinkers, not because they believed in and discoursed on magic but rather because they regarded texts and oral folklore from ancient times on with an uncritical eye, believing these sources to have absolute authority. Although Agrippa wrote repeatedly on the lodestone, all of his statements on that topic were copied from Pliny or others, and it is unlikely that he ever took a lodestone in his hand and conducted experiments on it. One can only concur with the observation that the "great enemy of scientific progress was not speculation and experiments in natural magic, but the old fossilized bookish learning."[35]

It was to take a great thinker of the 1500s to transcend this limitation.

Most historical surveys make no distinction between the magic of the 1400s and the 1500s, but in fact the magic of these two periods should be viewed as qualitatively different, or at least representing two distinct stages. As we have seen, neo-Platonism and Hermetism had a significant impact on the former, but the latter, as we will discuss later, was inspired by Roger Bacon and greatly influenced by Aristotle. For this reason, whereas the magic of the 1400s tended to be more religious and speculative, enmeshed in a world of words, the magic of the 1500s was empirical, mathematical, and practical in nature, and it tended to be paired with technology passed down by craftsmen. It is from this later type of magic that modern science, with its experimental methods, mathematical reasoning, and pursuit of technical application, is thought to have arisen—a subject that will be discussed in detail in later chapters.

At any rate, magical thought—the conception of nature as full of things that mutually interacted and of man as able, through observation, to understand nature's powers—was to prepare the way for the discovery of the law of universal gravitation, the key concept of the modern physics eventually initiated by Johannes Kepler, Robert Hooke and Isaac Newton. The idea that distant celestial bodies could influence terrestrial bodies, originally an astrological concept, was a part of Renaissance magical thought concerning the interaction between the macrocosm and the microcosm. However, before Europeans could become aware that this idea also pertained to the existence of a physical power acting between the Earth and the sun or the moon, they first had to reexamine and rediscover the Earth itself, and this required the expeditions of discovery and exploration that came with the age of great ocean voyages. In the next chapter, therefore, we turn our gaze to the events of this era and to the discovery of the Earth's magnetic field.

Chapter 11

# The Age of Exploration and the Discovery of Magnetic Declination

## Section 1—Magnetic Mountains

The fact that magnets drew iron was common knowledge even among the masses in late medieval Europe; this much is clear from more than one literary work of the day.

A love song probably written around the twelfth century and then included in the *Carmina Burana*, an anthology of popular songs from the Middle Ages, makes this analogy:

> Indeed like a magnet she draws me back,
> my wonderfully perfect girl,
> her grace of strong attraction
> directly overcomes me.[1]

And then there is *The Romance of the Rose*, written sometime between 1268 and 1278—just about the time that Petrus Peregrinus's *Letter concerning the Magnet* made its appearance—in which we find the observation that, just as a magnet so cleverly attracts iron, so too do gold and silver given to a man win his heart.[2]

In these two cases, the magnet is used as a simile for the ability of the opposite sex and of wealth, respectively, to charm the human heart. And in Bishop Richard de Bury's 1345 work *Philobiblon* we find reference to people who "run to us well pleased as listeners, just as the admant draws the willing iron."[3]

And about a half century later, in 1380, came this in Chaucer's *The Parliament of Birds*:

> Exactly like a piece of iron set
> Between two lodestones of force parallel,
> Which has no power to move this way or that—
> One magnet tries to pull, the other repel ...[4]

The association of the magnet with attraction was strong indeed. And if we look just a bit closer, things become even more interesting.

Let's return for a moment to the time of Richard de Bury and consider *Canzoniere*, a book of verse by Italian poet Francesco Petrarch, crafted somewhere around 1330. First, a small excerpt:

> Out in the Indian Ocean
> there is a stone so bold it naturally
> draws iron towards itself,
> and steals the nails from ships, so that they sink.[5]

Several lines later the romantic yearnings of a lover are described by means of another personification of lodestone:

I see myself being drawn
up on the shore by such a live lodestone!

Most important to our discussion is the poem's reference to the waters of the Indian Ocean. Marbode in the eleventh century and Gervase and Albert the Great in the thirteenth cited India as an area that produced lodestone. But until that time, Mount Ida had held distinction as the place where—according to Pliny's widely read *Natural History*—the shepherd Magnes had discovered the magnet accidentally.

*Natural History* also claims that iron had been discovered by a metalworking tribe called Dactyls on Mount Ida on the island of Crete.[6] Meanwhile, histories authored by Herodotus and Diodorus place Mount Ida in the region of Phrygia, near Troy.[7] From this, we may conclude that Mount Ida, where a shepherd named Magnes supposedly discovered the lodestone, was located either on Crete or else somewhere in Asia Minor. According to another account, in the seventh century Isidore of Seville mistakenly wrote "Ida" as "India" and this led to the belief that India was the world's source of lodestone. We have no way of judging the veracity of this specific chain of events, but it does indeed seem that at some point "Mount Ida" was replaced in the historical literature with "India." Two sixteenth-century works, for example, tell the story of Magnes the shepherd, citing Pliny as a source but placing the legendary mountain in East India. One of these is a technical work on seafaring written by a Spaniard named Martin Cortes in the mid-sixteenth century, and the other is *The Newe Attractive*, written in 1581 by Englishman Robert Norman.[8]

Before the age of exploration, however, the term "India" did not signify for Europeans a place so geographically specific as the

Indian subcontinent or the Indian Ocean. In the time of Herodotus, the term referred to everything from the Indus river eastward, and even as late as Columbus's day the "Indias" designated, rather vaguely, India and all of Asia to its east. Because of this lack of clarity, until the fourteenth century the word "India" conjured up images of a strange world that lay beyond Asia Minor and the world of Islam, seemingly as far away as the end of the Earth itself.

Moreover, it is said that "in the Middle Ages the unknown and the marvellous lay very close by"[9] and that the India of people's imaginations was filled with strange wonders. Even Marco Polo, who ventured as far as China, said that India contained "many marvels unlike anything to be found in the rest of the world."[10] Pliny's *Natural History* has it that "India and parts of Ethiopia especially teem with marvels," including humans with the heads of dogs, humans with no mouths, and satyrs.[11] Such tales multiplied in the Middle Ages, yet none was questioned. The *Hereford Mappa Mundi*, crafted around the year 1300, includes a drawing of a man with one leg and using it as an umbrella—that is, none other than a member of the so-called umbrella-foot tribe described by Pliny.[12] The Western world of the Middle Ages of course knew nothing of India, and Christians continued to propagate images of that land as a storehouse of strangeness.[13]

It was said that the seas off the coast of faraway India were home to a magnetic mountain (or, alternatively, a magnetic island) capable of sinking ships. But this claim did not originate with the fanciful poem by Petrarch cited earlier; rather we can trace it back to Pliny's *Natural History*, wherein is told the tale of two mountains near the Indus River, a tale continued in Ptolemy's *Geographia* of the second century. Book 7, chapter 2 of *Geographia* claims that there

> are said to be … ten … islands forming a continuous group called Maniolai, from which ships fastened with iron nails are said to be unable to move away, (perhaps on account of the magnetic iron in the islands) and hence they are built with wooden bolts.[14]

Although *Geographia* was not rediscovered by Europeans and translated into Latin until the fifteenth century, the claim itself had been passed down orally from ancient times and was well known both in Europe and in the Arab world.

We have already noted that Albert the Great also mentioned this story in the thirteenth century. Now, let us turn to a fourteenth-century author known as John Mandeville, and his book *Mandeville's Travels.* Written in 1356, shortly after the poem by Petrarch cited above, the book contains two references to a magnetic island or islands. One of these claims that in the land of "Prester John, the Emperor of India" (a legendary figure believed at that time to be a Christian king of the East),

> there are in many places in the sea great rocks of the stone that is called adamant,

and that ships built with iron nails dare not sail near for fear of being drawn in. The second reference is to an island called Chermes, said to be even further away than India. "In this isle," says Mandeville,

> are ships made without nails or bands of iron; and that is because of rocks of adamant that are in the sea, which would draw ships to them.[15]

This idea was perpetuated in literature right through to the late medieval period, as seen in a best-selling German folktale published in the fifteenth century, *The Voyage of Saint Brendan*. This work and also the story upon which it was based, *Herzog Ernst*, spoke of a faraway demonic sea whose waters were sticky. According to this version of Saint Brendan's tale, near this evil sea was a magnetic mountain to which many a ship, along with its seamen and cargo, had been lost.[16] And then there is *The Faerie Queene* by Englishman Edmund Spenser, which refers to "an hideous Rocke … / Of mightie *Magnes* stone" that pulled in travelers, sinking their ships.[17]

But this was still an age when the boundary between what we now think of as fiction and nonfiction was less than clear. Until the sixteenth century, Mandeville was known as the greatest traveler in Europe, and yet his book makes no distinctions between factual chronicle on the one hand and unreliable hearsay and lore of doubtful veracity on the other. And yet the people of his time believed what amounted more or less to a pack of lies, including his claims about magnetic boulders. Even the traditional tale of Saint Brendan's sea voyage was accepted as true. And so, to Europeans in the late Middle Ages, magnetic mountains that sank ships and demonic seas from which no one returned were facts. These concepts were significant, at the very least, as straightforward symbols of the fear of the unknown—of the terror that must surely lie at the unseen edge of the world's vast oceans.

Such tales of a ship-sinking mountain continued to incite the literary imaginations of European writers in subsequent periods as well. Even as late as the eighteenth century, Johann Wolfgang von Goethe mentions such a magnetic mountain in *The Sorrows of Young Werther*, noting how it was said that

> ships that sailed too close were suddenly stripped of all their ironwork [and] the nails flew to the mountain.[18]

Yet although the motif of sinking ships endured throughout the centuries, somewhere along the way to Goethe the Indian connection was lost. Magnetic mountains, yes, but no longer located in the waters surrounding the mystical land of India. The first time we see this change is actually in the mid-thirteenth century, when an Italian poet named Guido Guinicelli wrote:

> In those parts under northern skies are mountains of lodestone.[19]

It is clear that this bit of poetry was directly inspired by the fact that sailors had begun to use compasses on the Mediterranean Sea.

The voyages of Christopher Columbus and Vasco da Gama came and went, and by the sixteenth century the legendary magnetic mountain had been moved to the Arctic Circle. The *mappa mundi* created in Rome in 1508 by Johannes Ruysch places it to the north of Iceland and includes the remark that in that area

> the nautical compass becomes useless, and ships having iron in them cannot return.[20]

In 1555, a Swedish cleric named Olaus Magnus said in *A Description of the Northern Peoples* that

> in the extreme North are found magnets, by which sailors steer their course at sea.

Magnus also crafted a map of northern lands, in which we find pictured a magnetic island close to Nordkapp (North Cape), the northernmost point in Europe.[21]

On the other hand, Garcia da Orta, a Portuguese doctor who lived in Goa from 1543, explicitly denied the legend of the magnetic mountain in India. In his *Colloquies on the Simples and Drugs of India*, published in Goa in 1563, he responds as follows to the claim that ships in India have wooden nails because of a reputed magnetic mountain that would pull out nails of iron:

> These are fables, for the Portuguese never saw anything of the kind, and at Calicut, as well as all along that coast, the vessels have iron rather than wood. It is true that in the Maldive Islands they use wooden nails, but the only reason is that wood is cheaper than iron.[22]

Goethe, of course, must have understood the magnetic mountain he wrote of to be fictitious. Writing in the nineteenth century, Jules Verne introduces a fictitious mountain of his own in *An Antarctic Mystery*, only this time the scene in question is on the continent of Antarctica. And we can see why: from the fourteenth century to the sixteenth, what amounted to new and uncharted territory for Europeans shifted from India to the North Pole and its environs, and by the nineteenth century it had shifted again to the South Pole. The transition was huge, and what brought it about was extensive discovery across the seven seas. Over the course of this half millenium, Europeans conquered the Atlantic, discovered the Pacific, and brought the sea routes to and from India under their control. This accomplished, India was no longer a world of wonder and mystery, and the waters off its coasts were no longer demonic and frightening. Europeans had vanquished this terror by advancing out into the wider world with rifles in one hand and compasses in the other.

It bears remarking that Verne's placement of a magnetic mountain in Antarctica merely reflected the fact that in his day Antarctica was one of the few regions of the world that remained unexplored; it had nothing at all to do with the concept of magnetic north-south polarity. On the other hand, Ruysch's choice was based on the experience of sailors who had repeatedly found that their compasses ceased to function as they approached the Arctic Circle.

Olaus introduced the notion of a magnetic island "by which sailors steer their course at sea," and we can surmise that in so doing he was aware of the fact that magnets and magnetized needles aligned themselves geographically northward. More than this, it is likely that he introduced the magnetic island and the magnetic mountain in an attempt to explain what was already understood about magnetic directionality. The same line of thinking was expressed in other quarters in the sixteenth century, for example (according to Gilbert) by Girolamo Fracastoro, an Italian physician; and by Francesco Maurolico, a mathematician, astronomer, and refugee from the Byzantine Empire, who went so far as to claim that the point on Earth with which magnets aligned themselves was none other than Olaus Magnus's magnetic mountain itself.[23]

In direct contradiction of Peregrinus's and Roger Bacon's thirteenth-century view that it is from the poles of the heavens that the poles of the magnet receive their virtue, as well as of Ficino's claim in the fifteenth century that magnets acquire their power from the constellation Ursa Minor, the emerging facts strongly suggested that what attracted magnets was a point on the Earth itself. This represented a radical shift in the interpretation of magnetic force, which in turn led to a new understanding of the Earth and paved the way not only for the realization that the Earth itself is what influences magnets and magnetic needles but also for Gilbert's discovery that the Earth itself is a huge magnet.

## Section 2—The Magnetic Compass: Discovering the World

As discussed earlier, little is known about the way in which compasses first came into use by European sea voyagers beyond the fact that this transition took place sometime before the end of the twelfth century. But determining the timing of the compass's introduction is far less important to science history than investigating the consequences of its invention.

The magnetic compass—or, more accurately, the compass-equipped caravel, a type of sailing ship highly suited to ocean travel—is among the triad of technologies that demarcate medieval from modern times in Europe, the other two being gunpowder and mechanical printing. Because of these technologies, we look back and see on one side of this divide the history of Europe and on the other side the history of the world as a whole. These three powerful inventions were the prerequisites of the activities by which Europeans were able to widen their sphere of activity, previously confined mainly to the Mediterranean region and extending at most only to northern England and the Rhine River.

It has often been said that Francis Bacon (1561–1626) was the first to recognize the historical significance of the nautical compass. Let's take a look at the famous passage, from Bacon's 1620 work *The New Organon*, upon which this distinction is based:

> Again, it helps to notice the force, power and consequences of the discoveries, which appear at their clearest in three things that were unknown to antiquity,… namely the art of printing, gunpowder and the nautical compass. In fact these three things have changed the face and condition of things all over the globe: the first in literature; the second in the art of war; the third in navigation; and innumerable changes have followed; so that no

> empire or sect or star seems to have exercised a greater power and influence on human affairs than those mechanical things.[24]

These words have been quoted time and again—by Benjamin Farrington in *Francis Bacon*, by Paolo Rossi in *Francis Bacon: From Magic to Science*, and by Allen George Debus at the beginning of his *Man and Nature in the Renaissance*, to name just a few. But in fact the viewpoint this passage expresses did not originate with Bacon. Italian philosopher Tommaso Campanella (1568–1639) made the association between the compass and modernity in 1602 in *The City of the Sun*, where he refers to what people say about

> our present century [i.e., the sixteenth century], which has produced more history in a hundred years than the whole world did in the preceding four thousand! More books have been written in the last century than in the previous five thousand years. And what they say about our stupendous inventions—the compass, the printing press, the harquebus—mighty signs of the imminent union of the world.[25]

And Girolamo Cardano (1501–1576) made the same point still earlier, in the fifteenth century. Born in northern Italy, Cardano grew up in step with the expansion of European horizons to and beyond distant shores. In *The Book of My Life* (1576), he writes that "I was born in this century in which the whole world became known." In that same work he encapsulates the first three-quarters of the sixteenth century with observations like the following:

> For what is more amazing than pyrotechnics? Or than the fiery bolts man has invented so much more destructive than the lightning of the gods?

> Nor of thee, O Great Compass, will I be silent, for thou dost guide us over boundless seas, through gloomy nights, through

> the wild storms seafarers dread, and through the pathless wilderness.
>
> The fourth marvel is the invention of the typographic art, a work of man's hands, and the discovery of his wit—a rival, forsooth, of the wonders wrought by divine intelligence.[26]

Cardano calls the invention of typography the "fourth marvel," having in an earlier passage identified the first marvel to be none other than the discovery of the world by Europeans—or, in other words, world conquest by Europeans and European civilization. He thus grasped correctly that the significance of the compass and a small number of other inventions lay in their giving Europeans the power to discover and then to dominate the globe.

Given everything that Cardano had seen in his lifetime, this is not at all surprising. He and his contemporaries were shaped by a powerful thrust of events, from Portugal's conquest of the Indian Ocean in 1509 to the occupation of Goa and Malacca (1510–1511); to the discovery of the Pacific and circumnavigation of the globe by Ferdinand Magellan's Armada de Molucca (1519–1522); to Spain's forays into the New World, beginning with Christopher Columbus and continuing with Hernando Cortes's subjugation of the Aztec Empire in 1521 and Francisco Pizarro's conquering of the Inca Empire in 1533. In 1542, Jean Fernel, a French physician only four years older than Cardano, wrote of global circumnavigation, the discovery of the New World, printing technology, and firearms as the most representative developments of the new age.[27] While he did not mention compasses explicitly, their presence is obviously implied, and we can safely assume that their significance was likewise recognized by most or all intellectuals of the day.

In this context, Bacon's contribution was remarkable in that he grasped the particular significance that these new technologies had on European thought and learning. We can see more clearly what Bacon was really getting at if we reread the previously cited excerpt from *The New Organon* together with a passage that follows it:

> They [i.e., the ancient Greeks] knew only a fraction of the parts and regions of the world,... knew nothing in Africa beyond the nearest part of Ethiopia, nothing of Asia beyond the Ganges, much less the territories of the New World, even by report or consistent and believable rumour. In fact most climates and zones, in which uncounted nations live and breathe, were declared uninhabitable; and the travels of Democritus, Plato and Pythagoras, which certainly did not take them far from home, were celebrated as major undertakings. But in our time large parts of the New World and the farthest parts of the Old are becoming known everywhere.[28]

Because of the explosive increase in ocean voyages after Columbus and Da Gama, Europeans of the sixteenth century were able to set eyes on lands, peoples, plants, and animals—on an entire world, in fact—that had been completely unknown to ancients such as Aristotle and Ptolemy.

Indeed, ancient literature provides example after example of mistaken claims about geography that continued to be read and believed throughout the Middle Ages. Aristotle asserts in *Meteorology* that "Beyond the tropics no one can live."[29] Ovid, a Roman of the first century, claims in *Metamorphoses* that a god created the Earth with five zones—two polar, two temperate, and one tropical—but that the "central zone is so hot as to be uninhabitable."[30] Writing not long after that, Pliny states in *Natural*

*History* that "the middle portion of the lands, where the sun's orbit is, is scorched by its flames and burnt up by the proximity of its heat."[31] In the third century we find the claim by Origen, an early church father, that the ocean "is impassable to men."[32] And to take just one more example: Macrobius, writing in the fifth century in his *Commentary on the Dream of Scipio*, asserts as follows:

> The earth ... is divided into regions of excessive cold or heat, with two temperate zones between the hot and cold regions.... The belt in the middle and consequently the greatest, scorched by an incessant blast of heat,... is uninhabited because of the raging heat.[33]

But then came the accounts of travelers in the age of discovery. In 1522, an Italian named Antonio Pigafetta completed a round-the-world journey with Magellan, returning alive to tell the tale: "we sailed for sixty days of rain to the equinoctial line," he says, adding that this "was a thing very strange and uncommon, in the opinion of the old people."[34] And Gonzalo Fernández de Oviedo y Valdés, a Spaniard who ventured to the New World, made this comment in 1535: "Pliny said that the tropics and the equatorial region were uninhabitable, but he and others who at one time expressed this belief were mistaken."[35] Francisco López de Gómara observed in 1552 that

> all philosophers in ancient times believed that both the intense heat [of the tropical belt] and the vastness of the seas made it out of the question for people of our hemisphere to travel to the hemisphere of antipodal peoples. But now that traveling to those lands has become something ordinary for we Spaniards and we know the truth based on our ample experience of having set foot there, we find our firsthand experience to be in conflict with philosophy.[36]

And so, based on experiences accumulated throughout the sixteenth century, people of the age of exploration came to reject completely the beliefs that had been passed down from ancient times. Concerning cinnamon, a spice native to East Asia, Garcia da Orta, who lived many years in India, wrote that "ancient writers saw this spice after it had come from such a distance that they could not have had a correct notice of it, and as the price was very high there arose a thousand fables which Pliny and Herodotus repeat." And in 1563, Da Orta wrote the following about lodestone:

> A stone which God created with such strength may well have these other gifts. But to say that when one is put behind a woman's head unknown to her, and when she is asleep, she will embrace her husband if he [*sic*] hath been faithful, and if the contrary will avoid him, is a thing I am unable to believe, though they tell me that it is stated by writers of authority; for they say the same of certain herbs and we know that to be nonsense.[37]

José de Acosta, a Spanish priest and missionary to Peru who has been called "the Aristotle of the New World," relates the following in his 1590 masterpiece *Natural and Moral History of the Indies*:

> I will confess here that I laughed and jeered at Aristotle's meteorological theories and his philosophy, seeing that in the very place where, according to his rules, everything must be burning and on fire, I and all my companions were cold. For the truth is that no place in the world is there a calmer and more moderate region than that under the equator. But there is great diversity in it, and it is not the same in all places. Parts of the Torrid Zone are very temperate, as in Quito and on the plains of Peru; parts are very cold, as in Potosí; and parts are very hot, as in Ethiopia, Brazil, and the Moluccas.[38]

And so, facts gradually dealt a crushing blow to the illusion, held early in the Renaissance, that the ancients had been unequivocally superior human beings upon whom God had bestowed worldly truth. Underlying the words of Bacon cited above seems to be the confident assessment that the ancients were people of only scant wisdom. As Spaniard Juan Luis Vives put it in 1531, "if we only apply our minds sufficiently, we can judge better over the whole round of life and nature than could Aristotle, Plato, or any of the ancients."[39] Fernel expresses a similar sentiment: "This age of ours sees art and science gloriously re-risen, after twelve centuries of swoon. Art and science now equal their ancient splendour, or surpass it. This age need not, in any respect, despise itself, and sigh for the knowledge of the ancients."[40] Bacon follows suit when he declares Greek philosophy to be "the words of idle old men to callow youths."[41] *The New Organon*, created as part of a planned larger work titled *The Great Instauration*, contains in its 1620 edition an engraved title page depicting a ship setting sail for the Atlantic between the Pillars of Hercules—mythical pillars associated with two large promontories that straddle the Strait of Gibraltar—as a symbol of the move by Europeans out into the open sea after having been confined until then to the Mediterranean. The illustration bears a Latin caption taken from the Old Testament's Book of Daniel that can be translated as "Many shall pass through, and knowledge will be increased." In the book itself, Bacon elaborates that "it is in the fates, that is, in providence, that the circumnavigation of the world … and the increase of the sciences should come to pass in the same age."[42]

Moreover, Acosta points out that it is none other than the advent of the magnetic compass that marks the superiority of the moderns over the ancients:

> I find no mention made of the use of the lodestone and compass among the ancients, nor do I believe that they were aware of them; and without knowledge of the compass, it is obvious that there is no possibility of crossing the ocean.... It is astonishing that something with such excellent qualities as the lodestone should have been unknown to the ancients for so long and to have been discovered by modern men. That it was unknown to the ancients is clearly seen in Pliny, who, although he was a conscientious historian of natural things, when he tells wonders about the lodestone never says a word about the functions and uses that it has, the most remarkable of which is that any iron that it touches turns toward the north. Nor did Aristotle speak of it, or Theophrastus, or Dioscorides, or Lucretius, or any historian or natural philosopher that I have read, although they do discuss the lodestone. Nor does Saint Augustine mention it, although on the other hand he praises at length the wonderful qualities of the lodestone in the books of *The City of God*.[43]

The compass, firearms, and printing technology jolted Europeans out of their illusory, almost cult-like worship of the ancients, clearly demonstrating to them that new learning was possible and planting the seed from which the notion of limitless progress would sprout.

We can only be impressed by the timing of two parallel sets of events. Lorenzo de' Medici, patron of the Accademia Platonica (Platonic Academy in Florence), and two central figures in the academy, Marsilio Ficino and Pico della Mirandola, saw their last days on Earth—these three men passed away in 1492, 1499, and 1494, respectively—just as Columbus and shortly after him Vasco da Gama were pushing the Earth's horizons forward: Columbus first set sail in 1492, and Da Gama became the first to discover a sea route to India in 1498. The enormous impact of these voyages on Europeans, with people from all social strata recognizing the existence of and taking an interest in a whole new world, was not to be grasped until the second half of the sixteenth century. But be that as it may, the journeys by sea that Columbus and Da Gama undertook were pivotal, dividing the Renaissance into two distinct phases—the fourteenth century (quattrocento) and the fifteenth century (cinquecento)—and at the same time marking the transition from the Middle Ages to modern times.

## Section 3—Columbus and the Discovery of Magnetic Declination

As travelers ventured further out across the globe, they came to realize that in some locations compass needles failed to point due north—that is, that at certain points on the Earth's surface, compass direction veered from the geographical meridian and from the horizontal plane. These two types of deviation are referred to respectively as declination—meaning the angle of compass-needle deviation from the geographical meridional plane—and inclination, meaning the angle of the dip in the needle below the horizontal plane. The discoveries of declination and inclination were by-products of compass use itself, and in time they would lead to the understanding that compass directionality was caused by the Earth rather than derived from the heavens. This realization would lead in turn to the discovery of a whole new Earth—not the Earth known previously as an inert lump of soil but the Earth we now know to be itself a magnet, something active and full of power.

But precisely when it was that Europeans discovered declination—in other words, the fact that the degree of magnetic variation was some number other than zero—is something of a mystery.

Until the beginning of the twentieth century, Columbus was credited with both the discovery of declination itself and the knowledge that the degree of declination varied from one location to another. In 1879, for example, Johann Christian Poggendorff wrote in *Geschichte der Physik* (History of physics) that

> Christopher Columbus was the first person in Europe to observe the phenomenon of declination and differences in declination at different points on the Earth's surface.[44]

And Timoteo Bertelli in 1892 and Park Benjamin in 1895 likewise named Columbus as the discoverer of these phenomena. Even as late as 1962 (Charles Singer) and 1970 (Fritz Krafft), there were writers who claimed that Columbus had discovered declination on his first voyage.[45]

In contrast, in an 1849 work called *Cosmos*, Alexander von Humboldt writes that declination itself had been discovered before Columbus's time but adds that

> Christopher Columbus has not only the merit of being the first to discover *a line without magnetic variation*, but also of having excited a taste for the study of terrestrial magnetism in Europe, by means of his observations on the progressive increase of western declination in receding from that line.[46]

This understanding was more or less echoed by Paul Fleury Mottelay in 1922.[47]

But in the 1930s, A. Crichton Mitchell and N. H. de Vaudrey Heathcote took a critical look at the evidence for such claims, carrying out positive research that was extremely detailed.[48]

So what is the evidence both for and against a Columbus connection? Let me state the conclusion from the outset: it does in fact seem that the phenomenon of magnetic declination was known to manufacturers of sundials and compasses in the mid-fifteenth century.[49] The sundials I refer to here (see Figure 11.1) were produced in Germany in the fifteenth century. They were quite popular, were portable and designed for travelers, and were equipped with magnetic compasses. Using one of these sundials, it was theoretically possible to find true north by observing the direction of the shadow formed when the sun crossed the meridian (the time of day when shadows are at their shortest) and then determine the degree of magnetic declination by comparing true north with compass direction. In those days, however, discrepancies between compass direction and true north were usually attributed to variations in the way in which the compass needle had been magnetized; to differences in the location where the magnet used for magnetization had been mined; to nearby iron-ore deposits; or even to simple errors of measurement. In time, though, variations in magnetic declination became a recognized phenomenon, as they were observed at many locations. But there are no explicit records to inform our understanding of the specific process through which craftsmen became aware of this phenomenon, nor are there records indicating whether or not any actual numerical readings were taken.

Evidence concerning knowledge of declination prior to Columbus's first voyage is, as we've just seen, sketchy, and certainly there are no clues as to who might have been first to discover the phenomenon. Even so, Mitchell makes a convincing case for the conclusion that Columbus was not the one. Let's take a look at some of the details.

First, Mottelay's assessment relies upon a description of conditions on a particular day—September 13, 1492, to be exact—during Columbus's first voyage, when,

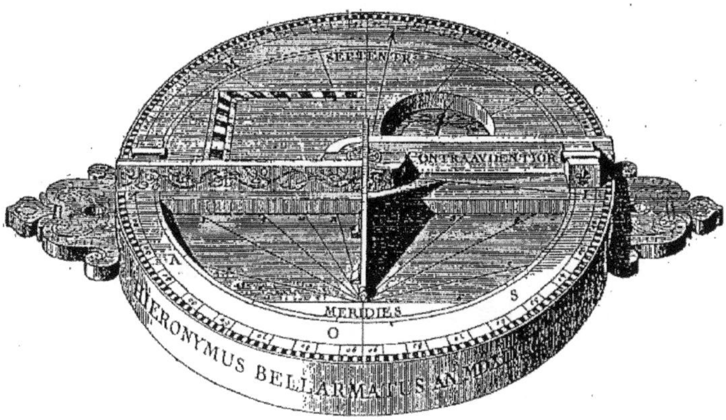


Figure 11.1: Portable sundial with magnetic needle. *Source*: Hellmann, "Die Anfänge der magnetischen Beobachtungen."

> 2 1/2 degrees east of the island of Corvo [in the Azores], the magnetic variation changed and passed from N.E. to N.W.

Humboldt, too, cites this description. The quote, however, seems to come from Washington Irving's 1828 biography of Columbus. In other words, it is not from a primary source.[50] The actual sea logs from that journey in fact have been lost; the closest we have to an authentic diary are, first, a summary prepared by Bartolomé de las Casas—using hand-copied fragments and secondhand excerpts from the original sea logs—as background research for his *History of the Indies*; second, quotes contained in a biography written by Columbus's son Ferdinando in the 1530s; and third, letters written by Columbus himself. It is thought that Ferdinando Columbus would have had access to his father's logs. Regarding the date in question, Ferdinando records as follows:

> On 13th September, he found that at nightfall the needles of the compass varied half a quarter to the northwestwards, and at break of day half a quarter more.[51]

Las Casas, however, simply notes the following:

> On this day, at the beginning of the night, the needles declined to the northwest, and in the morning they declined a trifle to the northeast.[52]

Ferdinando's reference is much the same, but he follows it by elaborating that his father

> understood that the needle did not point at the North Star but at some other fixed and invisible point. This variation no man had observed before and therefore he had occasion to be surprised at it.[53]

But the biography by Ferdinando was written when the fact of declination was already widely known. Moreover, his description is biased by his desire to magnify his father's achievements and is therefore unreliable.

In these three descriptions, the directions of the variation differ. While it is true that Columbus recorded the fact of declination, he does not seem to have recognized it as a new discovery. His attention was focused not on declination itself but rather on its variation with the movement of his ship.

Mitchell, in arguing that Christopher Columbus was not the first to discover declination, presents as one piece of evidence an excerpt by Ferdinando of his father's record for May 20, 1496—when Columbus was on his second voyage. According to Ferdinando, somewhat to the west of the Azores

> the Flemish compass-needles northwested a quarter as usual, and those of Genoa … did not except slightly.[54]

Let's pause here for a moment to discuss compass design. In general, a compass consists of a magnetized needle capable of spinning horizontally and affixed to a compass card that spins with it. On the compass card are perpendicular lines drawn east-west and north-south, forming four quadrants that are further divided into eight parts each, for a total of 32 radii, thus enabling the user to determine his bearing relative to the direction in which the compass needle is pointing. The term "quarter" as used by Ferdinando in the above quote signifies one of the angles on the compass card formed by two adjacent radii; in other words, $360° \div 32 = 11°1/4$.

The Genoa compass, used in the Mediterranean, was designed so that its magnetized needle aligned with the radius marked as north on the compass card. For this reason, a northward reading on the Genoa compass was skewed away from geographical true north to precisely the extent of magnetic declination. In contrast, the Flemish compass was designed so that the needle was affixed to its compass card pointing one "quarter" east of the north radius. As a result, a northward bearing as shown by the Flemish compass was in fact in almost perfect alignment with geographical true north, at least for sailors in northwest Europe. This of course meant that compass designers in this part of the world realized that magnetized needles in nearby waters did not point straight north but instead were always skewed approximately one "quarter" to the east.

One fact made certain by the above is that on his second voyage Columbus brought with him both Flemish and Genoa compasses. Since he left for that voyage in September of 1493, we also know that Flemish compasses were available in the Western world by that time, meaning that compass manufacturers in the Netherlands and surrounding areas already knew of magnetic declination. Now suppose instead, for just a moment, that Columbus had discovered declination on his first voyage. This knowledge would not have reached southern Spain until sometime after his return home on March 15, 1493. This would have left a mere six months before Columbus's subsequent departure, during which knowledge of magnetic declination would have had to have spread all the way to the North Sea coast, compass manufacturers would have had to have incorporated this knowledge into new designs, and these new compasses would have had to have arrived in Spain. Considering the speed of information transmission and the pace of technical innovation of the day, this scenario is highly implausible. And this is to say nothing of the fact that there is simply no evidence to suggest that Columbus had come back from that first journey trumpeting news of such a discovery. Actually, the details of that voyage were closely guarded and appear to have been kept as a Spanish state secret. And so it is only natural to conclude that sailors and compass makers in northwest Europe already knew of magnetic declination before Columbus first set sail in 1492.

But this is not to say that Columbus made no contribution to our understanding of declination. He understood that declination occurred and that its degree varied from one location to another, and he went far beyond mere understanding when he noted and recorded the specifics of this phenomenon at a particular point on the globe. He left to history the information that near the island of Corvo in the Azores archipelago declination shifted from northeast to northwest, with the implication that somewhere in the process compass needles pointed briefly to geographical true north. While the accuracy of what we know about Columbus's sea logs remains in question, we must credit the man with having provided the Western world's first record (or at least the first known and remaining record) of quantitative data concerning declination.

## Section 4—The Quantification of Magnetic Declination

Columbus's measurements of magnetic declination were few and imprecise, and his descriptions of what he observed were vague. It is therefore not particularly meaningful to attempt to decipher his work in this area in order to determine whether or not he was the first either to discover declination or to discover that it was variable. More important in the history of ideas about magnetism is an examination of how the understanding of declination changed and deepened from the time of its discovery to the point when it was discovered that the Earth itself was a magnet. For here, among these changes, we can trace the birth and growth of the science of geomagnetism.

We can say that the study of geomagnetism was born after Magellan's return to Europe in 1522 from his round-the-world voyage. The homecoming of Magellan's fleet marked the end of travel on the open seas as an adventuresome and somewhat reckless enterprise engaged in by a few ambitious individuals—the kind of enterprise that had begun with the likes of Da Gama and Columbus. Europe's sights were now set on the establishment of trade routes, with voyages to the New World and East Asia—that is, travel along the Atlantic sea route to the West Indies and the route to India around Africa's southern tip—taking place more and more frequently. Sea travel became serious business, and nations threw their weight behind it. It would no longer do to have sailors groping around hoping to find land; it now mattered a great deal how quickly and reliably ships could reach a certain destination, and innovations in navigational technology to improve safety and reliability were a pressing concern. For this reason, increasing the accuracy and performance of compasses was of utmost importance, and here is where geomagnetic research really came into its own.

Of particular interest was the fact that, when traveling west from Europe, compasses would start out showing declination toward the east, but the degree of this bias would fall steadily until it reached zero in the vicinity of the Azores and then westward bias would set in, increasing in degree as the voyage continued. At the time of its discovery, these peculiarities of the phenomenon convinced people that there was a simple connection between declination and longitude and that therefore measurement of magnetic declination was closely and directly bound up with determining longitude at sea. In particular, two of the greatest Atlantic sea powers of the day, Spain and Portugal, realized that carrying out precise and systematic measurements of declination across a wide expanse of ocean was both crucial and urgent. And both countries also had a strong additional motivation to conduct measurements of longitude. In 1494, the two nations signed the Treaty of Tordesillas, whereby they agreed to divide all overseas lands between them, designating the north-south meridian point 370 leagues west of Cape Verde as the border between their two claims. Executing this treaty would require an accurate measurement of longitude.

Francisco Falero (years of birth and death unknown) was the first to create in print a practical guide to the measurement of magnetic declination. Though Portuguese, Falero was a member of the Spanish navy, and early in the sixteenth century he wrote (in Spanish) the *Tratado del Esphera y del Arte del Marear* (Treatise on the sphere and the art of navigation), a work that can only be considered the first-ever true manual of seafaring. The handbook was printed in 1535 but may actually have been written sometime before 1519. Though it was produced by means of the printing press, only one copy is now extant, and this resides in Madrid, at the National Library of Spain. Copies of the manual appear to have been scarce even in its day, as there are almost no references to it in any other literature on nautical science; presumably it was a classified document of which only a small number of copies were produced, for the Spanish navy. Despite its low profile, Falero's book is valuable because of the extreme clarity with which it relates the importance that people of his day placed on awareness and knowledge of declination. In modern times, the section of Falero's manual devoted to declination was included in an 1898 work called *Rara magnetica*, edited by Gustav Hellmann; and an English translation was published in 1943 in the journal *Terrestrial Magnetism and Atmospheric Electricity.*[55] Let's begin with an excerpt from part 2, chapter 8, "On the northeasting of the needles," in which declination is explained:

> The northeasting and northwesting of the needles are nothing else than their deviation from the meridian in which they are. They do not show this exactly except when they seek accurately the pole. And they seek this exactly, according to navigators, only when they are in the meridian of the islands of the Azores, and the most precise seek it in that of the Island of Corvo, according to the experience of some.... Accordingly you are to know that sailing from the meridian of the Island of Corvo or any other of the Azores in which the needle points exactly towards the pole, going toward the west the needles decline to the northwest, and sailing from the same meridian toward the east, they decline to the northeast.

Obviously this explanation includes knowledge that had first been discovered by Columbus. But Falero goes further than this by offering the hypothesis that there is a straightforward relationship between declination and longitude. If one travels either east or west from the island of Corvo, maintaining a steady latitude, up until 90° in either direction, declination continues to increase (moving either eastward or westward as the case might be) at a constant rate. At 90°, declination peaks, thereafter gradually falls as the journey continues beyond that point, and then returns to zero—reliably pointing due north—when the traveler has gone 180° away from Corvo. Falero then makes this case for the importance of measuring declination:

> The northeasting of the needles causes navigators many doubts, from which they may be freed by knowing precisely how much the needles northeast or northwest. In addition to this, other advantages will follow, such as knowing exactly in what direction they are sailing. Knowing this they will follow exactly their courses without error or wandering, and also it will help much to a knowledge of the longitude in which they are navigating.

This is the first time we find in print the notion that knowledge concerning declination could help in determining longitude. Here we see the intimacy of the connection between navigational technology and the science of geomagnetism expressed in the most eloquent of terms.[56]

Falero follows this with a detailed explanation of his method for accurately measuring declination. This basically involves using a sundial to first get an accurate reckoning of true north and then measuring the angle between that direction and the direction of the compass needle. His explanation makes for difficult reading, the technical details centering on an elaborate discussion of the several ways to accurately determine true north. His first method is to assume that true north is the direction in which a shadow is cast at the time of day when the sun crosses the meridian, when shadows are at their shortest; the second is to assume that true north is to be found at the bisector between the direction of shadows cast in the morning and in the afternoon, at two times of day when the shadows are of equal length; and the third is to assume that true north is to be found at the bisector between the direction in which the sun rises and the direction in which it sets.

The next piece of literature we encounter concerning declination is the set of log books created from 1538 to 1541 by João de Castro, a man who in Renaissance Portugal displayed outstanding talent as a naval sailor, as a scientist, and as a writer. These log books form a record of very reliable declination measurements taken in a planned and systematic way. The measurements were taken at various points off the west coast of Africa, off the continent's southern tip, in the Indian Ocean, and near the Red Sea. Some of the locations included in this survey were in the waters near Lisbon, the Canary Islands, the Tristan da Cunha Group, Cape Agulhas, and Goa[57]

Castro based his method on almost the same principles as Falero. Castro defines $\theta_1$ as the angle formed by (a) north as indicated by the compass needle and (b) the direction of the shadow cast by the sundial's gnomon (pointer) when the morning sun has reached a certain height (see Figure 11.2). Similarly, the angle formed by compass north and the gnomon's shadow when the afternoon sun has fallen to the same height in the sky was defined as $\theta_2$. Then, declination $\delta$ is obtained by the formula $\delta = (\theta_1 - \theta_2) \div 2$. (With observations made in the southern hemisphere, north and south are reversed, the direction of the sun's movement across the sky is reversed, and accordingly $\theta_1$ and $\theta_2$ change places in the equation.) In each location observed, Castro takes eight measurements, two (morning and afternoon) for each of four points in the sky. He thus

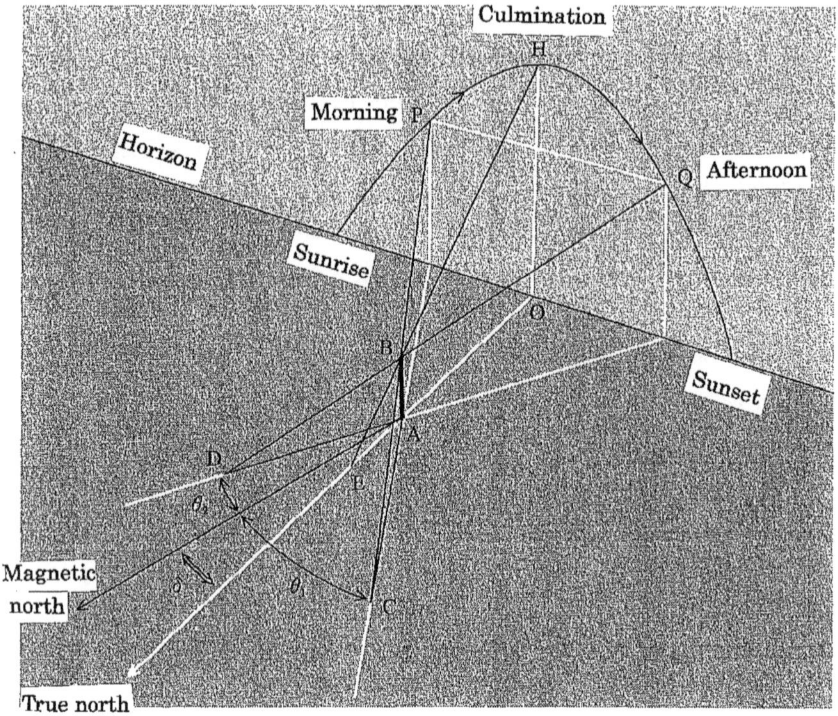


Figure 11.2: Determining declination.

obtains four separate values for declination in each geographical location, and he finds that the four are almost identical, varying by only a quarter of a degree at most. Castro, who was Portugal's greatest military hero and who rose to become governor-general of India, was quite definitely, as Hellmann puts it, "the most important representative of scientific maritime investigations at the end of the epoch of discoveries."[58]

On the basis of his own measurements, Castro discovers that from one latitude to another, "in one and the same meridian, the needle can northwest and northeast more or less." And so he is able to conclude that longitude and declination do not in fact have a simple, unique mathematical relationship. He muses that "such variations are caused by particular and inherent mysteries concealed by mighty Nature in its vast and secret workshops."[59]

This assessment dissuaded the Portuguese from further attempts to study the determination of latitude by means of declination.

Castro also relates in his log books an interesting episode about something he noticed after losing, near Goa, India, the magnetized needle from the compass fitted to the sundial he had brought with him from Portugal. He uses instead a needle made in Germany that just happens to be the right size.

> Before magnetizing this small needle I placed it on the pivot and socket where it should swing, straightening it along the meridian-line of the dial-plate, and taking note where the shadow of the style cut the circle, and promptly removed it, and the pilot magnetized it; after it was magnetized I replaced it, and straightening it as before, with the meridian-line or the north-south of the dial-plate, the shadow of the style cut the circle in the same point it was cast at first before its magnetization.
>
> This made me very thoughtful, for the watch from which I removed the needle was made in Germany, where certainly the needle was magnetized with the loadstones of that country; the stone now used by the pilot to remagnetize the needle was from this coast of India, and despite the fact that the regions are so different the property of the loadstones seems to be one and the same.[60]

This is history's first direct confirmation of two important facts. The first is that a chief property of the magnet (its ability to magnetize iron) is the same in both Indian and German magnets and so by extension must be the same everywhere in the world. And the second is that the degree of declination of a given compass needle is not influenced by the origin of the magnet used to magnetize that needle. Moreover, Castro's recognition of sameness between European and Indian lodestones was a sign that Westerners had outgrown the old tales depicting India as a mystical land populated by monstrous creatures.

We have just been discussing measurements of declination that were taken at sea, and as we have seen, these were mainly carried out by Spain and Portugal, two of the greatest sea powers of their day. But what is known about measures of declination taken on land? The first such measurement appears to have been taken by a man named Georg Hartmann (1489–1564) using a compass-fitted sundial. Hartmann was an instrument maker in Nuremberg who later became the vicar of Saint Sebaldus. In a letter to Duke Albrecht of Prussia, dated March 4, 1544, Hartmann writes that

> the magnet-stone is still more wonderful in that the needle stroked by it does not point due north, but turns away from the true south-north line and points eastwards, in some countries by 6°, as I myself have found and studied, at the time in Rome,... but here in Nuremberg I find that this deviation is 10°, and in other places more or less.[61]

In fact, this same letter mentions another scientific first: Hartmann also makes the first known reference to the phenomenon of inclination, or dip, and moreover discusses its measurement. But this historically valuable document disappeared for almost three centuries, hidden away in an archive at Königsburg until it was found in 1831. (In 1581, Robert Norman published *The Newe Attractive*, in which he records his 1576 discovery of inclination; his story will be examined in the next chapter.)

And so, urged on by the establishment of sea routes west from Europe to the New World and east to India and East Asia around the southern tip of Africa, knowledge of declination advanced remarkably far in the mere half century after the voyages of Columbus and Da Gama.

## Section 5—The Earth's Magnetic Pole: A New Concept

We have just seen how knowledge of declination continued to deepen during the first half of the sixteenth century. The concept of the Earth's magnetic pole developed as a direct extension, so to speak, of this line of inquiry, and it was decisively important in the understanding of the Earth. The starting point was Martin Cortes of Aragon and his originally Spanish-language work *Brief Compendium on the Sphere and Art of Navigating*, which was published in Seville in the mid-sixteenth century.[62]

Chapter 3, "On the virtue and property of the lodestone," does not surpass what Augustine or Pliny wrote on this subject, except for the addition of newer information on places in the world where lodestones were produced and of discussions of magnets' north-pointing directionality and the use of compasses for ocean navigation. Chapter 4, "Of making the needle or navigating compass," explains the method of manufacturing nautical compasses.

Chapter 5, on declination, is titled "On the northeasting and northwesting of the needle" and begins as follows:

> Many and various are the opinions which I have heard and read in some modern writers regarding the northeasting and northwesting of the needles and in my opinion none of them is exact and few of them hit the mark.

We can infer from this that the phenomenon of declination was already quite well known by that time to people involved in ocean travel. But here Cortes also implies something else. What he calls inexact and off-the-mark opinions are hypotheses that compass needles point toward a single, specific spot in the heavens. He takes an Aristotelian stance in refuting this belief. In Aristotle's view of the cosmos, the Earth sits in quiet repose at the center of a series of concentric spheres that house all of the planets, the sun, and so forth. The outermost of these spheres houses the fixed stars, and the prime mover of the universe causes this sphere to make one complete rotation per day. Therefore, Cortes argues, if we posit that the attractive point is located in the heavens and

> moves to the movement of the *primum mobile*, then the needle in consequence will make the same movement in 24 hours. This is not found to be the case; therefore, this point is not in the mobile heavens nor in the pole because, if it were in it, the needle would neither decline to the northeast nor to the northwest.

Reading this argument, one would expect Cortes's conclusion to be that the "attractive point" exists on Earth itself, but—by some extraordinarily difficult-to-follow logic—he instead conjectures that the attractive point is to be found in the opposite direction, outside the sphere holding the fixed stars. He explains it as follows:

> The attractive point is somewhat separated from the pole of the world and from it extends a thread which always has to pass through the north-south of the needle. And the needle being in the meridian of the attractive point which passes through the pole, it will indicate the pole, and outside of it, it will decline to the northeast or northwest, departing from the true meridian which starts from the pole of the world.

At all events, if one accepts the premise that compass needles on Earth all point to one particular and fixed point that exists at a specific distance from the pole of the world—that is, in the heavens—then it follows that magnetic declination should have a specifically fixed value at each point on the Earth and a predictable mathematical relationship to longitude. Thus, Cortes's arguments provide renewed support for the idea that declination measurements can be used to determine longitude.

The greatest turning point in the history of geomagnetic studies came with the work of Gerard Mercator (1512–1594), a Flemish mathematician and cartographer famous for his method of map projection. It was Mercator who first located on the Earth the spot to which magnetized needles pointed, proposing the concept of the *polus magnetis* of the Earth. In a letter dated February 23, 1546, he offers the following as his answer to the question "In what place … that point lies, which the magnet so greatly seeks":

> In the first place, it has been found by experience that at one and the same place, the magnetic needle declines from the true north by the same amount. The point, therefore, can certainly not be in the heavens, because, since every point in the heavens, except the poles, is subject to a rotational motion, the needle, owing to the diurnal rotation of such a point in the heavens, would necessarily wander now this way now that, and hence decline alternately to the east and to the west, which is contrary to experience. On the Earth, therefore, which remains fixed, this point is to be sought.[63]

Here was something very, very different. While Peregrinus and Roger Bacon had sought the place that influenced the behavior of magnets in the pole of the heavens, Thomas Aquinas and Ficino had sought it in the North Star, and Cortes had sought it outside of the heavens, now for the first time a new thinker sought this point—the point to which anything magnetic was attracted—on the Earth. In terms of human understanding of the Earth's magnetic field and, indeed, of the Earth itself, this was a decisive shift.

In arguing as he does, Mercator assumes *a priori* the diurnal rotation of the celestial sphere on the one hand and, on the other, the constancy of the declination. Although Nicolaus Copernicus's *On the Revolutions* had already been published in 1543, his heliocentric theory was still barely approved in Mercator's time. And until Henry Gellibrand (1597–ca. 1637) discovered the phenomenon of secular variation of declination in 1634, the constancy of declination at any given location was not questioned. Aside from these two issues, Mercator's argument is much clearer than Cortes's and is the superior of the two in terms of both its logic and its conclusion.

Moreover, Mercator goes so far as to specify a location for the magnetic pole on the basis of his quantitative observations. He takes 9° east, the declination measured on the island of Zealand in Denmark and on this basis estimates, according to some rather obscure logic, that declination in Danzig (Gdansk) is 14° east. He then takes two great circles—the circle that crosses its local meridian with 9° on Zealand and the circle that crosses its local meridian with 14° at Danzig—and finds that the two circles intersect at 168° longitude and 79° north latitude. However, because this longitude takes as the prime meridian the one passing through the island of Corvo—roughly 31° west of the Greenwich meridian and in terms of modern Greenwich longitude equivalent to about 137° east—the calculation locates the Earth's magnetic pole in the Bering Strait.[64]

Of course, this calculation presupposes that a magnetic pole exists and that a magnetized needle anywhere on Earth would point to such a pole—in other words, that there are magnetic meridians that, just like the geographic meridians, pass through two fixed points on the Earth, forming a great circle. These premises, however, had not been corroborated by actual experience. To the contrary, measurements taken by Castro contradicted them. In this connection, it is worth noting the contribution of an Austrian mathematician named Georg Joachim Rheticus (1514–1574). Rheticus went to Poland in 1539 in order to learn heliocentric theory directly from Copernicus. Rheticus also visited the city of Danzig, where he took a measurement of declination and came up with a figure of slightly greater than 13°. This is quite close to the 14° arrived at by Mercator,[65] despite the fact that Mercator's calculation did not have a sound basis.

Mercator later recalculated his value for the magnetic pole using a slightly different figure for declination at Zealand, and this time he located the magnetic pole at a longitude of 180°and a north latitude of 73°30'.[66] (See Figure 11.3) In maps produced by Mercator in 1569 and 1595, the cartographer drew in a so-called *polus magnetis*, a magnetic pole in the form of a mountainous island,[67] in the sea off northern Japan, between north latitudes 70° and 80°. The specific numerical values he came up with are themselves of no particular importance; what is instead historically valuable is his proposal of the concept that the Earth had something called a magnetic pole.

And so the mystical "magnetic mountain" was reinvented as the geoscientific "magnetic pole."

At first, both notions coexisted and overlapped. In a work published in 1580, Antwerp mathematician Michiel Coignet writes that

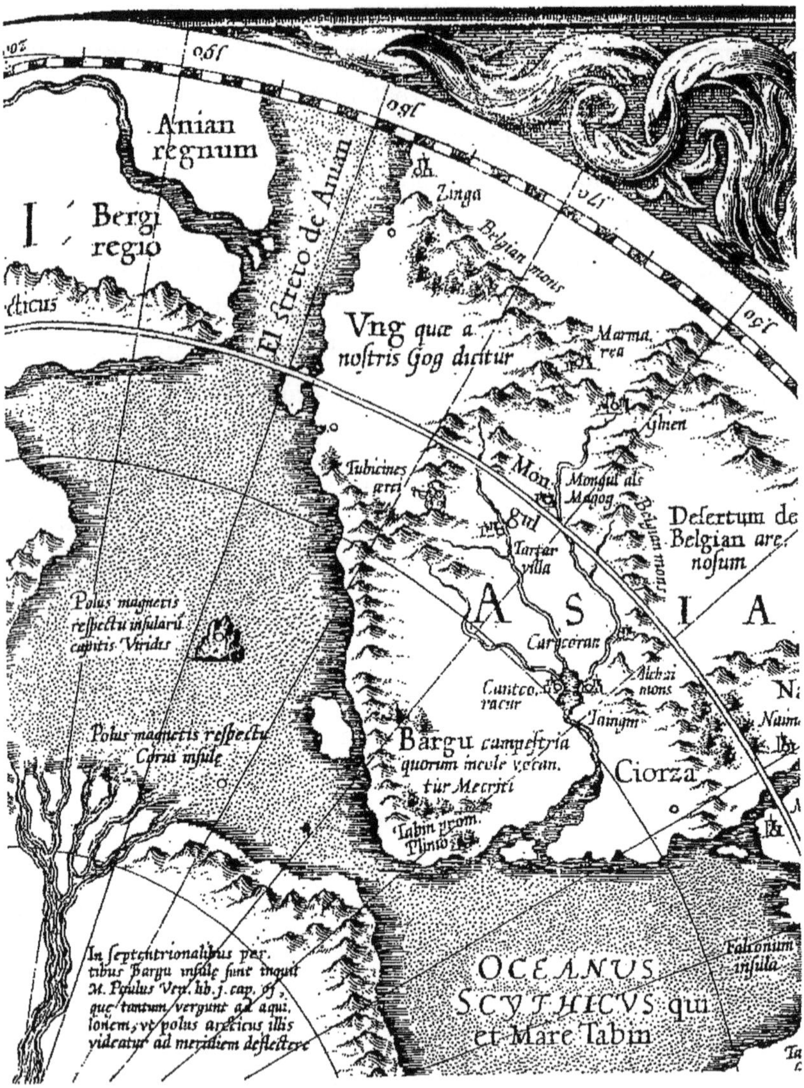


Figure 11.3: The magnetic pole (*Polus magnetis*) on Mercator's 1595 map. *Source*: Nordenskiöld, *Facsimile-Atlas to the Early History of Cartography.*

Mercator locates at 16°30' from the North Pole "a very prodigious rock and mine of loadstone, towards which all other loadstones in the whole world are attracted."[68] Clearly, the magnetic pole proposed by Mercator was regarded along the same lines as the older image of the magnetic mountain.

Mercator's research brought him very close to developing the concept of a terrestrial magnetic field; one might say that, at this point, science was just one step away from Gilbert's discovery of the fact that the Earth itself was a single giant magnet. In any case, it is baffling that in the year 1600 Gilbert (and, near the end of the nineteenth century, Park Benjamin as well) completely ignored this work of Mercator's.

* * *

The use of the nautical compass during the age of exploration brought about the discovery of magnetic declination, which, together with the discovery of inclination, had huge implications for our understanding of the Earth. (We will discuss inclination in the next chapter.) As knowledge unfolded, the ancient and medieval notions of a "magnetic mountain" in a place vaguely referred to as India gave way to the sixteenth-century idea that such a mountain existed in the Arctic Circle, which in turn led to the concept of a "magnetic pole." This discovery made it possible for Gilbert to realize that the Earth itself constituted a single gigantic magnet. Late in the sixteenth century, Giordano Bruno remarked as follows in *Essays on Magic*:

> It is not easy to explain why magnetic attraction occurs at the pole of the earth, especially if what some say is not true, namely, that in that region there are many large magnetic mountains.[69]

Although this excerpt implies skepticism about the existence of magnetic mountains per se, it nonetheless makes clear that by this point in history the shift to an understanding that magnets got their power from a terrestrial pole was more or less complete. In other words, some thinkers who came just before Gilbert were a hair's breadth away from anticipating his understanding of the Earth itself as a magnet.

Moreover, this entire intellectual process entailed a movement away from the medieval view of science as an exercise in textual interpretation—a kind of exegesis—to a view of science as an exercise in the observation and measurement of actual natural phenomena, with mariners and craftsmen and soldiers now participating in scientific activities. And that shift was to take even clearer shape with the work of the man we are about to meet, the British mariner and compass maker Robert Norman.

Chapter 12

# Robert Norman and *The Newe Attractive*

## Section 1—The Discovery of Magnetic Dip

The discovery of magnetic dip (magnetic inclination)—or, more precisely, of the fact that the angle made with the horizontal plane by a magnetized needle is not zero—equaled the discovery of magnetic declination in importance not only for furthering people's understanding of the Earth's magnetic field but also for forming a new understanding of the Earth itself. Dip was first identified by the German Georg Hartmann, introduced in the previous chapter, and the Englishman Robert Norman. Norman's account is particularly notable in that it goes beyond describing the circumstances of his discovery and recording the results of his measurements to discuss the notion of a "sphere of virtue," albeit in a very simple manner. It also includes an important consideration based on his observations that the force exerted on magnetized needles is not mere attraction. This idea would lead to the discovery of pairs of forces known as *couples*. Norman's work also represents a major shift in the nature of scientific activity. Here was a craftsman publishing the results of independent research into technical problems that arose in his work; and, what's more, he was presenting them in his native language. Let's take a closer look at this important work and the circumstances that led to its publication.

Of the two men, Hartmann must be credited with having discovered magnetic dip first, as the letter cited below was written in 1544, before Norman's treatise discussing the subject. In his letter, Hartmann states:

> I also find this about the magnet, that it not only turns from the north and deviates towards the east, by 9° more or less, as I have said, but also it dips downwards. This is now to be proved.

He then adds:

> I make a needle a finger's length, that stands level on a pointed rod, or level with a water-surface, so that it in no way inclines earthward, but both ends stand level in exact balance; but when I once stroke its ends, no matter which, then the needle no longer stands level, but dips downwards by 9° more or less. The cause why this happens I could not explain ...[1]

As mentioned in the previous chapter, however, Hartmann's letter remained in obscurity until 1831, when it was rediscovered and published.

Norman wrote *The Newe Attractive*[2] while working in London as a navigational instrument maker and salesman after a career of almost twenty years as a mariner. The book contains precise measurements and related observations concerning magnetic dip, which he made independently from, and recorded in much more detail than, Hartmann. Published in English in 1581, it is considered the first work devoted solely to geomagnetism. That this work was based entirely on Norman's own experiments and observations of the behavior of magnetized needles is evident from a statement in a section at the beginning titled "The Epistle": "I have heere set downe whatsoever I could finde by exacte triall, and perfect experimentes."

In the third chapter of *The Newe Attractive*, titled "By what meanes the rare and straunge *Declining* of the needle, from the plaine of the Horizon was first founde," Norman describes how he discovered magnetic dip and his reasons for measuring it. Although somewhat long, the account is of great interest to our discussion, so I include it here in its entirety (noting that by "Stone" Norman means lodestone):

> Havyng made many and diverse compasses, and using alwaies to finish and ende them, before I touched the needle, I found continually, that after I had touched the Irons with the *Stone*, that presently the North poinct thereof would bende or *Decline* downwards under the Horizon in some quantitie: in so much that to the Flie of the Compasse, whiche before was made equall, I was stil constrained to put some small peece of Waxe in the South part thereof, to counterpoise this *Declining*, and to make it equall againe.
>
> Whiche effect havyng many tymes passed my handes, without any great regard therunto, as ignorant of any suche propertie in the *Stone*, and not before havyng heard nor read of any suche matter: It chanced at length that there came to my handes an Instrument to be made, with a needle of sixe ynches long, whiche needle after I had pollished, cut of a just length, and made to stand levell upon the pin, so that nothyng rested but onely the touchyng of it with the Stone: when I had touched the same, presently the North parte thereof *Declined* downe, in suche sorte, that beeyng constrayned to cut awaie some of that

> parte, to make it equall againe, in the ende I cut it too short, and so spoyled the needle wherein I had taken so muche paines.
>
> Hereby beyng stroken into some choler, I applied my self to seeke further into this effect, and makyng certaine learned and expert men, my freendes, acquainted in this matter, they advised me to frame some Instrument, to make some exacte triall, how muche the Needle touched with the Stone would *Decline*, or what greatest Angle it would make with the plaine of the Horizon. (TNA 3:8–9)

As we can see from this passage, Norman's discovery of magnetic dip was triggered by a phenomenon that compass makers witnessed repeatedly during their daily work. This may also have been true for Hartmann, a sundial maker. This point is well worth examining in a little more detail.

Excluding such cases as the invention of the microscope or telescope, where technological advances in equipment dramatically enhanced human observational powers, "discoveries" of "previously unknown" natural phenomena do not generally arise from a sudden ability to see what was formerly invisible. Rather, such a discovery occurs when some chance event causes a frequently observed yet overlooked phenomenon to be viewed in a new and different light; a phenomenon formerly dismissed as unworthy of notice is now perceived as extraordinary, as deserving of thorough study and requiring a proper explanation.

With scientists and scholars, in many cases it is the introduction of a new theoretical framework that gives previously overlooked phenomena new relevance and meaning and focuses attention upon them. But this is not necessarily true for engineers and craftsmen. Rather, the truth is likely closer to what Norman describes: only after countless anonymous craftsmen repeatedly experience the technical and practical inconveniences associated with a particular phenomenon is someone finally compelled to pay attention to it and begin investigating it. This was probably how sundial and compass makers stumbled upon declination in the fifteenth century. Generally speaking, however, such "discoveries" made by craftsmen or engineers are rarely recognized as such, and it is even more unusual for a written record to be preserved. This is why, as we saw in the previous chapter, attempts to determine when and by whom such discoveries are made are relatively unproductive. Norman's testimony is thus invaluable as a vivid description of the kind of process leading to such achievements, a process that is normally buried in history without due credit given to the discoverer. In fact, *The Newe Attractive* deserves special mention as virtually the first work recording such a discovery by a craftsman from start to finish in his own hand and interpreted in his own words.

The details of Norman's experiments and measurements are recorded in the fourth chapter, "How to finde the greatest *Declinyng* of the Needle under the Horizon." To obtain his measurements, Norman made a special device consisting of an iron needle, uniform on both sides, that could swing freely on a horizontal axis and stop at any angle. The center of the needle is fixed to the center of a vertical disc, the diameter of which is the same length as the needle and the circumference of which is divided into 360 equal parts. (The original text states that the circumference is divided into 160 parts, but from the appended figures and results of his measurements, it appears that this number is a misprint.) The disc is aligned with the geomagnetic meridional plane. Norman then touches the iron needle with a lodestone and measures the needle's angle of incline from the horizontal plane. He thus determines

> the *Declination* [i.e., inclination, or dip, in modern usage] of the North poinct of the touched Needle, which for this citie of London, I finde by exact observations to be about 71 degrees 50 mi[nutes]. (TNA 4:10, Figure 12.1)[3]

The significant discrepancy between Norman's and Hartmann's results can be attributed to the imprecision of their measurements—particularly Hartmann's, which were very rough. Although the angle of dip in Germany at that time is of course not known, in comparison with an estimate based on the proven dip from a later date, Hartmann's figure is too small, even if we allow for changes that would have occurred over a long period of time.[4]

*The Newe Attractive* is not lengthy, but it is meaty: it not only describes the phenomenon of dip but also contains important observations about other experiments inspired by that discovery and about magnetism in general. In particular, Norman argues that the discovery of magnetic dip overturns previous views on the point to which lodestones and compass needles are attracted. Moreover, his descriptions and interpretations of other experiments clarify certain previously unknown properties of magnetism. Accordingly, let's take a more detailed look at the content of the work as a whole.

The first chapter is titled "Of the *Magnes* or *Lodestone*, where thei are founde, and of their colours, weight, and vertue in drawyng Iron, or Steele: and of other properties of the same stone." Here, Norman presents contemporary European knowledge and views on magnetism and magnets, none of which is particularly new. He lists regions where lodestone is obtained, including the coast of China and Bengal, the island of Elba, the Red Sea, the iron mines of Norway, and the mines of Caravaca in Spain. Clearly, by this time India was no longer an obscure, distant land. Whereas

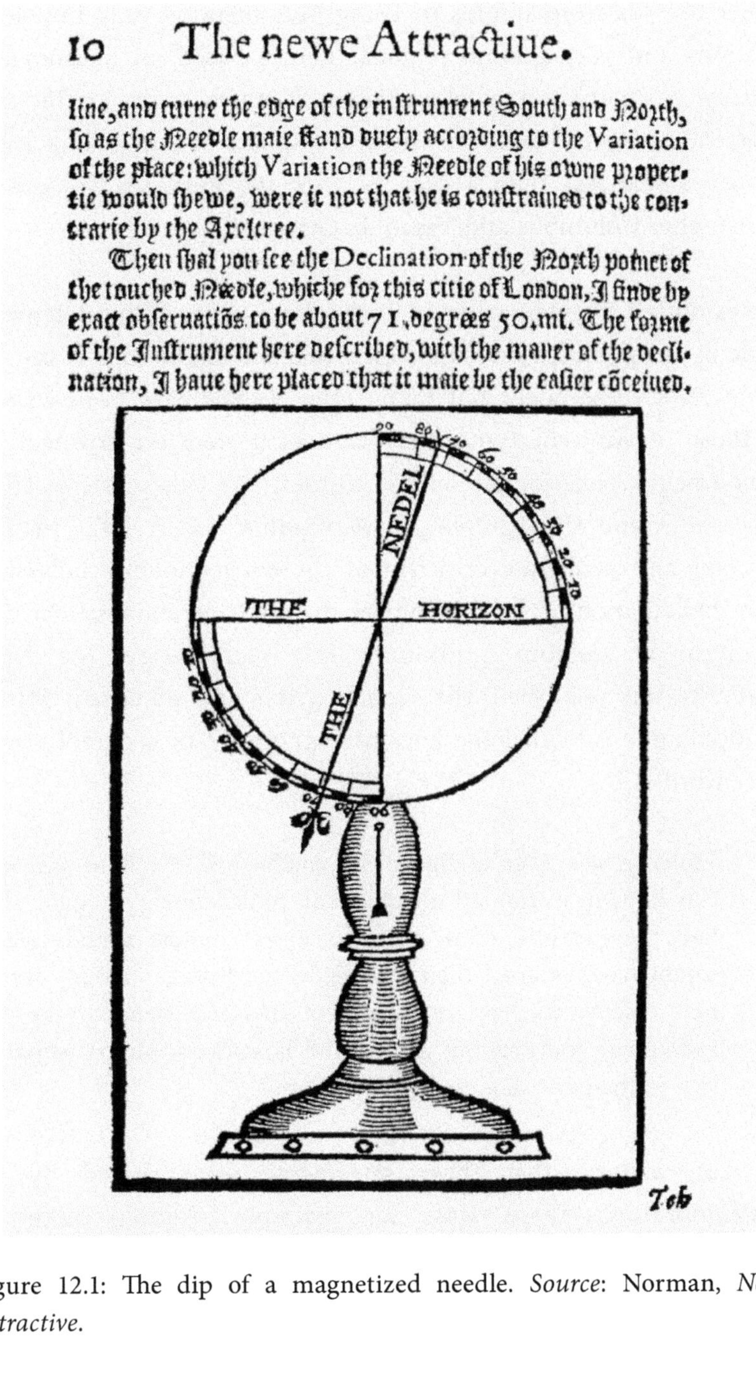

10 The newe Attractiue.

line, and turne the edge of the instrument South and North, so as the Needle maie stand duely according to the Variation of the place: which Variation the Needle of his owne propertie would shewe, were it not that he is constrained to the contrarie by the Axeltree.

Then shal you see the Declination of the North poinct of the touched Needle, whiche for this citie of London, I finde by exact obseruatiōs to be about 71. degrées 50. mi. The forme of the Instrument here described, with the maner of the declination, I haue here placed that it maie be the easier cōceiued.



Figure 12.1: The dip of a magnetized needle. *Source*: Norman, *Newe Attractive*.

earlier records from Plinius to Peregrinus mention only imprecise locations and very general regional names based on questionable hearsay, Norman's list of place names is much more specific and clear, indicating the extent to which the sphere of European activity had expanded less than a century after the great sea voyages of Christopher Columbus and Vasco da Gama.

This is not all Norman achieves in chapter 1. After citing statements made by his predecessors concerning magnetism, he then dismisses them: "Not these onely, but many other Fables have been written by those of auncient tyme, that have as it were set downe their owne imaginations for undoubted truthes, and this moste of all in *Geographie*, and *Hydrographie*, or *Navigation*" (TNA 1:4). This is a very apt and deliberate criticism of the conventional scholarship, with its adulation of the wisdom of the ancients and its uncritical parroting of dubious traditions. The year before *The Newe Attractive* was published, the French writer and statesman Michel de Montaigne ridiculed the ancients' ignorance of geography with these words:

> Ptolemy was a great figure; he established the boundaries of the known world; all the ancient philosophers thought they had the measure of it, save for a few remote islands which might have escaped their knowledge.... But ... in our century new discoveries have revealed, not the odd island or the odd individual country, but an infinite land-mass, almost equal in size to the part we already knew.[5]

We can assume that these sentiments were shared by his contemporaries. There was, for example, André Thevet, a Frenchman who had traveled to Brazil and who published a report about South America in 1557 that not only became very popular in France but was also published in English translation in London in

1568 under the title *The New Found Worlde, or Antarctike.* At the end of that work we find the following:

> By this ye may see how long tyme we haue bene ignorant of many countreys, as well Ilands as maine landes, beleuyng onely that, whiche our Elders hadde set forth, untyll that since of late yeares men haue ieoparded them selues in the Nauigation: So that nowe all our Hemispherie is discouered and founde inhabited; of the which *Ptolomeus,* and others knewe not the halfe.[6]

Then there was an account by Jacques Cartier, another Frenchman, of an expedition to New France (Canada) that was likewise published in English translation in London in 1580.[7] It is probable that Norman, a sailor-turned-craftsman, was well acquainted with matters as described in these works.

Norman, however, goes beyond criticizing to propose, in clear and simple language, a new approach for investigating the truth that would replace the excessive emphasis on book learning: "Therefore I wishe experience to bee the leader of Writers in those artes, and reason their rule in settyng it downe, that the followers bee not ledde by them into errours, as oftentymes it is seen" (TNA 1:4). This statement indicates that Norman, the craftsman, had a quite clear understanding of scientific and technological activity as an endeavor shaped by and dependent upon rational argument while at the same time based on experience and experimentation.

Norman maintains that the magnet's power of attraction is a quality of the stone itself rather than a quality belonging to some fixed and distant point or to the iron attracted to it. Concerning the magnetization of iron, he asserts that

> the iron hath no *Attractive* vertue, nor power of it self, untill it hath received it of the Stone. But Iron havyng a certaine affinitie, or naturall qualitie agreable to the Stone, doeth aptly and freely receive his vertue,... whiche when it hath received by touchyng the Stone, it is endewed with the verie same propertie and operation in all respects (though not in so greate force) as the Stone it self. (TNA 1:3)

On this basis, he describes the action of the magnet as follows:

> *The vertue of this stone is distributive,...* so the Iron that hath received this vertue of the Stone, will extende, and give that same to an other, and that Iron to an other, and so to many.
>
> And in this poincte the stone is mervailous, that notwithstanding you touche ten thousande Irons or nailes with him, every one of them cariyng awaie as muche virtue as will lifte up an other his like... yet the stone it self wilbe nothyng diminished of his strength, but continue of one force. (TNA 1:4–5)

Cornelius Agrippa von Nettesheim had already noted the unlimited nature of the magnet's capacity to magnetize. Obviously, this quality appeared just as marvelous to a craftsman such as Norman.

Next, Norman describes the lodestone's northward directionality and the phenomenon of declination, noting that

> if you put it in a drie Dishe, and sette it to swim in a tub of water, it will turne the Dish about, and the North part of the Stone after many swarvinges to and fro, will rest, and directly shewe the line of *Variation*, or imagined *Attractive* poinct. (TNA 1:4–5)

These facts were already well known, but in chapter 2, "Of the divers opinions of those that have written of the *Attractive* poinct, and where they have imagined it to be," Norman proceeds to unfold his critique of previous notions concerning the point to which magnets and magnetized objects are drawn. He concludes that neither Martin Cortes's theory—that this point was farther away than the celestial sphere—nor Olaus Magnus's theory—that it was some enormous magnetic mountain located near the North Pole—was correct. He saves the development of his criticism of Cortes's theory for chapter 6. Concerning Magnus's theory, he makes the very practical observation that if Magnus's proposal were true, compass needles in the Gulf of Lepanto should point to the island of Elba, with its large iron mine, but that in fact no such phenomenon had ever been reported. (Gilbert later made the same argument in *On the Magnet.*)

Norman's main point was that these men erred because they did not know about magnetic dip, and that leads into his discourse on the topic.

## Section 2—Norman's Observations concerning Magnetism

Norman touches upon his interpretation of magnetic dip in chapter 5, titled "That in the vertue of the *Magnes* or *Lodestone*, is no ponderous or weightie matter, to cause any suche *Declining* in the Nedle." He begins by noting the view held by some of his contemporaries that

> this Declinyng of the Needle, is caused by some ponderous substance that it receiveth from the *Stone*, and not (as I take it) to proceede of the simple vertue, and secrete influence thereof, because the Stone it self, wherein the vertue remaineth and is nourished, is weightie.

In other words, such people claimed that magnetic dip resulted from an increase in weight—the inflow of a "ponderous substance"—as a result of magnetization of the needle's north-pointing end. Norman refutes this claim in the following manner.

First, he presents counterevidence based on direct experimentation. Using a balance scale to weigh pieces of iron or steel against lead weights, he found that, even after being rubbed with a lodestone and sufficiently magnetized to raise other scraps of iron, the pieces still balanced against the lead weights exactly as they had before being magnetized. This was the first known experimental verification that magnetization causes no change in weight in either iron or steel.

He also provides logical argument for his view, making the very reasonable point that

> if the North poinct of the Needle doe *Decline* by any ponderous or weightie matter, in the vertue received by touchyng the *Stone*, why then should not the South poinct of the Needle, beeyng touched with the contrary ende of the *Stone*, have the same *Declinyng* Southwardes …?

In fact, he says,

> touche the Needle with what part of the *Stone* you list [i.e., like], that ende of the Needle that sheweth the North will alwaies *Decline*. (TNA 5:11–12)

That is, regardless of which end of the needle is rubbed or which pole of the magnet it is rubbed by, it is invariably the northward pointing end of the needle that dips and not necessarily the end that is rubbed. Not only did this experiment demonstrate that magnetic force was unrelated to gravity; it also suggested that magnetism *differed qualitatively* from gravity, which actually draws objects with weight in a single direction.

Norman then goes on to record his observations from a further experiment in chapter 6, titled "A confutation of the common received opinion of the poinct *Attractive*." From the results of his experiment in chapter 5, it could still be surmised that the dip in the needle's north-pointing end is caused by an attraction to a point somewhere on the Earth or by the attraction of the needle's south end to a point somewhere in the heavens. Norman, however, asserts that neither is the case.

To prove this, he sticks an approximately two-inch length of iron or steel wire into a piece of cork, which he floats in water, and conducts the following experiment:

> Then you shall take a deepe glasse, bole, cup, or other vessell, and fill it with faire water, setting it in some place, where it maie rest quiet, and out of the winde. This done, cut the Corke circumspectly by little and little, untill the wier with the Corke be so fitted, that it maie remaine under the superficies of the water two or three inches, both endes of the wier lying levell with the superficies of the water, without ascendyng or descending, like to the beame of a paire of ballance beyng equally poised at both endes.
>
> Then take out the same wier, without moving the Cork, and touch it with the *Stone*, the one ende with the South of the *Stone*, and the other ende with the North, and then set it againe in the water, and you shall see it presently turne it self upon his owne Centre, shewyng the aforesaied *Declinyng* propertie, without descendyng to the bottome, as by reason it should, if there were any *Attraction* downewards, the lower parte of the water beeyng nerer that poinct, then [*sic*] the superficies thereof.

> And as this maie prove no *Attraction* or drawing downewards; in like maner, the Corke beyng so made, that it maie sincke very slowly to the bottome, and then taken out, and touched with the *Stone*, and put in againe downe to the bottome with your finger, if any *Attractive* drawyng were upwards, it would ascende, and come up to the superficies of the water, beeyng nerer to that poinct then [*sic*] the bottome. But I finde by diligent and exact triall, that it hath no suche effect. (TNA 6:13–14)

In the same way, he then sticks a magnetized needle into a cork and floats it on the water. He proposes that

> if there were any suche *Attractive* poincte as hath been imagined, either in the earth by vertue of huge Rockes of the *Magnes* stone neere the Pole, or otherwise in the heaven,… it should of necessitie bee drawen in tyme to some side.

As Norman points out, however, although the needle exhibits a particular declination,

> you shall finde that it will not be drawen from his place, neither to the one side, nor the other. (TNA 6:15)

On the basis of these experiments, Norman concludes

> that the *Attractive* poincte before imagined, is no where, nor no suche thyng: and therefore, as moste proper, I will call the poincte, whereunto the Needle inclineth by vertue of the Stone, *The poinct Respective*, and attribute the whole power of shewyng that poinct, to bee in the *Stone*, and in the Needle, by the vertue received of the *Stone*. (TNA 6:15)

Norman appears to say that, judging from the results of his experiments, the faculty exhibited by the magnet and the magnetic needle is an inherent characteristic of these objects rather than the result of an external action. In modern terms, the essence of his discovery is that the effect exerted by the Earth's magnetic field on a magnetized needle is not a single attractive force that pulls in one direction but rather what physicists call a couple—that is, a pair of equal and opposite forces whose combined result is that the needle is merely turned to face in a certain direction, rather than actually pulled.

In that sense, William Gilbert puts it most aptly when he notes that

> direction does not arise from attraction, but from the disposing and turning power which exists in the whole earth.

Therefore, according to Gilbert:

> This assertion of the Englishman, Robert Norman, is plausible and appears to do away with attraction.[8]

Norman's discovery provided the first correct understanding of the nature of the force exerted upon a magnetic needle—in other words, upon a magnetic dipole—in a nearly uniform magnetic field.

In the following passage, Norman presents another opinion concerning magnetism that is also worthy of attention, namely, that

> if this vertue could by any means be made visible to the eye of man, it would be found in a Sphericall forme extendyng rounde about the *Stone* in great compasse. (TNA 8:19)

Although simply expressed, this statement marks the beginning of the extremely important concept of *orbis virtutis*, or "sphere of virtue," which would be passed down through Della Porta and subsequently Gilbert. This is a subject I will touch upon again in a later chapter. Suffice it to say that the seeds of the modern view of force were beginning to sprout.

I will add here that when Norman refers to the "point respective," he is not claiming that there is a single, fixed pole on the Earth to which all compass needles point. Rather, as he states in chapter 9:

> This *Variation* is judged by divers travailers to bee by equall proportion, but herein they are muche deceived, and therefore it appeareth, that notwithstandyng their travaile, they have more followed their bookes then [*sic*] experience in that matter. True it is, that *Martin Curtes* doeth allowe it to be by proportion, but it is a moste false and erroneous rule. For there is neither proportion nor uniformitie in it, but in some places swift and sudden, and in some places slowe. (TNA 9:21)

Indeed, William Bourne, an Englishman, had already proposed, in 1574, that there was no simple relationship between declination and longitude; and astronomer Thomas Digges (1546–1595) showed in 1576 that the directions indicated by compass needles in different locations did not converge in a single point, and rejected the idea of uniform variations in declination.[9] Norman's view was an extension of this line of thinking. In 1599, English mathematician Edward Wright (c. 1558–1615) asserted in *Certaine Errors in Navigation* that since declination was highly irregular, measurements were necessary on a global scale. In the same year, Wright also translated *The Haven-finding Art*, by Flemish mathematician and engineer Simon Stevin (1548–1620), into English. The book notes in its opening lines that

> some people, hoping to find [an accurate knowledge of longitude] through the variation of the compass, ascribed a pole to the said

> variation, calling it magnetic pole, but it is found upon further experience that these variations do not obey a pole.[10]

This lays to rest Mercator's 1546 hypothesis that geomagnetic meridians consist of great circles that pass through a single magnetic pole.

Whatever the case, the discovery of magnetic dip combined with Petrus Peregrinus's method for discovering the polarity of spherical magnets would seem to almost necessitate the conclusion that the Earth itself is a magnetic sphere. This conclusion, however, would have to wait until Gilbert's work in 1600. Petrus's *Letter concerning the Magnet* was first printed in 1558 at Augsburg, but Norman, either because he did not know of it or because he was not versed in Latin, appears not to have read it.

## Section 3—The New Proponents of Natural Science

*The Newe Attractive* is infused throughout with a consistent logic and rational methodology. Norman presents precise measurements of dip made with a device that, though crude, was specially crafted for its purpose; he provides a clear, quantitative outline of the results; he formulates a hypothesis on the nature of the force that causes dip; and he undertakes carefully designed experiments to test that hypothesis.

At that time, Europe stood on the threshold of the modern age, and nowhere was this transformation more conspicuous than in England. As one scholar points out: "In the eighty years before 1640 England, from being a backward country in science, became one of the most advanced."[11] Behind this lay England's rapid rise from an industrially undeveloped nation in the mid-sixteenth century to a nation boasting one of the best mining and manufacturing industrial bases in Europe by the mid-seventeenth century. During that period, England's population almost doubled, and the country experienced a transformation that can only be described as a precursor to the industrial revolution. The changes that took place in the fifty years from 1575 were particularly rapid.[12] The development of a commodity economy and the intensification of capitalist competition along with dramatic technological advancements meant that craftsmen and engineers could no longer rely on the antiquated skills and know-how that had been passed down through the conventional apprenticeship system within the medieval guilds. This in turn drove them to acquire rational and promising knowledge that offered greater possibilities for open-ended expansion and also to collaborate with scholars.

In the latter half of the sixteenth century, England was eager to overtake the world's greatest naval powers of the day, Spain and Portugal. Accordingly, advancing the art of navigation and related science and technology was a particularly pressing need. In 1578, Queen Elizabeth issued letters patent authorizing the English explorer Humphrey Gilbert to establish a colony in North America; and in 1584, English writer Richard Hakluyt published his *Discourse of Western Planting*, in which he strongly advocated the building of colonies on the North American continent. In 1580, the year before *The Newe Attractive* was published, Sir Francis Drake returned to England from the two-year, ten-month voyage that had made him the first Englishman to circumnavigate the globe. Eight years later, this same man was to lead the English navy in the destruction of Spain's so-called Invincible Armada. Nationalism had reached its zenith, launching England into a period of rapid territorial expansion. England's navy is said to have actually more than quadrupled in size between 1580 and 1640.[13] According to economist and sociologist Werner Sombart, the English fleet, which had a total tonnage of 7,110 tons when Queen Elizabeth ascended the throne in 1558, grew almost tenfold, to 62,594 tons, in the century from then until 1660.[14]

Long sea voyages required constantly updated geographical knowledge and new, more advanced shipbuilding technology to produce ever larger and faster vessels, as well as astronomical observation instruments and compasses, maps, and charts to determine one's location and route. Knowledge of astronomy and mathematics was essential not only to produce and upgrade such tools but also just to use and maintain them. Upgrading the quality and skills of engineers, craftsmen, and crewmen was thus a top priority both for the nation and for the financiers of the expeditions. Appended to *The Newe Attractive* was a slim booklet titled *A Discourse of the Variation of the Cumpas, or Magneticall Needle* and written by English navigator William Borough. In its preface, addressed "To the Travelers, Seamen, and Mariners of Englande," Borough states: "I wishe all Seamen & Traveilers, that desire to bee cunnyng in their profession, firste to seeke knowledge in Arithmetik & Geometrie, whiche are the groundes of all Science and certain artes."[15]

Such was the professional climate surrounding highly enterprising engineers, who now studied mathematics to round out their technological expertise and took a scientific approach to technical problems. Norman was well aware of this. In his introduction, titled "To the Reader," he states with obvious pride that

> albeeit it maie bee saied by the learned in the Mathematicalls, as hath been alreadie written by some, that this is no question or matter for a Mechanician or Mariner to meddle with, no

> more then [*sic*] is the finding of the longitude, for that it must bee handeled exquisitely by Geometicall demonstration, and Arithmeticall Calculation, in whiche Artes they would have all Mechanicians and Seamen to bee ignorant, or at least insufficiently furnished to performe suche a matter, alledgyng against them the Latine Proverbe of *Apelles, Ne sutor ultra crepidam*.... yet there are in this lande divers Mechanicians, that in their severall faculties and professions, have the use of those artes at their fingers endes, and can applie them to their severall purposes, as effectually and more redily, then [*sic*] those that would moste condemne them. For albeit thei have not the use of the Greeke and Latine tongues, to searche the varietie of Aucthours in those artes, yet have they in Englishe for Geometrie *Euclides* Elementes with absolute Demonstrations: and for Arithmeticke *Recordes* woorkes ... and divers others, bothe in Englishe and in other Vulgar languages, that have also written of them, whiche bookes are sufficient to the industrious Mechanician, to make hym perfecte and ready in those Sciences.... And therefore I would wishe the learned to use modestie in publishyng their conceites, and not disdainfully to condemne men that will search out the secretes of their Artes and Professions, and publishe the same to the behoofe and use of others.

It was mariners and compass makers from the time of Columbus up to the end of the sixteenth century who elucidated the phenomena of magnets and magnetism—men who worked with lodestones on a daily basis. Practitioners of emerging occupations, including craftsmen, soldiers, and seamen, who were neither clergy nor academic intellectuals, were beginning to express what they knew and understood about magnetism and about the new and now wider world, foreshadowing tectonic shifts in the realm of knowledge. A clear indication of this was the fact that the majority of them, including Norman, Falero, and Cortes, penned their reports and books not in Latin, the common tongue of intellectuals in medieval Europe, but in their respective vernaculars.

## Section 4—Robert Recorde and John Dee

Of course, a mere craftsman like Robert Norman would not have had the confidence to express his views on his work as early as 1581 unless the ground had already been prepared. As he himself testifies in the above quotation, it was a series of mathematical texts written for engineers in English by Robert Recorde in the mid-sixteenth century and the publication of Euclid's *Elements* in English in 1570 that made it possible. The latter was the first English edition of Euclid's work as well as the first ancient scientific treatise to be published in English.[16] Around the same time, in 1576, Thomas Digges wrote a commentary on Copernican heliocentric theory in English, including an English-language translation of certain crucial chapters from the first book of Copernicus's *On the Revolutions*; indeed, he ventured even further than Copernicus by proposing an infinite universe.

Humphrey Gilbert had already suggested the establishment of "Queene Elizabethes Achademy" in 1562, and in his prospectus he declared, "wheareas in the vniuersities men study onely schole learninges, in this Achademy they shall study matters of accion meet for present practize, both of peace and warre."[17] The academy was not to focus solely on mathematics; rather, the professor of geometry would be required to teach the theory and practice of gunnery, and the professor of astronomy, the art of navigation and the use of navigational instruments. Moreover, English was the designated language of instruction as opposed to Latin, the language of the intellectual elite used in universities.[18] In 1579, Thomas Gresham (c. 1519–1579), a highly successful merchant and financier known for Gresham's law, passed away, leaving instructions in his will for the establishment of a college to educate ordinary citizens in new branches of science. He specifically stipulated that college lectures at this institution (Gresham College) were to be offered free of charge to the general public in English and Latin. Thus, by Norman's time, both the spiritual climate and the material circumstances required to provide urban-dwelling bureaucrats, merchants, engineers, and craftsmen access to new branches of study were already in place.

Robert Recorde (1510–1558) became a medical doctor at Cambridge after graduating from Oxford. A true Renaissance man, he was well versed in mathematics, astronomy, medicine, metallurgy, mining, the art of minting coins, theology, and law. Aside from being the first person in England to comment on Copernicus's hypothesis as well as being the inventor of the equals sign (=), he was also known as "the first mathematician of any note to publish works in the English language."[19] His works include *The Grounde of Artes*, first printed in 1542; *The Pathway to Knowledge*, first published in 1551; *The Castle of Knowledge*, first published in 1556; and *The Whetstone of Witte*, published in 1557. *The Grounde of Artes* was a manual of elementary mathematics specifically written for engineers and craftsmen and the first mathematical text to be written in English using Arabic numerals. Widely read, it was reprinted numerous times between 1542 and 1699 and had a tremendous impact.[20] Similarly, *The Pathway to Knowledge* and *The Castle of Knowledge* were the first works written in English in the fields of geometry and astronomy, respectively. The latter was basically a discussion of Ptolemaic geocentric astronomy, but it also included the first favorable mention of Copernicus's heliocentric theory in England.[21] Finally, *The Whetstone of Witte* was not only written in English but was also the first book on algebra written in England.

And it was the erudite, multitalented John Dee (1527–1608), a son of Elizabethan England who published enlarged, revised editions of *The Grounde of Artes* in 1561 and 1570 and wrote the outstanding *Mathematicall Praeface* to the first English translation of Euclid's *Elements*. This translation was completed in 1570 by Henry Billingsley, who would later become mayor of London. Concerning the *Praeface*, one scholar on Dee's work has gone so far as to state that "it is probably almost impossible to estimate its true influence on the development of scientific and philosophical thought in England during the Renaissance."[22] The fact that England produced mathematicians who had never attended university, such as Hamfrey Baker and William Bourne, owed much to the works of both Recorde and Dee.

In 1542, Dee entered St. John's College at Cambridge, where he studied liberal arts and classical languages. Between 1547 and 1550 he traveled to Louvain, Brussels, and Paris and came into close contact with such figures as Flemish mathematician Gemma Frisius and cartographers Gerard Mercator and Abraham Ortelius. During those travels, Dee is said to have acquired his knowledge of mathematics and mastered the new branches of knowledge being pursued on the Continent, including geography, the art of navigation, Hermeticism, Kabbalah, mechanical engineering, and architectural theory. As none of these new sciences had yet taken root in England, it was Dee who introduced and propagated them.[23]

Dee was a mathematician, philosopher, and geographer, but he was not an academic. He was the early Englishman to recognize the importance of mathematical and scientific technology, but his primary interest in mathematics and astronomy was in their application, and the importance of science for him lay in its ability to enhance the prosperity and power of England as a nation. In fact, he served as Queen Elizabeth's royal naval adviser and guided her in many other matters. He was a passionate promoter of English imperialist policies, which explains his extraordinary interest in improving and advancing the art of navigation. In fact, for thirty years during the late sixteenth century, he acted as an adviser to such English colonialists and explorers as Humphrey Gilbert and Walter Raleigh and trained and advised navigation officers. He also kept up a tireless correspondence with famous scholars in Europe and, at the same time, was England's greatest book collector. He gathered around him those who were interested in the sciences and arts and gave them unlimited access to his personal library, creating a sort of private academy. Among the products of this institution were Phillip Sidney, a central poet of the Elizabethan Renaissance, and Thomas Digges, the most ardent advocate of the Copernican theory in that era and a forerunner to declare that the universe was infinite.

Looked at in this way, John Dee can be seen not only as a leader of England's scientific enlightenment and a point man of the scientific revolution but also as someone who advocated technocracy two centuries before the French Revolution. Yet, at the same time, he was a Hermetic and a believer in Renaissance magical thought, much influenced by Agrippa. This earned him the ignominious title of the greatest magician in sixteenth-century England, and his scientific achievements went virtually ignored until the early twentieth century. Indeed, one commentator notes: "There is perhaps no learned author in history who has been so persistently misjudged, nay, even slandered, by his posterity."[24] Dee's notoriety as "a magician deeply immersed in the most extreme forms of occultism"[25] was perpetuated in large part by the sensationalized posthumous publication of his diary, which recorded his attempts to communicate with angels. Of course, neither of these portrayals is the sole truth. As Frances A. Yates puts it: "In him [i.e., Dee] Prospero and Sir Francis Drake meet and are one."[26]

The most noteworthy fact about Dee, for our purposes, is that he looked to engineers and craftsmen engaged in the practical application of technology as the driving force behind new branches of learning. Disdaining university students engrossed in Scholastic disputation, he washed his hands of the two universities in England at a very early date. In fact, conventional universities devoted solely to preserving, elaborating, and teaching Scholasticism were falling far behind the times. Incredibly, the basic textbooks used at Oxford in the mid-sixteenth century were, according to one source, Ptolemy for astronomy and Strabon and Pliny for geography; as far as Oxford was concerned, Columbus and Da Gama, let alone Copernicus, simply did not exist.[27] Small wonder that such institutions were accused of worshipping the authority of old, worm-eaten books. European universities originated in the twelfth century as student-teacher guilds intended to mine the wealth of knowledge made accessible through the prolific translation movement of that time, and in the beginning they were liberal and progressive. By the fifteenth century, however, they had degenerated into exclusive, upper-class, conservative institutions, and, compared to the rest of Europe, the ill effects of this situation were particularly noticeable at Cambridge and Oxford.[28]

In a letter written in 1563, Dee bemoaned the fact that, despite the proliferation of university scholars versed in theology, Hebrew, and Latin, England had no one skilled in the sciences of "number, weight and measure."[29] For that reason, he wrote the *Mathematicall Praeface* to an English-language edition of Euclid's *Elements* primarily for the engineers and craftsmen of the burgeoning middle class—in his own words, for "vnlatined people, and not Vniuersitie

Scholers" and also for the "Common Artificer … that dealeth with Numbers, Rule, & Cumpasse."[30] Accordingly, he deliberately discussed his subject in English. "This Preface," observes one commentator, "with its utilitarianism and its vigorous defence of experiment, made a great impact on young men of the middle class, sons of the traders and craftsmen."[31]

Naturally, the *Praeface* must have served as an inducement and an encouragement to Norman, a mere tradesman, to write and publish *The Newe Attractive*. But, of course, this was not the only work that influenced him. During the same period, many other English-language publications were produced in England to provide mariners with the knowledge they required. At the back of Eva Germaine Rimington Taylor's *The Mathematical Practitioners of Tudor & Stuart England* is a list of sixteenth-century books on mathematics, including applied mathematics such as astronomy, navigation, and geography. According to this source, over thirty texts on mathematics, navigation, and geography were published in English between 1550 and 1600, not including translations and handwritten manuscripts.

Looking at the most notable works on navigation, we find that William Bourne's *A Regiment for the Sea*, about navigational techniques and practices, was published in 1574 and reprinted several times thereafter. In addition, Digges's *A Treatise on the Arte of Navigation*, published in 1579; William Borough's *A Discourse of the Variation of the Cumpas, or Magneticall Needle*, published in 1581; and Edward Wright's *Certaine Errors in Navigation*, published in 1599, were all written in the English language.

And then there were numerous translations into English: Martin Cortes's work, which was discussed in the previous chapter, was published in English under the title *The Arte of Navigation*; and *De natura magnetis et ejus effectibus* (The nature of the magnet and its effects), written in Latin by the Belgian Jean Taisner and published in 1562, was later published in English under the title *A Very Necessarie and Profitable Booke concerning Navigation*. These two works were translated by Richard Eden in 1561 and 1579, respectively. Moreover, the English version of *Aguja de marear y de sus inventors*, completed in 1539 by the Spaniard Antonio de Guevara, was published as *The Invention of the Art of Navigation* in 1578, and *Arte de Naviguar* by Pedro de Medina was translated from the Portuguese and published as *The Arte of Navigation* in 1581. This succession of English translations demonstrates the enormous demand for such works by craftsmen and mariners who could not read Latin or other languages. Simon Stevin's *De havenvinding*, which was mentioned earlier, was published in 1599 and in the same year appeared in an English-language translation as *The Haven-finding Art*, the title page of which proclaims: "And now translated into English, for the common benefite of the Seamen of *England*."

Within this milieu, craftsmen, mariners, and soldiers were forging a path toward a new kind of science through experience and practice. Although it is dangerous to generalize, this was true at least in the case of the magnet. The observations and measurements of craftsmen and mariners were opening the way toward the study of magnetism as a modern science, taking a separate route from magic.

* * *

The discovery of magnetic dip by the sundial maker Hartmann and mariner and instrument maker Norman, which followed on the heels of the discovery of declination, set the stage for the realization that the Earth itself was a magnet. These discoveries were clearly by-products of the remarkable development of long-distance travel, with its use of the magnetic compass. At the same time, the discovery of dip heralded the emergence of craftsmen and mariners as the new agents of progress in natural science.

The voyages of exploration went hand in hand with a shift from the former preoccupation with book learning to a zeal for knowledge based on experience. Above all, this had the effect of increasing man's understanding, particularly of the magnet's relationship to the Earth, but it also led to the emergence of a new type of scholar. Within this context, both technology and magic underwent a metamorphosis that would lead to their convergence in modern science. Through the efforts of such unique men as John Dee, the road to modern science would be paved by, on the one hand, the reconciliation of an emerging experimental science practiced by craftsmen and engineers and, on the other, revived magical thought. Indeed, Dee and other believers in magic had a very strong interest in technology. The nature of the reconciliation they helped to effect will become clearer in subsequent chapters when we examine Girolamo Cardano's research into static electricity and Della Porta's study of the magnet.

Chapter 13

# Mining and the Continued Peculiarity of Magnetism

## Section 1—The Cultural Revolution of the Sixteenth Century

In the previous chapter we met Robert Norman, who deliberately penned his work, *The Newe Attractive*, in the English language. In 1580, a year before that book was published, a French potter named Bernard Palissy wrote, in French, a book on geology that includes the following:

> I have advanced this proposal [i.e., that practice has begotten theory, not vice versa] to silence those who ask how it is possible for a man to know something and to speak of natural things without having seen the Latin books of the philosophers? Such a statement could be made about me, since in many instances I prove by practice that the theory of several philosophers is false.[1]

Around the same time, Ambroise Pare was revolutionizing the profession of surgery, writing a number of textbooks in French based on his long experience and clinical observations as a military surgeon—this despite his having in fact begun his career as a barber-surgeon and therefore been scorned by the medical profession as a mere tradesman. And then there was Agostino Ramelli, a military engineer in the service of the Kingdom of France, whose *Various and Ingenious Machines* was written in French and Italian and published in 1588. In Italy, Giovanni Ventura Rosetti, a Venetian dyer, wrote *Instructions in the Art of the Dyers* in Tuscan (then the most popular Italian vernacular) as early as 1548; and Benvenuto Cellini, a goldsmith and sculptor, composed the treatise *On Goldsmithing* in Italian in 1568, printing it at his own expense the same year. In 1574, a Bohemian metallurgist named Lazarus Ercker produced, in German, his *Treatise on Ores and Assaying*, the world's first monograph on the art of assaying. Another work produced in the vernacular of Germany had actually appeared a half century earlier, in 1525; namely, *The Painter's Manual*, a work on the principles of applied geometry that was written by artist Albrecht Dürer for the benefit of painters, stonemasons, engravers, furniture builders, and goldsmiths. And in 1539, Jerome Bock (Hieronymus Bock), a Lutheran pastor, published his *New Herbalism Book*, in which he focuses on native German herbs and plants and provides descriptions based on his own experiments and observations.

In the 1580s, the United Provinces of the Netherlands was fighting for independence from Spain. Simon Stevin, an engineer who worked for the Dutch army, published in Dutch a series of textbooks on dynamics and mathematics. Stevin also made a strong case for the significance of writing about science in vernacular languages:

> In the first place, what we lack is a large body of data obtained by practical experience, on which the sciences can be firmly founded. In order to arrive at such a body of data, a great many people would have to apply themselves jointly to this task.
>
> ...
>
> To arrive at so great a number of men as is needed for this, the aforesaid experiences and pursuit of the sciences would have to be practised by a nation in its own native language.[2]

In Stevin's view, if science needed to be founded upon experience, then it followed that the doors of science needed to be thrown open to a wider class of individuals. In the halls of academia all across Europe, a seismic shift in the realm of knowledge—a kind of cultural revolution—was taking place.

And this shift—the appearance of books written by engineers and craftsmen in their native languages in the mid-sixteenth century—was due in part to the appearance of new book-printing technology. In the mid-fifteenth century, Johannes Gutenberg, a German, invented a printing press using movable type made of cast metal; and around 1455, he and his associates used this new device to produce the now-famous Gutenberg Bible. The invention was more or less perfected over the next decade, allowing publication of printed books to emerge as a commercial business; and underlying this development was the technology of papermaking that had come to Europe from China in the late Middle Ages via the Islamic world. In the remaining fifty years or so of the fifteenth century, book publishing spread like wildfire throughout Western Europe.

At first, many of the books printed were religious texts written in Latin, mere reproductions of the handwritten manuscripts formerly created in monasteries and at universities: the medieval status quo in literary culture had yet to be challenged. From around 1480, however, the volume of books produced rose sharply, as did the number of individual collectors. No longer were books seen as objects solely by, for, and of the monastery or the university.[3] By the early sixteenth century, the true power of the printing press—namely, mass production—had been amply demonstrated, helping to create a mass readership. Over the forty years from 1520, book printing completely distanced itself from any resemblance to the art of hand-copied book production.[4] Incidentally, from the outset, both printing and publishing were established as profit-making businesses ruled by market principles.[5] And while there was incentive for the production of books written in vernacular languages, the same could not be said of books written in Latin, whose readership was more limited. For this reason, printing technology actually spurred the trend toward using vernaculars.[6]

Figure 13.1 shows the change over the course of the sixteenth century in the number of medical books either written in French or translated into French.[7] We know that significant change began around 1530. Figure 13.2 shows the total number of books published in Paris in the sixteenth century as well as the proportion thereof published in French. Both the number and the percentage of books published in French continued to rise, and in 1575 more than half of all books published in Paris were written in the vernacular. Books originally written in Latin were translated into numerous languages in keeping with demand. Thus the monopoly on knowledge previously enjoyed by clerics and academics crumbled, and a whole new literary culture emerged, one completely apart from the world of the university, where scholars did little but produce literal interpretations of the works of Thomas Aquinas and lecture on ancient writings.

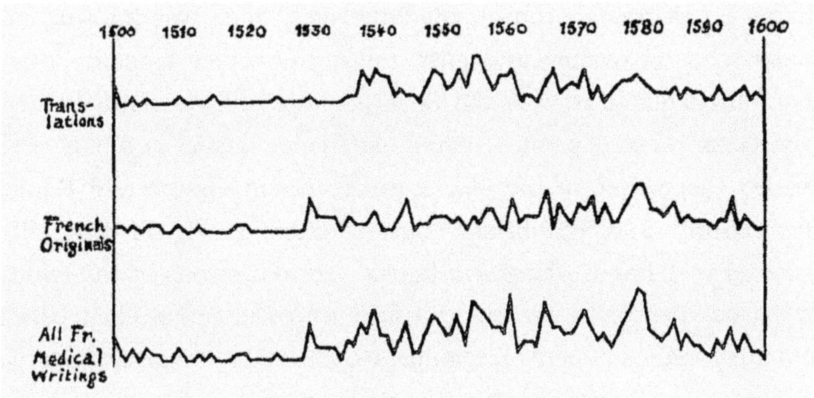


Figure 13.1: French medical writings in the sixteenth century. *Source*: Stone, "The French Language in Renaissance Medicine."

Table 13.1: No. of Books Printed in Paris and the Percentage Printed in French. *Source*: L. Febvre and H. Martin, *The Coming of the Book*, chapter 8.

| Year | Total | No. in French | Percentage |
|---|---|---|---|
| 1501 | 88 | 8 | 9% |
| 1528 | 269 | 38 | 14% |
| 1549 | 332 | 70 | 21% |
| 1575 | 445 | 254 | 55% |

University scholars, of course, were not pleased and stubbornly railed against these changes. In Europe, resistance to the use of books in the vernacular was very strong indeed throughout the sixteenth century. As Table 13.1 shows, publishing in the vernacular had become mainstream by 1575, and yet in that same year Ambrois Paré was attacked by embittered doctors at the University of Paris when a complete anthology of his writings—in French—was published. Paré was compelled to take the defensive: "For my part," he remarked, "I esteem nothing in my book pernicious because it is written in our vulgar tongue. Thus the divine Hippocrates wrote in his language, which was known and understood by women and girls, talking no other language than that." In England as well, the Royal College of Physicians derided physicians who depended upon experience alone and had not availed themselves of the teachings of ancient Greece and Rome; the Oxford- and Cambridge-educated membership of this elite believed that English-language books were dangerous in the hands of various people in the medical field who had not studied Latin, and they deemed such practitioners useless.[8] As historian George Sarton notes, "Latin was the esoteric language used to prevent the dissemination of learning to people who were deemed unworthy of it, or who might make a bad use of it."[9]

In sixteenth-century European society, discussing academic topics and lofty theories in a vernacular language was much harder than one might imagine. Such languages had not yet been refined to the point where they were suited to such subjects, and their lexicons were still poor. More importantly, Latin—the common language of the ruling classes—and religious unification through the church had in medieval Europe been the only means of bringing together peoples and regions of diverse cultures and histories to form a single entity. In addition, as the language of the intellectual elite, Latin could be acquired only at an educational institution controlled by the church. Thus the church controlled not only religious affairs but also all of academic life as well as political and administrative circles. By insisting that Latin alone be the lingua franca of the upper echelons of society in all areas, the clergy maintained hegemony over all European territory, wielding power far greater than that held by secular authorities. Authorship in vernacular languages dealt this order a direct blow.

From the perspective of those in power—those who had a

monopoly on learning—there was a very real danger that the use of the vernacular, particularly in religious contexts, could lead directly to heresy. The church had seen this happen numerous times and so had some basis for its fears. For example, there was the case of Peter Waldo, a merchant from Lyon who founded the Waldensians in the late twelfth century. This sect was suppressed by the church after recognizing discrepancies between the teachings of Christ and the preachings of the Vatican, discoveries its members made upon reading a Bible they had had translated into French. In the fourteenth century, John Wycliffe, a pioneering religious reformer, defied church orders by attempting to translate the Bible into English and was condemned by the church as a heretic. Similarly, Jan Hus of the Kingdom of Bohemia, who had translated the Bible into Czech and attempted to reform the church, was excommunicated and executed in 1415. And then in the sixteenth century came Martin Luther, who wrote *To the Christian Nobility of the German Nation*, maintained that the Bible was the ultimate authority, and translated it into German.

To prevent such incursions, in 1543 authorities in England "forbade the reading of an English Bible by women, artificers, prentices, journeymen, serving-men of the rank of yeoman or under, husbandmen and labourers."[10] In France, where Catholicism held sway, a French-language translation of the Bible appeared in the second half of the sixteenth century, but "for several decades the doctors of theology (strongly backed by secular law) tried to defend their monopoly on its interpretation by denying the right of the uneducated to read it."[11] Meanwhile, successive generations of popes throughout the fifteenth century gave tacit consent to early Renaissance humanists who toyed with heretical opinions—but only so long as their writings were confined to Latin so that their influence did not penetrate beneath the upper echelons.

Eventually, the rendering of religious texts into vernacular languages did lead to religious reform. Moreover, the use of vulgar tongues had an even wider impact in academia: not only did the trend serve to oppose university Scholasticism; it also went beyond the efforts of well-born humanists of the early Renaissance, who had attempted to revive the refined and elegant ancient Latin. The trend toward writing in the vernacular characterized what we now think of as the late Renaissance period. And this development was significant indeed, because the "liberation from medieval Latin, the gradual construction and development of the *volgare* as an independent scientific form of expression was the necessary prerequisite for the free development of scientific thought and its methodological ideals."[12] Books written in the native tongues of their authors—by people such as England's Robert Recorde, John Dee, Robert Norman, and William Bourne; Flanders's Simon Stevin; Germany's Albrecht Dürer, Jerome Bock, and Lazarus Ercker; France's Bernard Palissy and Ambrois Paré; and Italy's Giovanni Ventura Rosetti, Benvenuto Cellini, and Vannoccio Biringuccio, who we will be discussing very shortly—had an impact whose importance went beyond scientific enlightenment and the popularization of academic study: before long, such works had set a new standard for the very manner in which scientific research was to be pursued. And so, the cultural revolution of the sixteenth century laid the groundwork for the scientific revolution that was to take place in the seventeenth century.

## Section 2—Biringuccio's *The Pirotechnia*

From the second half of the fifteenth century through the first half of the sixteenth, Germany and Italy were far ahead of other regions in the nascent book-publishing industry as a whole and led them by about fifty years in the production of scholarly works written in or translated into vernacular languages. We have already taken a brief look at Luther, Bock, Ercker and Dürer of Germany. Italy, too, had its literary pioneers, including Dante Alighieri, Francesco Petrarch, and Giovanni Boccaccio. Another leader was Pietro Bembo, whose *Prose della volgar lingua* (Discussions of the vernacular language), published in 1525, championed the Italian language. Around the same time, political theorist Niccolò Machiavelli and historian Francesco Guicciardini also wrote in Italian. In fact, Italians were if anything first among Europeans to develop the conviction that their language was in no way inferior to Latin. Almost from the start of the commodity economy's expansion, authors and publishers were coming out with works in Italian specifically for the benefit of urban merchants and craftsmen.

*The Pirotechnia of Vannoccio Biringuccio* stands out among such works for being published in Italian on a technical subject by someone who was himself an engineer. Vannoccio Biringuccio (c. 1480–c. 1539) was born in the central Italian province of Siena. His father was an architect. As a young man, Biringuccio traveled in Italy and Germany, and in 1513 he took a job at a weapons factory, managing iron mining and production operations. Twice, in 1515 and 1526, Biringuccio left Siena amid citizens' revolts. He traveled to Naples and Sicily and also to Germany and made it his business to observe mining operations in all three locations. In 1530 he returned to Siena, where he was involved in both construction and weapons manufacturing. In 1538 he was given responsibility for overseeing both foundry operations and ammunition production for the Vatican, and he passed away a short time later. Thus, Biringuccio received all his professional training directly on the job, without any connection whatsoever to the clergy or to university life. *The Pirotechnia* was published posthumously, in 1540.

The Italian word *pirotechnia* generally signifies the art of using gunpowder and other explosives, but Biringuccios's great work was no mere manual of gunpowder manufacturing. Instead, the series was a veritable treatise on the wider subjects of mining, metallurgy, and any technology involving the use of fire. The table of contents makes this clear:

> Book 1: Every Kind of Mineral, in General
> Book 2: The Semiminerals
> Book 3: Assaying and Preparing Ores for Smelting
> Book 4: The Separation of Gold from Silver
> Book 5: The Alloys that are Formed between Metals
> Book 6: The Art of Casting in General and Particular
> Book 7: Methods of Melting Metals
> Book 8: The Small Art of Casting
> Book 9: The Procedure of Various Works of Fire
> Book 10: On Certain Artificial Combustible Materials and the Procedures Followed in Making Fireworks to be Used in Warfare and for Festivals

Biringuccio gave particularly thorough and superb treatment to the subject of casting, which was then a cutting-edge technology. The third and fourth chapters of book 3 include a discussion of the blast furnace, which was then just coming into use, and the eighth chapter of book 9 goes into the manufacturing of metal type, also a new technology at that time. Numerous editions of *The Pirotechnia* were published, and the work was widely read during the sixteenth and seventeenth centuries.

In the western Europe of the late Middle Ages, the use of metal increased sharply with the development of industries and the expansion of the monetary economy, and in the thirteenth and fourteenth centuries mining flourished. The industry stalled in the mid-fourteenth century as the Black Death took its toll on the population, but it revived around a century later as trade with the East and increased use of firearms spurred demand for metal. By the sixteenth century, the mining industry was again growing rapidly and began to develop on a large scale in the New World as well. We have already seen how navigation and printing technology flourished during the Renaissance, and now we can add to the list military and mining technology as well.

The rise of the mining industry brought about an increase in the labor population and in free competition. As a result, the previously insular guilds began to relax their restrictive practices, and demand arose for technologies and knowledge that surpassed the old, outdated traditions that the guilds had developed and preserved. In order to embark on a large-scale mining venture in those days, a plethora of minute details had to be meticulously planned, requiring ample investments of capital and comprehensive administrative frameworks. Even before excavation and mining could begin, it was essential to first select suitable locations from which to begin tunneling, to secure sources of water for motive power, to construct waterwheels as well as a variety of equipment to process mined metals, to make ready the necessary materials and manpower, to construct dwellings and other facilities needed for the welfare of the miners employed, and to put infrastructure in place for the transporting of mined products.

In short, getting metals out of the ground and moving them through to the smelting process was an enormous endeavor bringing together motive power, machinery, and labor on a vast scale. Setting up an enterprise of this magnitude required having in advance a full understanding of all the technologies involved. And this is exactly where *The Pirotechnia* came in. It was the earliest compilation and publication of all knowledge and technological information concerning assaying, digging, smelting, and casting—in short, everything there was to know about mining and metallurgy—and as such the work is one of the great highlights in the history of technology.

In the preface to book 1, Biringuccio writes that "the mountains are the matrices of all the most prized riches and the repository of all treasures," and then goes on to draw the following conclusion:

> For this reason all men who wish to have wealth should turn their attention to the excavation of mines rather than to warfare, with all its annoyances, or to commerce, which goes about outwitting the world and perhaps doing other tiresome things which may be illicit for honest men, or to going on long and weary journeys over land and water—journeys full of annoyances and discomforts, among strange and unknown peoples who are often of animal-like natures—or to applying oneself (as many do) to the fabulous philosopher's stone in the hope of enslaving its elusive service to make fixed silver or to perform magic rites, or to other things vain and without foundation.[13]

We see here unmistakably the ethos of a rational, ambitious, progressive industrial capitalist. Biringuccio, rather than writing a technical treatise for mining engineers, produced something that was both a manual and a set of guiding principles for mining as a business. Here is one passage that makes this intention very clear:

> When it is known that there is an ore and what quantity of what metal it contains, and it is found by calculation that you will recover enough value in it to justify the expense, I exhort you to begin courageously and to continue the undertaking with every care, and to start mining. (PVB, 16–17)

What is most striking about *The Pirotechnia* is its detailed coverage of what was then the most up-to-date knowledge of relevant machinery. For example, perhaps the greatest innovation of that period was the use of a water-driven bellows to increase the force of air, paving the way for the development of the blast furnace. Historically, this led to the production of cast iron, which in turn led to the use of iron on a larger scale; it also made possible the manufacture of larger iron weapons. In chapter 7 of book 7, Biringuccio includes eight illustrations (see Figure 13.2 for one example), with accompanying text, explaining the mechanisms of several types of bellows in use at the time. This was also the beginning of an indirect smelting process in which pig iron was manufactured in a blast furnace and then converted into wrought iron.

The second striking feature of *The Pirotechnia* is its heavy emphasis on the importance of on-site experience in mining and metal production; it paints a starkly real picture of what such experience must have been like.

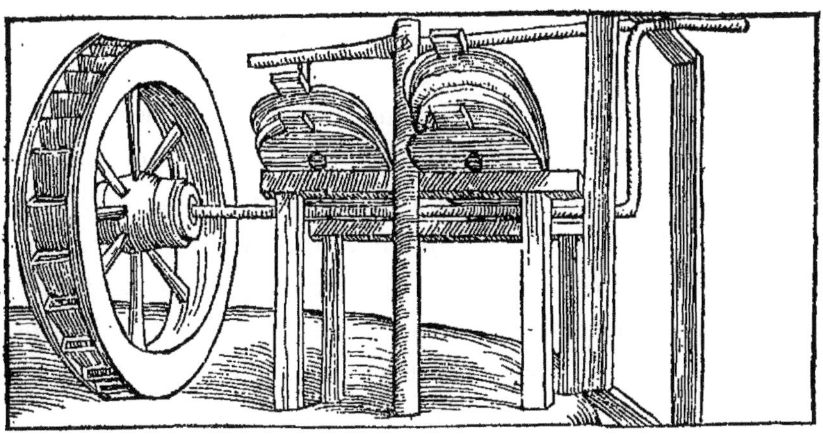

Figure 13.2: Water-driven bellows. *Source*: Biringuccio, *Pirotechnia*.

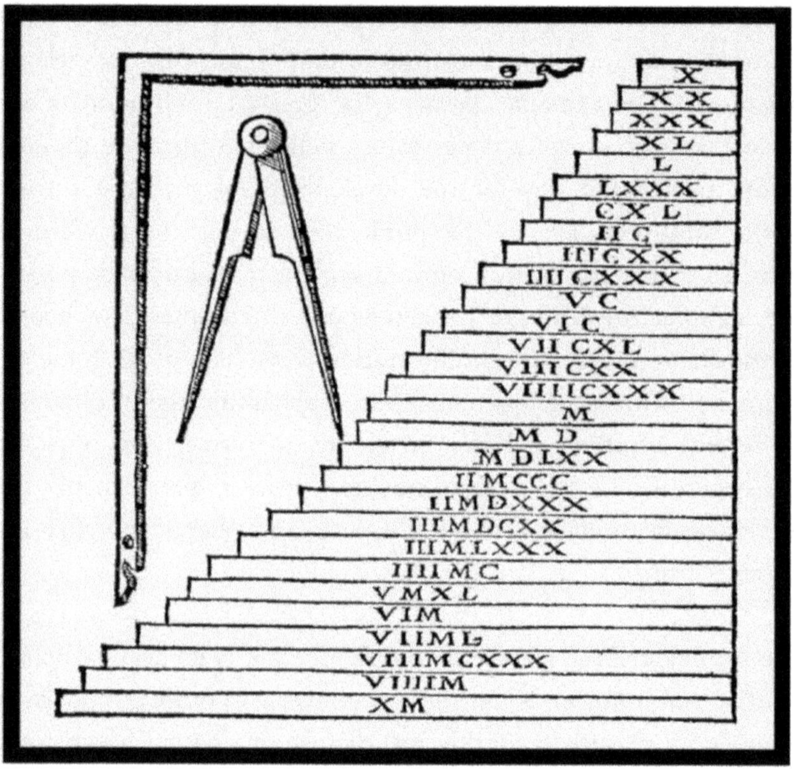


Figure 13.3: The bell scale. The length of each line represents the thickness of the rim of a bell of the designated weight. *Source*: Biringuccio, *Pirotechnia*.

In the preface to book 6, for example, Biringuccio describes the "extraordinary obstacles" involved in the casting process, work that he says requires "the bodily labors heavy as a stevedore's" (PVB, 214). His graphic description of this type of occupation—what is now sometimes called "3D" work (dangerous, difficult, and dirty)—could not have been crafted by anyone who had not seen it firsthand. His practical advice, too, of course reflects long and direct experience. Consider, for instance, his discussion of the production of a mold for a hanging bell:

> It has been discovered by skilled bell founders, more through experience than from geometrical calculation (although calculation does enter), that a certain relationship of dimensions in both large and small bells makes the tone and the weight almost certainly what is desired. (PVB, 260)

In addition, he includes a graph showing in quantitative terms the ideal thicknesses for the molds of bells of different sizes (see Figure 13.3). Clearly, this graph is the work of an author who himself was also an experienced caster. Moreover, it is the earliest example in the history of technical writing of the use of a bar graph.

These detailed accounts of the workplace and what went on there were of a kind that until Biringuccio's time had been related only among people working in the trade; they were not the sort of discussions recorded for posterity by intellectuals, who in the first place took no notice of such business. And yet here they were being published. The magnitude of this change almost defies the modern imagination.

## Section 3—Georgius Agricola

Albeit few in number, there were some university-educated people who welcomed the technical works being published by the likes of Norman and Biringuccio, men who had never set foot in the hallowed halls of academia. The business of scholarship as conducted by the universities from the time of Thomas Aquinas had by the fifteenth century become quite detached from reality, having rigidified into a formula consisting of dogmatic dialogues and interpretative commentaries on existing literature. University education had become a superficial exercise in acquiring the skills of rhetoric and argumentation necessary to serve scholarship within this limited framework. The humanism of the fifteenth-century Renaissance, which operated outside of academia and aimed to revive ancient texts, was one example of a backlash against this rigidity. Another example was the search for a new type of scholarship as the Renaissance moved into the sixteenth century—one that recognized the need to learn from the experience and accomplishments of craftsmen, engineers, and merchants at work. In *Gargantua*, published in 1534, author François Rabelais wrote of the importance of including within the scope of education the study of technical crafts such as metal casting, gemstone polishing, printing, dyeing, and goldsmithing.

Thus in the sixteenth century, intellectuals gradually began to voice the importance of scholarship that grew out of man's interaction with nature and the need for practical education in related technologies. People in many fields began to insist that the technical know-how and abilities of craftsmen and engineers were valuable assets that could lead to progress in human understanding of the natural world and that therefore these needed to be made subjects of scholarly inquiry. Some went so far as to claim that this type of academic pursuit was far and away more valuable than the contemplative, pedantic scholarship and intricate-yet-sterile theorizing that came out of the monasteries and universities.

A cosmopolitan figure born in Spain and educated in various countries throughout Europe, Juan Luis Vives associated with prominent figures, including Desiderius Erasmus and Thomas More, and is considered a pioneer in the field of psychiatry. In 1531 he wrote that a student

> should not be ashamed to enter into shops and factories, and to ask questions from craftsmen, and get to know about the details

> of their work. Formerly, learned men disdained to inquire into those things which it is of such great import to life to know and remember, and many matters were despised and so were left almost unknown to them.[14]

Similarly, Paracelsus, who was a year Vives's junior, had this to say:

> The physician does not learn everything he must know and master at high colleges alone; from time to time he must consult old women, gypsies, magicians, wayfarers, and all manner of peasant folk and random people, and learn from them; for these have more knowledge about such things than all the high colleges.

We will be taking a close look at Paracelsus in the next chapter; for now, suffice it to say that he was an itinerant doctor, was the first person in history to lecture in German at a German university, and also wrote in German. He advised that medicine should be taught "cleanly and clearly in the language of the homeland."[15]

One contemporary who put into practice the advice of Vives and Paracelsus to learn directly from tradesmen—in this case, from those involved in mining and metallurgy—was Georg Bauer (1494–1555), known by his Latin name, Georgius Agricola. Two years younger than Vives and one year younger than Paracelsus, Agricola was born in Glauchau in Saxony, near what today is the border between Germany and the Czech Republic. Agricola studied classical languages at the University of Leipzig, and in 1518 he became vice principal at a public school in the city of Zwickau, where he taught Greek and Latin. In 1522 he returned to Leipzig. As a student, this time taking up medicine, he ventured to Bologna and Padua, and after applying himself to the study of medicine and natural philosophy, he returned to Germany in 1526. It is said that while studying abroad Agricola gained favor with Erasmus and other humanists. Upon returning to his native land, he began practicing medicine in the silver-mining town of Joachimsthal and took the opportunity presented by his time there to acquire both theoretical and practical knowledge of mining and geology.

Agricola's *Bermannus*, which appeared in 1530, was his first book on mining studies. After completing this work, he moved to the mining town of Chemnitz, also in Saxony, and then proceeded to turn out a series of books on mineralogy, mining, and metallurgy, beginning with *De natura fossilium* in 1546. His magnum opus, *De re metallica* (On metals), consists of twelve books and took over twenty years to complete; it was published posthumously, in 1556. Most of Agricola's works were published in Basel by Froben, a publishing house founded by Johann Froben. Froben was an important cultural figure in the Renaissance and a leader in the history of printing and publishing. His publishing house had close ties with Erasmus, who was the first to read the manuscript of *Bermannus* and, deeply impressed, had been eager to see it published.

Like *The Pirotechnia*, *De re metallica* is a compilation of what were then the latest theories and techniques concerning mining and metallurgy. The series of twelve books is organized as follows by topic (though no actual chapter titles are given).

> Book 1: The importance of mining and metallurgy
> Book 2: Mining as a business, the location of mine shafts, and the surveying of veins
> Book 3: Veins
> Book 4: The establishment of mining areas and the responsibilities of mining officials
> Book 5: Digging techniques

Book 6: Tools and machinery
Book 7: Assaying
Book 8: Ore dressing
Book 9: Smelting
Book 10: Techniques for separating gold and silver and for separating lead
Book 11: Techniques for separating copper and silver
Book 12: Techniques for manufacturing salt, soda, and alum

The work as a whole has less to say about iron and steel than it does about gold, silver, and copper, and it also fails to mention the blast furnace, which at the time of publication had recently come into use along the Rhine; these omissions demonstrate how constrained *De re metallica*'s content was by the place and time in which Agricola wrote it. Agricola had of course seen *The Pirotechnia*, and *De re metallica* shows signs here and there of having been based on that earlier work. And yet Agricola did not touch upon iron casting, despite Biringuccio's having gone deeply into this subject in *The Pirotechnia*. In contrast, *De re metallica* includes more thorough discussions than Biringuccio's work on the subjects of tunneling techniques and devices for drainage, pumping, ventilating, and lifting and crushing ore.

*De re metallica* was published in Italian in 1563 and in German (in Basel and Frankfurt) in 1580. In 1621, new editions came out in both of these languages, so the book was no doubt read by engineers and tradesmen who were unschooled in Latin. Furthermore, a copy was apparently placed upon a church altar, as though it were a Bible, near the mine of Potosi, in Bolivia, which was started by Europeans.[16]

*De re metallica* is one of several books of that period that stand out for their many beautiful and realistic illustrations. In that respect, it shares the company of Agostino Ramelli's *Various and Ingenious Machines*; Ercker's *Treatise on Ores and Assaying*; Leonhart Fuchs's book on botany, *De historia stirpium commentarii insignes* (Notable commentaries on the history of plants), published in 1542 at Basel; and *Fabrica*, a 1543 work written by Andreas Vesalius that is said to have revolutionized the study of anatomy. Agricola wrote that he hired an illustrator for *De re metallica* using his own funds. The large number of illustrations and the meticulous, vivid quality in which they are rendered are evidence of a move away from contemplative scholarship, with its narrow focus on the refinement of terminology and logic, to a practical scholarship that relied on actual observation and visual understanding. To put it another way, the illustrations convey the distinct impression of an intellectual shift away from dependence on the written word and toward an emphasis on experience. The reproduction of a page from *De re metallica* shown in Figure 13.4 is one example of how Agricola attempted to reveal details that would normally be hidden from view and to faithfully convey meaning so that readers could understand at a glance that to which words alone could not do justice. Incidentally, Figure 13.5 also demonstrates that *De re metallica* is primarily a series of books about machinery on a large scale—machinery powered not only by man but also by beasts and water.

Book 1 begins with this observation:

> Many persons hold the opinion that the metal industries are fortuitous and that the occupation is one of sordid toil, and altogether a kind of business requiring not so much skill as labour. But as for myself, when I reflect carefully upon its special points one by one, it appears to be far otherwise.[17]

Figure 13.4: Large-scale machinery used in a German mine. *Source*: The 1561 Latin edition of Agricola's *De re metallica*, Libri XII (reprinted by Edindutria Editoriale S.p.A, Rome, 1959).

In this work, Agricola does battle with the prejudiced view that "the scum of the miners exist wholly by fraud, deceit, and lying" (DRM, 21). Again and again he emphasizes, for example, that "the gain derived from mining is not sordid" (DRM, 22), that "mining is a calling of peculiar dignity" (DRM, 24), and so on. Statements like these tell the tale of how outsiders of that era tended to view the mining industry and those who toiled in it; until Agricola came along, no one would have dreamed that a respectable scholar might actually get to know this supposedly base, despicable business firsthand, study it, and record his findings for posterity. Of course, the method of research that we now call fieldwork was not yet known. For this reason, it is no exaggeration to say that the greatest significance of *De re metallica* lies precisely in its having taken the mining industry as a whole and turned it into a subject of intellectual inquiry.

Moreover, Agricola saw mining not just as a single general subject but rather as a field whose specialists required multidisciplinary knowledge. To begin with, they needed everything from geology and mineralogy to chemistry for digging and smelting. Beyond this, they also needed physics for locating mineral deposits and understanding their various characteristics; medicine for seeing to the welfare of mining laborers; astronomy for judging the direction in which mineral veins extended; surveying for the purpose of tunneling; arithmetic for managing finances; architecture for constructing necessary facilities; and drafting for manufacturing tools and materials. Furthermore, mine operators also needed knowledge of mining laws pertaining to land claims (DRM, 3–4). Indeed, the bulk of this knowledge was a prerequisite for any entrepreneur wishing to go into the mining business, and it was precisely this laundry list of subjects—everything from the technology to the bookkeeping—that subsequently became a new set of academic disciplines.

In the preface to *De re metallica*, Agricola declares as follows:

> I have omitted all those things which I have not myself seen, or have not read or heard of from persons upon whom I can rely. That which I have neither seen, nor carefully considered after reading or hearing of, I have not written about. (DRM, xxx–xxxi)

Agricola's writing, just like Biringuccio's before him, is filled with the spirit of unequivocal empiricism, with an elevation of the concept of firsthand, on-site learning. For these two men, ancient literature and the great philosophers were now utterly without authority.

## Section 4— Attitudes toward Alchemy

From what we have seen so far, we can say that both Biringuccio and Agricola write like rational men of the modern age—like industrial capitalists. They reject outright most of the superstitions that had been passed down among both alchemists and miners. Biringuccio rejects as a delusion the alchemists' notion of a "philosopher's stone" and asserts that gold produced through alchemy is something "which I believe not only does not exist but also, in truth, has never been seen by anyone, although many claim to have seen it (DRM, 42). Instead, he says, "I am drawn … to follow the path of mining more willingly than that of alchemy, even though mining is a harder task, both physical and mental, is more expensive, and promises less at first sight … than does alchemy" (DRM, 40).

In the preface to *De re metallica*, Agricola even remarks rather sarcastically that since "there have been so few who have written on the subject of the metals, it appears to me all the more wonderful that so many alchemists have arisen who would compound metals artificially, and who would change one into another"; and that

> seeing that we do not read of any of them ever having become rich by this art, nor do we now see them growing rich, although so many nations everywhere have produced, and are producing, alchemists, and all of them are straining every nerve night and day to the end that they may heap a great quantity of gold and silver, I should say the matter is dubious." (DRM, xxvii–xxviii)

He also dismisses claims long passed down among miners about the efficacy of the divining rod—specifically, the *virgula divina*, a type of divining rod made from the forked twig of a hazel tree. Regarding people who used such rods, folklore in those days held that "the moment they place their feet on a vein the twig immediately turns and twists, and so by its action discloses the vein" (DRM, 39). Agricola's unequivocal advice to miners tempted by such stories was that "the divining rod passed to the mines from its impure origin with the magicians," and that "a miner, since we think he ought to be a good and serious man, should not make use of an enchanted twig, because if he is prudent and skilled in the natural signs, he understands that a forked stick is of no use to him" (DRM, 41). In *The Pirotechnia*, Biringuccio likewise dismisses similar claims as fairy tales (PVB, 14).

Despite these views, however, neither Agricola nor Biringuccio was able to completely transcend his particular social milieu. In ancient times, Pliny, writing in *Natural History*, had claimed that mines left untouched for a time would grow more productive.[18] And Origen, a church father from the third century, includes the following remark in the context of discussing the animate versus the inanimate:

> Other things have the cause of their movement in themselves, as animals, trees and all things which are held together either by natural life or by soul. Among these some think that even the veins of metals should be reckoned.[19]

This idea—that veins of ore were living things, the outgrowths of minerals generated within the bellies of mines—had been passed down throughout Europe without interruption from ancient times. Early in the sixteenth century, the Italian humanist Pietro Martire d'Anghiera likewise asserted that "the vayne of golde is a lyuinge tree."[20] And Biringuccio echoes this idea, albeit with some reservations and qualifications, in a passage in *The Pirotechnia* about a mine on the island of Elba:

> [Though the ore of Elba] has been mined in such quantities for so many centuries, and is still mined even today so that not only the mountains but even two islands like that one should have been leveled, yet nevertheless it is still mined today and is of better quality than ever before. Thus it is the opinion of many that within a certain time the ore is regenerated anew in that soil which has already been mined. If, indeed, this were true, it would be a great thing and would demonstrate a great provision of Nature or a great power of the heavens. (PVB, 62)

Biringuccio goes further still in the preface to book 1 of *The Pirotechnia*, painting the following portrait of a typical mineral deposit:

> They [i.e., ores] show themselves almost like the veins of blood in the bodies of animals, or the branches of trees spread out in

> different directions. Indeed, careful investigators of minerals, wishing to show by analogy how ores are located in mountains, have drawn a large tree with many branches, planted in the middle of the base of a mountain. From its principal trunk extend various branches, some thick, some slender, exactly like real trees in mature forests. They think that these grow and enlarge continually and draw themselves toward the sky, ever converting into their own nature the most disposed adjacent materials so that finally the tips arrive at the summit of the mountain and emerge with clear sign, sending forth, in place of leaves and blossoms, blue and green fumosities, marcasites with small veins of heavy mineral, or other composition of tinctures. (PVB, 13)

Although it is difficult to guess how literally Biringuccio took this notion of *planta sepulta*, or "buried plants," we do know that in his day these deposits were generally referred to as *arbores subterraneae*, or "underground trees," and were thought to be gigantic living things that continually grew. In 1525, Paracelsus had asserted that "the older its veins, the richer and more abundant in metals is the mine," and that "all metals, so long as they remain in their matrix, so long do they continually increase.[21] And in 1574, Ercker had claimed that "God, the Almighty Creator, when He first created the world, also deposited the minerals in the rocks with their fissures and fractures and made them grow like other creatures."[22]

Robert Boyle, writing in the second half of the seventeenth century, recorded testimony by mine owners to the effect that mines housing various types of metals had made great recoveries, the veins regaining their abundance after being left alone for a number of years.[23] In 1735, physician, botanist, and zoologist Carl von Linné went so far as to state flatly that "*Minerals* grow." And even as late as 1759, German physicist Franz Aepinus wrote in his *Essay on the Theory of Electricity and Magnetism* that "as many Physicists think, metals are still growing in the earth today."[24]

Furthermore, it was thought that metals went through a process in which they matured from baser to nobler materials. It was believed that, as nature always perfected whatever it sought to produce, the end product of this process, unless thwarted by some external obstacle, was gold. In this sense, all metal was gold in its potential form. But when nature met with resistance or with some obstacle, the process was either aborted or perverted, resulting in a baser metal. This thinking formed part of the theoretical basis for alchemy: alchemists thought that by creating appropriate conditions they could accelerate the sluggish natural process by which metals grew in the womb of the Earth, artificially inducing an early completion. Geber (Jabir Ibn Hayyan), an Islamic alchemist of the eighth century, held that alchemy "treats of the *Imperfect Bodies* of *Minerals*, and teacheth how to perfect them."[25]

Biringuccio has no objection to alchemy as such. He maintains, for example, that "the goldsmith's craft" includes certain secrets that "in truth are parts of alchemy" (PVB, 363). He distinguishes between alchemy that is "sophistic, violent, and unnatural" and therefore criminal and fraudulent (PVB, 337), and a different kind of alchemy whose practitioners he calls "imitators and assistants of Nature" (PVB, 336). He approves of the latter and praises them for being "manipulators and true physicians of mineral bodies, purging them of superfluities and assisting them by augmenting their virtue and freeing them from their defects" (PVB, 336). Biringuccio also values the process of alchemy for the many by-products it offers, from new medicines to perfumes and dyes and countless other compounds, as well as derivative technologies such as a certain distillation technique. His overall conclusion is thoroughly pragmatic:

> Thus … this art is the origin and foundation of many other arts, wherefore it should be held in reverence and practiced. But he who practices it must be ignorant neither of cause nor of natural effects, and not too poor to support the expense. (PVB, 337)

An engineer to his very core, Biringuccio rejected anything that claimed to surpass nature. Nonetheless, he shared with alchemists an idea that had been around since Roger Bacon: that there dwelled in the natural realm forces and forms of life that people, though unable to fathom them, could nonetheless imitate and utilize.

## Section 5—Biringuccio's and Agricola's Understanding of Magnetism

Bearing in mind the background that we've just been discussing, let's turn our attention back to the magnet. Both magnets and compasses were used in the digging of mines and in smelting. In the preface to *The Pirotechnia*, Biringuccio writes about the use of magnetic compasses for the purpose of digging tunnels that did not veer from a straight line. This use also comes up in book 5 of Agricola's *De re metallica*. Moreover, both book 7 of *De re metallica* and book 4 of Ercker's treatise refer to the use of the magnet in assaying mineral samples for iron ore.[26]

We can of course assume that both Agricola and Biringuccio had firsthand experience with magnets. In fact, Biringuccio writes that "we know very well that kind of lodestone that is in our regions, that we see every day," and he demonstrates his knowledge of the magnet thus:

> It is a mineral material although it does not melt, and since it does not melt it cannot be said that it contains metal. Wherefore I will say, regarding the question of its beginnings, that it is generated from the common composition of stones with a good deal of metallic elemental mixture. This is clearly shown by its color as well as by its great weight and by the fact of its birth among iron ore. For the same reasons that Nature wishes that like always hunger for like, it appears that this hungers for iron, and for other things if it be true. I once saw a piece of this that had such a great desire for iron that, since it could not draw the disproportionate quantity of the iron to itself by its own force, it became almost alive, moved toward it and itself started off to visit the iron in order to be joined with it. (PVB, 115)

With the second half of this excerpt we see the first written record of the observation that not only is iron drawn to lodestone, but lodestone is likewise drawn to iron. In terms of intellectual history, this remark is significant because it divests the magnet of the status that Aristotle had granted it—that of a so-called first mover, an entity that moves of its own accord, without being impelled by anything external. Biringuccio's observation is also interesting in that we know without doubt that it was a product of his firsthand experience.

Nonetheless, in parts of *The Pirotechnia* Biringuccio shows no hesitation in recording mere hearsay. He relates, for example, that

> it is said that all [lodestones] lose their power and virtue every time that a rough diamond comes near them or when the lodestone and the thing it attracts are bathed in goat's milk or garlic juice, or are greased with oil. (PVB, 116)

He later elaborates on the legendary relationship between the diamond and the magnet, demonstrating a lack of critical discrimination concerning a claim that had been passed down from the time of Saint Augustine:

> With its presence it [i.e., diamond] stops the virtue of lodestone so that it cannot attract iron, and if it has already attracted and taken hold of it, the diamond makes it let go. (PVB, 123)

And this mindset is even more conspicuous in Biringuccio's decision to include in *The Pirotechnia* the following bit of hearsay, which was commonly related among ordinary people of his day and which we in modern times can only regard as superstition:

> They say that some lodestone is found that has a white color and that this has the property of attracting flesh, because it is very efficacious in facilitating the parturitions of women if it is tied to the left thigh of the pregnant in childbirth. They also say that whoever wears it so that it touches the skin has the power of inducing great benevolence in the minds of persons. This is particularly true if it has engraved on it the conjunction of Venus and Mars with their symbols and images. (PVB, 115)

We cannot fail to remark on the vast gulf that thus clearly existed between Biringuccio's expressed views on the magnet and the factual and empirical statements he made on other subjects.

In that respect, there was scant difference between Biringuccio and Agricola. Let's take a look at the principal work in which Agricola's thoughts on the magnet are expressed, which, as it happens, is not *De re metallica* but *De natura fossilium*, published in 1546. The latter work is considered the earliest "textbook" on mineralogy, and it includes a detailed list of all minerals known at that time, along with their characteristics—the first known such compilation since Theophrastus and Pliny.

*De natura fossilium*[27] begins as follows: “Mineral substances vary greatly in color, transparency, luster, brilliance, odor, taste, and other properties which are shown by their strength and weakness, shape, and form” (DNF, 5). And a few pages later we find this:

> Minerals differ even more in other qualities which are perceived by touch and which reveal the position of some, and in the well-known qualities due to strength and weakness, namely, unctuousness and meagerness, density and porosity, hardness and softness, roughness and smoothness, heaviness and lightness, and many others. (DNF, 9)

These passages show that Agricola’s criteria for classifying minerals were chiefly sensory, in particular involving appearance, texture, taste, and odor; he seems to have given almost no consideration to chemical characteristics. His criteria were, in fact, the accepted standard for identifying and classifying different types of ore in the mining industry at that time.

Agricola distinguishes between what he calls noncomposite and composite mineral bodies. The former is divided into four subtypes: earth, congealed juice, stone, and metal. Agricola defines earth in this context much the way we think of it in common-sense terms, namely, as “a simple mineral body which can be worked in the hands when it is moistened and from which mud can be made when it is saturated with water” (DNF, 17–18).

The category of “congealed juice” is divided into “harsh” and “unctuous” kinds. Examples of the former include alum, halite (rock salt), *nitrum* (saltpeter), sulfur, realgar (red arsenic), and bitumen, as well as such metal oxides as aerugo (copper rust) and iron rust. On the other hand, amber and jet are included among the "unctuous" kind, that is, as forms of congealed juice that is smooth. Agricola makes the following assertions about jet and amber:

> Having been warmed by rubbing [jet] will pick up small light objects as does amber. It will not do this unless it has a natural brilliancy or has been polished. (DNF, 68)
>
> …
>
> The Greeks call [amber] ἤλεκτρον [*electron*] because when it is rubbed it will attract chaff and other small light objects. (DNF, 70–71)

This is thought to be the first time that the "amber phenomenon" was attributed not to frictional heat but rather to friction itself. Several pages later, however, the distinction is blurred:

> Having been warmed by rubbing amber will draw and support feathers, chaff, balls, leaves and other small light substances in the same manner as lodestone attracts iron.… It will pick up metal shavings. (DNF, 77)

Clearly, Agricola does not deliberately distinguish between friction itself and friction-induced heat as factors contributing to static electricity. This distinction would not gain recognition until after Gilbert had made it.

Agricola defines the fourth subtype of mineral body, metal, as "a natural mineral body which is either liquid or solid and will melt in a fire"; and, contrastingly, he defines stone as "a dry, hard mineral body that may soften a little after standing in water for a long time and is reduced to a powder in fire or is not softened in water and melts in only the very hottest fire" (DNF, 18–19).

Agricola further divides stones into four subcategories: stones without a special name, gems, marbles, and rocks. Among rocks he counts limestone and sandstone. Marble is, in fact, a type of limestone, but Agricola's criteria in establishing these categories are based on external appearance and other sensory qualities; he distinguishes marble from other rocks because it shines when polished.

Agricola places magnets, along with hematite, among stones without a special name. His decision not to place these among metals has a historical basis in Theophrastus and Albert the Great, who did likewise in their mineral classifications. Agricola believes, along with Biringuccio, that magnets cannot melt.

Book 5, on stones without a special name, begins with a discussion of magnets that includes the following:

> Theologians attribute the powers this mineral possesses to divine origin, scientists to natural origins, the nature of which cannot be interpreted. (DNF, 84)

Not that even Agricola attempts to explain magnetic force; most of his knowledge about it comes straight from Pliny, and for the most part he offers no new knowledge or thoughts on the subject. Nonetheless, *De natura fossilium* provides a record of what people in Agricola's time were saying about the magnet and magnetism. Let's look at two more related statements:

> Many people know that lodestone will pick up iron but only a few know that this power is weakened and destroyed if the stone is immersed in an acid. (DNF, 5)
>
> ...
>
> Magnetite removes excess fat. (DNF, 13)

It is not clear what specifically he means by "acid" and "fat," but that is beside the point; what we must note here is Agricola's uncritical attitude and his extreme credulity with regard to these tales about the magnet. In *De re metallica*, he echoes Biringuccio by claiming that

> a magnet smeared with garlic juice cannot attract iron. (DRM, 39)

And in *De natura fossilium*, he makes the following claims:

> Lodestone will not attract iron that is covered with rust, impure iron or iron that has been smeared with the juice of onion or garlic.
>
> To no less degree, diamond resists its power. If a diamond is placed next to iron a lodestone cannot draw it or if iron has been attracted to the stone as soon as a diamond is placed along side of it the iron will drop. (DNF, 85)

And he goes on to declare that

> *Theamedes* has the opposite power to lodestone since the latter attracts iron, the former repels it.... Even today stones are found composed in part of lodestone, in part of *theamedes*. (DNF, 85–86)

Paolo Rossi asserts that Agricola's work is characterized, and distinguished from its predecessors, by an "ideal of clarity and the conscious departure from the fabulous."[28] That this same Agricola nonetheless put forward such views about the magnet and magnetism testifies to just how elusive this force of nature was to even the keenest minds of that era.

* * *

Technologies related to mining, to navigation, and to warfare expanded rapidly from the second half of the fifteenth century and on into the sixteenth. In the sixteenth century, mining engineers and seamen, as well as soldiers and military surgeons, began to engage in substantive inquiry into and discourse about these technologies and to take up questions that expressed related theoretical problems. Moreover, such discourse took place in the native languages of the participants rather than in Latin. The emerging book-printing and publishing industries accelerated this trend and broadened the urban population's intellectual interest in the natural world and technology. The seismic shift this effected in the world of knowledge was nothing short of a cultural revolution. Among the university-educated as well, there now began to appear those who took up as topics of scholarly research the very issues of technology that academics had previously ignored.

Biringuccio's *The Pirotechnica* and Agricola's *De re metallica* were two products of this trend. In both works, the authors present consistently empirical discussions based on real-life experience, marking a clear departure from the argument-based scholarship of the universities, which was dependent on past authority and literature. Yet, when it came to the magnet, even these two comparatively modern figures unquestioningly repeated the superstitious tales passed down in Europe over several millennia. Despite their empiricism and positivism, these thinkers appeared to suddenly take a giant step backward when dealing with the subject of magnetic force.

As things thus stood in the mid-sixteenth century, the puzzle of action at a distance as manifested by magnetic force remained a major impediment to the advancement of modern science.

Chapter 14

# Paracelsus and Magnet Therapy

## Section 1—Paracelsus

The process by which modern empirical science developed and displaced Scholasticism was anything but straightforward, and the physician Paracelsus (1493–1541; known also as Philippus Aureolus Theophrastus Bombastus von Hohenheim) embodied the contradictions inherent in that transition. He was born in Einsiedeln, near Zurich, Switzerland, and moved to Villach, a mining town in Kärnten, Austria, at the age of nine. It seems that this exposure to the environment in which miners lived and worked directly contributed to his later discovery of what he termed "miners' sickness," and that life among laborers under the rule of the Fugger family, which virtually monopolized the mining industry, nurtured his sympathy for the lower classes. It was in Kärnten that he obtained his first lessons in medicine from his father, a doctor, and acquired some knowledge of science and metallurgy, which at that time were closely tied to alchemy. He then left home to wander Europe as an impoverished student of medicine.

The outdated medicine taught at universities relied solely on the written word in the form of works that constituted a sort of medical canon—works by such ancient and medieval authorities as Galen and Avicenna. Lecture content was not just conservative but openly contemptuous of practical training and clinical experience in anatomy and pathology. Reading and recitation of the ancient texts was considered almost the sole means for training physicians. On the rare occasion that a dissection of a human body was performed, the students would merely observe a lowly surgeon making hasty incisions in the cadaver, following the instructions of a professor at the podium reading aloud from a text by Galen.[1]

Paracelsus studied at a number of universities in Europe, and although there is no proof, some claim that he obtained a degree at the University of Ferrara. After completing his studies, he continued to roam all over Europe, at times as an army surgeon, traveling from one battle to the next. The year 1526, however, represented a major turning point in his life. While in Basel, he successfully cured the famous publisher Johann Froben of gangrene without amputating his leg, and this catapulted Paracelsus to fame. With the support of the Dutch humanist Desiderius Erasmus and of Johannes Oecolampadius, a leader of the Protestant Reformation, in June 1527 Paracelsus was appointed municipal physician, a post that involved lecturing at the University of Basel. This was the first time he was permitted the privilege of lecturing at a university.

In introducing his lectures, Paracelsus announced that his educational approach would be based on his long years of clinical experience:

> What a doctor needs is not eloquence or knowledge of language and of books, illustrious though they be, but profound knowledge

> of Nature and her works.... The doctor must know the causes and symptoms of the disease and use his judgment to prescribe the right medicine.
>
> Thanks to the liberal allowance the gentlemen of Basle have granted for that purpose, I shall explain the textbooks which I have written on surgery and pathology, every day for two hours, for the greatest benefit of the audience, as an introduction to my healing methods. I do not compile them from excerpts of Hippocrates or Galen. In ceaseless toil I created them anew upon the foundation of experience, the supreme teacher of all things.[2]

On the one hand, this statement was a manifesto proclaiming his intent to transform the sterile, impractical, convention-bound study of medicine into a practical, empirical, clinical discipline. At the same time, it represented an outright rejection of Galen's humoral pathology, which the universities considered infallible. "I do not believe," he asserted, "in the ancient doctrine of complexions and humors which has been falsely supposed to account for all diseases." Moreover, he went so far as to claim: "Who does not know that most contemporary doctors fail because they slavishly abide by the precepts of Avicenna, Galen, and Hippocrates, as though these were Apollo's oracles from which it is not allowed to digress by a finger's breadth." In essence, he was declaring war on the medical education system of the universities.

Incensed, university officials announced that any students who attended Paracelsus's lectures would not be permitted to sit their exams and refused to give him access to the university's lecture halls. In response, Paracelsus held his lectures off campus and, in a further challenge to academism, delivered them in vernacular German rather than Latin. His cantankerous nature was also a source of trouble, although lack of understanding among the general public and jealousy on the part of his colleagues certainly contributed. With the death of Froben, Paracelsus lost the last of his few patrons, and in February 1528 he left Basel after being forced out of his post as municipal physician.

He resumed his travels, continuing to record his medical philosophy until his death in 1541 at the age of 48. He died in poverty in Salzburg, a foreign land. Although he was a prolific writer, the majority of his works never saw the light of day during his lifetime, partly because many people dismissed him as a charlatan or megalomaniac, and partly because he filled his pages with inflammatory abuse aimed at academic medicine and university physicians. Posthumously, however, rumors of his supposedly miraculous cures spread, spurring greater interest in and appreciation for his work. Forty-three years after Paracelsus's death, Italian philosopher Giordano Bruno wrote in *Cause, Principle and Unity* that "a man who knows no Greek, nor Arabic, nor perhaps Latin, like Paracelsus, can have a better knowledge of the properties of drugs and medicine than Galen, Avicenna and all those who communicate with the Roman tongue."[3] Soon people were vying to find and print his unpublished works, which were scattered around Switzerland and Germany.

Paracelsus was a very religious man, but this does not mean he supported any of the religious sects formed during the Reformation. While outwardly he remained a Catholic to his death, his beliefs deviated considerably from church doctrine and bore not a shade of resemblance to the Scholasticism espoused by the universities. He appears to have been critical of the activism engaged in by Martin Luther, his contemporary, yet he sympathized with the impoverished peasants who rebelled during the Peasants' War, begun in 1524. The persecution he experienced everywhere he went in Germany essentially stemmed from class discrimination. Late in life he wrote in his *Seven Defensiones* (Seven defenses) that "what one received in youth sticks to one all one's days"; and that "those who are brought up in soft raiment and in the women's apartments, and who grow up among fir-cones do not understand one another well."[4] True or not, this certainly seems to be how Paracelsus felt.

A study of medieval medicine states:

> The university educated physician took care of only a comparatively small proportion of the sick or injured in the medieval period. He was the medical adviser to the king or pope, leading nobility, and bishops, but if he entered practice he did so in the larger cities where the greatest opportunities existed. Only rarely was he found in the countryside, the villages, or even in the smaller cities.[5]

The situation was almost the same in the time of Paracelsus, and his indictment of the status quo was directed at the privileged class of established physicians and university scholars. Behind his criticism lay his ethical views on the class system.

The *Great Surgery*, Paracelsus's chief work and one of the few published during his lifetime, was printed in Augsburg in 1536. In it, he recalls his travels throughout Europe in quest of medical knowledge:

> I eagerly and diligently investigated and sought after the tested and reliable arts of medicine. I went not only to the doctors, but also to barbers, bathkeepers, learned physicians, women, and magicians who pursue the art of healing; I went to alchemists, to monasteries, to nobles and common folk, to the experts and the simple.[6]

Regarding his reference to "learned physicians," note that in his day "physician" referred to university-educated doctors of internal medicine. In Germany, the dirty, manual work of actual clinical procedure was entirely relegated to "barbers" trained under an apprenticeship system. The latter were held in contempt as "illiterate surgeons," mere tradesmen capable only of surgical interventions. Similarly, bathkeepers performed cupping and bloodletting at public baths. For the masses, medical treatment rested entirely in the hands of barbers, bathkeepers, midwives, and the like. Moreover, Paracelsus records in his *Great Surgery* that tinsmiths used slag to stop bleeding and to dry festering sores; that blacksmiths used powdered iron oxide, called "crocus of Mars," to treat wounds; and that potters used litharge (lead monoxide) for the same purpose.[7] Tradesmen thus practiced and passed down methods of medical treatment particular to their respective professions. Paracelsus in his travels actively sought out, learned, and mastered precisely these practices—the folk medicine that had not merely been dismissed but often completely ignored by learned physicians as lowbrow and unscholarly.

To learn not from the books of the ancients but from the experience and practice of those who actually provided medical treatment—this was the starting point of Paracelsus's medical reform.

It must be noted that Paracelsus's criticism of the training of physicians and his doubts concerning the medical content taught at universities were provoked primarily not by theory but rather by the sober truth of contemporary medicine's futility in actual practice. In a passage of the *Great Surgery* that precedes the one quoted above he observes that "in my time, there has been no physician who could even cure a toothache with certainty, to say nothing of more severe illness."[8] Even allowing for some exaggeration here, we can be certain that Paracelsus was correct in finding it impossible to ignore the miserable state of academic medicine with its sterile theories badly in need of replacement by practical and clinical experience. His motto was, “The patients are your textbook, the sickbed is your study.”[9]

After he left Basel, Paracelsus went to Colmar and Nuremberg. His treatment and clinical observations of syphilis, or “French disease” as it was known, are still highly respected. This disease swept across Europe during the age of discovery and was beyond the capacity of university physicians, whose training largely began and ended with commentaries on Galen and Avicenna, to whom syphilis was unknown. In *On the Origin and Causes of the French Disease*, written in 1529, Paracelsus states his position clearly: “The right path does not consist in speculation, but leads deep into experience. From experience the physician receives his help, and upon it rests all his skill.”[10]

In 1533 Paracelsus visited a mining region along the Inn River, a tributary of the Danube in Tirol. There he focused his attention on the harsh, hazardous working conditions in the mining and smelting industry and on lung ailments prevalent among men employed in those trades. The result was his discovery of “miners’ sickness,” which he recorded in *On the Miners’ Sickness and Other Miners’Diseases.* This work represents a groundbreaking achievement in the history of medical theory, as it was the first to recognize any disease as being of occupational origin. It contains a record of the symptoms of “miners’ sickness” and identifies it as a lung disease caused by inhalation of dust (a condition now called “pneumoconiosis”) and exposure to mineral pollutants in toxic steam, such as lead, arsenic, and mercury. Moreover, the work was inspired not only by its author’s ethical perspective as a physician but also by his sympathy for the poor, the weak, and the oppressed—by his class consciousness. In terms of scientific methodology, particularly noteworthy are his conclusions, which are that "nothing is found about these diseases in the old writers"; that this is why "they remained undescribed until now, and the cure too was omitted"; and that "the most fundamental thing is to hasten to experience."[11] This position could have been arrived at only from Paracelsus's underlying conviction that doctors must learn by observing their patients' actual conditions and symptoms and by accumulating clinical experience.

Four years later, Paracelsus wrote *Seven Defensiones* and *The Doctor's Labyrinth*. In the former he remarks that anyone "who would explore [nature], must tread her books with his feet," and in the latter he declares: "Practice should not be based on speculative theory; theory should be derived from practice."[12] His study on "miners' sickness" was a product of this motto and a tremendous triumph.

But while Paracelsus placed great importance on "experience" and "practice," he did not necessarily see these terms in what would pass as modern, rational concepts within a logical framework. Francis Bacon noted at the beginning of the seventeenth century that "in all times, in the opinion of the multitude, witches and old women and impostors have had a competition with physicians"[13]; and indeed, folk medicine coexisted with academic medicine in Europe until the beginning of the modern period. In fact, folk medicine was the only medicine available to the masses everywhere, not just in rural or mountainous areas. Paracelsus observed the treatment methods passed down among tradesmen, the balneotherapy practiced in hotspring areas, the herbology of village elders, midwifery, and incantations by magicians. The majority of this was folklore, much of which appears to us today as being more akin to magic and superstition than to medicine.

While he did offer a logical basis to explain these practices, one that criticized Scholasticism, this was not modern empiricism but rather a combination of Hermeticism, astrology, and alchemy—in a nutshell, natural magic. And unlike his contemporary Agricola, who likewise practiced medicine in a mining region but had already embraced the ethos of industrial capitalists, Paracelsus's sympathies lay with the disintegrating peasant class and the newly emerging proletariat, and, consequently, he was more strongly influenced by premodern mysticism.

## Section 2—Paracelsus's Medicine

In *Paragranum* (1530), which presents the theoretical basis for his "new medicine," Paracelsus mentions four pillars of medicine: philosophy, astronomy (astrology), alchemy, and (moral) virtue.

> The first pillar is the philosophical knowledge of earth and water; the second, astronomy, supplies its full understanding of that which is of fiery and airy nature; the third is an adequate explanation of the properties of all the four elements—that is to say, of the whole cosmos—and an introduction into the art of their transformations; and finally, the fourth shows the physician those virtues which must stay with him up until his death, and it should support and complete the three pillars.[14]

Paracelsus's first pillar is natural philosophy, the study of nature and all that exists on this planet, while the fourth pillar, virtue, is in modern terms a philosophy focused on the professional ethics and responsibilities of a physician—or, indeed, on the physician's *love* for his patients.

The second pillar is that of astronomy and astrology. According to scholars on the subject, Paracelsus was known in his day not as a medical reformer but rather for his discourse on prophecies and astrology.[15] Having studied in Italy during his youth, he was greatly influenced by magical thought and Hermeticism. Although he was not inclined to praise others, he called Ficino—the author of *Three Books on Life*, a treatise on astrological medicine—the "best Italian doctor."[16]

Like Ficino, Paracelsus views man as a "microcosm" that corresponds to the "macrocosm." In *Paragranum* he asserts, "Heaven is man, and man is heaven."[17] In other words, "the sun and the moon and all planets, as well as all the stars and the whole chaos, are in man." "The inner stars of man," however, "are, in their properties, kind, and nature, by their course and position, like his outer stars, and different only in form and in material." Man, therefore, "can be understood only as an image of the macrocosm."[18] In fact, in *Volumen paramirum*, which preceded *Paragranum* and concerns the causes of disease, he maintains that the sun, moon, and planets of the "macrocosm" correspond to organs in the human body. For example, the heart corresponds to the sun; the brain, to the moon; the spleen, to Saturn; the liver, to Jupiter; the gallbladder, to Mars; the kidneys, to Venus; and the lungs, to Mercury.[19]

This view naturally led to the belief that the sun, the moon, and the planets influenced the human body in some way. It was believed, for example, that the planets caused infectious diseases and that that explained why the latter occurred in cycles. In short, at that time the macrocosm was believed to influence both the mind and the body—that is, the microcosm—of man, and for Paracelsus this meant that the causes of diseases, and consequently, the proper treatment and medication for them, were inevitably dictated by these influences. "Everything that astronomical theory has profoundly fathomed by studying the planetary aspects and the stars," he maintains, "can also be applied to the firmament of the body."[20] Thus knowledge of astronomy (astrology) is essential to medicine. In *Paragranum* Paracelsus writes: "Know that philosophers are of two kinds, the philosopher of heaven and the philosopher of the earth. Each sphere constitutes only one aspect of the physician, and in itself neither makes the whole physician."[21]

In considering the third pillar, alchemy, we must remember first of all that in Paracelsus's day there was no clear distinction between this discipline and chemistry. Second, we must note that "alchemy" in the sense that Paracelsus used the term did not mean merely making precious metals from base metals. Paracelsus, like his contemporaries, believed that metal-bearing veins grew within the Earth and that not only such growth but also the processes by which living creatures turned food into nutrients and separated out toxins, as well as such phenomena as the fermentation of grape juice into wine and the decomposition of organic matter, were all forms of alchemic metamorphosis. In this broad sense, "alchemy" meant the discovery of the natural principles governing such transformations through experimentation and observation, and the art of using this knowledge to accelerate such changes. Among the tools at the alchemist's disposal were such physical and chemical means as heating, distillation, dissolution, precipitation, and filtration, all means of making the imperfect perfect and extracting pure and useful elements. As Paracelsus puts it, "alchemy is nothing but the art which makes the impure into the pure through fire."[22] Accordingly, all those who artificially promote the process of perfection are alchemists.

> Nature is so keen and subtle in her operations that she cannot be dealt with except by a sublime and accurate mode of treatment. She brings nothing to the light that is at once perfect in itself, but leaves it to be perfected by man. This method of perfection is called Alchemy. For the Alchemist is a baker, in that he bakes bread; a wine merchant, seeing that he prepares wine; a weaver, because he produces cloths. So, whatever is poured forth from the bosom of Nature, he who adapts it to that purpose for which it is destined is an Alchemist.[23]

Paracelsus rejects Galen's view that illnesses are caused by imbalance in the body's four humors and proposes instead that disease is a localized dysfunction due to breakdown in the body's alchemic ability to sort out foreign substances and transform them into something beneficial.

In *Volumen paramirum*, Paracelsus discusses in detail what he sees as five causes of illness. One of the causes he cites is toxicity (*ens veneni*). Just as the body of a pig extracts nutrients from the refuse it eats, or a peacock isolates and eliminates the poisons from the lizards in its diet, so too "every animal has food adapted to it and which has been ordained for him by his alchemist who separates the proper materials." Similarly, people need their own food, and the alchemist in the human body separates the toxins from the nutrients. "When the meat reaches the stomach, there is the alchemist to divide the good from the bad. What does not belong to health he casts away, and sends the good to where it is needed." Clearly, the alchemist he refers to is the stomach.[24] But when this alchemist is weakened and its ability to separate and extract deteriorates, the body absorbs toxins along with necessary nutrients. These accumulate in and damage certain organs, causing illness.

Paracelsus concludes that every illness has its own cause and that, accordingly, as he states in *Seven Defensiones*, "each disease has its own physic."[25] In treatment, therefore, he places great importance on the use of specific, chemically refined medicines (*specificum*) that can act as a substitute for the body's ailing alchemic action. Thus alchemy in the narrow sense of the word as it applies to the practice of medicine entails studying the effects and efficacy of different drugs and chemically preparing effective remedies for each type of disease or injury. As he states in the *Great Surgery*: "God did not choose to give us the medicines prepared. He wants us to cook them ourselves." In this same work he writes, "I praise the art of alchemy because it reveals the mysteries of medicine and because it is helpful in all desperate illnesses." Therefore, "never must … medicine and alchemy be separated from each other."[26]

Ultimately, this meant learning to refine and extract so-called *arcana*, secret and powerful remedies believed to transmit the power of the cosmos. In other words: "It is not, as some madly assert, that Alchemy makes gold and silver. Its special work is this—to make arcana, and direct these to disease."[27] "With Paracelsus," in fact, "alchemy had shifted from a search for gold to a quest for a better pharmacopoeia."[28] Because Paracelsus's medicine was closely connected with chemistry in the form of alchemy, he is often considered to be the founder of iatrochemistry.

"A physician," Paracelsus concludes, "should first of all be an *Astronomus*"; but, at the same time, "it is also necessary that the physician be an alchemist."[29]

## Section 3—Paracelsus's Views on Magnetism

My purpose in introducing Paracelsus's philosophy is to define the landscape at this stage in our journey as we trace the development of man's understanding of magnetism. Let's see what Paracelsus has to say about this mysterious force of nature.

Paracelsus appears to have left no writings addressing head-on the question of exactly what magnetic force is. In one place he states that the magnet is "an iron stone, and so attracts iron to itself."[30] In *Astronomia magna,* completed in 1537, he writes that the magnet "has no effect on either lead or tin" because "things which are the same come together while those that are not the same, do not."[31]

In short, Paracelsus's assertion was based on the same logic that had been around since Plato's *Timaeus*—that "all kindred substances are drawn towards one another."[32]

Elsewhere, however, Paracelsus asserts that

> the spirit of iron in the magnet attracts the body of Mars to itself; and this happens not only in the magnet but in all other natural things, so that the foreign spirit which is in an alien body, which is not of its own nature, always attracts a body agreeing with its own nature.[33]

In this case, Mars means both our solar system's fourth planet and the mineral iron. This statement is based on the belief that the planets correspond not only to the organs of the body, as previously mentioned, but also to metals, as shown in Table 14.1. We have evidence that this correspondence between metals and planets was, most likely, commonly accepted throughout Europe during the Middle Ages. It comes up in "The Canon's Yeoman's Tale" in Chaucer's *The Canterbury Tales*, as well as in John Gower's *Confessio amantis* (Lover's confession), both works from the late

Table 14.1: Correspondence between Astronomical Bodies, Metals, and Human Organs[35]

| Astronomical Body | Metal | Organ |
|---|---|---|
| Saturn | Lead | Spleen |
| Jupiter | Tin | Liver |
| Mars | Iron | Gallbladder |
| Venus | Copper | Kidney |
| Mercury | Mercury | Lungs |
| Moon | Silver | Brain |
| Sun | Gold | Heart |

fourteenth century.[34] The sun's correspondence to gold and to the heart and the moon's correspondence to silver and to the brain make some sense, at least in a metaphorical, literary way; but the relationships between the other planets and metals is mystifying. There is no point, however, in trying to force these abstract relationships to conform to modern common sense or logic. Regardless of what we may think, at that time, they were accepted as having some sort of foundation.

Paracelsus seems not to have troubled himself about why iron was attracted to the magnet or what the nature of magnets was. What was important to him was the magnet's practical utility in medicine, its effects and efficacy as a modality of treatment. In *Herbarius*, written around 1525, he makes a very interesting comment concerning this point. After asserting that the power of the magnet is "above all natural understanding," he remarks as follows:

> Now this is my argument: that this power in the magnet lies open to the eyes of all physicians and yet no one has given any thought to whether the magnet can also be used with other

> things. Rather, all physicians have neglected further experience and have made do with their [peculiar sort of] kitchen gossip. This cannot be answered with honor. They have a subject like this, see plainly its effect, and [nevertheless] deal with it in a way that might advance experience no further.[36]

Paracelsus's great interest in magnets and magnetism can be inferred from this statement. And here is something intriguing: In the early summer of 1525, just before he wrote this, he had been forced to leave Salzburg due to his involvement in the Peasants' War and his association with peasant insurgents. In the process of fleeing under the scrutiny of local authorities, he abandoned a number of his belongings. Among the items later confiscated was a magnet[37], leaving us to wonder just what he used it for. Did he carry it around to conduct experiments? Or was it a necessity that travelers often carried? Whichever the case, it seems certain that Paracelsus was deeply interested in magnets. In *Herbarius* he also writes:

> I say on the basis of experience which proves all things, including that which flows from it, namely theory, that the magnet is a stone in which there is obviously an attractive power *vis a vis* iron and steel. Moreover, there is also in it an attractive power in regard to all martial illnesses in the entire body.[38]

And more specifically:

> Experience proves that the magnet draws to itself all martial illnesses from their place [in the body] to another, and draws the excrement to itself and everything to its place.[39]

Paracelsus claims here that the magnet acts on certain types of illnesses—or perhaps more accurately, on the causes of certain illnesses–and also on the secretions they generate. But just what are the "martial illnesses" he refers to as being so susceptible to magnetic influence? Paracelsus gives some concrete examples: "all fluids of women, and all fluxes of the stool." And elsewhere he writes:

> This is also the method to draw away spasm and when it is diverted into its origin, then anoint with *oleo salis*. The procedure is the same with *tetanus* with the back of the magnet. And it is the best medicine [also] for pregnant women, especially [those] with cramps.[40]

The use of powdered magnets in internal medicine and in ointments had been discussed even earlier. In the thirteenth century, a Franciscan encyclopedist, Bartholomaeus Anglicus wrote:

> Its [i.e., the magnet's] dust, perfected with *apostolico*, an application for wounds, is especially valuable for wounds. Indeed the powder of a magnet in a quantity of two drams … [and] with the juice of fennel is valuable against dropsy, spleen, and fox mange, as Avicenna says, for it draws phlegm and black bile.[41]

But the therapeutic use of magnets externally as described above, which has become quite popular in modern times, essentially started with Paracelsus. Although others had mentioned the idea before him, Paracelsus is generally credited as its originator for being the first physician known to apply it in his practice.[42]

We cannot help but wonder how a man who repeatedly emphasized the importance of experience and the need to abide by it could possibly make such preposterous claims about magnetism. The concept underpinning Paracelsus's assertions was the correspondence, detailed above, between the macrocosm and the microcosm, and specifically the supposed relationships between celestial bodies, minerals, and bodily organs. Correspondences like those said to exist between Mars, iron, and the gallbladder were a key issue in the medicine of his day, and Paracelsus spoke of the medical benefits of magnetic influence on the human body solely from that perspective.

This understanding of Paracelsus's intellectual context does not make his claims for the efficacy of magnetic therapy any more defensible in modern terms, and in fact he held other views that defy comprehension. For example, *Concerning the Nature of Things*, Paracelsus's work on alchemy, contains an odd method for increasing magnetic potency:

> Take a magnet, and heat it in the coals to such a degree that it may be at a high temperature, but still not red hot. Extinguish this immediately in the oil of the crocus of Mars.

"Thus," he goes on to explain,

> you will make a magnet so powerful that with it you can pull out the nails from a wall, and do other wonderful things which a common magnet could never accomplish.[43]

In the same work, Paracelsus explains how to make "crocus of Mars" (powdered iron oxide). He further asserts that in "the conservation of metals the first thing to learn is what are their enemies"; and also that

> the magnet should be guarded and kept from Mercury, for it exerts hostility on it as on Mars. For every magnet which common mercury touches, or which is anointed with mercurial oil, or only placed in Mercury, never afterwards attracts iron. Let no one be surprised at this; there is a natural cause for it,

> seeing that Mercury extracts the spirit of iron which the magnet holds latent in itself.[44]

Clearly, this view is derived from the belief in the correspondence between the planets and minerals. In another variation on this theme, Paracelsus writes that "metals exercise hostility amongst each other, and mutually hate one another from their inborn nature." Accordingly, he believes that whereas mercury destroys the power of the magnet, iron conversely protects it because of the supposed sympathetic relationship between the latter two. Therefore, he claims,

> for the conservation of the magnet nothing is better than filings of iron or steel. If the magnet be placed in these, not only does not its force decrease, but it grows more and more every day.[45]

In other words, Paracelsus's interest in the magnet derived from his interest in natural magic, and his claims concerning the magnet's medical and other effects were grounded in the view, still prevalent during the Renaissance, that nature was full of relationships of sympathy and antipathy, as well as in magical thought, which had held since the time of Ficino that man could exploit the powers of heaven.

## Section 4—Weapon Salve and Paracelsus's Posthumous Influence

Paracelsus died around the same time Copernicus did, the former in 1541 and the latter two years later. In retrospect, Copernicus had by far the greater—indeed, a decisive—influence on the revolution that took place in man's view of the world and nature. Interestingly, however, whereas Copernicus's theories generated little controversy for the first fifty years or so after his death, the controversy surrounding Paracelsus began much sooner and raged for a century.[46]

Montaigne regarded the conflict between Ptolemy's geocentric theory and Copernicus's heliocentric theory dispassionately, writing in 1580: "What lesson are we to learn from that, except not to worry about which of the two opinions may be true?" He then turned his attention to Paracelsus, praising him as follows:

> How long has medicine been in the world? They say that some newcomer called Paracelsus is changing or reversing the entire order of the old rules, maintaining that, up to the present, medicine has merely served to kill people. He will be able to prove that easily enough, I believe.[47]

Around that time, Copernican theory was mentioned by only a rather small number of people, and even then, at best, mainly as a technological and mathematical improvement to Ptolemaic astronomy. In contrast, the debate between Paracelsus's fervent admirers, who hailed him as a great reformer, and his detractors, who denounced him as the devil's pawn, intensified after his death as rumors of his miraculous cures spread.

In 1575, Joseph Duchesne, a French Calvinist who received his medical degree from the University of Basel while in exile, extolled Paracelsus as someone who "teacheth many things almost divinely, in Phisicke, which the thankfull posteritie can neither commend and praise sufficientlie."[48] In 1602, the English surgeon William Clowes wrote, "I can say by experience, that I have practised certaine of his inventions Chirugicall, the which I have found to be singular good, & worthy of great commendations."[49] In the seventeenth century, Flemish physician Jean Baptiste van

Helmont described him in these glowing terms: "Paracelsus was a forerunner of the true medicine. He was sent by God and endowed with knowledge. He was an ornament for his country, and all that has been said against him is not worthy to be listened to."[50] Van Helmont himself wrote a pamphlet entitled *De magnetica vulnerum curatione* (Magnetic treatment of wounds). Incidentally, it has been claimed that Van Helmont was the first to use the word "chaos" to denote gas, but Paracelsus had already used "chaos" to refer to vapor in *On the Miners' Sickness and Other Diseases of Miners*, so it would not be unreasonable to surmise that Van Helmont's new usage was influenced by Paracelsus.[51]

These are just a few examples that clearly demonstrate Paracelsus's widespread fame. And his influence extended beyond physicians. His name was used as a synonym for a good doctor in plays by John Webster and Ben Jonson, both English dramatists and contemporaries of Shakespeare.[52]

Today we understand a key part of the scientific revolution to be the process whereby Scholasticism reached a dead end and was superseded by mechanistic philosophy in the seventeenth century. However, mechanism was not the only substitute proposed: Paracelsian theory was also touted as a replacement for the outdated medicine of Galen and Avicenna taught at universities. Paracelsus's followers and disciples in England saw his chemical philosophy, which was at the heart of Paracelsian thought and which was rooted in Christianity and Hermeticism, as a new, innovative, and truly Christian alternative to the heathen Aristotelian philosophy underpinning Galen's medicine. The need to reassess the outdated theories and methods of medical education in universities was already pressing, since by then the newer approaches were "by no means fringe phenomena limited to small intellectual groups."[53]

Thus, in the century following Paracelsus's death, chemical philosophy, which viewed chemistry and alchemy as the keys to unlocking the secrets of nature, wrestled with the newly emerging mechanistic philosophy for supremacy as the "new science."

Chemical philosophy was basically Hermeticism. It held that there was a correspondence and harmony between the macrocosm (the heavens) and the microcosm (human beings), and that the latter was capable of manipulating and exploiting the powers of the former. To us, this appears to be nothing more than a delusion preceding the establishment of modern science, but Hermeticism's influence on the medicine of that day was far greater than we could ever imagine. The greatest medical discovery of the early seventeenth century, for example, was the discovery of blood circulation by William Harvey. This same Harvey declared that the heart "is the beginning of life; the sun of the microcosm, even as the sun in his turn might well be designated the heart of the world." The Hermetic thinkers, in turn, regarded Harvey's discovery as proof that man was indeed a microcosm.[54]

For our purposes in discussing the concept of force, it is noteworthy that Hermetic chemical philosophy accepted the idea of action at a distance, which had been rejected by both Aristotelianism and mechanism. Chemical philosophers saw relationships of sympathy and antipathy as the real-world manifestations of the correspondence between macrocosm and microcosm and, furthermore, saw these relationships as being none other than examples of action at a distance.

In this connection, let me fast-forward slightly to take a look at weapon salve, also known as weapon ointment, which was a topic of significant debate in the seventeenth century. Interest in weapon salve demonstrates the peculiar influence of Paracelsianism, particularly chemical philosophy.

Weapon salve was a mysterious ointment believed to produce healing effects when rubbed not on a wound but on the weapon that inflicted it. Although it seems a highly bizarre and even ridiculous treatment, there are actually reports of cases in which it appeared to work. These apparent successes have since been attributed to the fact that, in an age with little concept of hygiene, there was more chance of an open wound's healing naturally if it were simply washed and left alone rather than rubbed with contaminated ointments.[55] Be that as it may, weapon salve was referred to as a "magnetic cure" and a "cure by sympathy," and its advent was attributed to Paracelsus.

We find this treatment method bizarre and ridiculous because we are convinced that ointments have no effect on an injury unless they are directly applied to it and that medicine cannot act at a distance. At that time, however, people who accepted the validity of weapon salve were convinced of its efficacy for two reasons: they believed that magnetism could heal external wounds, and they believed that magnetism acted at a distance. Moreover, whether or not Paracelsus directly mentioned weapon salve, its supporters claimed that he originated these two concepts.

Concerning the idea that magnetism is force acting at a distance, we find the following passage in Paracelsus's treatise on mental illness, *The Diseases that Deprive Man of His Reason*:

> The stars have the power to hurt and weaken our body and to influence health and illness. They do not fall into us materially

> or substantially, but influence reason invisibly and insensibly, like a magnet attracting iron.... Such power of attraction is possessed by the moon, which tears reason out of man's head.[56]

Paracelsus believed that just as the faraway planets influenced human beings, so too magnetic force acted at a distance. Or, more accurately, he believed the opposite: that the effect of the planets operated in the same way as magnetism. This idea did not seem particularly far-fetched at that time in history. In 1502, a doctor in Siena described the effect of the moon on the tides as that "the moon attracts the sea as the magnet attracts iron."[57] In the case of weapon salve, the early seventeenth-century Paracelsian Daniel Sennert likewise believed that "the cure is done by the magnetique attractive power of this Salve, caused by the Starres."[58]

If people believed this, it is hardly surprising that they would think a medicine based on magnetic power, regardless of whether an actual magnet was used in the process, could cure a wound even when not directly applied.

Of course, this bizarre method of treatment also met with loud criticism. Moving forward just a little in history, we find, for example, the English Aristotelian William Foster denouncing weapon salve in 1631. He begins by describing Paracelsus as "the first inventer of this Magicall oyntment" and notes that "*Paracelsus* saieth, if the weapon be annoynted, the wounded partie may be cured, though 20 miles absent." People who defend the use of weapon salve, Foster explains, claim that "this naturall Balsame by the influence of the Starres, causeth a sympathy betwixt the weapon and the wound: and so the application of the Medicine to the one, effects the cure upon the other." Having thus summed up the arguments of his opponents, the Paraclesians, Foster adopts a contrary position: "For 'tis a rule amongst both Divines and Philosophers that; *Nullum agens agit in distans*. Whatsoever workes naturally, workes either by corporall or virtuall contact." From this premise, Foster condemns weapon salve as "witchcraft which comes from the Divell."[59]

Most critics of weapon salve used the same argument. In other words, the debate focused not on the effectiveness of magnetism in treatment but rather on whether or not action at a distance was possible. Weapon salve was rejected and denounced on one or both of two grounds: that natural action at a distance was impossible or occult and that the employment of such action was nothing but witchcraft. While Foster was a conservative in his outlook, his rejection of action at a distance was basically the same as that of the mechanists, whose philosophy we will discuss in the next chapter. Thus, albeit for subtly different reasons—one group rejected weapon salve as evil magic and the other as unscientific fraud—the old-school Aristotelians and the new-school mechanists formed a united front against chemical philosophy.[60]

Virtually everyone today would reject the idea of weapon salve on similar grounds, yet somehow we do not deny that our Earth exerts gravitational force upon the moon and is pulled by the sun. In contrast, the mechanists of the seventeenth century, without exception, branded Newton's universal gravitation as occult force just as Paracelsus's weapon salve had been so branded. At that point in time, only those who believed in Hermetic magical thought or astrology accepted the idea of action at a distance. The riddle of action at a distance hindered the development of modern science, which had to wait for Newton to come along and find a way to integrate the concept of action at a distance into a coherent theory—a development we will take up in a later chapter.

* * *

The sixteenth century saw much technological development coupled with a widening chasm between Scholasticism and the facts of reality. Out of these changes emerged the modern approach to scientific research, which values experience and is informed by the productive labors of engineers and tradesmen. Yet the road to empirical modern science was far from straightforward. Some form of theoretical framework was needed in order to judge, articulate, and systematize observed facts and experiences and make them more readily comprehensible. As people's experience expanded and Scholasticism's lack of usefulness became apparent, an intellectual vacuum developed and was soon filled by astrology and alchemy, as well as by magical thought, all of which together provided the logic for understanding nature based on experience and practical activity. For a time, these disciplines served in place of a new foundation for scientific endeavor. "To the sixteenth century world at large," indeed, "the scientist was mage rather than sage.[61]

This magical view of nature nonetheless engendered a perspective that defeated the Scholastic view. The alchemist saw the Earth as a living, productive entity with the power to grow metals in its womb, and this led to the view that the Earth was an agent that moved of itself and could influence other things. This perspective ultimately destroyed the Aristotelian model of the cosmos in which the Earth was a lowly, inert lump of soil and the superior celestial spheres revolved independently of it. Coupled with the astrological view that celestial bodies affect objects on Earth, this new view of the Earth made possible the concept that the planets influenced each other. Certain key modern scientific concepts, including that the Earth is a magnet and that gravitational pull exists between planets, were, in fact, discovered within this context. We will soon see how that process unfolded through the ideas of Gilbert and Kepler, but in preparation, let us first take a brief look at the changes in magical thought that occurred in the sixteenth century.

Chapter 15

# Changes in Magical Thought during the Late Renaissance

## Section 1—Magic Sheds its Mysticism

With the beginning of the Renaissance in the 1400s, magical thought made a comeback, and it underwent a huge transformation during the century that followed. Roger Bacon had hinted at the changes to come as early as the thirteenth century. Although magic had, from ancient times, always maintained some sort of a presence in western Europe, medieval Christianity accorded it the status neither of orthodox theology nor of orthodox philosophy, and church authorities offered it no cloak of legitimacy. Hugh of Saint Victor, who taught at the monastery-based School of Saint Victor near Paris, had said in the twelfth century that magic "is not accepted as part of philosophy, but stands with a false claim outside it."[1]

But in reality things were not so simple. Comparing the two with a critical eye, what difference was there between the recitation of prayers to exorcise demons and the casting of spells to summon them? Likewise, the idea that the bread and wine consumed as part of the Catholic mass could become the body and blood of Christ was no different from a magical rite.[2] The Bible describes a number of miracles, including the transformation of water into wine and—in the presence of many witnesses—of a staff into a serpent. The sacred relics used in Catholicism were said to have the power to cure the sick, and many with serious illnesses or physical handicaps would at the expense of great pain travel from one holy site to another in hopes of supernatural relief—in hopes of a miracle. The difference between all of this and so-called magic was no wider than a hair's breadth, and on the basis of factual evidence alone there was really no difference at all. Marvelous deeds were considered "miracles" when, and only when, performed by figures such as Christ or Moses or one of the saints, while the same deeds performed by pagans were invariably termed acts of "magic." Keith Thomas observes in *Religion and the Decline of Magic* that the church "did not deny that supernatural action was possible," and that the "difference between churchmen and magicians lay less in the effects they claimed to achieve than in their social position, and in the authority on which their respective claims rested."[3]

According to Christian ideology, all natural phenomena were examples of divine revelation—in other words, nature was completely subject to God's will. As long as the masses remained under the spell, so to speak, of this doctrine, then all was well. But as soon as people began to accept that nature followed intrinsic laws of its own, the distinction between the miraculous and the magical suddenly broke down. Among the first to see this coming was Roger Bacon.

Bacon was an adherent of Aristotle's rational physics and a pioneer of experimental science. In his *Opus majus*, Bacon dismisses attempts to explain mysterious natural phenomena in magical terms. Referring to a curious effect of apparent attractive force in tree saplings, for example, he notes that

> magicians perform this experiment, repeating different incantations, and they believe that the phenomenon is caused by virtue of the incantations. I have disregarded the incantations and have discovered the wonderful action of nature, which is similar to that of the magnet on iron.[4]

A little further on in the same work, we find the following:

> This [experimental] science, moreover, knows how to separate the illusions of magic and to detect all their errors in incantations, invocations, conjurations, sacrifices, and cults. But unbelievers busy themselves in these mad acts and trust in them, and have believed that the Christians used such means in working their miracles.... [This science] is the only branch [of philosophy] that considers matters of this kind, and is able to overcome all falsehood and superstition and error of unbelievers in regard to magic, such as incantations and the like already mentioned.[5]

And in a work titled *Roger Bacon's Letter concerning the Marvelous Power of Art and of Nature and concerning the Nullity of Magic*, which, according to one theory, may have been written by Bacon himself, the author asserts that "it is unnecessary for us to aspire to magic since Nature and Art suffice."[6] Baconian experimental science thus declares magic to be a delusion based on ignorance. But this attack has implications for the Christian concept of miracles as well. Citing "experiments with the magnet, not only on iron, but on gold and other metals," Bacon argues that "if the experiment on iron were not known, it would be viewed as a great miracle"—that is, counted among the "miracles of nature."[7] In other words, that something appears to be a miracle is the result of ignorance rooted in insufficient experience. While Bacon recognizes that vast powers dwell hidden within nature itself, he asserts the impossibility both of so-called magic based on the art of incantation and of so-called miracles in the form of supernatural phenomena.

When Pico della Mirandola and Marsilio Ficino revitalized magic in the fifteenth century, they emphasized that they were speaking of natural magic, which depended solely upon the inherent powers of nature, as distinguished from ritual magic and demonic magic, which invoked transcendental powers thought to be held by certain symbols (icons) and incantations (words). Framing Bacon's thought in these terms, we can conclude that although Bacon did not himself use the term "natural magic," this was in fact the proper term for the only kind of magic he thought possible. Any other kind of so-called magic or miracle was, in his view, impossible.

This brings us to Cornelius Agrippa von Nettesheim, whose life illustrates how Bacon's position on this issue gained ground at the turn of the sixteenth century.

In chapter 10 we examined Agrippa's *Three Books of Occult Philosophy*, a veritable encyclopedia of early Renaissance magic. Agrippa completed the work around 1510, but in fact publication was delayed until 1533, perhaps out of fear that the book's strongly expressed mysticism and pagan elements would be seen as heretical. Three years before the release of this work, Agrippa had published another, *De incertitudine et vanitate scientiarum et atrium, atque excellentia Verbi Dei, declamatio invectiva* (On the uncertainty and vanity of the arts and sciences: An invective declamation), in which he dismisses and disavows certain aspects of magic, especially concerning demonic magic. At least one scholar has suggested that, seen in light of the three-volume publication that was to appear just several years later, this was a preemptive ploy to reduce the likelihood that *Three Books of Occult Philosophy* would be condemned.[8] We cannot be certain of Agrippa's intentions, but to some extent the views he expresses in *De incertitudine et vanitate* must have been genuine. Let's take a look at one passage from that work:

> Therefore natural magic is that which having contemplated the virtues of all natural and celestial things and carefully studied their order proceeds to make known the hidden and secret powers of nature in such a way that inferior and superior things are joined by an interchanging application of each to each; thus incredible miracles are often accomplished not so much by art as by nature, to whom this art is as a servant when working at these things. For this reason magicians are like careful explorers of nature only directing what nature has formerly prepared, uniting actives to passives and often succeeding in anticipating results so that these things are popularly held to be miracles when they are really no more than anticipations of natural operations; ... therefore those who believe the operations of magic to be above or against nature are mistaken because they are only derived from nature and in harmony with it.[9]

By thus maintaining that magic did not contradict nature but merely predicted natural operations and helped them along, Agrippa accomplished two things. First, he emphasized the role that nature played in magic. Second, he relegated to the background the mystic worldview and philosophy in which magical thought was rooted. Agrippa, who had always held natural magic to be subordinate to demonic magic and numerology, was now instead pushing natural magic to the fore.

## Section 2—Pietro Pomponazzi and Reginald Scot

In the Renaissance of the 1500s, magical thought was characterized by an emphasis on the idea that magic—understood as natural magic—did not contradict nature but rather was controlled by nature's intrinsic powers and laws. In this section, we'll be taking a close look at two notable proponents of this view, one from the first half and the other from the second half of the sixteenth century: respectively, Italian philosopher Pietro Pomponazzi (1462–1525) and English writer Reginald Scot (1538–1599).

Born to an aristocratic family in Mantua, Pomponazzi was a dyed-in-the-wool academic philosopher. He studied medicine at the University of Padua, then lectured on philosophy at the universities in Padua, Ferrara, and Bologna. Though a year younger than the precocious and short-lived Pico della Mirandola, Pompanazzi was essentially a thinker of the sixteenth century, and Pico one of the fifteenth. Like Ficino and Pico, Pomponazzi placed human beings at the center of his ideology, and in this respect he must be considered a Renaissance philosopher. It has been said, however, that "Renaissance Aristotelianism culminates ... in Pomponazzi,"[10] indicating Pomponazzi's position as a representative of the neo-Aristotelianism that flourished in northern Italy. Pomponazzi based his philosophy on the ideas of Aristotle and Aristotelian commentators such as Alexander of Aphrodisias and Averroes; thus he drew on a different academic legacy from that followed by fifteenth-century Renaissance thinkers such as Ficino. In fact, Pomponazzi was a pragmatist who did not believe in the immortality of the soul. This view was of course heretical given that in 1513 the Fifth Lateran Council had established the concept of the immortality of the soul as official doctrine. This was more than merely a matter of theology, as it signified Pomponazzi's emphasis on life in the present world: For Pomponazzi, salvation in the next world was neither the purpose of life nor the ideal that one should strive for; rather, life's purpose and ideal were the realization of virtue in the here and now.

We have already seen how acceptance of Aristotelian philosophy in the thirteenth century hinged upon the commentaries of Averroes and how his theory of twofold truth was condemned as heresy by the Bishop of Paris. In northern Italy, and particularly in Padua, however, Averroism survived, for two reasons. First, bad relations between the church and the throne meant that municipal governments retained greater autonomy. Second, Italian universities, unlike the University of Paris, were established for practical study in the fields of medicine and law, and therefore theology had a relatively low profile at these institutions.

Thus Pomponazzi's fundamental stance vis-à-vis the contradictions between revelation and philosophy was, in essence, the theory of twofold truth. While on the one hand he believed, in the manner of Catholicism, in a transcendental world, on the other hand his discussions of philosophy reveal an indifference to church doctrine. He believed that a theological foundation was unnecessary to the pursuit of academic learning. His position, at bottom, was that scholarship had its own independent foundation; philosophy, he believed, held a status equal to theology and could therefore be discussed on its own merits without reference to religion. Pomponazzi's Aristotelianism differed in two ways from the model espoused by the philosophers of Paris: first, it was less dependent on theology than the Aristotelianism of the Thomists; and second, it was based on the Aristotelian commentaries of Alexander of Aphrodisias. Consequently, Pomponazzi's thoughts on natural philosophy were characterized by a naturalism in which the measure of truth was to be found in rationality, experience, and natural law. Accordingly, he did not believe in the kind of miracles or magic that were thought to occur through the caprice of angels or the selfish whims of demons.

Pomponazzi developed his theory of magic in *De naturalium effectuum causis sive de Incantationibus* (On the causes of natural effects, or On incantations; hereafter, *De Incantationibus*). Though written in 1520, this work did not see print during Pomponazzi's lifetime. It was finally published in 1556 in Basel by a Protestant in exile. But this is not to say that *De incantationibus* remained unknown all that time. To the contrary, it was widely circulated and read as a handwritten manuscript, and its contents are said to have become the subject of controversy even before the book was published in printed form.[11] What made the work controversial was Pomponazzi's frequently repeated thesis that, whereas "it is the custom of ordinary folk to place blame or credit with demons or angels whenever the cause of something is unknown," in fact miracles and "all magic can be reduced to natural causes."[12] He rejects transcendental explanations for anything magical or miraculous, allowing only explanations according to which all phenomena are part of the fabric of nature. His discussion, however, is based on his assumption that these natural causes consist of both manifest qualities and occult forces. He counts among obvious examples of occult forces the force of magnetism, which acts, he says, with *qualitas insensibilis* (qualities not perceivable by the senses):

> Magnets attract iron, and diamonds interfere with this action. Sapphires chase away ulcers and heal the eyes. Such occult forces (*virtus occulta*) are numerous.... When ordinary people see such things (because they cannot perceive the causes for

> these various actions), they do not understand their causes and therefore fall into believing that they are brought about by God or angels or the devil.[13]

As further examples, Pomponazzi mentions that diamonds interfere with the power of magnets and that garlic smeared on magnets has the same effect; that a small fish called an echeneis can stop ships 200 feet or more in length, even when the ships are carrying cargo and require either wind or oars to move forward; and that the torpedo (electric ray) can cause people to faint even from a distance.[14] As we know today, of the various examples Pomponazzi cites, only the magnet and torpedo have the powers he claims they have, powers known to us of course as magnetism and electricity. But no matter: Pomponazzi's point is that forces should not be considered supernatural just because their causes are not understood.

Pomponazzi also includes among natural causes the astrological effects that celestial bodies were thought to exert on terrestrial bodies and on human beings. More than that, he believes that it is precisely these influences that cause the various phenomena generally seen as supernatural and therefore referred to as "miracles." To Pomponazzi, so-called magic was, "like medicine and most other sciences, a practical and genuine science in itself with its foundation in natural philosophy and astrology."[15]

Furthermore, Pomponazzi supposed that miracles had a natural cause in the form of the psychological phenomenon of human imagination. According to Pomponazzi, stories of medical cures effected using sacred relics such as the remains of deceased saints were not supernatural miracles. They were instead the result of the natural working of imagination aroused in patients who placed faith in those sacred objects. In this view, the objects used in treatment might just as well have been the bones of deceased dogs, for the result would have been equally powerful as long as the patient believed himself or herself to be undergoing treatment with articles considered holy. One critic notes that "it is difficult for us now, after reading the *De Incantationibus*, to believe that Pomponazzi was a Christian," and though that might be putting it rather too bluntly, it is difficult to disagree with the thrust of that assessment.[16] Another commentator observes more concisely that Pomponazzi "denied consecrated truth."[17] Without going that far, I prefer Cassirer's description of Pomponazzi's writings as "Enlightenment in Scholastic garb."[18]

And now, let us move on to the contribution of English protestant Reginald Scot. In 1584, Scot published *The Discoverie of Witchcraft*, in which he rejects demonic magic from the point of view of reason and common sense, dismissing the idea of sorcery practiced by witches in pacts with the devil as foolish superstition. Scot is known for having vociferously criticized the practice of witch-hunting conducted on church authority. *The Discoverie of Witchcraft* was a courageous and frank indictment of the witch-hunting that was prevalent in Europe in Scot's day. It was the author's means of taking decisive action against the practice by, in his own words, offering "a thorough-going demonstration of the magical elements in medieval Catholicism and their affiliation with other contemporary kinds of magical activity."[19]

Scot, like Pomponazzi, takes "a position tantamount to atheism"[20] by denying the efficacy of both the devil and angels. And yet he also recognizes so-called natural magic. In *The Discoverie of Witchcraft*, he asserts that natural magic "of it selfe is not evill," elaborating as follows:

> In this art of naturall magicke, God almightie hath hidden manie secret mysteries; as wherein a man may learne the properties, qualities, and knowledge of all nature. For it teacheth to accomplish maters in such sort and oportunitie, as the common people thinketh the same to be miraculous; and to be compassed none other waie, but onelie by witchcraft. And yet in truth, naturall magicke is nothing else, but the worke of nature.[21]

Under the topic heading "What strange things are brought to passe by naturall magicke," Scot cites the examples of wild bulls becoming docile when tied to fig trees and of remora and echeneis stopping cargo-laden ships with full sails merely by passing in front of them. He states his own position on the validity of such frequently mentioned phenomena as that "it seemeth unpossible." "And yet," he continues,

> it is affirmed by so manie and so grave authors, that I dare not denie it; speciallie, bicause I see as strange effects of nature otherwise: as the propertie of the loadstone, which is so beneficiall to the mariner.[22]

Although Scot refused to jump onto the antiwitch bandwagon—which, again, he saw as foolish and superstitious—he did accept the notion of relationships of sympathy and antipathy, typified by magnetism, as belonging to the category of natural action. He believed that the practice of natural magic entailed discerning sympathy and antipathy in the midst of natural phenomena and manipulating these dynamic relationships through human intervention. There was, he argued, nothing about any of this that should be considered supernatural.

And so, here in the sixteenth century, at the threshold of the scientific revolution, thinkers were openly discussing natural magic with natural causes—the rational basis for miracles and magic that had been anticipated centuries earlier by Roger Bacon. But even though the supernatural in the form of miracles and magic performed by angels, demons, or the devil had now been rejected, the natural causes offered as an alternative explanation were still dependent on the concepts of astrological cause and effect, in which the celestial world influenced terrestrial affairs, and on the notion of occult forces expressed as sympathy and antipathy. Roger Bacon, too, had held that "celestial things are the causes of terrestrial"[23] and that these astrological relationships were a part of nature and not anything supernatural.

## Section 3—Magic and Experimental Methods

The view that magic was a practice that obeyed nature rather than contradicted it, using human intervention to augment natural activity for human ends, only became stronger and more vocally expressed in the second half of the sixteenth century. The key conceptual framework applied during this period involved a series of ideas such as occult properties (*proprietas occulta*), occult forces (*virtus occulta*), and occult action (*opus occultus*). Alistair Cameron Crombie, a historian of medieval science, characterizes the belief of the time thus: "'Marvels', when not the work of demons and therefore evil, might be produced by occult virtues resident in certain objects in nature, that is by 'natural magic.'"[24] In other words, it was commonly thought that natural magic, as opposed to demonic magic, was simply a technology entailing the use of occult forces as studied in the laboratory of direct experience.

It is necessary here to emphasize that although the word "occult" as used in English today (as well as the words *occulte* in French and *occulto* in Italian) carries connotations of the mysterious or supernatural, this was not so for the medieval Latin word *occultus*, which was merely an antonym for *manifestus* ("overt" or "plainly perceivable"). In actual use, though, the word *occultus*, which rightfully translates as "hidden," had two separate connotations. The first signified something incomprehensible, as when Thomas Aquinas asserts in his *Summa theologica* (Summary of theology) that "in the physical order things have certain occult forces, the reason of which man is unable to assign."[25] The second signified something imperceptible to the five senses, as in the case of magnetism. These two meanings, however, were intimately related, as is demonstrated by Ficino's reference to things "hidden from our senses, and hence only with difficulty known to our reason."[26]

This view originated in the Aristotelian concept that a quality that could be apprehended by any of the five human senses–a manifest property–could be understood in logical terms based on the four-element theory. To put it the other way around, anything that could not be apprehended by the five senses was by definition a thing that defied explanation. In other words, there were powers in nature that, while they could not be directly perceived and were therefore incomprehensible, could still be assumed to exert action on material bodies due to their directly observable effects; these were called "occult forces" (*virtus occulta*).

This understanding of occult force was, in the sixteenth century, held not only by philosophers and scholars of magic but also by engineers and craftsmen. In a 1540 work, Italian engineer Vannoccio Biringuccio, for example, makes a comparison between the peculiarity of magnets and the qualities of more commonplace things. Most things, he observes,

> reveal themselves to us through the sense of sight by their splendor and variety of colors, through the sense of smell by their odors, through the sense of hearing by the melody of the harmonious or unpleasantly discordant repercussions of the air. They likewise manifest themselves by their softness or harshness.

The lodestone, in contrast,

> is one of those things that show their apparent effects to our sight by a particular virtue hidden in such a way that its cause, as far as I know, has never been known to the very wise natural speculators who have spoken on every other subject.

In other words, magnetism was regarded as an occult force—a phenomenon that could not itself be perceived by the five senses and whose cause was unclear but whose existence could nonetheless be taken for a fact because of the repeated experience of witnessing the result, namely, magnets drawing iron. Hence Biringuccios's conclusion:

> This action [of lodestone] is certainly great enough to bear certain witness to the virtues possessed by other created things that do not manifest their powers to the external senses.[27]

In other words, the fact that magnetism existed did not merely hint at the existence of numerous other occult forces; rather, it positively testified to their existence. And so, even dyed-in-the-wool engineers like Biringuccio could accept the validity of natural magic, which had the notion of occult force as its theoretical core.

Moreover, as the Middle Ages gave way to the Renaissance, it was this very issue that, by extension, formed a basis for establishing methods of researching natural phenomena—methods of what we today call experimentation. Crombie explains:

> In medieval literature … the term *experimentum* (whence *experimentalis*, *experimentatio*, *experimentator*) and its synonym *experientia* were linked again with *mirabile*, *mirandum*, *stupendum*, *miraculum*, *magicum*, *secretum*, with the whole world of natural magic with its astonishing and theatrical effects produced from the hidden secrets of nature.[28]

In other words, *experiments* and *experience* were considered the means for investigating the properties of the occult forces we have been discussing. Given this context, we can see why, at the dawn of the modern age, the term "occult science" meant something very close to what we today would call "experimental science." "Indeed," notes Peter French, author of a biogaphy of John Dee, "almost every magician of the sixteenth century advocated some sort of methodological experimentation."[29]

Here we must touch upon a distinction so convoluted as to be highly perplexing from a modern vantage point. With Scholasticism still holding sway at this point in history, science (in Latin, *scientia*) as it had long been understood meant the deductive demonstration of the behaviors and attributes of each object of inquiry from arguments concerning the true nature of that thing—that is, its definition.

According to this perspective, discerning scientific truth required an exacting process of deductive reasoning, and this process was considered superior to arguments based upon the evidence of human senses, which was subjective and therefore uncertain and unstable. In a fifteenth-century work, the then bishop of Sardinia candidly expresses this distinction as follows:

> Moreover, to assert that such experimental knowledge is science or a part of natural science is ridiculous, wherefore such magicians are called experimenters rather than scientists.... [But] natural science in itself and all its parts is purely speculative knowledge.[30]

Occult forces, however, were of unknown nature and operated on unknown principles, their existence being inferred solely from their effects. Therefore, occult forces could not be known in the academic, properly scientific sense; instead, one was forced to rely on the second-best strategy of experimentation and experience. Indeed, Roger Bacon, the founder of the experimental method, spoke of the convergence and refraction of light by means of lenses as "a wonderful thing in the eyes of men of science," adding: "the causes are hidden. We need not now investigate the causes, since we know this marvel by means of a very certain test."[31] Similarly, in the sixteenth century Agrippa said that "the occult properties in things ... can no otherwise but by experience and conjecture be inquired into by us." And with regard to magnetism and similar activity, Agrippa spelled out his position thus:

> And they are called *Occult Qualities*, because their causes lie hid, and man's intellect cannot in any way reach and find them out. Wherefore philosophers have attained to the greatest part of them by long experience, rather than by the search of reason.[32]

It is clear that experimental method was closely connected to the investigation of occult forces and qualities.

Let us take up the example of one more proponent of this position, Albert the Great. A work titled *The Book of Secret of Albertus Magnus* is thought to have been penned after Albert's death by his disciples to record what he had taught them about magic. The work was published in numerous editions and probably altered somewhat along the way. Be that as it may, the main body of the work is thought to have been written near the end of the thirteenth century, around the same time as Petrus Peregrinus's *Letter concerning the Magnet* and Bacon's *Opus majus.* Here is an excerpt of particular interest from *The Book of Secrets of Albertus Magnus*:

> Therefore certain things must be believed by only experience, without reason, for they be hid from men; ...
>
> For although we know not a manifest reason wherefore the Loadstone draweth to it Iron, notwithstanding experience doth manifest it so, that no man may deny it. And like as this is marvellous, which only experience doth certify, so should a man suppose in other things. And he should not deny any marvellous thing although he hath no reason, but he ought to prove by experience; for the cause of marvellous things are hid, and of so diverse causes going before, that man's understanding, after Plato, may not apprehend them.... So marvellous things are declared of Philosophers to be in things by experience.[33]

When Petrus said in his *Letter concerning the Magnet* that studying the hidden action of magnetism required one to be "clever in the use of his hands," he was drawing basically the same conclusion. In all of this we can see the point of departure for a new method of experimentation blending the skills of technology and magic.

In this way, natural magic was given its own distinct category: labeled "occult science," it was positioned in between Scolasticism's *scientia*, which crafted formal arguments about natural phenomena based upon first principles, and *magia*, with its spells, symbols, and belief in the power of demons. Natural magic, thus distinguished as a discipline in its own right, was defined essentially as the art of manipulating occult forces whose properties had been discovered through experience and experimentation. Natural magic differed from *scientia* by being based on an experiential and experimental methodology rather than on rational theorization. And it differed from *magia* by remaining within the framework of nature and its uniformity, as opposed to incantation-based (demonic) magic, which presupposed numerous supernatural forces.

And so, in the sixteenth century, natural magic foreshadowed the coming of the scientific revolution. It certainly contained elements of modern empirical science, based as it was on both observation and experimentation, and as such it was seen in contrast to the excessive rationalism of Scholasticism. As science historian Marie Boas puts it, "nascent experimental science was popularized as natural magic."[34] Lynn Thorndike similarly notes that, already in the Middle Ages, "under the guise … of magic experimental science is at work," and that "natural science, originally philosophical and speculative, took over experimental method in a crude form, as well as the conception of occult virtue, from magic."[35] It is therefore safe to say that, in the sixteenth century, as natural magic came to be more openly discussed, so the validity of experimental methods was gradually approved. I would add, however, that the experiments of natural magic differed from modern experiments in one particular sense: whereas in the former the purpose was to confirm the efficacy of techniques used to manipulate occult forces, or to improve upon such techniques by trial and error, in the latter—in modern times, that is—the purpose is to test hypotheses and discover new laws by using deliberately constructed devices.

In any case, magnetic force was most certainly the model for occult forces and properties. And for this reason the magnet—which so readily drew to itself heavy, inert lumps of iron and held them suspended, and which through rubbing gave directionality to iron needles—was an ideal subject for the experiments of natural magic. It was no meaningless coincidence that the mysterious magnet was the subject of the European earliest formal dissertation on experimental physics, the *Letter concerning the Magnet.*

## Section 4—John Dee and the Mathematization of Magic

Not only experimental science but also mathematical reasoning became what seems an unlikely bedfellow to magic.

Another figure of the sixteenth century who was greatly influenced by Roger Bacon's writings was the English mathematician John Dee (1527–1608), discussed briefly in chapter 12. Dee was a noted book collector, and his biographer, Charlotte Smith, notes the following about Dee's selection of reading material: "Of English authors ... the most eagerly sought after ... were Roger Bacon and Robert Grosseteste." Nicholas Clulee, another scholar who studied Dee's life, observes that Dee considered Bacon "a source of inspiration for his own natural philosophy."[36] Having discovered Bacon's works in 1556, Dee proceeded the following year to defend them against the criticism that Bacon's research had been aided by the devil. And in fact, the views Dee expressed on natural magic seem to closely parallel Bacon's research.[37] The chief lesson that Dee took from Bacon was in drawing a distinction between, on the one hand, the legitimate technological application of hidden natural powers and, on the other hand, magic, which was rightfully denounced for seeking demonic assistance.

Early in his career, Dee's interest in natural philosophy focused heavily on astrology. In a 1558 work, *Propaedeumata aphoristica*,[38] he paints this portrait of the universe:

> XI. The entire universe is like a lyre tuned by some excellent artificer, whose strings are separate species of the universal whole. Anyone who knew how to touch these dextrously and make them vibrate would draw forth marvelous harmonies. In himself, man is wholly analogous to the universal lyre.
>
> XII. Just as the lyre is an arrangement of harmonious and disharmonious tones, … so the universe includes within itself parts among which a most close sympathy can be observed, but also other parts among which there is harsh dissonance and a striking antipathy. The result is that the mutual concord of the former and the strife and dissension of the latter together produce a consent of the whole and a union eminently worthy of admiration.

Like Ficino and Paracelsus, Dee saw the universe as an organic, integrated whole consisting of a network of sympathetic and antipathetic relationships, and he saw human beings as microcosms replicating all parts of this network in smaller form.

Dee's aim in *Propaedeumata aphoristica*, one that is consistently developed throughout the work, is to develop a mathematical foundation for astrology. He maintains that everything in the universe, including all of its activity, can be understood in strictly quantitative terms, and in this he differs decisively from both Ficino and Paracelsus. At this still early stage of his career, Dee is fascinated not only by astrology but also by mathematics. While he accepts the premise that the celestial world influences Earth, he also embraces Bacon's concept of *multiplication of species* as a model of the way in which celestial action is transmitted to the terrestrial realm. It is said that Bacon's *On the Multiplication of Species* was among the works in Dee's collection[39]; and in Dee's understanding, both astronomy and optics are essentially mathematical disciplines.

Moreover, he sees Baconian multiplication of species as the mechanism not only by which light travels but also by which celestial bodies act upon terrestrial ones. Taking this understanding a step further, it follows that celestial influences are governed by geometric principles and can be described mathematically and measured in quantitative terms, just as the transmission of light, as explained by Grosseteste and Bacon, can be so described and measured.

According to Dee, God created the world out of nothing. While natural laws do not apply to creation itself, they do govern everything from the time the world came into being until such time as God may choose to extinguish it. Although these laws, in Dee's view, reveal relationships of astrological cause and effect, they are also imbued with a precise mathematical logic, and it should therefore be possible to understand them mathematically. Dee believes that just as the strength of light emitted by any given body in the heavens can be measured on the basis of that body's size and distance from Earth, the particular degree to which celestial bodies influence human beings and other bodies on Earth at any given time should also be measurable. Moreover, Dee goes so far as to argue that celestial influences can be focused in a particular direction and amplified with the use of lenses and reflecting mirrors. This understanding is the essence of Dee's particular brand of natural magic. In his own words:

> LII. If you were skilled in "catoptrics," you would be able, by art, to imprint the rays of any star much more strongly upon any matter subjected to it than nature itself does. This, indeed, was by far the largest part of the natural philosophy of ancient wise men.... By this means obscure, weak, and, as it were, hidden virtues of things, when strengthened by the catoptric art, may become quite manifest to our senses.

"Catoptrics" (the study of optical reflection) here refers to the art of focusing light rays using a spherical mirror. Clulee explains as follows:

> The essential components of Dee's early natural philosophy can therefore be identified as mathematics, astrological physics, natural magic, and optics ... [It] involved a naturalistic conception of the world and of man, in which there is little dualism between the natural and the supernatural, and all phenomena are natural and self-sufficiently explainable without appeal to intervention of extra-natural demons, spirits, or intelligences.[40]

To this extent, Dee clearly started out as a follower of Roger Bacon. In a nutshell, both the theoretical and the practical foundations of Dee's natural magic consisted of optics and mathematical astronomy, and his *Propaedeumata aphoristica* represents a transitional period from the pursuit of magical predictions based on astrological principles to the pursuit of scientific predictions based on mathematical principles.

As he grew older, Dee came to embrace neo-Platonism and particularly, under Agrippa's influence, numerology; but at the same time his magical philosophy became both more practical and more technical. This evolution in Dee's thought is evidenced by *The Mathematicall Praeface*, the preface he wrote for an English translation of Euclid's *Elements*.[41] We touched upon Dee's *Praeface* in chapter 9 and are now ready for a closer look.

At this point in his career, Dee develops the idea that the world is formed according to mathematical principles and therefore can be understood solely in mathematical terms—an idea that stems both from neo-Platonism and from a Pythagorean numerology. Dee clearly expresses his convictions clearly in this regard, following the Roman philosopher Boethius: "*All thinges … do appeare to be Formed by the reason of Numbers. For this was the principall example or patterne in the minde of the Creator*" (TMP, *jr); and again when he quotes Pico della Mirandola: "*By Numbers, a way is had, to the searching out, and vnderstandyng of euery thing, hable to be knowen*" (TMP, *jv).

According to Dee, all numbers exist in one of three states, namely, within the Creator, within Creation, and in a so-called third state. Physical bodies can all be classified into one of three corresponding groups, which Dee calls "Thinges Supernaturall," "Thinges Naturall," and "a third being." Dee obviously bases this framework on Agrippa's designation, in *Three Books of Occult Philosophy*, of an "Intellectual World," an "Elementary World," and a "Celestial World."[42]

Dee characterizes his framework thus:

> Thinges Supernaturall, are immateriall, simple, indiuisible, incorruptible, & vnchangeable. Thinges Naturall, are materiall, compounded, diuisible, corruptible, and chaungeable. Thinges Supernaturall, are, of the minde onely, comprehended: Thinges Naturall, of the sense exterior, ar hable to be perceiued. In thinges Naturall, probabilitie and coniecture hath place: But in things Supernaturall, chief demonstration, & most sure Science is to be had.

The things of his "third being," which he also refers to as "Thynges Mathematicall," are "immateriall." Dee further explains that "a meruaylous [marvelous] newtralitie haue these thinges *Mathematicall*, and also a straunge participation betwene thinges supernaturall … and thynges naturall" (TMP, p.*v).

One of Dee's chief ideas is that numbers occupying the third state are situated between the supernatural and the natural and can be applied to the other two states in order to help understand or affect them. Dee saw number as, on the one hand, "a thyng so Immateriall, so diuine, and aeternall," but on the other hand as applicable "to thynges sensibly perceiued" (TMP, *jv). "Thus," he explains, "can the Mathematicall minde, deale Speculatiuely in his own Arte: and by good meanes, Mount aboue the cloudes and sterres: And thirdly, he can, by order, Descend, to frame Naturall thinges, to wonderfull vses" (TMP, ciijv).

In other words, numbers not only have importance for understanding the celestial realm but can also be used as a technical means of achieving purposes here on Earth; numbers therefore hold the promise of widening the bounds of technology. This is Dee's particular contribution to philosophy. While it is true that he borrows much directly from Boethius and Pico, he differs from those thinkers in his conviction that numbers are important as a basis not only for comprehending the Platonic Ideas behind the perceived world but also for grasping the nature of terrestrial things and acting upon them.

In the *Praeface*, after touching upon the general significance of numbers, Dee defines arithmetic as "the *Science that demonstrateth the properties, of Numbers, and all operations, in numbers to be performed*" (TMP, *ijr); and geometry as the "Science of *Magnitude*, his properties, conditions, and appurtenances" (TMP, aijr). And he explains his main aim in writing the *Praeface* as follows:

> Mathematicall Speculations, both of Numbers and Magnitudes: are meanes, aydes, and guides: ready, certaine, and necessary. From henceforth, in this my Preface, will I frame my talke …

> to such, who well can, (and also wil,) vse their vtward senses, to the glory of God, the benefite of their Countrey, and their owne secret contentation, or honest preferment, on this earthly Scaffold. To them, I will orderly recite, describe & declare a great Number of Artes, from our two Mathematicall fountaines, deriued into the fieldes of *Nature.* (TMP, aiijr)

The list of these sciences and mathematical arts that Dee presents at the end of the *Praeface* is shown in Figure 15.1. Where he uses the term "derivative," we can understand this to mean "application." The concept of application was very important to Dee; for him, the aim of intellectual inquiry was no longer to grasp the Platonic realm of Ideas. Note, in particular, that for Dee the purpose of learning was no longer solely the glorification of God but also the promotion of England's national interests and the earthly advancement of the individual. This shift indicated that a new citizenry was emerging in this newly formed nation-state, a citizenry which considered the prosperity of the nation and increased national power to be directly linked to their personal interests, and it was to such people that Dee addressed his discussion on science in the *Praeface.* It was written for engineers and craftsmen who had had little or no formal academic education. These were the very first days of the modern era, and until this time people in technical fields had relied not on rational principles but rather on instincts and skills developed over many years of experience. Dee was awakening these individuals to the news that mathematics was something they could actually find both useful and important.

For example, Dee counts "*Architecture*, among those *Artes Mathematicall*, which are Deriued from the Principals," even though this discipline "to many may seme not worthy, or not

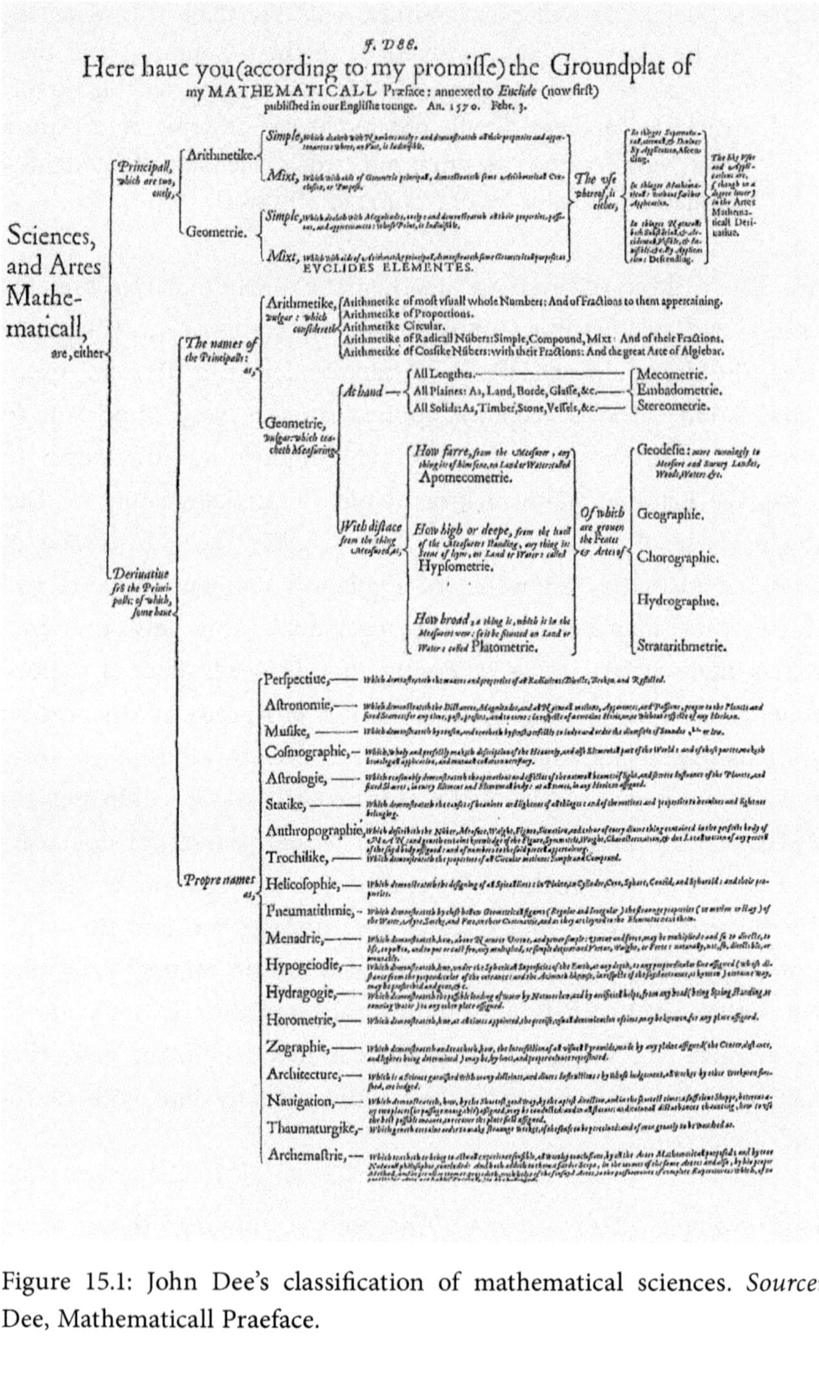


Figure 15.1: John Dee's classification of mathematical sciences. *Source*: Dee, Mathematicall Praeface.

mete, to be reckned" as belonging to that category (TMP, diijr). Moreover, he declares the writings of Vitruvius, an architect of the ancient Roman Empire, to be a "Storehouse of all workmanship." Vitruvius's view was that architecture requires not only mere craftsmanship but also academic disciplines such as (in Dee's words) "Geometrie, Arithmetike, Astronomie, Musike, Anthropographie, Hydragogie, Horometrie &c." (TMP, diiijr).

In addition, Dee coins the term "Zographie" to denote a mathematical craft of painting: "Great skill of *Geometrie*, *Arithmetike*, *Perspectiue*, and *Anthropographie*, with many other particular Artes, hath the *Zographer*, nede of, for his perfection." Accordingly, says Dee, to be highly skilled a painter has to be also a "Mechanicien of Zographie" (TMP, dijv).

Another coinage of Dee's was "Menadrie," which he describes as "an Arte Mathematicall, which demonstrateth, how, aboue Natures vertue and power simple: Vertue and force may be multiplied" (TMP, djr). In modern terms, this is the discipline of statics—the principles behind levers, pulleys, bevels, and the like. Here again, Dee is proposing that technical crafts, to which academics were previously considered irrelevant, be properly supported with mathematical and scientific principles in the form of theoretical mechanics.

In the *Praeface*, Dee refers to the mathematical technique of "Thaumaturgike" (the art of miracle working), which he defines as an "Art Mathematicall, which giueth certaine order to make straunge workes, of the sense to be perceiued, and of men greatly to be wondred at" (TMP, Ajr). As Dee explains it, these strange works include, for example, the inventions of Heron of Alexandria, which made use of aerodynamics or other devices, such as a mechanical doll and a clock with revolving puppets made with springs, wheels, and so forth. According to Dee, such things are a part of natural magic.

The last technique Dee presents in the *Praeface* is "Archemastrie":

> This Arte, teacheth to bryng to actuall experience sensible, all worthy conclusions by all the Artes Mathematicall purposed, & by true Naturall Philosophie concluded,... to the performance of complet Experiences, which of no particular Art, are hable (Formally) to be challenged.... *Science* I may call it, rather, then [sic] an Arte: for the excellency and Mastershyp it hath ... And bycause it procedeth by *Experiences*, and searcheth forth the causes of Conclusions, by *Experiences*: and also putteth the Conclusions them selues, in *Experience*, it is named of some, *Scientia Experiementalis*, The *Experimentall Science*. *Nicolaus Cusanus* termeth it so, in his *Experimentes Statikall*. (TMP, Aiijr–v)

Dee not only credits the term *scientia experimentalis* to Cusanus but also places the initials "R.B." in the margin at this point. According to Clulee, "R.B." stands for Roger Bacon, indicating that Dee's use of the term is synonymous with Bacon's and therefore does not necessarily signify experimental science in the modern sense.[43] In the introduction to a reprint of Dee's *Praeface*, historian Allen G. Debus explains that Dee's methodology does not encompass experimentation in the modern sense of controlled experiments designed to test hypotheses. Instead, he argues, Dee's "experience" signifies something more akin to "observation."[44] Boas corroborates this with her interpretation that Dee is referring to "genuine observation of nature."[45] We cannot, therefore, view Dee as a proponent of experimentation in the modern sense. Nonetheless, we can see clearly that he not only emphasized the importance of mathematics but also advocated linking natural science with technological endeavor, with the qualification that natural science was not to follow in the path of Scholasticism. That is, instead of proceeding by textual analysis and argumentation, natural science should be based on and guided by experience and practice.

In his later years, Dee became immersed in mysticism, believing that numbers could be used as a tool for communicating with angels. There is no doubt, however, that *The Mathematicall Praeface* paints his view of magic as something much more down-to-earth. Dee—at least at the time he wrote the *Praeface*—saw magic as a body of mechanical and practical techniques that could be verified by experience and backed up and validated by mathematical theories. In the *Praeface* itself, he clearly distinguishes his brand of magic from demonic magic, lamenting that "for these, and such like marueilous Actes and Feates, Naturally, Mathematically, and Mechanically, wrought and contriued: ought any honest Student, and Modest Christian Philosopher, be counted, & called a Conjurer?" (TMP, Ajv).

Dee's view of science, then, was supported by an absolute faith in the value of mathematical and mechanical technique. In his hierarchy, science was first of all a mathematical pursuit; second, it was something that presupposed applicability to technological pursuits; and, third, it needed to withstand the scrutiny of experience. In short, Dee's science was to some extent modern in character.

In this regard, Peter French's assessment is no exaggeration: "Since Dee's philosophy … looked forward to the scientific spirit of the Royal Society, he parallels the French academicians in attempting to transform medieval magic into Renaissance science."[46] At the very least, we can concur with another historian, Frances Amelia Yates, when she applauds Dee and his *Praeface* for the monumental impact they had on seventeenth-century English science:

> As a manifesto for the advancement of science, Dee's mathematical Preface is of greater importance than Francis Bacon's *Advancement of Learning*, published thirty-five years later, for Dee fully understood and emphasized the basic importance of mathematical studies for the advancement of science, whereas, as everyone knows, Bacon underrated mathematics, which was the chief reason why his methods did not yield results of scientific importance.[47]

And so, with this examination of John Dee and *The Mathematicall Praeface*, we arrive at the conclusion that theoretical mathematics and magical thought are not necessarily mutually exclusive.

## Section 5—Cardano: Magic and Electromagnetism

The preceding discussion outlines how Renaissance magical thought was transformed into what could be called a premodern form of natural science in the second half of the sixteenth century.

Two men in particular personified the new magic of the mid- to late sixteenth century: Girolamo Cardano (1501–1576) and Giambattista della Porta (c. 1535–1615). Their philosophies of magic were strikingly different in several ways from those that had come before. First, they broke from the tradition of unqualified praise and uncritical approval of ancient writings. Second, as a consequence of this new attitude, the new magic emphasized experience and observation akin to experimentation. Third, it concerned itself with technological application. Cardano and Della Porta picked up where Petrus Peregrinus had left off, carrying out what amounted to experimental research into magnetism and electricity. Della Porta's contributions to experimental research on magnetism were especially important, so let's leave them for separate examination in the next chapter and focus for the moment on the work of Cardano.

Cardano was the son of a Milanese lawyer and is known for his work in solving cubic equations and for having originated probability theory. Cardano was also a physician, trained at universities in both Pavia and Padua. Influenced by this background, he was an Aristotelian, aligned philosophically with Pomponazzi and a bearer of much of Pomponazzi's intellectual legacy. But we cannot do justice to Cardano with such simple labels as "Aristotelian" or "physician." He is credited, for example, with having written the first known monograph on algebraic equations in Europe and the leading mathematical work of his era, *Artis magnae sive de regulis algebraicis* (also known simply as *Ars Magna* or *The Rules of Algebra*), published in 1545. Aside from his accomplishments in mathematics, he had strong interests in natural philosophy, music, theology, astrology, palmistry, and machinery. He was also a confirmed gambler, and in fact his probability theory was a by-product of this particular hobby. He lived free and unrestrained, a true Renaissance man. Even while engaged in this dazzling array of endeavors, he managed to author numerous books. In 1576, just before he passed away, Cardano wrote *The Book of My Life*, a work of tremendous interest to anyone wanting to learn about the mentality of Renaissance Italian intellectuals.[48]

In this memoir, Cardano relates, for example, that one of the rules he set himself was "not to think that anything happened fortuitously in nature" (BML 23). Reading only this by itself, one might conclude that Cardano was oriented toward modern science and believed strongly that nature controlled events through strict cause-and-effect principles. And yet throughout the book he repeatedly makes rather unscientific observations like the following: "Does there not seem to be that about dreams—the fact that they have been so real—which can be worthy of admiration?" (BML 37). And referring to his birth, he relates that "since the sun, both malefics, and Venus and Mercury were in the human signs, I did not deviate from the human form. Since Jupiter was in the ascendant and Venus ruled the horoscope, I was not maimed, save in the genitals" (BML 2).

His observation that nothing is fortuitous—that nothing is without cause—means that, in his view, all events out of the ordinary and seemingly supernatural are omens of one kind or another. When the stars align themselves in a peculiar way, when one has a strange dream or palpitations, when a room shakes in the absence of an earthquake—such events are not to be considered as having occurred for no reason; they should be taken as unmistakable signs that some misfortune is about to occur.

Cardano believed not only in astrology but also in oneiromancy (divination by means of dreams) and in the idea of a guardian spirit. He thus exemplifies the fact that experience- and observation-based inquiry alone did not lead directly to modern science, and that the significance ascribed to individual experiences and observations varied according to the perspective of the observer who integrated and articulated them. In the case of Cardano, however, we know that he "attempted so far as possible to give a natural explanation for magical phenomena,"[49] and therefore we can conclude that his understanding of magic was close to that of Pomponazzi.

In *The Book of My Life*, Cardano writes: "I have taught that a contemplation of the natural universe leads to an artistic expression and to creative labor, although no others before my time have undertaken to offer such a suggestion" (BML 44). This is, in effect, a statement of what Cardano regards as his own greatest achievement. To Cardano, the study of nature is important and praiseworthy because of its practical applicability.

Cardano displays the same orientation in a 1551 work called *De subtilitate rerum* (On the subtlety of things), a book that deserves to be classified with works on natural magic.

In *De subtilitate rerum*, Cardano considers the world from the perspective of five principles: primal matter, form, soul, position, and movement. From his choice of words, we can see that Cardano was strongly influenced by Aristotelian philosophy. But even more important than this metaphysical framework is Cardano's discussion of the forces and movements observable in the natural world. By this I mean that Cardano unveils his understanding of nature primarily in the course of discussing theories about the way machinery and mechanical instruments make use of various forces and about the design of such devices. In volume 1 of *De subtilitate rerum*, in fact, he devotes page after page to explaining the principles governing the operation of machines that make use of atmospheric pressure, hydropower, and the force that we know today as gravity.

Cardano proposes that there are four types of natural movement (*motus naturales*).[50] The first is movement into a space with less air, as in Aristotle's concept that "nature abhors a vacuum." The second is essentially the reverse: movement away from an area of strong pressure. Third comes what Aristotle had referred to as "natural movement": the downward movement of heavy bodies and the upward movement of light bodies. The fourth is the movement seen between lodestone and iron or between amber and straw. Since Cardano holds that movement derives from force, these four categories can be considered four types of force. In modern terms, we would restate Cardano's classification system to say that there are three types of force: atmospheric, gravitational, and electromagnetic, Cardano's first two types being essentially the same.

In volume 5 of *De subtilitate rerum,* Cardano attributes the amber phenomenon (static electricity) to amber's having

> a fatty and glutinous humor which, being emitted, the dry object desiring to absorb it is moved toward the source, that is the amber. For every dry thing, as soon as it begins to absorb moisture, is moved toward the moist source, like fire to its pasture; and since the amber is strongly rubbed, it draws the more because of its heat.[51]

Here again we see static electricity attributed not to friction itself but to the heat generated by friction. More relevant to our inquiry, however, are two facts: first, Cardano does not mention the principle of like attracting like; and second, he leans utterly on the notion of action through an intermediary. In his explanation, the attracting force is generated by "a fatty and glutinous humor" emitted by the amber. As a mechanism for the propagation of force, this was a revival of the kind of mechanistic theories that had not been seen in the thousand years since Plutarch, and at the same time it was also pathbreaking, leading the way toward modern mechanistic theories of electricity. Cardano's "fatty and glutinous humor" was the origin of William Gilbert's *ab humore effuluvia*, also known as "electric effluvia"; and in the theories of the seventeenth and eighteenth centuries it would be transformed into various types of ethereal materials.

Cardano also recognizes that "the magnet stone and the amber do not attract in the same way," enumerating the ways in which static electricity and magnetism differ as follows:

> 1. The amber draws everything that is light; the magnet, iron only.
> 2. The amber does not move chaff when something is interposed: the magnet nevertheless will attract iron.
> 3. The amber is not mutually attracted by the chaff: the magnet is drawn by the iron.
> 4. The amber does not attract at the end: the magnet attracts the iron sometimes at the North and sometimes at the South.
> 5. The attraction of the amber is greatly aided by heat and friction: that of the magnet, by cleaning the attracting part.[52]

That magnets acted selectively and that they were capable of acting through a physical barrier had been known since the time of Alexander of Aphrodisias, and given that Alexander's writings were well known and highly regarded at the University of Padova, we can assume that Cardano studied them there. Nonetheless, Cardano deserves credit for originality on three accounts: he was the first to note that amber attracted all light things, that its effect was obstructed by shielding, and that in these two respects static electricity differed qualitatively from magnetism. The above excerpt, therefore, represents the earliest codification in scientific history of the differences between these two mysterious phenomena.

It is probable that Cardano drew most of his conclusions from actual experimentation. In volume 7 of *De subtilitate rerum,* he describes how he physically carried about each of a variety of gemstones in order to confirm their medicinal properties, their value as amulets, and so forth. Of course, most of these so-called experiments had to have been quite simplistic, if not childish, but we must acknowledge Cardano's efforts to take a critical and empirical look at claims that had been passed down through many generations from ancient times forward. With regard to inquiries into the causes of natural bodies and phenomena, Cardano writes that Pliny and Albert the Great "cannot be trusted in this field because they clearly misrepresent the facts."[53] Indeed, whereas Albert the Great had said that there were two types of magnets—those that attract iron and those that are attracted by iron—in *De subtilitate rerum* Cardano becomes the first to recognize that iron and magnets act reciprocally.

As we will see later, Cardano's model of static electricity as a form of action through an intermediary, as well as his observations concerning the differences between electricity and magnetism, would be adopted in their entirety by Gilbert. In particular, Gilbert would borrow two of Cardano's more original observations—that amber cannot attract through shielding and that amber's force is one-sided—as fundamental principles setting magnetic force apart from static electricity.

## Section 6—Giordano Bruno's Understanding of Electromagnetism

Although Cardano saw amber as a fertile subject of study within the framework of natural magic, in the end he explained amber's force as something less than mysterious: as a form of action through an intermediary (in this case, some sort of "humor"). In other words, with respect to force Cardano was a reductionist through and through. This is noteworthy because, in seeking out an explanation for the mechanism by which forces were transmitted, Cardano pointed the way to mechanist theory and therefore to the erosion of the theoretical foundation of natural magic. Natural magic was a discipline based on the belief that certain powers of nature had to be accepted as occult forces, the results of their action being the only possible evidence of their very existence. In natural magic, further explanations were not to be sought.

Roughly half a century after Cardano came Giordano Bruno (1548–1600), an even more ardent proponent of magic than Cardano, and one who went even further in the direction Cardano had indicated.

Bruno was born in southern Italy, on the outskirts of Naples. He entered a monastery for training as a cleric at the age of 17, and at first was a faithful adherent of traditional Catholicism and a boy who worshipped Thomas Aquinas. In the 1570s, however, he was drawn to Renaissance humanism and came under scrutiny as a dangerous figure who had drunk the poison of heretical thought. Near the end of the decade, Bruno left Naples, condemned to a life of exile from his native region. At that time, southern Italy was under the control of Spain, a nation steeped in Catholic dogma. The Counter Reformation was raging at full force, granting heresy no tolerance. From 1564 to 1567, more than thirty people were burned at the stake in Naples for this crime, and the smoke from those fires threatened to asphyxiate Renaissance humanism in its infancy.

Bruno moved from place to place throughout northern Italy and crossed the Alps in 1579. From there he roamed about Europe, writing as he went, until his return to Italy years later. Despised by Catholics and Protestants alike, Bruno found permanent safety nowhere, and the writings he left were all penned during this unsettled period of wandering. After his return to Italy in 1591, he was arrested, incarcerated for eight years, and eventually burned at the stake in Rome's Campo de' Fiori in 1600.

Bruno's *On the Infinite Universe and Worlds* did more than lend support to Copernicus's heliocentric theory; it also surpassed Copernicus's thinking with a new view of the cosmos that included an infinite universe containing multiple worlds. Indeed, one of the charges against Bruno in his prosecution for heresy was that he held this particular view. Bruno wrote *On the Infinite Universe and Worlds* in London in 1584, along with another book, *Concerning the Cause, Principle, and Unity.* It goes without saying that his inspiration in writing these books came partly from Nicolaus Cusanus, but it is important to emphasize that his inspiration also came from the ancient atomists Democritus and Lucretius. Note, for example, his remark that, as physical beings,

> we receive a perpetual flow of fresh atoms, and those that we have received previously are ever leaving us [and the] body attaineth form and growth when the influx of atoms exceedeth the efflux. Moreover this same body is of a certain consistency when the efflux equals the influx, and finally declineth when the efflux exceedeth the influx.[54]

Bruno thus considered the indivisible *atom* to be the ultimate formative principle of all things in nature, and he explained natural phenomena as a cycle of atoms coming together and then breaking apart. This atomistic theory also formed the basis for Bruno's understanding of magnetic force.

Bruno discusses magnetic force in *Essays on Magic,* a short treatise published around 1590. He begins by listing the various types of magic, including natural magic and other types of magic that make use of words, spells, numbers, images, symbols, and so on. Mathematical magic is among the types Bruno writes about in some detail. In what follows, we will restrict our discussion to Bruno's thoughts on natural magic only, although it appears that he did believe in other varieties as well. Bruno writes that "'natural magic' in the proper sense" is

> what happens as a result of the powers of attraction and repulsion between things, for example, the pushes, motions and attractions due to magnets and such things, when all these actions are due not to active and passive qualities but rather to the spirit or soul existing in things.[55]

In saying that the actions of a magnet are "due not to active and passive qualities," Bruno asserts that magnetism cannot be comprehended in terms of Aristotle's four-elements theory.

As it was for many thinkers who came before him, for Bruno the mysterious power of magnetism was not only the first and foremost example of natural magic but virtually its sole representative.

There is, however, a decisive difference between Bruno and his predecessors who had written about natural magic. With regard to force, Bruno is a reductionist, and he attempts to explain magnetic force in terms of atomistic theory and the concept of action through an intermediary. He holds that there are two types of attractive force. One is attraction between like, or identical, bodies, which act in the manner of a part being pulled by the whole. The other force Bruno explains as follows:

> There is another type of attraction which is not perceived by the senses. This is the case of a magnet attracting iron. The cause of this cannot be attributed to a vacuum or to any such thing, but only to the outflow of atoms or parts, which occurs in all bodies. For when atoms of one type move towards and mutually encounter other atoms of a similar type or of a congenial and compatible nature, the bodies develop such an attraction and impulse for each other that the overpowered body moves towards the whole of the stronger body. For since all the parts experience this attraction, then so must the whole body also be attracted.[56]

We can see clearly here the influence of ancient atomism. Moreover, what emerges next is an explanation of force that is mechanistic in the extreme:

> The fact that this attraction is caused by the outflow of parts from such bodies also indicates that when a magnet or amber is rubbed, it attracts iron or straw much more strongly. For the heat causes more parts to be emitted, since it opens the pores and rarefies the body.

Although this is not a reiteration of Cardano's observations about the differences between static electricity and magnetic force, we do hear echoes of Cardano's break with the past: in this excerpt, Bruno presents magnetic force and static electricity as phenomena whose existence calls for an explanation. This was a far cry from the position of natural magic, which saw magnetic attraction solely as an "occult force" and abandoned further inquiry. We can therefore say that in the history of thought regarding the concept of force, if Cardano is to be credited with having revived the Platonic and Plutarchian theories of action through an intermediary, then Bruno can be acknowledged for having resurrected the ancient theory of atomism.

I should add, however, that Bruno's atomism is in substance different from both ancient atomism and the kind that would be propounded in the seventeenth century. From the following excerpt it is clear that Bruno does not see the characteristics of four simple elements as an explanation for the cause of magnetic attraction:

> A sign of this is that when a piece of iron touches a magnet, it acquires the same power of attracting other pieces of iron. This could not happen if this were due to an elementary quality. For when heat and coldness are accidentally present in a subject, they quickly disappear when the source of heat is removed. Therefore, one must explain this in terms of the emission of parts or of a spiritual substance which flows from the magnet into the iron....
>
> This same explanation and cause accounts for the fact that diamonds are said to block such an attraction, and similar types of explanation account for various other things.[57]

In speaking of "parts," although Bruno does mean some sort of atom, he does not mean the sort specified by the simple atomistic and mechanistic theories of older times, according to which atoms consist of inert matter whose properties depend solely upon shape and movement. As indicated in the passage quoted at note 56, Bruno holds that like atoms act on one another. Behind this thinking is his belief that the universe itself is a single living organism,[58] and that within this organism are atoms, each of which has a soul. Or, to put it another way, each atom has a spirit that endows it with the ability to act in a particular way—an active nature derived from the workings of this spirit, also known as the "ether."

In *On the Infinite Universe and Worlds*, Bruno describes not only the existence of the four elements but also of this ether, which he posits as an active principle that envelops and permeates all created things and within which all created things move, live, and grow. To clarify, let's take a look at what Bruno himself has to say on the subject of ether:

> In so far as [the ether] entereth into and formeth part of the mixture of the elements, it is commonly named air—the word applying to the vaporous layer around the waters and within the land,… [And] when it is in some part of us or at least hath a part in our composition—as when it is found in our lungs, our arteries and other cavities and pores of our body—it is called spirit. The same, when around a cold body, becometh condensed into vapour, but around a hot star it is attenuated like flame,… Thus the ether is of his own nature without determined quality, but it receiveth all the qualities offered by neighbouring bodies, and carrieth them with his own motion to the furthest limits of the horizon wherein such active principles have efficacy.[59]

In sum, Bruno asserts that magnetic force and static electricity are forms of action through an intermediary—namely, atoms—and that this action takes place because of a living, active ether, or spirit.

But it is clear, at any rate, that Bruno had a preference for speculation. Whereas Cardano had emphasized observation, experimentation, and technological application, Bruno emphasized none of this. The man who truly took the torch from Cardano and carried it further in the quest to understand magnetic force was a contemporary of Bruno's, a native of southern Italy, and a magician. And it is to that figure—Della Porta—that we will next turn our attention.

* * *

When the Platonists of Florence, influenced by both neo-Platonism and Hermeticism, revived magic in the 1400s, they recognized two types of magic—demonic and natural. For religious reasons, Marsilio Ficino and others hesitated to discuss demonic magic at all, although they did believe in its existence. But the thinkers of the 1500s, influenced by Roger Bacon and by Aristotle, dismissed demonic magic on philosophical grounds. They rejected the transcendental perspective, which held that magic and miracles occurred at the whim of demons and angels, and therefore thought demonic magic to be bogus.

With this philosophical orientation, in the 1500s experience, observation, and experimentation led the way, shedding light on occult forces that were understood to be amenable to manipulation in accordance with natural laws. Natural magic practiced in this manner took on the form of what we can call premodern physics, evolving into something both secular and technological. John Dee imbued natural magic with a mathematical and technological character, and others used it to develop experimental techniques and to inform their observations of real-world phenomena. And because magnetism and static electricity were seen as exemplars of the so-called occult forces that were at the heart of magical thought, these two phenomena in particular became the subject of much observation and experimentation in the field of natural magic.

But the most important development for our purposes here was the realization that mathematical reasoning and experimental technique were not in themselves incompatible with magic. Moreover, the demand arose for rational explanations of unseen forces once figures such as Cardano and Bruno had smuggled the reductionist theories of mechanism and atomism into the arena of magical thought. "In early modern Europe," notes Paolo Rossi, "magic and science formed a web not so easily unraveled."[60] Driven by seafarers, tradesmen, and military men on the one hand and by magicians, astrologers, and alchemists on the other, during the latter half of the sixteenth century, research into magnetic force was on the verge of a momentous confluence with the intellectual streams of experimental physics and rational reasoning.

Chapter 16

# Della Porta's Investigations into Magnetism

## Section 1—Della Porta's *Natural Magick* and its Background

The late-Renaissance Italian scholar Giambattista della Porta (c. 1535–1615) undertook a general, experimental study of magnets and magnetic force as part of his research into natural magic. Born to a noble family in Naples, Della Porta was educated at home and thus remained free from university academism throughout his life. In 1560 he founded a private learned society known as the Accademia dei segreti (Academia secretorum naturae; Academy of the Mysteries of Nature), the forerunner of later academies of modern science, such as the Accademia dei lincei (Academy of the Lynxes) and the Accademia del cimento (Academy of Experiment), both founded in Italy in the seventeenth century. A true Renaissance man, he wrote not only scholarly works such as *De refractione optices* (On optics; 1593) but also numerous plays, including both comedies and tragedies.

Della Porta's most famous work, however, is *Natural Magick*, which enjoyed remarkable popularity from the mid-sixteenth to the mid-seventeenth century, when modern science was in its infancy. The first edition, published in four volumes in 1558, reportedly went through over five Latin printings and was also published in Italian, French, Dutch, Spanish, and even Arabic. In 1589, after acquiring new knowledge through travels in Italy, France, and Spain, Della Porta published an extensively revised and expanded edition of *Natural Magick* comprising twenty volumes. The revised edition also became immensely popular throughout Europe, being published twelve times in Latin, four in Italian, seven in French, two in German, and two in English—astounding figures considering the low literacy rate at the time. As I noted earlier (see chapter 13), market principles governed the print publishing industry from its inception and publishers eagerly sought out likely bestsellers.[1] Judging by the number of translations and printings *Natural Magick* enjoyed, publishers evidently had great confidence in the work's lasting popularity. It continued to be read in the seventeenth century, including by Johannes Kepler, Francis Bacon, and René Descartes, and was among the works in the respective personal libraries of Isaac Newton and Thomas Browne.[2]

But despite being renowned for his extensive knowledge during his own lifetime, Della Porta sank rather inexplicably into obscurity thereafter. Neglect of his contributions is particularly evident in works on the history of science. Joseph Priestley, for example, makes no mention of him in *The History and Present State of Electricity*, published in 1767. Nor does Della Porta appear in William Whewell's voluminous *History of the Inductive Sciences*, published in 1837, or in either J. D. Bernal's *Science in History* or Herbert Butterfield's *The Origins of Modern Science*, both of which were written yet another century later. While not quite banished to oblivion, Della Porta was thus certainly underestimated by subsequent generations. Perhaps the age of enlightenment simply buried his achievements along with everything else that had dealt with "magic," especially as *Natural Magick* was the last work on the subject to attract interest in the intellectual mainstream. With the emergence of mechanism, and particularly after the defeat of Paracelsian theory in England, magic was no longer of primary concern in the intellectual world. Nonetheless, *Natural Magick* played a major role in the birth of modern science.

*Natural Magick*, at least from book 2 onward, is neither a discussion of the esoteric mysteries of magic nor a systematic, scholarly treatment of the topic. It neither expounds upon the delicate relationship between magic and religion nor provides a particularly thorough treatment of natural philosophy. Rather, in descriptive approach it is akin to the first-century *Natural History* by Pliny; in fact, it is essentially a sixteenth-century version of that work, a hodgepodge of strange tales, instructive know-how, and other anecdotal information thrown together more or less at random. For the most part *Natural Magick* can best be regarded as a compilation of practical knowledge and skills rooted in the daily life of the rising urban population of Della Porta's day. "The idea of building up a personal library," explain historians Lucien Febvre and Henri-Jean Martin, "was popular among wealthy merchants, among the comfortable bourgeoisie, and even among the successful artisans ... in the 16th century."[3] *Natural Magick* was therefore probably more popular among the new urban class of physicians, merchants, craftsmen, and bureaucrats, who valued the practical and the utilitarian, than among either university scholars or high-ranking clergymen.

From book 3 to book 20, *Natural Magick* covers a plethora of practical subjects. These include how to improve plant species and different methods of cultivation (book 3); winemaking, fruit preservation, and bread making (book 4); methods for coloring gemstones (book 6); various kinds of medicines (book 8); cosmetics (book 9); instructions for making perfumed powder (book 11); methods of tempering steel (book 13); cooking recipes (book 14); and how to lure and catch animals (book 15). While book 5 explicitly deals with "Alchymy," in the preamble to that volume Della Porta remarks on this choice of topic that "I do not here promise any golden mountains, as they say, nor yet that Philosophers stone"; and he flatly dismisses the "golden liquor, whereof if any man drink, it is supposed that it will make him to be immortal" as being a "meer dream." Instead, the "alchemy" presented consists mainly of simple metallurgy for metals with low melting points, such as tin and lead. Overall, *Natural Magick* is thus at best a work of popular science, or a sort of household encyclopedia, which is precisely why it was so well received.[4]

*Natural Magick* is also a catalogue of the strange, the rare, and the mysterious, phenomena that fascinated the "virtuosos" (people who studied just as a hobby) among the leisure class of the time. In Della Porta's day there were many science lovers, particularly in Italy, who enjoyed observing, experimenting, collecting, and so on, not for the lofty aim of investigating truth, nor for any particular practical purpose, but merely out of interest in the mysteries of nature or simply to show off.[5] Of course, such dilettantes would also have made up a considerable portion of Della Porta's readership.

Although *Natural Magick* was clearly influenced by and in many places directly modeled on works by Marsilio Ficino and Giovanni Pico della Mirandola, in tone it differed considerably from their speculative and argumentative texts. Ficino and Pico took great pains to reconcile Christianity with neo-Platonism, Hermeticism, and magical thought, all of which the church viewed as heretical. Indeed, this effort consumed much of their energy and comprised the bulk of their writings. The objective of the Accademia Platonica (Platonic Academy) launched in Florence by Ficino was, after all, "the reconciliation of the spirit of antiquity with that of Christianity"[6] and, therefore, "religious and theological interests determined the whole attitude and development of philosophical thought in the Academy."[7] In contrast, this attitude is noticeably absent from Della Porta's work, in which he shows little interest in religious issues.

Della Porta's approach reflects the changes that occurred in political and social conditions from the fifteenth to the late sixteenth century. As Jacob Burckhardt points out, the humanism espoused by Ficino and Pico was "in fact pagan";[8] but during the fifteenth century, in which those two scholars lived and wrote, humanism was limited to the elite, and successive popes therefore tended to be somewhat tolerant of heretical opinions expressed from that quarter. Beginning in the mid-sixteenth century, however, the situation changed radically. On the one hand, the trend toward publishing in vernacular languages resulted in a dramatic increase in the general readership, giving the masses much greater exposure to new modes of thought. On the other hand, the Catholic church, which had let the print publishing industry go virtually unchecked for just short of a century, began tightening its control over the printed word, having finally recognized its enormous role in promoting and spreading the Protestant ideas of Martin Luther and his ilk in Germany, Switzerland, and other countries. In 1540 the pope officially sanctioned the Jesuits' tough stance toward the Reformation, and in 1542 the Holy Office of the Inquisition was reestablished in Rome to keep a stricter watch over intellectual activities. Then came the Council of Trent, first convened in 1545, which sought to reinforce church discipline and administration and definitively clarify church doctrine. It should be noted here that the council had the underlying purpose of protecting the Catholic faith from magic; unlike countries north of the Alps, Italy had had little exposure to Protestantism, and the church therefore regarded magic, rather, as the primary threat. The process continued with the Treaty of Cateau-Cambresis in 1559, whereby the majority of Italy came under the jurisdiction of Spain, a dogmatically Catholic state. That was also the year in which Pope Paul IV issued the first *List of Prohibited Books.*

Thus, by the time *Natural Magick* appeared in the latter half of the sixteenth century, Italy was in the grip of the Counter Reformation and the Inquisition was condemning countless souls to imprisonment or death at the stake. Many heretical thinkers fled into exile. Della Porta himself was summoned before the Inquisition in 1574 and 1580 and commanded to disband his private academy. At a time when even an offhand remark connecting religion with magic could put one's life at risk, in *Natural Magick* Della Porta quite understandably emphasized only the secular aspects of magic—that is, natural philosophy and technology—and relegated the religious aspects to the background.

Changes had also occurred in the readership. Many of the city dwellers who had recently joined the ranks of the literate were more interested in secular life than in religion. The publication of *Natural Magick* coincided with this social shift. As one scholar of Renaissance magic puts it, "In no treatise of the Renaissance can we examine more favorably what natural magic was like at its least mystical level."[9] This "natural magic" was the end product of a process whereby early Renaissance thinking about magic was cleansed of its mystical and religious elements and magic was secularized and popularized through a shift of emphasis to its practical and recreational aspects.

At the same time, however, *Natural Magick* was more than a lowbrow how-to book. As symbolized by the lynx depicted on the title page of the 1589 Latin edition, with its penetrating gaze fixed on its prey, and in keeping with the maxim expressed in the preface that "the most Majestick Wonders of Nature are not to be concealed," Della Porta infused this work with his passion to expose the hidden secrets of nature and share them with the world. It was precisely such voracious curiosity about the natural world and this spirit of investigation that led to modern science. In fact, *Natural Magick* includes what amount to the first steps toward experimental physics in the fields of optics and magnetism, and therefore must be given due recognition in any thorough history of physics.

## Section 2—From Textual Magic to Experimental Magic

Book 1 of *Natural Magick* presents a general overview of what magic was essentially thought to be. It is almost entirely based on Ficino, Pico, and Agrippa, with very little original content. Let's look at how Della Porta addresses the concept of magic in this first volume.

In chapter 2 of book 1, Della Porta explains that "there are two sorts of Magic." One is "infamous, and unhappie, because it hath to do with foul spirits, and consists of Inchantments and wicked Curiosity." This, he says, is called "sorcery" and "stands merely upon fancies and imaginations." The other is "natural magic," "which all excellent wise men do admit and embrace, and worship with great applause." In fact, "the most noble Philosophers that ever were, Pythagoras, Empedocles, Democritus, and Plato … do call this knowledge the very highest point, and the perfection of natural Sciences."

Della Porta states his own view clearly:

> I think that Magick is nothing else but the survey of the whole course of Nature … it openeth unto us the properties and qualities of hidden things, and the knowledge of the whole course of Nature; and it teacheth us by the agreement and disagreement of things. (NM 1.2)

He later explains the terms "agreement" and "disagreement" as follows:

> By reason of the hidden and secret properties of things, there is in all kinds of creatures a certain compassion, as I may call it, which the Greeks call Sympathy and Antipathy, but we term it more familiarly, their consent, and their disagreement. (NM 1.7)

This is, of course, the same conceptual framework, the same dichotomy of sympathy and antipathy, that had been in place since well before his time.

Up to this point, then, Della Porta is merely adhering to long-standing traditions in the thought of natural magic. This organic view of nature as a network of sympathy and antipathy among all things—including the concept of astrological causality, according to which celestial bodies influence terrestrial bodies—had been reiterated through the ages. In Della Porta's own words,

> the parts of this huge world, like the limbs and members of one

> living creature, do all depend upon one Author, and are knit together by the bond of one Nature; therefore as in us, the brain, the lights, the heart, the liver, and other parts of us do receive and draw mutual benefit from each other, so that when one part suffers, the rest also suffer with it; even so the parts and members of this huge creature the World, I mean all the bodies that are in it, do in good neighbour-hood as it were, lend and borrow each others Nature; for by reason that they are linked in one common bond, therefore they have love in common; and by force of this common love, there is amongst them a common attraction … And this indeed is Magick. (NM 1.9)

Della Porta thus seems to view the universe as a unified organic body, or even, rather, as a single living organism. In any case, his observation a little later in the same chapter that

> the Load-stone draws iron to it, Amber draws chaff or light straws, Brimstone draws fire, the Sun draws after it many flowers and leaves …

is copied directly from Ficino. In that sense, although Della Porta's work contributed to a more thorough understanding of magical nature than had existed before, it did not offer anything particularly new in terms of the basis of that understanding.

What does distinguish Della Porta from the writers who preceded him, however, is that he achieved the pivotal shift from speculative, purely textual magic to experimental magic built around demonstrable proof. Whereas Ficino, Pico, and Agrippa unquestioningly accepted past literature and based their discussions of natural magic on texts written in antiquity, Della Porta tried to actually perform magic in several fields of inquiry and in the process verify the truth of received views experimentally.

In book 1, in which Della Porta discusses magic in general, he maintains, as had Agrippa before him, that there exists a "deadly hatred" between coleworts and vines and an enmity between hemlock and rue and between cane and fern, and that apes cannot abide snails and that wolves fear the urchin. In the same place, he reports a

> notable Disagreement that is betwixt Garlick and the Loadstone; for being smeared about with Garlike [*sic*], it will not draw iron to it. (NM 1.7)

Insofar as he simply compiled these examples from ancient and medieval literature without holding them up to any critical scrutiny, Della Porta certainly deserves the criticism that "such associations derive … from naive thought-processes not searchingly questioned because of a reverence for traditional wisdom."[10]

On the other hand, Della Porta himself frankly admits that, of the various relationships of sympathy and antipathy, "there can be rendred no probable reason" (NM 1.7); and indeed such hollow theorizing has virtually no bearing on the individual magical practices he presents in subsequent chapters. In the preface, he explains his approach as follows: "In our Method I shall observe what our Ancestors have said; Then I shall shew by my own Experience, whether they be true or false, and last of all my own Inventions."

As we will see later, Della Porta disproves the purported "disagreement" between garlic and lodestone through experiments described in book 7, which deals specifically with magnets. Thus, regardless of what was written in the ancient texts or of the fact that he himself repeated those very statements in the first book of *Natural Magick*, in presenting the specifics of magic and its actual practice in subsequent volumes, Della Porta puts proof by experiment before traditional lore, at least in his discussion of magnets.

As Della Porta understands it, magic is not about employing incantations or symbols to call forth demonic powers, conjure supernatural phenomena, or perform miracles that defy nature. Rather, "the works of Magick are nothing else but the works of Nature, whose dutiful hand-maid Magick is" (NM 1.2). Magic is the technology for artificially either generating or promoting the workings of nature in obedience to nature's laws. Generating such effects means applying nature's power through, for example, experiments on optical phenomena using mirrors and lenses, or using steam to siphon water. Promoting natural processes, meanwhile, refers to such tasks as separating and extracting pure components from natural substances through distillation; hastening animal and plant growth and mineral metamorphism; and improving animal and plant breeds through crossbreeding, grafting, and so on. There is no calling on the devil for supernatural effects. Accordingly, Della Porta makes it clear that "superstitious, profane, and wicked men have nothing to do with this Science; her gate is shut against them" (NM 1.2).

Such a view of natural magic had already appeared in general discourse, but it was Della Porta who drove the point home. His uniqueness lies in the fact that he places far more emphasis than his predecessors on experimental, empirical methods and practical, utilitarian objectives. As he puts it, magic is "a practical part of Natural Philosophy" (NM 1.3). For this reason, he insists that a magician must be a flawless philosopher, a skilled physician, a learned herbalist, an adept in the nature of metals and gems, a master of the art of distillation, well versed in alchemy, skilled in the mathematical sciences, especially astrology, and proficient in optics. In other words, the magician must excel in every branch of practical knowledge and skill. He caps this list of qualifications with a thoroughly down-to-earth observation: “Lastly, the professor of this Science must also be rich: for if we lack money, we shall hardly work in these cases: for it is not Philosophy that can make us rich; we must first be rich, that we may play the Philosophers” (NM 1.3). Della Porta probably used his own money to purchase all the materials and equipment he needed for his experiments. In any case, we should bear in mind that in his time a single magnet, lens, or measure of chemicals would have been an extremely valuable and therefore expensive item, as indicated by his observation that an Ethiopian magnet, the kind considered the finest in quality, “costs the weight in silver” (NM 7.1).

We can see from the above that, while Della Porta explores the same topic of natural magic as had others before him, his emphasis differs from theirs in key respects. This is evident, for example, in book 18, which is devoted to the study and application of what today we call hydrostatics and describes, among other things, the mechanism of a water clock. In the preamble to the volume, Della Porta says of its subjects that “these things may be applied to very necessary and profitable uses, and if any man shall more deeply consider these things, he may invent many new things: that may be employed for very profitable ends.” As he repeatedly emphasizes, knowledge is important precisely because it leads to the discovery of things that are beneficial and useful.

Thus, with Girolamo Cardano and Della Porta, in the sixteenth century the natural magic of the Renaissance arrived at the threshold of the modern era of technical practice characterized by thoroughgoing instrumental rationality.

## Section 3—*Natural Magick* and Experimental Science

Della Porta's diverse and disorderly exposition on magical practice covers topics that nowadays would fall under elementary physics, including empirical descriptions of lodestone (book 7); observations and experiments on lenses, mirrors, and other optical equipment (book 17); hydrostatics and its applications (books 18 and 19); and observations and experiments on the physics and dynamics of air (book 20).

The experiments and theoretical considerations on optics presented in book 17 are particularly noteworthy. Although experimental studies of optics in the Middle Ages are generally still regarded as belonging to the field of natural magic outside the standard natural philosophy of the schools,[11] the observations and experiments contained in book 17 merit praise even from the perspective of modern physics. As early as 1882, Ferdinand Rosenberger in *Die Geschichte der Physik* (History of physics) stressed that the most important part of *Natural Magick* is optics,[12] and early in the twentieth century Ernst Cassirer also acknowledged that "Porta certainly did make decisive contributions to the foundation of the science of optics."[13]

In chapter 6 of book 17, Della Porta describes an experiment with a camera obscura. He constructed a darkened room with a small hole in one wall into which was fitted a convex lens. With light entering the room through the lens only, the scene outside the room was projected as an inverted image on a white sheet hung on the opposite wall. For people in that day and age it must have been quite amazing to see, inside an otherwise completely dark room, a clear image of the view outside it. The camera obscura was already known by that time but, as far as we know, it was Della Porta who first added a convex lens to the pinhole and who, in *Natural Magick*, first compared it to the optical structure of the eye. Della Porta's camera obscura experiment formed the basis for Kepler's later theory of vision, and thus deserves special mention in any history of optics.[14]

Chapter 10 of book 17 contains what appears to be the earliest known written reference to the device that later came to be known as the Galilean telescope, which comprised both concave and convex lenses:

> Concave Lenticulars will make one see most clearly things that are afar off; but Convexes, things neer hand; so you may use them as your sight requires. With a Concave you shall see small things afar off, very clearly; with a Convex, things neerer to be greater, but more obscurely; if you know how to fit them both together; you shall see both things afar off, and things neer hand, both greater and clearly. I have much helped some of my friends, who saw things afar off, weakly; and what was neer, confusedly, that they might see all things clearly. (NM 17.10)

Ever since Galileo's comment in *The Sidereal Messenger* of 1610 that he had heard rumors that a Dutchman had created such a spyglass some ten months earlier,[15] people have attributed the invention to lensmaker Hans Lippershey. In fact, however, the above passage preceded Galileo's remark by about twenty years, which makes it impossible to refute Della Porta's role in the telescope's invention. Indeed, in *Kepler's Conversation with Galileo's Sidereal Messenger*, Kepler declares that the Galilean telescope had been invented not recently by a Dutchman but rather much earlier by Della Porta, in evidence of which he quotes the above passage from *Natural Magick*.[16] And in 1609, after hearing that Galileo had used a telescope for astronomical observations, Della Porta himself laid claim to priority in the invention in a letter to Federico Cesi, the founder of the Academy of the Lynxes.[17]

Book 17 of *Natural Magick* thus occupies a unique place in the history of optics. Notwithstanding the connotations of the overall work's title, book 17 reads much like a textbook on elementary experimental science. It must be admitted that Della Porta's experiments fail to qualify as modern science because "there is no theoretical framework in *Magiae naturalis* [*Natural Magick*], and no program to [Della] Porta's observations."[18] Nor can it be denied that the main aim of his experiments was not proving hypotheses or discovering physical laws but something like recreation and entertainment: he experimented primarily for the sheer pleasure of it and to amaze people, and his experiments were little more than trial-and-error attempts to artificially reproduce strange natural phenomena. Nonetheless, as Marie Boas asserts, Della Porta's "natural magic" must be understood in terms of a "magico-experimental approach to nature."[19]

With the exception of Petrus Peregrinus's *Letter concerning the Magnet*, written three centuries earlier, the earliest real forays into experimental research on the lodestone were *The Newe Attractive* by instrument maker Robert Norman and *Natural Magick* by magician Della Porta, both of which appeared in the 1500s. In most histories of physics, however, the credit for this is awarded to William Gilbert's *On the Magnet*, first published in 1600. Gilbert himself offered the following assessment of Della Porta's research on magnets:

> Most recently, Baptista Porta, no ordinary philosopher, in his *Magiae Naturalis*, has made the seventh book a custodian and distributor of the marvels of the loadstone; but little did he know or ever see of magnetick motions; and some things that he noted of the powers which it manifested, either learned by him from the Reverend Maestro Paolo, the Venetian, or evolved from his own vigils, were not so well discovered or observed; but abound in utterly false experiments, as will be clear in due place: still I deem him worthy of high praise for having attempted so great a subject ... and having given occasion for further research.[20]

One can only suppose that Gilbert deliberately sought to underrate *Natural Magick*. Although he may not have wanted to admit it, in truth his *On the Magnet* owes much to Della Porta. This debt is not apparent from reading *On the Magnet* alone, but it becomes obvious when one compares that work with book 7 of *Natural Magick*.

## Section 4—Overview of the Study of Magnetic Force in *Natural Magick*

As suggested by its title, "Of the wonders of the Load-stone," book 7 of *Natural Magick* is devoted to the study of magnets and magnetic force. It begins with these words:

> We pass from Jewels to Stones: the chief whereof, and the most admirable is the Load-stone, and in it the Majesty of Nature doth most appear.

Della Porta's interest in magnets derived not from any concern for their application in navigation (compasses) or medicine but purely from his curiosity about the mysterious powers they manifested.

Book 7 consists of fifty-six chapters. I list all of the chapter titles here to convey an overall picture of the volume's content, though in some cases it is difficult to surmise the content of the chapter from the title alone.

Chapter 1. What is the Name of this Stone, the kind of it, and the Countrey where it grows.
Chapter 2. The natural reason of the Loadstones attraction.
Chapter 3. That the Loadstone hath two opposite Poles, the North and South, and how they may be known.
Chapter 4. The force of the stone is sent by a right line from North to South, through the length of it.
Chapter 5. That the polar line in the Loadstone is not stable, but moveable.
Chapter 6. That the force of North and South is vigorous in the points.
Chapter 7. That by the touching of other stones, those points will not change their forces.
Chapter 8. That a Loadstone will draw a Loadstone, and drive it from it.
Chapter 9. A sport of the Loadstone.
Chapter 10. The greater the Loadstone is, the greater is the force of it.
Chapter 11. That the force of this Stone will pass into other Stones, that sometimes you may see as it were a rope of Stones.
Chapter 12. That in the Loadstone that hairyness is contused.
Chapter 13. The attractive part is more violent then [*sic*] the part that drives off.
Chapter 14. The contrary parts of the Stones are contrary one to another.
Chapter 15. How to know the Polar points in the Loadstone.
Chapter 16. That the force of drawing and driving off, can be hindred by no hindrance.

Chapter 17. How to make an Army of Sand to fight before you.

Chapter 18. The Situation makes the Vertues of the Stone contrary.

Chapter 19. How the attractive force of the Loadstone may be weighed.

Chapter 20. Of the mutual attraction, and driving off of the Loadstone, and of Iron.

Chapter 21. The Iron and Loadstone are in greater amity, then [*sic*] the Loadstone is with the Loadstone.

Chapter 22. The Loadstone doth not draw on all parts, but at certain points.

Chapter 23. That the same Loadstone that draws, doth on the contrary point drive off the iron.

Chapter 24. How iron will be made leap upon a Table, no Loadstone being seen.

Chapter 25. That the vertue of the Loadstone, is sent through the pieces of Iron.

Chapter 26. The Loadstone within the sphere of its vertue, sends it forth without touching.

Chapter 27. How the Loadstone can hang up iron in the air.

Chapter 28. The forces of the Loadstone cannot be hindred, by a wall or table coming between.

Chapter 29. How a man of wood may row a little Boat; and some other merry conceits.

Chapter 30. A Loadstone on a plate of iron, will not stir iron.

Chapter 31. The position of the Iron, will change the forces.

Chapter 32. That the iron rubbed with the northern point of the Loadstone, will turn to the south, and with the south point to the north.

Chapter 33. That iron touched by the Loadstone, will impart that force to other iron.

Chapter 34. The vertue received in the iron, is weakned by one that is stronger.

Chapter 35. How in a stone the south or north point is discerned.
Chapter 36. How to rub the iron needle of the Mariners Compass.
Chapter 37. Of the divers uses of Mariners Compasses.
Chapter 38. How the Longitude of the world, may be found out by help of the Loadstone.
Chapter 39. If the Mariners Needle stand still, and the Loadstone move, or contrarily, they will move contrary ways.
Chapter 40. The Loadstone imparts a contrary force to the Needle.
Chapter 41. Two Needles touched by the Loadstone, obtain contrary Forces.
Chapter 42. That the force of the Iron that draws, will drive off Iron by diversity of Situation.
Chapter 43. The Needle touched by the Loadstone on one part, doth not alwayes_receive Vertue on both parts.
Chapter 44. The Needle touched in the middle by the Loadstone, sends forth its Force at both ends.
Chapter 45. An Iron Ring touched by a Loadstone, will receive both Vertues.
Chapter 46. An Iron Plate touched in the middle, will diffuse its forces to both ends.
Chapter 47. How filings of Iron may receive force.
Chapter 48. Whether Garlick can hinder the vertues of the Loadstone.
Chapter 49. How a Loadstone astonished may be brought to itself again.
Chapter 50. How to augment the Loadstones vertue.
Chapter 51. That the Loadstone may lose its vertue.
Chapter 52. How the Iron touched with the Loadstone loseth its force.
Chapter 53. It is false, That the Diamond doth hinder the Loadstones vertue.

Chapter 54. Goats blood doth not free the Loadstone from the inchantment of the Diamond.
Chapter 55. The Iron touched with a Diamond will turn to the North.
Chapter 56. The forces and remedies of the Loadstone.

These chapter titles are taken from the second edition, published in 1589. For the sake of comparison, let's look briefly at the treatment of magnets in the first edition, published in 1558.[21] Book 1 of the first edition deals with magic in general, book 2 with unusual tales of nature, book 3 with alchemy, and book 4 with experiments with light. Magnets are mentioned under the heading "How to prove a wife's fidelity" in book 2, chapter 30, titled "Of several amazing experiments." As the heading suggests, in regard to magnets the first edition of *Natural Magick* has little in the way of actual observations or experiments and mainly regurgitates folklore. The section in question claims, for example, that "if a lodestone be placed under a married woman's head while she sleeps, she that's chaste will embrace her husband but she that's a whore will fall out of the bed"; and that "as garlic hinders the virtue of the lodestone, if a seaman who has eaten garlic or onions approaches a compass, it will make the poles drunk."

When he wrote the first edition, Della Porta apparently accepted such ancient lore about magnets without question. But here is where the comparison with the second edition is telling: the headings from book 7 of the later work reveal that, between the first and second editions, Della Porta had radically revised his understanding of the magnet in a way that marked a crucial shift from purely textual magic to experimental magic.

Chapter 1 of book 7 in the second edition appears to have been copied almost verbatim from texts by Dioscorides and Pliny. Chapter 2 is devoted to considering the causes of magnetic force and begins by introducing the various theories of Aristotle, Galen, Avicenna, Epicurus, and Lucretius in that regard. It contains the first mention in well over a thousand years of the atomic theory proposed by Epicurus to explain magnetic force, though Della Porta's account here appears to be based on Galen's critique of that theory.

Della Porta then presents his own theory, as follows:

> I think the Loadstone, is a mixture of stone and iron, as an iron stone, or a stone of iron. Yet do not think the stone is so changed into iron, as to lose its own Nature, nor that the iron is so drowned in the stone, but it preserves it self; and whilst one labours to get the victory of the other, the attraction is made by the combat between them. In that body, there is more of the stone, then [*sic*] of iron; and therefore the iron, that it may not be subdued by the stone, desires the force and company of iron; that being not able to resist alone, it may be able by more help to defend it self. For all creatures defend their being: Wherefore, that it may enjoy friendly help, and not lose its own perfection, it willingly draws iron to it, or iron comes willingly to that. The Loadstone draws not stones, because it wants them not, for there is stone enough in the body of it; and if one Loadstone draw another, it is not for the stone, but for the iron that is in it. (NM 7.2)

Gilbert subsequently denounced this rather obscure explanation as "the ravings of a babbling old woman, not the inventions of a distinguished mage,"[22] but essentially it is just another variation on the long-standing theme of attraction between things of like kind or nature. What is noteworthy, however, is that, whereas previous thinkers, from the ancients right through to Albert the Great and

Georgius Agricola, had classified magnets as stone and not metal, in this passage Della Porta becomes the first to take lodestones to be comprised of iron, or at least to have some properties of iron, and to propose that magnetic force originates in those properties.

Chapter 2 of book 7 also has the following:

> I shall not pass by the opinion of Anaxagoras, set down by Aristotle in his Book *De Anima*, who by a similitude calls it a living stone, and that therefore it draws iron.[23]

Della Porta's own view of magnetic force was similarly hylozoistic.

## Section 5—Della Porta's Experiments with Magnets

In *Natural Magick*, Della Porta presents his experiments with magnets in book 7, beginning with chapter 3. The detailed nature of his reports makes it almost certain that he conducted the experiments himself, although the majority of them were aimed at checking those of Peregrinus.

Chapter 3 concerns the fact that magnets have two opposing poles, pointing north and south, respectively, and how to tell them apart. Chapters 4 and 5 explain that when a magnet is split apart, each of the separate pieces also has two poles, demonstrating that magnets are always dipolar. Chapter 6 records that the magnetic force of a magnet is strongest at its poles, chapter 7 that the force does not change when two magnets touch each other, and chapter 8 that opposite poles attract each other while identical poles repel each other. Up to this point, Della Porta thus simply reconfirms the observations that Peregrinus had already made.

But Della Porta's experiments on magnets do not stop there. In chapter 28 of book 7 he confirms that physical bodies do not obstruct magnetic force, but in chapter 30 he notes the exception of iron, whereby

> if you put the filings of iron upon a plate of iron, and with your hand underneath, you carry the Loadstone, the filing will not stir, but stand still upon the plate.

Similarly, in chapter 16, he notes that, whereas a lodestone above a table made of, say, wood can be drawn by another lodestone under the table,

> if the Table is made of Loadstone or Iron, the vertue is hindred, and can do nothing.

Not even Cardano had noticed this.

In chapter 20, Della Porta, like Cardano and Vannoccio Biringuccio before him, asserts that the attraction between lodestone and iron is mutual rather than a case of one pulling the other. According to Della Porta, lodestone and iron

> are so mutually in love [that] the lighter of these two will stir when the heavier approaches neer it.

In principle, of course, both should move, but in experiments on frictional surfaces often only the lighter object moves. This implies that Della Porta's findings were based on actual experiment. Although his experiments were not strictly controlled, his account nonetheless ranks as an early pointer toward the law of action and reaction in magnetic force.

But what truly sets book 7 of *Natural Magick* apart from previous discourse on magnets is the fact that Della Porta methodically tested, through simple but sound experiments using actual magnets, the veracity of each popular belief about magnetism that had been inherited without critical assessment for almost two millennia. To some, his methods may seem childish, but it should not be forgotten that practically he was the first person ever to conduct such experiments. Without this process, man's understanding of magnetism could never have risen above the level of baseless rumor to that of scientific theory. It is not known what prompted Della Porta, between the publication of the first and second editions of *Natural Magick*, to undertake this task of experimentally verifying those long-standing beliefs, but whatever the reason, by doing so he laid to rest many superstitions that had persisted unchallenged since antiquity.

Among them was the claim that a magnet would lose its power if smeared with garlic juice. As noted previously, Della Porta, in both the first edition of *Natural Magick* and also book 1 of the second edition, uncritically reiterates this notion of the "notable Disagreement that is betwixt Garlick and the Load-stone," and specifically that "being smeared about with Garlike [*sic*], it [lodestone] will not draw iron to it" (NM 1.7).

But in book 7 of the second edition, after listing various beliefs concerning the effects of garlic on magnets (including that seamen were forbidden to eat it because the fumes were thought to cause compasses to malfunction), Della Porta declares:

> But when I tried all these things, I found them to be false: for not onely breathing and belching upon the Loadstone after eating of Garlick, did not stop its vertues: but when it was all

> anoynted over with the juice of Garlick, it did perform its office as well as if it had never been touched with it. (NM 7.48)

While it is unclear whether or not the contradiction between the statements in book 1 and book 7 was intentional, it is quite clear, as far as we know, that Della Porta was the first to finally dispel, through actual experiments, the traditional belief that garlic hinders a magnet's action. Despite this, the majority of science historians—Paul Fleury Mottelay, Lynn Thorndike, and Marie Boas are among the few exceptions[24]—have instead attributed this achievement to Gilbert. Frankly, it is nothing less than amazing that Della Porta's contribution in this respect could be so overlooked despite the clear evidence in what he wrote. It is therefore gratifying to see his accomplishment explicitly acknowledged in physicist Robert S. Elliott's 1993 textbook *Electromagnetics*.[25]

Della Porta also tested the theory that diamonds impede magnetism, a claim dating back to Pliny and Augustine and perpetuated even in the sixteenth century by the philosopher Pietro Pomponazzi as well as the engineer Biringuccio. Della Porta, however, concludes: "I tried this often, and found it false; and that there is no Truth in it" (NM 7.53). And that is not all. In Della Porta's time, this superstition appears to have acquired a further dimension of strangeness by its association with another ancient belief, passed down even as far as Paracelsus, that the blood of a goat could break diamonds. That is, it was thought that, if diamonds could render magnets powerless and goat's blood could destroy diamonds, then goat's blood should be able to restore a magnet's power as well, on the principle that an enemy of one's enemy is one's friend. In chapter 54, Della Porta puts the proposition thus:

> Since … there is an Antipathy between the Diamond and the Loadstone; and there is as great Antipathy between the Diamond and Goats blood, as there is sympathy between Goats blood and the Loadstone; We are from this Argument proceeded thus far, that when the vertue of the Loadstone is grown dull, either by the presence of the Diamond, or stink of Garlick, if it be washed in Goats blood it will then recover its former force, and be made more strong.

But this, too, he then immediately refutes, declaring,

> but I have tried that all the reports are false.

At the beginning of *On the Magnet*, published in 1600, Gilbert criticizes those who came before him for their unquestioning acceptance of tall tales and lies and their failure to confirm for themselves the validity of propositions, and immediately cites the persistence of the old lore that garlic and diamonds nullify magnetic force and that goat's blood restores it.[26] Yet here again he makes no mention whatsoever that already Della Porta actually tested these erroneous claims, and this is yet another achievement that subsequent historians have instead attributed to Gilbert.

But Gilbert does not fail to mention Della Porta by name when pointing out the latter's mistakes. In book 7, chapter 55 of *Natural Magick*, for example, Della Porta asserts as an experimentally verified fact that iron rubbed with a diamond points north. We can only surmise that the iron used in his experiments had been magnetized at some point before the experiments were conducted. In any case, Gilbert elatedly denounces Della Porta's mistaken findings:

> Here one must marvel at a demonstrable error of B. Porta, who, while rightly opposing a very old falsehood about the diamond, in speaking of a power contrary to that of the loadstone, introduces another still worse opinion; that forsooth iron, when touched by a diamond, turns to the north.[27]

But Gilbert himself observes:

> A long piece of Iron, even though not excited by a loadstone, settles itself toward North and South.[28]

He then points out the possibility that iron could be naturally magnetized during casting, and supposes that Della Porta, "being ignorant of this, thought it was done by the diamond."[29] If so, the phenomenon Della Porta observed presumably did occur, and even though he erred in his interpretations, it cannot be said that he botched the experiments themselves.

Della Porta's work of verification by experiment was not limited to ancient traditions. He also turned his critical attention to the views of fellow sixteenth-century thinker Paracelsus. Regarding Paracelsus's strange belief that one could increase a magnet's force by heating it and immersing it in a kind of "iron oil" called "oil of crocus of Mars," Della Porta experimentally demonstrated not only that this was false but also that such treatment in fact extinguishes magnetic force:[†]

---

[†] According to both Mottelay and Park Benjamin, the first person to show that magnetism is lost through heat was Pietro (Paolo) Sarpi (1552–1623), a little before Della Porta. In the passage I quoted at the end of section 3, Gilbert refers to Sarpi as "the Reverend Maestro Paolo, the Venetian." Sarpi's work was destroyed, however, when the building containing his writings burned down. Mottelay, *Bibliographical History of Electricity and Magnetism*, 111; Benjamin, *Intellectual Rise in Electricity*, 225–26.

> I made the stone red hot, and quenched it often in Oyl of Iron: but it was so far from getting more strength, that it lost what it had: and fearing I had not done it right, I tried it often: so I found the falsity of it [i.e., Paracelsus's claim], and I warn others of it also. For a Loadstone made red hot in the fire, will lose all its vertue. (NM 7.50)

Gilbert similarly points out in *On the Magnet* that

> when a piece of iron has been touched by a loadstone, if it be placed in a hot fire until it is perfectly red hot and remain in the fire some considerable time, it will lose that magnetick strength it had acquired

and that

> even a loadstone itself through a longish stay in the fire, loses the powers of attracting implanted and innate in it, and any other magnetick powers.[30]

He repeats the latter point later in the book, noting that "a loadstone loses some of its virtue by too great a heat."[31] Gilbert thus declares Paracelsus's method for strengthening magnetism an "imposture,"[32] but nowhere does he mention that this error was already pointed out by Della Porta. In fact, Gilbert goes so far as to mark this point with an asterisk, his method of indicating what he considered his own discoveries.

Many modern historians also attribute to Gilbert the discovery that heat causes magnets to lose their power of attraction, but it must be pointed out that this claim is unjustified. It is unlikely that Gilbert was unfamiliar with the experiments of Della Porta. The Englishman William Barlow, a friend of Gilbert's, noted fairly in *Magneticall Advertisements* (1616) that it was Della Porta who, through actual experiment, proved Paracelsus's theory to be erroneous.[33] In the 1611 Italian edition of *Natural Magick*, Della Porta himself complains that Gilbert "took the whole seventh book of my *Natural Magic* [*sic*] and split it into many books, making some changes."[34] In Della Porta's view, Gilbert was using slander and insults to conceal his plagiarism.

Gilbert mentions Della Porta by name in *On the Magnet* mostly when finding fault with him: of the sixteen times Della Porta is mentioned, ten are to point out his mistakes. At the same time, Gilbert disregards the many experiments and discoveries that Della Porta made, presenting them as if they were his own. In fact, however, he owes much to Della Porta, and indeed Della Porta's name is second only to Aristotle's in number of appearances in *On the Magnet*. According to a study by Lois Irene Abromitis, Della Porta is mentioned—either quoted or referred to in connection with specific facts and ideas—more than any other person in Gilbert's book.[35]

Despite this, Della Porta has been largely ignored. He does not appear, for example, in Duane H. D. Roller's *The "De Magnete" of William Gilbert* (1959), still considered a standard reference on Gilbert's famous work. In histories of science, Gilbert has been consistently credited with effecting the shift from medieval misconceptions to modern scientific theory about magnetic force, while Della Porta's role has remained inordinately disregarded. Edmund Whittaker's *History of the Theories of Aether and Electricity* (1910), for example, maintains that the modern history of electricity and magnetism began in 1600 with Gilbert, and includes no reference at all to Cardano or Della Porta. The same tendency is evident even in W. James King's 1959 paper "The Natural

Philosophy of William Gilbert and His Predecessors." Unswayed by the popular view of Gilbert as the father of experimental science, King conducts an unbiased reading of *On the Magnet* and justly notes that "Gilbert added few novel empirical facts of a fundamental nature to previous observations on the loadstone"; but he nonetheless completely overlooks Della Porta when he goes on to assert that

> Gilbert's experimental work was in large part an expansion of Petrus Peregrinus' *De magnete* of 1269, and a development of works like Robert Norman's *The new attractive* ... and like William Borough's *Discourse on the variation of the compass, or magnetized needle.*[36]

The English physician-scholar Thomas Browne (1605–1682), who set out in the mid-seventeenth century to purge science of the superstitions passed down from ancient times and the Middle Ages, presents a generally much fairer assessment of Della Porta in *Pseudodoxia epidemica* (subtitled in some editions as *Or, Inquiries into Commonly Presumed Truths*; 1646), describing him as the "famous Philosopher of Naples, Baptista Porta; in whose Works, although there be contained many excellent things, and verified upon his own Experience; yet are there many also receptary, and such as will not endure the test."[37] In regard to *Natural Magick* as a whole, Browne's account is adequate, but at least as far as Della Porta's discourse on magnets is concerned, the former category—things "verified upon his own experience"—far outweighs the latter.

Thus, it was the second edition of *Natural Magick*, published by the magician Della Porta in 1589, that marked the watershed between ancient and medieval superstition and modern science, at least in regard to magnets. Early Renaissance thinkers rediscovered the works of Hermeticism and neo-Platonism and drew on the authority of ancient texts to oppose Scholasticism. Although magical thought was fraught with inherent religious and philosophical difficulties, its emphasis on experience and practical utility took knowledge beyond ancient authority and entrenched lore and encouraged greater attention to experimentation, thereby preparing the ground for modern science. As part of that process, book 7 of *Natural Magick* represented the first comprehensive experimental research on magnetism since Peregrinus.

## Section 6—Della Porta's Theoretical Discoveries

Della Porta's overall significance to science history notwithstanding, it must be admitted that he made few original discoveries concerning magnetic phenomena, the principal one being iron's shielding effect against magnetic force. Most of the facts he confirmed experimentally had already been revealed by Peregrinus three centuries earlier. Since he refers to Peregrinus in book 7, chapter 27, it is probable that he had studied Peregrinus's *Letter concerning the Magnet*, most likely the Augsburg edition, published in 1558, the same year in which the first edition of *Natural Magick* appeared.

The most important contributions to magnetic theory that Della Porta presents in book 7 of *Natural Magick* are his clear explication that the attraction between lodestones and iron and also the magnetization of iron (magnetic induction) are cases of action at a distance; his observation that magnetic force decreases with distance; and his formulation of the concept of the "compass of vertue" or "sphere of vertue," which he calls the *orbis virtutis*.

In chapter 16 of book 7, Della Porta declares that

> the force of the stone [lodestone] for attraction and repelling, can be included in no bounds, can be hindered by nothing, or held back; but it will penetrate invisibly, and will move and stir those stones that are sympathizing with it, if they be put to it, and will exercise its forces, as if there were nothing between.

This much—the lodestone's power of action at a distance—was already known. But in chapter 26 Della Porta makes the further observation that

> it is clear, that without touching, it [the lodestone] can impart its virtue to the iron.

Prior to Della Porta, it was generally believed that iron was magnetized only by rubbing or other direct contact with a magnet. In the face of that received view, with this passage Della Porta became the first to state explicitly that even magnetization (magnetic induction) was action at a distance. This is another discovery that has frequently and unfairly been attributed to Gilbert.

In chapter 26, a little before the sentence just quoted, Della Porta puts it even more clearly:

> And the Loadstone doth not onely impart its vertue to the iron, by touching it, but, which is wonderful, within the compass of its vertue, it will impart vertue to the iron, if it be but present, to draw another iron.

Just after the above passage from chapter 16, he qualifies the point about the lodestone's action at a distance with the condition that

> this must be within the compass of its vertue.

A somewhat more in-depth treatment of the same point appears in chapter 15:

> And as the light of a Candle is spread every way, and enlightens the Chamber; and the farther it is off from it, the weaker it shines, and at too great a distance is lost; and the neerer it is, the more cleerly it illuminates: so the force flies forth at that point; and the neerer it is, the more forcibly it attracts; and the further off, the more faintly: and if it be set too far off, it vanisheth quite, and doth nothing. Wherefore for that we shall say of it, and mark it for, we shall call the length of its forces the compass of its vertues. (NM 7.15)

This passage proves without a doubt that Della Porta was the originator of the important concept of the "compass (or sphere) of virtue". Moreover, it represents a significant refinement on Norman's observation that magnetic force extends around a magnet in a spherical shape. This refinement alone makes book 7 of *Natural Magick* a great contribution to physics.[‡]

Della Porta's reporting of the attenuation of magnetic force over distance also deserves special note. Even though his expression of the fact remained qualitative ("the neerer it is ... doth nothing"),

[‡] Science historian Fritz Krafft concludes that Gilbert took both the concept and the term *orbis virtutis* from Della Porta. Even Charles D. Kay, who in a dissertation that otherwise tends to downplay Della Porta's influence on Gilbert, criticizing the former's experiments as lacking in theoretical consideration and designed purely for entertainment, recognizes that Della Porta's *Natural Magick* was likely "the immediate source of Gilbert's notion of 'orbis virtutis'." Krafft, "Sphaera activitatis—orbis virtutis," 129; Kay, "William Gilbert's Renaissance Philosophy of the Magnet," 139.

falling short of a quantitative understanding (subsequently formulated as the law that the force is inversely proportional to the square of the distance between the bodies concerned), it was nonetheless an important contribution to magnetic theory.

Furthermore, Della Porta's likening of the propagation and attenuation of force to the emanation of light from a candle can be regarded as evincing his understanding that force, too, behaves like an emanation. In this analogy we can see the influence of thirteenth-century thinkers Robert Grosseteste and Roger Bacon, who proposed that all action was propagated by "multiplication of species." As for the influence of Della Porta's idea on subsequent thought, it would be too great a leap to claim that it foreshadowed the inverse-square law of force, but it easily qualifies as a forerunner of Kepler's concept of the emanation of gravity—or what Kepler termed the emanation of *anima motrix* (moving soul), which he later renamed *vis motrix* (moving force).

Della Porta's contributions do not end there. In chapter 19 of book 7 he declares:

> We can also measure that attracting or repelling vertue of the Loadstone

and proposes a method, albeit a simple one, for quantitatively measuring the strength of the attractive force. This involves placing a magnet into one pan of a balance and enough weight in the other pan to bring the balance to equilibrium. A piece of iron is then placed on the table under the pan holding the magnet so that the pan sticks to the iron. Sand is then poured gradually into the other pan until the one with the magnet is pulled free of the iron. Thus,

> by weighing the weight of the sand, we have the Vertue of Loadstone we sought to finde.

This was an improvement on the method proposed by Nicolaus Cusanus in 1450. Although it is unclear whether or not Della Porta actually conducted the measuring procedure he proposed, the fact that he explicitly raises the possibility of quantitatively assessing and measuring the strength of magnetic force is significant.

With the combination of the two ideas just presented—that magnetic force is measurable and that it decreases with distance—Della Porta's approach to magnets in *Natural Magick* represented a decisive step in the transition from treating magnetism as a magical, qualitative action to studying it as a physical, quantifiable force, and paved the way for subsequent research on force as part of the science of physics. In fact, the concept of force acting at a distance with an intensity that decreases with distance, coupled with the idea of a sphere of virtue, differs little from the modern concept of central force expressed as a mathematical function. In the formation of modern physics, Della Porta's work thus holds great significance as a steppingstone toward the subsequent achievements of Kepler and Newton.

In *Concepts of Force*, which traces the historical development of these ideas, Max Jammer rates Della Porta's *Natural Magick* as "one of the early manifestoes proclaiming the scientific spirit of the Newtonian era."[38] In light of the foregoing discussion, this seems a reasonable assessment. Others, however, have not been so fair in their judgment. Writing in the same year as Jammer (1957), George Sarton, a leading figure in the field of science history, while lauding Gilbert's *On the Magnet* and praising his concept of *orbis virtutis* as "one of his most remarkable intuitions,"[39] harshly criticizes Della Porta, claiming that "he lacked common sense; he was as erratic and mystical as Cardano" and that *Natural Magick* "contains many good ideas and good intentions, which, alas, are drowned in an ocean of nonsense." Indeed, Sarton denies *Natural Magick*, along with Cardano's *De subtilitate rerum* (The subtlety of things) and *De varietate rerum* (The diversity of things), any real scientific significance, panning them all as "typical of the incredible fermentation of the ideas ... which to a large extent was unsound and morbid."[40] Given that Gilbert's concept of *orbis virtutis* was undoubtedly derived from Della Porta, we can only conclude that Sarton was motivated by prejudice in so unjustly awarding Gilbert's work such importance while dismissing Della Porta's as worthless. New concepts in science are often incomplete, disordered, and mixed with impurities when first proposed, and usually it is only later that they are refined and perfected, but that is hardly cause for denying the achievement of those who first proposed them.

## Section 7—Magic and Science

Paolo Rossi maintains that the decisive difference between modern science and magic is that the former is open and democratic while the latter is secretive and elitist. That is, whereas science uses precisely formulated methods and clearly defined terminology and is, in principle, transmittable and accessible to anyone with a certain degree of ability, magic and alchemy are kinds of esoteric knowledge the secrets of which are not to be recklessly revealed but rather passed on only to select, highly talented initiates. Scientific theories are publicly available and conveyed in clear concepts to avoid misunderstanding, while magic is cryptic and its mysteries are only hinted at in ambiguous words open to various interpretations.[41]

In fact, however, modern scientific practice became transparent and democratic, at least in principle, rather late in its history. Throughout Europe, with the exception of Gresham College in England, there was little openness or equality of access to higher education in science until after the French Revolution. Prior to that, and particularly in the sixteenth and seventeenth centuries, before the advent of scientific journals and the concepts of copyright and intellectual property, a predisposition toward secrecy and concealment, including the use of anagrams, was common in the world of science as a way of protecting priority to discoveries.

The mudslinging between Cardano and mathematician Niccolo Fontana Tartaglia in the sixteenth century over the solution to cubic equations concerned precisely this issue of openness. To Tartaglia, the method he discovered for solving the equations was a weapon for winning mathematical contests, not something to be freely shared with others. In this case, however, Cardano, who was more of an advocate of magic, was instrumental in making Tartaglia's solution public. The Danish astronomer Tycho Brahe (1546–1601), who was eleven years younger than Della Porta, likewise guarded his enormous body of observational data as private property, sharing it with only his most trusted apprentices. When Kepler first began working under him, he complained that "Tycho is very stingy as to communicate his observations."[42] And when the Flemish engineer Simon Stevin (1548–1620), a contemporary of Tycho's, noted on the importance of collaboration in astronomical observations that "if the branch of science is practiced by few people, each of them will keep his findings to himself and conceal them,"[43] it was to Tycho that he alluded. But such secrecy notwithstanding, no one regarded Tartaglia's mathematics or Tycho's astronomical observations as magic. Hence, given the stage scientific knowledge was at in those days, it is practically impossible to pigeonhole specific sixteenth-century studies of nature as either science or magic purely on the basis of differences in their style of scientific activity, least of all by their degree of openness or secrecy.

So it deserves to pay attention that Charles Webster proposes a distinction between "exoteric" and "esoteric" expressions of natural magic.[44] In the sixteenth century, both magic and science had both open and secretive aspects. In Della Porta's *Natural Magick* the openness is more evident than the secrecy. Della Porta does note at one point in *Natural Magick* that "if you would have your works appear more wonderful, you must not let the cause be known ... for he that knows the causes of a thing done, doth not so admire the doing of it" (NM 1.3); yet throughout the work he in fact lays his subject bare, recording his own experiments in plain, straightforward language. When he fails to reveal the causes of certain phenomena, it appears to be because he himself does not know them, and his occasional lapses into ambiguous expression seem to derive more from his lack of expertise and understanding than from any intention to obfuscate. Ernst Mach points out that Della Porta, though he did not invent the camera obscura, was "the first to make it generally known".[45] When all these elements are taken into account, Webster seems justified in ranking Della Porta alongside Francis Bacon as one of the leading promoters of the "tendency towards the democratisation of magic."[46]

*Natural Magick* was never intended as an esoteric text for an exclusive, secret society. On the contrary, it was written for ordinary people, printed in large quantities, and sold as one product among others on the market; and as a work of popular science it enjoyed a tremendous reception among the general public. Moreover, much of the content is generally regarded as the reports of experiments and observations by members of the private academy based at Della Porta's home. In that sense, the science historian William Eamon rates *Natural Magick* "an important step toward establishing formal channels for the communication of scientific research. Indeed, it might be regarded as the 'proceedings' of one of the first scientific societies."[47]

Nonetheless, in medieval times it was widely accepted not only in the field of magic but in all branches of learning that the secrets of nature should not be shared openly. In *Opus majus* Roger Bacon frequently quotes from *The Secret of Secrets*, a work erroneously attributed to Aristotle and widely judged "the most popular book of the Middle Ages."[48] One such passage quoted by Bacon is the following, presented as part of a letter from Aristotle to his pupil, Alexander the Great:

> I am revealing my secrets to you figuratively, speaking to you with enigmatic examples and signs, because I fear lest the present book fall into the hands of infidels and arrogant powers, whereby they, whom God on high has deemed undeserving and unworthy, might arrive at that ultimate good and divine mystery. I would then surely be a transgressor of divine grace and a violator of the heavenly secret and occult revelation.[49]

Influenced by this view, the author of *Roger Bacon's Letter concerning the Marvelous Power of Art and of Nature and concerning the Nullity of Magic* describes several methods by which the secrets of nature have been kept hidden from the unworthy and then notes, "Perhaps I shall make use of certain of them because of the magnitude of our secrets."[50]

Even in the fifteenth century, Pico wrote that

> to disclose to the people the more secret mysteries, things hidden under the bark of the law and the rough covering of words, the secrets of the highest divinity, what was that other than to give what is holy to dogs and to cast pearls among swine? Consequently it was not human prudence but divine command to keep these things secret from the people, and to communicate them to the perfect, among whom alone Paul says that he spoke wisdom.[51]

Secretiveness was thus characteristic not only of the field of magic but of the pious Middle Ages in general. The difference between the two opposing attitudes—on the one hand that the secrets of nature should be revealed only to a select few and never to the ignorant masses, and on the other that they should be available to all and expressed in clear, unambiguous language—does indeed correspond to the difference between the medieval and modern approaches[52]; but the distinction between those two views does not correspond in every respect to the distinction between magic and science.

In the evolution of science, as the modern system of research societies and journals and the concept of copyright were gradually established, the medieval tendencies toward secrecy and closed transmission of knowledge were gradually replaced by greater disclosure and open education. Likewise, magic shed its mysticism as it was popularized, secularized, and exposed to public curiosity and critique with the rise of the printing industry. Hence, the significance of sixteenth-century magical practice to the formation of modern science cannot be denied simply because it retained medieval elements. Indeed, the modern scientific attitude of experimentally investigating the types and effects of natural forces and applying them in technology was essentially the upshot of the demystification of magic.

Except for the differing views of nature (organic vs. physical) and causality (astrological influence vs. scientific laws) that inform their theoretical foundations, essentially there is little difference between Della Porta's natural magic and modern technology. "Of all the mystical and occult sciences," as Boas puts it, "natural magic was, ultimately, the most fruitful."[53]

* * *

In *Natural Magick*, Della Porta shows that both the iron-attracting power of magnets and their magnetizing effect are cases of action at a distance, and observes that these forces diminish with distance. By proposing, as Cusanus had done, that magnetic force can be quantitatively measured, and furthermore by developing the concept of "sphere of virtue," Della Porta helps pave the way toward the understanding of force expressed in modern physics as mathematical functions.

Moreover, by experimentally testing and thereby refuting many long-standing myths about magnets, Della Porta relegates many magnet-related superstitions to the past and facilitates the shift from textual magic to experimental magic. *Natural Magick* further prepares the ground for modern science by breaking away from the medieval esoteric approach to understanding nature and by encouraging the demystification and popularization of magic.

Thus, through Della Porta, Renaissance magical thought reached the threshold of modern scientific and technological thought. Although his writings on magnetism remained hylozoistic, Della

Porta freed the discourse on magnets from many superstitions and misconceptions, thereby illuminating the path toward a more scientific understanding. What he lacked, crucially, was the broad perspective and theory necessary for articulating and systematizing the many separate empirical phenomena that he investigated.

# Part 3
# The Dawn of the Modern Age

# Chapter 17

# William Gilbert's *On the Magnet*

## Section 1—Gilbert and his Times

The treatise *On the Magnet* by English physician-philosopher William Gilbert (1544–1603) is hailed as the inception of modern electromagnetic science, yet little is known of its author. There are two main reasons for this: in 1648 civil war destroyed Colchester, the town in Essex, southern England, that was Gilbert's birthplace and childhood home; and in 1666 the Great Fire of London razed both his London residence and the College of Physicians (later the Royal College of Physicians), which had preserved his books, manuscripts, and experimental equipments. The few more or less reliable facts about him that have survived the centuries are as follows.

Gilbert was born in 1544 as the son of a wealthy municipal court magistrate. In 1558 he entered Cambridge University, from which he received his doctorate in medicine in 1569. He began practicing medicine in London in the early 1570s and subsequently held prominent posts in the College of Physicians, eventually being elected its president in 1600. In 1577 he received a family crest from Queen Elizabeth I, and in 1601 he was appointed as one of her court physicians. He remained a bachelor all his life and died at the age of 59 in 1603, just a few months after the queen's death.

It is reported that, in the years between graduating from university and setting up his London practice, Gilbert traveled abroad for study, particularly in Italy, and made the acquaintance of some of the leading philosophers of the day, though there is no evidence to confirm this. Other reports maintain that meetings with friends that he held in his home were essentially a private science academy that foreshadowed the Royal Society, and that he received special favor and even an inheritance from the queen. Such tales, however, appear to be embellishments and exaggerations concocted by later generations to enhance Gilbert's stature.[1] All we know for certain is that in his later years he was a prominent member of Elizabethan society and was considered one of the foremost physicians of his time.[2]

The year of Gilbert's birth was previously thought to be 1540. This opinion was based on an inscription on a monument erected by his younger half-brother that ends with the words *Obiit anno Redemptionis Humanae 1603, Novembris ultimo, aetatis suae 63* ("Died on the last day of November in the year of Human Redemption 1603, in the 63rd year of his age"). However, that would make Gilbert's age when he entered Cambridge eighteen, rather too late for that era. The 1544 date is based on an inscription on a portrait of Gilbert that was painted while he was still alive and that he himself presumably saw, which reads *1591: aetatis 48* ("1591: age 48"). The later year of birth is currently regarded as the correct one following the subsequent discovery by Silvanus Thompson, who translated *On the Magnet* into English, of a document in the

Bodleian Library that "specifically gives as the date of his [Gilbert's] birth the 24th of May 1544, at 2 hours 20 minutes p.m."[3]

But far more important than Gilbert's exact date of birth is the general nature of the age in which he lived. The year 1544 was seventeen years after the birth of John Dee, nine years after Giambattista Della Porta is thought to have been born, and just one year after the publication of Nicolaus Copernicus's *On the Revolutions of the Heavenly Spheres* and Andreas Vesalius's *De humani corporis fabrica* (On the structure of the human body). This was thus a period in which Aristotelian natural philosophy, Galenic medicine, and Ptolemaic astronomy were beginning to teeter while Hermeticism and magical thought, which had kept their vigor, were attracting intellectual interest as possible substitutes. Many other pioneers of modern science were also born around the same time as Gilbert, including Italian mathematician and mechanics scholar Guidobaldo del Monte (1545); Thomas Digges (1546), author of England's first book on heliocentric theory; Danish astronomer Tycho Brahe (1546); Italian philosopher and theologian Giordano Bruno (1548), who toured Europe preaching heliocentrism and the concept of an infinite universe; and Flemish engineer and mathematician Simon Stevin (1548).

It was also an age of great exploration and discovery. The first Europeans in Japan (the Portuguese) arrived there in 1543, thereby expanding the sphere of European activity to the eastern limits of the world by the time of Gilbert's birth. Through their great voyages, Europeans had discovered new regions of the globe unknown to their forebears, and it was from this time forward that Europeans in general began to take a real interest in that wider world. In England, the year of Gilbert's entrance into Cambridge University (1558) also saw the death of the Catholic Queen Mary

I and the crowning of the Protestant Queen Elizabeth I, and his subsequent years in London coincided with both the establishment of Elizabeth's absolute monarchy and the remarkable build-up of power among the emerging bourgeoisie. After its defeat of Spain's "Invincible Armada" in 1588, England entered a period of dramatic economic development and rapid overseas expansion, laying the foundations for the rise of the British Empire in the seventeenth century.

The creation of the modern Church of England (Anglicanism) following the Pope's excommunication of Elizabeth I in 1571 illustrated the fact that the authority of the nation state had overtaken that of the Vatican. Though vilified by both Roman Catholicism and Calvinist fundamentalism, Anglican theology was humanistic and moderate, emphasizing the authority of reason in religious belief, and culturally this climate of thought nourished the English Renaissance and the great flowering of the Elizabethan age. The era saw the births of mathematician and logarithm inventor John Napier in 1550, poet Edmund Spenser in 1552, novelist Philip Sidney in 1554, and playwrights William Shakespeare and Christopher Marlowe in 1564. Gilbert thus reached his prime just as England reached the height of its cultural vigor. Furthermore, with the establishment of Gresham College in 1597, the prosperous times in which Gilbert practiced medicine also witnessed the emergence of new sciences spearheaded by London's bourgeoisie, physicians, engineers, and artisans.

In its original Latin edition, the full title of the work that secured Gilbert lasting fame is *De magnete, magneticisque corporibus, et de magno magnete tellure: Physiologia nova, plurimis & argumentis & experimentis demonstrata* (*On the Magnet, Magnetick Bodies Also, and on the Great Magnet the Earth: A New Physiology, Demonstrated by Many Arguments and Experiments*).[4] First published in 1600, it was followed by a second edition in 1628 and a third in 1633. Gilbert also wrote *De mundo nostro sublunari philosophia nova* (New philosophy about our sublunary world)—commonly referred to as *De mundo*—which was posthumously edited by his younger brother and published in Amsterdam in 1651. As is suggested by the terms "new physiology" and "new philosophy" in the titles of these works, Gilbert's main purpose in writing them was not only to present his research on magnets but to delineate a whole new vision of the world—his *philosophia magnetica* (magnetic philosophy)—so that, as he put it in his preface to *On the Magnet*,

> the noble substance of the great magnet, our common mother (the earth), hitherto quite unknown, and the conspicuous and exalted powers of this our globe, may be the better understood. (OM, p. ij)

To Gilbert, the Earth was not the cold, inert lump of clay that Aristotle had described but rather a noble, living thing endowed with "conspicuous and exalted powers," and it was to bear this fact out that he wrote *On the Magnet*.

## Section 2—Overview of *On the Magnet* and its Place in the History of Science

Before we examine the details of *On the Magnet*, it is worth noting two seemingly minor but significant innovations that Gilbert made to scholarly practice.

The first was that, breaking with established custom, he did not include in his work any dedication to a patron. This was quite extraordinary for his time. Five years later, in 1605, Francis Bacon wrote in *The Advancement of Learning* that "neither is the modern dedication of books and writings, as to patrons, to be commended: for the books ought to have no patrons but truth and reason"[5], yet even Bacon began that very book with the words "To the King." Presumably, Gilbert enjoyed not only a high social position but also a great deal of self-confidence.

Gilbert explains the other scholarly innovation in his preface to *On the Magnet*: "To our own discoveries and experiments we have affixed asterisks, larger and smaller, according to the importance and subtlety of the matter." Hence, there are over two hundred large and small asterisks in the book's margins, indicating the author's claim of priority regarding specific discoveries and experiments. Yet despite this assiduous attention to safeguarding his own rights, he is quite insensible to the same rights of others. To give just one example, he affixes an asterisk to his statement that magnetism is destroyed by fire and heat (OM 2.4:66) but makes no mention anywhere that this fact was previously verified through experiment by the Italians Pietro (Paolo) Sarpi and Della Porta.

Indeed, many science historians note that Gilbert was reluctant to admit any debt to those who had come before him. Sydney Chapman observes that Gilbert "mentions Petrus [Peregrinus] and [Robert] Norman several times, though more often to condemn their few errors than to praise their great merits."[6] Edgar Zilsel likewise points out that, despite Norman's tremendous influence on Gilbert's work, "Gilbert himself does not emphasize it at all, but rather hides it."[7] And according to Park Benjamin, Gilbert "rarely mentions an earlier writer except to dispute conclusions."[8]

The preface of *On the Magnet* begins with these words:

> Clearer proofs, in the discovery of secrets, and in the investigation of the hidden causes of things, being afforded by trustworthy experiments and by demonstrated arguments, than by the probable guesses and opinions of the ordinary professors of philosophy … (OM, p. *ij)

Gilbert also declares there that "to you alone, true philosophizers, honest men, who seek knowledge not from books only but from things themselves, have I addressed these magnetical principles in this new sort of Philosophizing." He reiterates the point in the main text: "One may very easily fall into mistakes and errors when one is searching into the hidden causes of things, in the absence of real experiments" (OM 4.11:169). For this apparent focus on experiment and "things themselves," *On the Magnet* has been widely regarded as the first work to elevate magnetics to the level of modern empirical science. In 1837, for example, William Whewell wrote in his *History of the Inductive Sciences* that "Gilbert … repeatedly asserts the paramount value of experiments. He himself, no doubt, acted up to his own precepts"; in 1876 Peter Guthrie Tait praised Gilbert as "the originator of true experimental science"; and in his 1935 *A History of Science, Technology and Philosophy in the 16th & 17th Centuries* Abraham Wolf similarly noted that *On the Magnet* "is characterized almost throughout by its reliance upon the results of experiment."[9]

To be sure, Gilbert did record numerous experiments on magnets, but in many cases he was corroborating or refining experiments already performed by Petrus Peregrinus, Robert Norman, or Della Porta.[10] In the connection with Norman, Richard F. Jones is well within reason in his estimation that "if anyone has claims of priority in this field, it is not Gilbert but Norman, whose work truly prophesies Gilbert's greater achievement."[11] And in regard to Della

Porta, the truth of the matter is aptly expressed in Marie Boas's observation that "Gilbert's method was not very different from that of Porta"[12] and in the following assessment by Paolo Rossi:

> He [Gilbert] gave no measurements in the book, and its experiments were typically qualitative. In fact, his methods were basically like those of Giambattista della Porta, though the quality of his experiments was far superior: they were more ingenious, more detailed, and more carefully carried out.[13]

On the other hand, Della Porta surpassed Gilbert to the extent that he discussed the quantitative measurement of magnetism.

Gilbert's true originality, and what really distinguished *On the Magnet* from what had come before, lay not in the experiments themselves but rather in his motives for conducting them and in the meanings and interpretations that he gave to the results. Although many before him, such as Robert Norman and William Borough, had tried their hand at experimenting with magnets, "neither of these authors, or any of the many others who remarked on the lodestone in their works, attempted to build a general philosophy of the magnet or provide a general explanation for magnetic phenomena."[14] Gilbert was the first to treat magnetic phenomenon as part of a comprehensive view of nature, that is, his "magnetic philosophy."

From a modern perspective, Gilbert's greatest and truly pioneering achievement in the field of magnetics was his discovery that the Earth is a giant magnet. This was not, however, a conclusion that he arrived at through Baconian inductive reasoning from numerous experiments and observations. Rather, it was a hypothesis rooted in Gilbert's peculiar views on matter, and his experiments were aimed at proving the hypothesis indirectly using a replica of the Earth in the form of a spherical magnet that he called a "terrella" ("little Earth"). In developing his picture of the Earth from this premise, Gilbert applied himself to speculative or, if you prefer, Scholastic discourse—his "magnetic philosophy"—an approach he justifies in the preface to *On the Magnet* by claiming that "it is permitted to us to philosophize freely." Francis Bacon subsequently criticized Gilbert for thus indulging in a debate that exceeded the scope of his experiments.[15] Certainly, as A. Rupert Hall points out, "Gilbert's magnetic philosophy was founded, not on massive experiments, but upon an analogy."[16] Indeed, it is not experimentation but rather this "magnetic philosophy"—what Gilbert describes in his preface as the "new sort of Philosophizing"—that forms the basic theme of *On the Magnet*, and it was through that new philosophy, more than anything else, that Gilbert made his greatest impact on the seventeenth century.

## Section 3—Gilbert and the Founding of Electrical Science

In chapter 2 of book 2 of *On the Magnet*, Gilbert states that

> the causes of magnetical motions are widely different from the forces of amber. (OM 2.2:47)

He then digresses into a discussion of the "amber effect." Because the amber effect had no place in Gilbert's magnetic philosophy, his aim here was to deal with it at the outset and thereby avert any confusion in his subsequent elaboration of that philosophy. As it turns out, however, he thereby launched electrical science as an independent field of study. Gilbert was the earliest to recognize that the amber effect and the lodestone effect (magnetism)—until then lumped together under the name of "attraction" and treated indiscriminately through the use of such magical terms as "occult powers" and "sympathy and antipathy"—were distinct phenomena requiring separate study and explication, and to propose that the science of electricity was separate from magnetics. His way of explaining electric force was to exert a lasting influence on the development of electrical theory from the seventeenth century on. Furthermore, the experiments on static electricity presented in the second chapter of book 2 of *On the Magnet* were based on original ideas and conceived virtually single-handedly; for these reasons alone they hold great value. It is not surprising, then, that Gilbert asterisked many items in this chapter. From this perspective, if Gilbert was the founder of experimental physics, then it was really in the field of static electricity that he earned that epithet.

To conduct his experiments on the amber effect, Gilbert conceived a device he called a "versorium," a term derived from the Latin intransitive verb *versor*, which means "to rotate." He describes its application as follows:

> In order that you may be able clearly to test how such attraction occurs, and what those materials are which thus entice other bodies, ... make yourself a versorium of any metal you like, three or four digits in length, resting rather lightly on its point of support after the manner of a magnetick needle, to one end of which bring up a piece of amber or a smooth and polished gem which has been gently rubbed; for the versorium turns forthwith. (OM 2.2:48–49)

In short, a versorium is a simple apparatus consisting of a metal needle that rotates freely on a pivot (Figure 17.1). When a charged object is brought near the tip of the needle, the opposite charge is induced at the needle's tip, causing the object and the tip to attract each other, regardless of the type of metal from which the

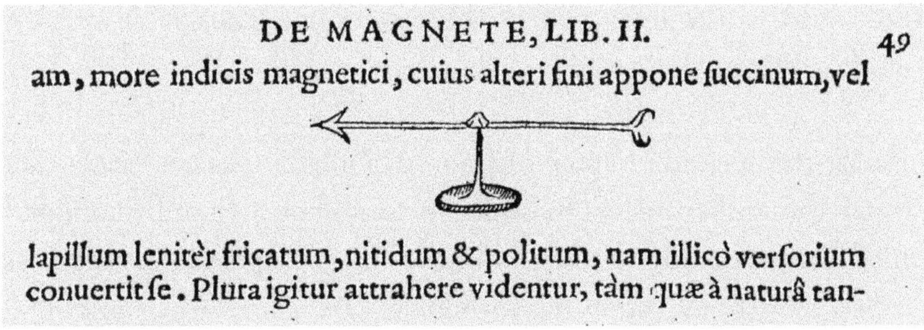

DE MAGNETE, LIB. II. 49

am, more indicis magnetici, cuius alteri fini appone ſuccinum, vel lapillum lenitèr fricatum, nitidum & politum, nam illicò verſorium conuertit ſe. Plura igitur attrahere videntur, tàm quæ à naturâ tan-

Figure 17.1: Versorium. *Source*: Gilbert, *De magnete* (1600/1967).

needle is made. At a time when the principle of static electrical induction was still unknown, Gilbert may have stumbled upon this effect by happening to place a piece of rubbed amber near a compass needle before the needle was magnetized. At any rate, his device was the first manmade electroscope. Despite the simplicity of its structure and operation, it was sensitive enough to detect even slight attraction. Furthermore, it even enabled the strength of that attraction to be ascertained, albeit only qualitatively, by observing the rapidity of the needle's rotation. It was presumably in this way that Gilbert came to describe the needle's being "attracted more when near" "electricks" (things with amber-like attractive properties) (OM 2.2:54), the first recorded observation that the strength of static electricity increases as distance decreases. Of course, such a simple device could not provide more precise quantitative measurements, nor did Gilbert himself grasp the significance or importance of such measurement of force. Nonetheless, these experiments were deliberate, systematic explorations of static electricity, and Gilbert's use of the versorium as an electroscope was groundbreaking, the first means of artificially enhancing the effects of weak static electricity so that they could be visually observed. As such, this was arguably the first significant achievement in experimental electrical science, one that would lead to the invention of the Leyden jar (capacitor) and the Voltaic cell (primary battery).

Also in the second chapter of book 2, Gilbert lists materials that exhibit the amber effect. In addition to amber, jet, and diamond,† which were already known to have this property, he includes materials such as sapphire, carbuncle, iris gem, opal, amethyst, vincentina, spar, beryl, crystal, glass, false gem, belemnite, sulfur, mastic, and hard sealing-wax (OM 2.2:48) and marks the list with a large asterisk. He also lists gems and other substances that he finds do not exhibit attraction no matter how hard they are rubbed, including emerald, agate, carnelian, pearl, jasper, chalcedony, alabaster, porphyry, coral, marble, touchstone, flint, bloodstone, emery, bone and ivory, the woods ebony, cypress, juniper, and cedar, and the metals gold, silver, brass, and iron (OM 2.2:51). His failure to detect static electricity in certain of these materials was most likely due either to his experimental conditions or to the conductivity of the materials in question, as in the case of metals. He lists chalcedony, for example, under materials that show no attraction, but the conditions under which he experimented on it presumably differed from those under which Marbodus came to the opposite conclusion about that material.

These extensive lists are clearly the result of systematic and sweeping experiments conducted by Gilbert himself. He seems to have experimented on whatever he could lay his hands on. Interestingly, he notes that for some materials clearer observations

---

† The fact that the amber effect occurs with diamond, which, when rubbed, attracts hairs and twigs in the same way that amber does, was first noticed by Girolamo Fracastoro in 1546. See Gilbert, *On the Magnet* 2.2:50; Roller and Roller, "The Prenatal History of Electrical Science," 353; and Benjamin, *Intellectual Rise in Electricity*, 241.

could be made "under a suitable dry sky" or "when the air is sharp and clear and rare" (OM 2.2:48). We now know that this is because humidity increases conductivity, diminishing frictional electricity, but in Gilbert's time that fact had yet to be discovered. We can therefore confidently surmise that his accounts are based on actual observations.

From the results of these experiments on numerous materials, Gilbert concludes of the amber effect that "this is not so much a singular property of one or two things (as is commonly supposed), but the manifest nature of very many" (OM 2.2:49). He coins the term "electricks" (from the Latin *electricum/electrica*) for materials endowed with the same property of attraction as amber (Lat. *electrum*), and the term "magneticks" (Lat. *magneticum/magnetica*) for materials that are or can be magnetized. Then, through a consideration of what each of these material types is and of what makes them different from each other, he proceeds to expound his own singular views on the nature of matter and the Earth.

According to Gilbert, all matter on this planet can be broadly divided into two types: "moisture" substances (sometimes referred to as "fluid" substances) and "earth" substances. Of the former he states that "all things ... which have sprung from a predominant moisture and are firmly concreted, and retain the appearance of spar and its resplendent nature in a firm and compact body, allure all bodies" (OM 2.2:52). Such substances—specifically, "those substances which have received their chief growth from moisture, whether aqueous or fatty, or have taken on their form by a simpler compacting from them, or have been compacted from these same materials in long ages" (OM 2.2:51)—are the electricks. He posits that "amber has been compacted of moisture, and jet also. Lucid gems are made of water; just as crystal" (OM 2.2:51); and he notes that all of these exhibit the amber effect. Accordingly, he concludes that the true cause of static electricity is a "humour": "it pleases us to call that force an electric force, which hath its origin from the humour" (OM 2.2:52). This is, incidentally, the first recorded use of the term "electric force" (Lat. *vis electrica*).

Regarding the other type of matter, Gilbert observes that

> those ... which partake of the true earth-substance, or are very little different from it, are seen to attract also, but from a far different reason, and (so to say) magnetically. (OM 2.2:52)

As we will see later, he considered such "earth" substances, which exhibit magnetism, to be the Earth's true and main constituents.

Gilbert postulates that, on the Earth's surface, efflorescence has caused true earth-substance to mix with moisture-substance, rendering the resulting matter nonmagnetic. When "moisture" materials are compacted they become electricks, but materials formed through the mixture and degradation of original "moisture" and "earth" exhibit neither electric nor magnetic force. In such mixed matter, the inherent magnetism of earth-substance is distorted and buried while the abundance of "earth" prevents the "moisture" matter from compacting. According to Gilbert, this is the natural origin of the differences evident among magneticks, electricks, and substances that are neither.

Gilbert further believed not only that magnetic force and electric force inhered in different substances but also that they were different effects even in their fundamental principles. We will examine the distinction in more detail later, but it is worth noting here that in the above quotations Gilbert states plainly that electricks "allure" objects but for magneticks he uses the more roundabout expression "are seen to attract." This shift in phrasing implies that the phenomenon commonly *perceived* as "attraction" in magnetic bodies is in reality something else. Without getting too far ahead of ourselves, suffice it say that Gilbert's basic view was that the amber effect resulted from external force applied to various other objects by electricks, whereas the magnetic effect was a kind of self-motion arising through interaction between impulses inherent in the magneticks themselves.

Gilbert's consideration of static electricity was simply a preliminary to his main project, the study of the Earth's magnetism. Nonetheless, along the way, through his invention of an electroscope (versorium), he devised a means for experimental study of static electricity, and with it he laid the groundwork for the static electricity research of the seventeenth and eighteenth centuries.

## Section 4—Gilbert's Explanation of Electric Force

In Gilbert's day, magnetism was still synonymous with the strange and the wonderful. Early in book 2, chapter 2 of *On the Magnet* he writes:

> Loadstone and also amber do some philosophers invoke when in explaining many secrets their senses become dim and reasoning cannot go further. Inquisitive theologians also would throw light on the divine mysteries set beyond the range of human sense, by means of loadstone and amber. (OM 2.2:46)

Later in the same chapter is this passage:

> Our own age has produced many books about hidden, abstruse, and occult causes and wonders; in all of which amber and jet

> are set forth as enticing chaff; but they treat the subject in words alone, without finding any reasons or proofs from experiments, their very statements obscuring the thing in a greater fog, forsooth in a cryptic, marvellous, abstruse, secret, occult way. (OM 2.2:48)

Clearly, Gilbert feels that adjectives such as "hidden" and "secret" explain nothing.

Because the concept of force was based on an understanding of muscular action, attraction could be represented only by the image of manually grasping and pulling an object. From ancient times, people had therefore believed that, as Gilbert puts it, "no action can take place by means of matter unless by contact" (OM 2.2:57). "Electric force" being no exception, Gilbert seems convinced that, as this phenomenon is observed in many common materials, it can be accounted for by just such an ordinary notion as mechanical grasping. He therefore regards explaining electricity as a matter of clarifying "by what force and with what hands (so to speak) electricks seize upon kindred natures" (OM 2.2:52). In framing the question in this way, he already hints at his overall image of the force he hopes to identify.

His answer to that question is that "all bodies are united and, as it were, cemented together in some way by moisture" (OM 2.2:57). More specifically, he maintains that electricks emit "electrical effluvia" that are "highly attenuated humours much more subtile [sic] than the ambient air" (OM 2.2:55) and that every electrical effluvia has "a singular tendency toward unity, a motion toward its origin and fount, and toward the body emitting the effluvia" (OM 2.2:57). Gilbert clearly owes this conception of attraction occurring through "humours" to Girolamo Cardano, but he rejects

Cardano's hypothesis that attraction is the result of the attracted dry objects absorbing "a fatty and glutinous humour."[17] Instead he proposes that the electrical effluvia released by electricks attract matter because the highly attenuated effluvia "embrace them [material bodies] as if with arms extended" (OM 2.2:59); that "as if they were material rays, they [effluvia] hold and take up chaff, straws, and twigs" (OM 2.2:60); and that through this process such bodies are united with the electricks attracting them:

> the one body lying near the other within the peculiar radius of its effluvia makes one out of two; united, they come together into the closest accord, and this is commonly called attraction. (OM 2.2:56)

Gilbert thus depicts electric force as action not at a distance but through a medium, that is, through the direct contact of the medium with the object acted upon. Like Cardano, Gilbert bases this idea of effluvial action largely on the shielding effect observed with electric force. As Gilbert puts it, the amber effect "is impeded by the interposition of a body (as paper, leaves, glass, or the like) when that way is impeded and obstructed, so that that which exhales cannot reach the corpuscle to be allured" (OM 2.16:86).

Gilbert also echoes Cardano by asserting that "an effluvium … is emitted by rubbing" (OM 2.2:55) and "effluvia are excited by a calorifick motion from attrition and attenuation" (OM 2.2:60). But in this connection Gilbert further points out that "amber truly does not allure by heat, since if warmed by fire and brought near straws, it does not attract them" (OM 2.2:49), thereby refuting the received view that heat is the true cause of the amber effect. In fact, electricks no longer exhibit the amber effect when burnt or roasted, and in Gilbert's explanation this is because "their primordial supplies of moisture perish by heat, and are changed and exhaled" (OM 2.2:52). In another historical first, Gilbert thus clearly identifies friction itself—not the heat generated by friction—as the only basic prerequisite for the amber effect.

Gilbert adds that it is the "bodies themselves, not the intermediate air" that are attracted by the effluvia from amber (OM 2.2:55). This argument appears to refute the Platonic and Plutarchian model of cyclical propulsion, according to which fluid exuded by the amber pushes the intervening air, causing action upon the object of attraction. To support his claim, Gilbert cites his observations that a candle flame does not move even when rubbed amber or jet is brought close to it, demonstrating that electric effluvia have no effect on flame or air.

From this, Gilbert reasons that "all things" are attracted to electricks "except such as are aflame or are too highly rarefied, as air, which is the universal effluvium of this globe" (OM 2.2:51). Specifically, the category of things thus attracted includes "not straw and chaff only, but all metals, woods, leaves, stones, earths, even water and oil, and everything which is subject to our senses, or is solid" (OM 2.2:48). Whereas Theophrastus and Cardano, among others, had already shown that a wide range of substances were subject to the amber effect, Gilbert extends the category to include "all things." On the basis of this salient fact, Gilbert also refutes the conventional wisdom that attraction is the result of similarity:

> Similarity … is not the cause [of attraction]; because all things around us placed on this globe of the earth, similar and dissimilar, are allured by amber and bodies of this kind … But neither do similars mutually attract one another, as stone stone, [or] flesh flesh. (OM 2.2:50)

This argument, too, had already been made—by Alexander of Aphrodisias—but in Gilbert's thinking it becomes the grounds for dispensing with terms like "sympathy" altogether.

Gilbert furthermore unequivocally denies the existence of electrostatic repulsion (as we now know it), declaring that "all electricks attract all things: they never repel or propel anything at all" (OM 2.39:113). Presumably he was unable to detect this phenomenon in his versorium experiments, and this becomes his grounds for rejecting "antipathy" along with "sympathy." Indeed, he flatly denounces the grouping and categorizing of the stuff of nature under magical concepts like "sympathy" and "antipathy" as erroneous folklore that leads to the ruin of true philosophy (OM 2.39:112).

Gilbert's explanation of electrostatic attraction was a classic theory of action through a medium, and apart from some minor differences in the details, his arguments concerning the amber effect were essentially a return to the reductionism espoused since Plato and Plutarch and revived by Cardano. This model of electric force transmitted through effluvia struck people as a concrete, readily visualizable explanation of force. That conceptual accessibility appealed strongly to early modern proponents of the mechanistic view of nature, and the theory was widely adopted, with a number of variations, during the formative period of electromagnetic science. Gilbert thus provided the paradigm for the study of static electricity in the seventeenth and eighteenth centuries. It is fair to say that, coupled with the invention of the electroscope (versorium), chapter 2 of the second book of *On the Magnet* launched the new science of electricity. Joseph Priestley, writing in the eighteenth century, was in that sense justified in designating Gilbert "the father of modern electricity."[18] But since the study of static electricity was secondary to Gilbert's primary purpose, the study of magnetism, Charles D. Kay was also correct when he pointed out that Gilbert "was hardly aware of the fact himself."[19]

## Section 5—Iron, Magnets, and the Earth

This brings us to Gilbert's investigations into magnetism. His treatment of this subject in *On the Magnet* is rather disorganized and repetitive, making it extremely difficult to follow the thread of his arguments or distinguish his premises from his conclusions. For this reason, science historians have tended to quote only those parts of *On the Magnet* that were convenient for their purposes, usually resulting in a rather selective representation and one-sided image of it. In the remainder of this chapter, therefore, I will attempt to present a more comprehensive picture of Gilbert's treatise, with as little modern interpretation as possible.

Gilbert begins book 1 of *On the Magnet* with a critical examination of views put forward about magnets since ancient times and then sets about demonstrating that lodestone and iron ore are essentially the same thing. Noting that, as is widely known from experience, "a strong loadstone … presents the appearance of unpolished iron, and is mostly found in iron mines," he argues that this is because iron and lodestone are "uterine brothers" (OM 1.2:8–9), conceived in the same womb within the Earth:

> At the highest part of the terrestrial globe, or at its perishable surface & rind, as it were, these two bodies usually originate & are produced in one and the same matrix, as twins in one mine. (OM 1.16:36)

Accordingly, "even metallurgists cannot decide between them, because they agree together in all respects" (OM 1.16:37).

Gilbert goes on to argue that in fact they are more than twins.

> The majority of loadstones ... yield in the furnace a very excellent iron. But iron ore also agrees in all those primary qualities with loadstone. (OM 1.16:38)

This suggests that they are identical in substance as well, and from this Gilbert concludes that "the loadstone & iron ore are the same" and that "the loadstone is by nature and origin of iron, and it and magnetick iron are both one in kind" (OM 1.16:36, 1.16:39).

Although this fact now seems so obvious as to be hardly worth noting, nonetheless Gilbert must be credited with its discovery. In fact, *On the Magnet* was the first work to completely discard the separate classification of iron as metal and lodestone as stone, a distinction that had persisted since the time of Theophrastus and was still espoused by Vannoccio Biringuccio and Georgius Agricola in the sixteenth century. The basis of this division had been that, since iron melted when heated but lodestone, like stone, did not, under the four-element theory iron was classified as a water element and lodestone an earth element. Gilbert, however, confirmed for the first time that in fact lodestone did melt when heated—in his words, that "loadstone also pours forth iron in the furnaces" (OM 1.16:39)—thereby removing the grounds upon which the distinction had rested. This discovery probably owed much to marked improvements in furnaces in the sixteenth century. A type of blast furnace capable of smelting pig iron at high temperature was developed along the banks of the Rhine and introduced to Sussex by Germans around 1500. By the end of the sixteenth century, the quality of English pig iron was as good as the German.[20]

Gilbert regards the material he calls "earth" as the primary substance constituting almost all of the Earth:

> The thews and dominant stuff of the globe is that terrene body which far exceeds in quantity all the volume of flowing streams and open waters ... and which takes up most of the whole globe and almost fills it internally, and by itself almost suffices to endow it with sphaerick shape. (OM 1.17:40)

He maintains that "native iron which is *sui generis* ... is formed when homogenic parts of the earth grow together into a metallick lode" (OM 1.17:42); that iron is the most abundant of all things buried in the Earth (OM 1.8:25); and that "in iron, especially in the best iron, there is earth in its own nature true and genuine" (OM 1.7:21). Thus, in Gilbert's theory, iron and lodestone, essentially two species of the same thing, are the true stuff of the Earth. It is notable in this connection that in his posthumously published *De mundo* Gilbert refutes Aristotle's four-element theory of the sublunary sphere, arguing that fire is not a primary element, that air and water are the Earth's effluvia, and that only "earth" can be considered a primordial element. This underscores the fact that, to Gilbert, iron and lodestone, being made of that earth-substance, occupy a special, superior, and fundamental place in the sublunary world.

Following this line of thinking, in *On the Magnet* Gilbert maintains that "a strong loadstone shows itself to be of the inward earth" (OM 1.17:43) and that this is why "magneticks are conformable to and are regulated by the earth, and are subject to the earth in all their motions" (OM 1.17:42). In other words, the nature of the magnet is the nature of the Earth itself:

> The loadstone and all magneticks, not the stone only, but every magnetick homogenic substance, would seem to contain the virtue of the earth's core and of its inmost bowels, and to hold within itself and to have conceived that which is the secret and inward principle of its substance; and it possesses the actions peculiar to the globe of attracting, directing, disposing, rotating, stationing itself in the universe, according to the rule of the whole. (OM 1.17:41)

In short, this means that the Earth itself must therefore be a magnet, which is Gilbert's fundamental contention and claim, though he does not always put it so definitely.

Up to this point, however, his argument has been based solely on the fact that natural lodestone is found in veins of iron ore; it has not been supported by any other form of evidence nor directly deduced from experiment.

The main tools used in Gilbert's experiments were the spherical magnet he called a terrella and a versorium-like instrument for testing magnetism (a freely rotating magnetic needle mounted on a pivot). The idea of using a spherical magnet was clearly borrowed from Peregrinus. In fact, however, a sphere is not always the most suitable shape for investigating the behavior of magnets or the general physical properties of magnetic force. Gilbert himself recognized that rod-shaped magnets exhibited greater force and more pronounced effects than spherical ones (OM 2.14, 2.15, 2.31). Other than to demonstrate the polarity of magnets, the only purpose for using a spherical magnet in experiments would be to reproduce and observe the Earth's behavior in miniature, on the assumption that the Earth were indeed a magnet. This implies that Gilbert's experiments in fact proceeded from the presuppositions that the Earth was a giant magnet with north and south magnetic poles and that a magnetic needle on the Earth's surface would point to either of those poles.

In fact, in the third chapter of book 1, prior to discussing his experiments, Gilbert states firstly that

> these poles [of the lodestone] are observant of the earth's poles, move toward them, and wait upon them

and secondly that because

> the spherical form, which is also the most perfect, agrees best with the earth, being a globe, and is most suitable for use and experiment, we accordingly wish our principal demonstrations by the stone to be made with a globe-shaped magnet as being more perfect and adapted for the purpose. (OM 1.3:12)

Concerning the second point in particular, Gilbert describes a procedure, drawn once again from Peregrinus, in which he places a compass needle (magnetic versorium; hereafter simply "versorium") at numerous points on a terrella, maps the many lines (magnetic meridians) indicating the needle's direction at those points, and thereby identifies the terrella's two poles as the points where the many lines intersect. This also enables him to identify the terrella's equator as the line equidistant from both poles. Thus he shows that the terrella is naturally endowed with an equator, meridians, and two poles, just as the Earth is. Accordingly, he asserts that

> the stone thus prepared is a true, homogeneous offspring of the earth and of the same shape with it: artificially possessed of the

> orbicular form which nature granted from the beginning to the common mother earth. (OM 1.3:12)

This is the basis for his claim that his experiments using terrella models accurately reproduce the behavior of magnets on the Earth's surface.

Via these considerations, Gilbert concludes in the final chapter of book 1 that

> the globe of the earth is magnetick, & a magnet. (OM 1.17:39)

As noted above, however, this proposition is not arrived at by inductive reasoning from unbiased experiments. Rather than proceeding from a blank slate, Gilbert *starts* with the hypotheses that the Earth is a magnet and that all terrestrial magnets are governed by it, and his experiments are designed to verify that the relationship between the terrella (as the Earth miniature) and the magnetic versorium is the same as that between the Earth and terrestrial magnets (compasses).

In the early nineteenth century, Thomas Thomson hailed *On the Magnet* as "one of the finest examples of inductive philosophy that has ever been presented to the world" and one that "is the more remarkable because it preceded the 'Novum Organum' of [Francis] Bacon, in which the inductive method of philosophizing was first explained."[21] For the reasons just noted, however, it is clear that such assessments fail to grasp the true nature of Gilbert's work. Peregrinus had already experimented with a spherical magnet and a compass needle and identified the magnet's poles in exactly the same way that Gilbert did, yet Peregrinus had concluded that compass needles on the Earth pointed to a celestial pole. This was because Peregrinus regarded his spherical magnet as a model of the celestial sphere, not the Earth. Thus, even with such a simple phenomenon, two people conducting the very same experiment to study the exact same effects can nonetheless give quite different interpretations of the results if they start from different hypotheses or theoretical premises. In this connection, Mary Hesse hits the nail on the head when she points out that "Gilbert's theories determine throughout the interpretations he makes of his experiments."[22]

## Section 6—Concerning "Magnetical Motions"

In the opening chapter of book 2, Gilbert summarizes, organizes, and classifies everything previously known about the behavior of magnets and compass needles under the banner of "magnetical motions." "These motions," he explains, "are incitements of homogeneal parts either among themselves or toward the primary conformation of the whole earth" (OM 2.1:45), and he classifies them into five categories:

> Five movements or differences of motions are then observed by us: Coition (commonly called attraction), the incitement to magnetick union; Direction towards the poles of the earth, and the verticity and continuance of the earth towards the determinate poles of the world; Variation, a deflexion from the meridian, which we call a perverted movement; Declination, a descent of the magnetick pole below the horizon; and circular motion, or Revolution. (OM 2.1:45–46)

(Note that Gilbert's term "variation" signifies what today is called "declination" while his "declination" denotes what is now known as "dip.")

All of the motions cited here are forms of magnetic phenomena observed in the behavior of terrestrial magnets and compass needles, and all were already known by Gilbert's time. But Gilbert adds three new aspects to the study of them. First, he distinguishes "coition" from "attraction"; second, he regards the five motions as natural motions also of the Earth itself; and third, he attempts to provide a unified explanation of them, particularly the last four, not just as observable facts demonstrated by compass needles at any point on the Earth but as the result of the conformity of all magnets to the enormous spherical magnet of Mother Earth, or in other words as a result of the action of the Earth itself.

Let's begin with "coition." As the word implies, this is the tendency of magnets to draw together and unite with other magnets or with iron. Much was already known about this phenomenon and its characteristics. Della Porta, for example, had already reported properties that Gilbert observes, such as change in magnetic strength—"the force of a magnet is always the stronger the nearer it is" (OM 2.24:91–92)—and the phenomenon of magnetic induction at a distance—that is, that a piece of iron is altered when it falls within a magnet's sphere of influence, even when it is at some distance from the magnet (OM 2.4:68). Della Porta had also known that, as Gilbert puts it

> the magnetick power excels light, in that it is not hindered by any opaque or solid substance, but proceeds freely, and extends its forces on every side (OM 2.7:77)

and likewise that

> thick boards do not obstruct, nor earthen vessel nor marble vases, nor the metals themselves; nothing is so solid as to carry away or impede the forces excepting an iron plate. (OM 2.16:83)

But one novel aspect of Gilbert's experimental research was his discovery that "to arm" a magnet—meaning to cover it with an iron cap—strengthened its magnetic force (OM 2.17 to 2.22), and his experiments in this area represented pioneering work on magnetic circuits. *On the Magnet* also contains the first mention of magnetic saturation, as follows:

> Magneticks can restore a certain soundness to magneticks … ; some can even exalt them beyond their proper strength; but when a body is at the height of perfection in its own nature, it is not capable of being strengthened further. (OM 2.25:93)

But perhaps more significant is Gilbert's discriminating preference for the terms "magnetical motions" and "coition." He occasionally uses such expressions as "attract or allure" (*trahere sive allicere*), as when, in book 1, he declares:

> It is trite and commonplace that loadstone draws iron; in the same way too does loadstone attract loadstone. (OM 1.5:15)

But he does so only because these are the terms commonly used to describe magnetism. They are not the ones he himself prefers:

> We say coition, not attraction. The word attraction unfortunately crept into magnetick philosophy from the ignorance of the ancients [but] if ever there is talk about magnetick attraction, we understand thereby magnetick coition, or a primary running together. (OM 2.3:60)

Thus, to Gilbert, the phenomenon known in unreflective, everyday language as the magnet's "attraction" is more accurately described as "coition."

The first implication of this is that coition is a kind of "motion." That is, whereas attraction is a species of force—"there seems to be force applied where there is attraction" (OM 2.3:60)—coition is motion that occurs spontaneously, not as a result of the application of force. In Aristotelian natural philosophy, the falling of a stone or a drop of water is not the result of gravity but rather the natural movement of elemental matter (earth or water) seeking to return to its original position. Presumably, Gilbert insisted on using the term "coition" instead of "attraction" because he shared more or less the same understanding, namely, that coition was the magnet's natural motion.

The second implication of coition is its mutuality. Before beginning the main text of *On the Magnet*, Gilbert provides a glossary of some of his terms, where, in explaining "magnetic coition," he states that

> in magnetick bodies, motion does not occur by an attractive faculty, but by a concourse or concordance of both. (OM, p. *vj)

The prefix "con" already implies "together," but to make sure there can be no misunderstanding, in the main text Gilbert adds that

> the magnetick coition is a motion of the loadstone and of iron, not an action of one. (OM 2.4:68)

Whereas "attraction" suggests an active agent unilaterally pulling a passive object toward itself, "coition" implies a spontaneous mutual approach and conjoining of two magnets, or a magnet and iron, as the result of an inherent impulse arising from the peculiar nature common to both.

Most importantly, in Gilbert's view the difference between "coition" and "attraction" is precisely the difference between magnetism and static electricity:

> The difference between Magneticks and Electricks is that all magneticks run together with mutual forces; electricks only allure. (OM 2.2:60)

This is why Gilbert frequently counterposes the "amber force" with "magnetic motion." His reason for making this distinction is presumably the empirical fact that magnetism works only between magnets or between magnets and iron whereas electricks, such as amber, attract most other objects. As we will see later, to Gilbert this shows that magnetism and static electricity are different both in essence and in principle.

The next kind of motion Gilbert mentions is "direction," referring to magnets' characteristic of aligning in a north-south orientation. Gilbert's treatment of this topic is not only the first elucidation of the connection between magnetic motion and the Earth but also a very lucid one.

But first let's look briefly at the related term "verticity of the earth," which Gilbert coined. The following is from chapter 13 of book 2:

> Let the line be called the axis which is drawn in the earth (as in a terrella) through the center to the poles [which] by the Latins [are] called *Cardines* or *Vertices*; because the world rotates and is perpetually carried around them. We are about to show, indeed, that the earth and a terrella are turned about them by a magnetick influence. (OM 2.13:81)

*Cardines* (the plural of *cardo*) meaning "axes of the heavens" and *vertices* being the plural of *vertex* (apex), here Gilbert is clearly referring to the Earth's axis of rotation. In the title of the second chapter of book 3 he qualifies what he calls "the directive virtue" with the aside "which we call verticity" (OM, p. 119). Furthermore, in enumerating the five magnetical motions in the abovequoted passage, he refers to "the verticity and continuance of the earth towards the determinate poles of the world," from which we can surmise that the "verticity of the earth" means the directionality of the Earth itself—that is, the nature of the Earth to turn in a fixed direction around its axis of rotation.

In Gilbert's view, the directionality of a magnet on the Earth's surface is precisely its self-alignment with the Earth's axis of rotation. Rejecting all previous theories that compass needles point to the north pole of the heavens, the North Star, or some point beyond the celestial sphere, he argues instead for seeking the cause of this phenomenon closer to home:

> The crowd of philosophizers, in order to discover the reasons of the magnetical motions, called up causes lying remote and far away ... But we study the earth itself and observe in it the cause of so great an effect. The earth, as the common mother, has these causes inclosed in her innermost parts. (OM 3.1:116)

Accordingly,

> true direction is the motion of a magnetick body in regard to the verticity of the earth, the natures of both agreeing and working together toward a natural position and unity. (OM 3.1:118)

But the direction to which terrestrial magnets point is not determined by any attractive or repulsive power of the poles. As Gilbert puts it,

> direction does not arise from attraction, but from the disposing and turning power which exists in the whole earth. (OM 4.6:162)

Furthermore, from the fact that the ends of magnets point to the Earth's poles, Gilbert also corrects a confusion in terminology that had persisted since the time of Peregrinus. First he asserts that

> all who before our time wrote of the poles of the stone [lodestone], and all the craftsmen and navigators, have been very greatly in error in considering the part of the stone which tended to the north as the north pole of the stone, and that which verged toward the south, the south pole. (OM 1.4:15; see also OM 5.1:187)

Later he clarifies what he means:

> the true austral pole of a loadstone, not the boreal (as all before us used to think), if the loadstone is placed in its boat on the surface of water, turns to the North. (OM 3.1:115)

And a little further on he puts it more precisely, explaining that

> a piece of a iron when touched by a loadstone acquires a disposition to motion toward the opposite pole of the earth, and not toward that pole of the earth to which that pole of that loadstone turned by which it was excited. (OM 3.4:125–26)

This fact is well known. The end of a compass needle rubbed with the "south" pole of a lodestone should be the needle's "north" pole, but if that end of the needle points northward, that would mean that the needle's "north" pole was being drawn to the Earth's North Pole, which would make the Earth's geographical North Pole its magnetic south pole. Considering this awkward, Gilbert sought to clarify matters by naming the north-pointing pole of a magnet its "south" pole and the south-pointing one its "north" pole. That way, the Earth's geographical North Pole was also its magnetic north pole. Before Gilbert, no one had noticed the problem because the directionality of the compass needle was not thought to be caused by the Earth's being a magnet.‡

Concerning magnetic force, Gilbert acknowledges that the magnet's poles pull iron more strongly than do its other parts, as he repeatedly indicates with statements such as the following:

> [It has] been ordained by nature that … at the pole itself should be the seat, the throne, as it were, of a consummate and splendid virtue, to which magnetical bodies on being brought are more vehemently attracted, and from which they are with utmost difficulty dislodged. (OM 1.5:17)

> The highest power of alluring exists in the pole, and … it is weaker and more languid in the parts adjacent to the aequator. (OM 2.14:81)

> Magneticks … incline languidly towards magneticks in the parts bordering on the aequator, but quickly in the places near the pole. (OM 2.28:97)

He maintains, however, that

> this is not so because there is really greater virtue residing at the pole, but since all the parts are united in the whole, they direct their forces towards the pole. (OM 2.5:72)

‡Today the pole located near the Earth's geographical North Pole is known as the Earth's magnetic south pole, while that located near the geographical South Pole is the Earth's magnetic north pole. Accordingly, the end of a magnet that points north is now called its "north-seeking pole" and that pointing south its "south-seeking pole," which conforms to the earlier terminology used by Peregrinus.

His explanation for this is as follows. If you take a single magnet and split it into two pieces of equal size, each piece will have two poles of equal strength regardless of whether the pieces were cut from the magnetically stronger poles or from near the magnetically weaker equator of the original magnet. In other words, magnets contain "all in every part" (OM 2.4:68). Accordingly, the fact that the force of the magnet appears more strongly at the poles is not because the magnet's strength lies solely in its poles but rather because all the strength of its parts accumulates there.

Also in connection with this point, Gilbert asserts that the strength of coition in the magnet is proportional to its "size" or "mass," that is, that the lodestone

> which excels in size shows greater powers, since it seizes greater weights and has a wider orbe of virtue. (OM 2.29:97)

Accordingly, Gilbert surmises, if a good-quality magnet weighing 1 drachm would seize 1 drachm of iron, then a 1-ounce magnet of the same quality would seize 1 ounce of iron (OM 2.29:97). This is not generally the case, however, so it is unlikely that Gilbert verified the hypothesis through actual experiments. Rather, his claim is probably based on the following inference:

> Magneticks in conjunction make one magnetick. Wherefore as the mass increases, the magnetick vigour is also augmented. (OM 2.22:90)

As we will see in the next chapter, Johannes Kepler adopted the same argument. It is certainly conceivable that Kepler was directly influenced by Gilbert's reasoning when he spoke of gravity being proportional to mass.

Regarding the phenomenon Gilbert calls "variation"—"declination" in modern terminology—Gilbert attributes it to the unequal distribution of matter (land) over the Earth's surface (OM 4.1:153). His explanation is that

> on the earth the verticity is perturbed by great continents, which are mostly elevated above the depths of the seas and make the versorium deviate sometimes from the right tracks (that is, from the true meridians). (OM 4.2:158)

Gilbert claims that this could be proved by experiments using a terrella with an irregular surface. In fact, however, the irregularities of land on the Earth's surface are miniscule in comparison with the Earth's radius (6,400 km): the lofty peaks of the Himalaya Range (8 km above sea level) would be only 1 millimeter in height on a terrestrial globe with a radius of 80 centimeters. It would be virtually impossible to build a terrella with such minute surface irregularity, and even if one could be made, it is hardly credible that such miniscule effects could be observed. In any case, in Gilbert's model the Earth has no magnetic poles separate from its geographic poles, and since magnets in principle point northward, in this view "variation" (declination) is simply an irregular perturbation effect and therefore no simple relationship could exist between "variation" and longitude. As Gilbert puts it,

> the variation is ... always uncertain ... nor does it ... follow the rule of any particular meridian. (OM 4.9:167)

Gilbert then turns to what he calls "declination"—now termed "dip"—which he identifies as the "remarkable motion of magnetick bodies dipping below the horizon" (OM 5.1:184). He describes the effect as follows:

> The end of the needle which looks toward the north dips below the horizon in northern regions, whilst in southern regions the end of the needle which looks toward the south tends toward the center of the earth, in a certain proportion … to the latitude of the district in question, from the aequator on either side. (OM 5.1:184)

Gilbert regards "declination" (dip) as, unlike "variation" (declination), a phenomenon inherent in the Earth itself, and asserts that the former

> arises in truth, not from any motion from the horizon toward the center of the earth, but from the turning of the whole magnetick body toward the whole of the earth. (OM 5.1:184)

"Declination" (dip), in other words,

> does not arise from the attraction of the loadstone, but from a disposing and rotating influence. (OM 5.5:195)

This much had already been pointed out by Robert Norman. Gilbert's uniqueness here lies in his statement that

> if the versorium dipped by reason of an attractive force, then a terrella made from a very strong magnetick stone would cause the versorium to turn toward itself more than one made out of an average stone, and a piece of iron touched with a vigorous loadstone would dip more. This, however, never happens. (OM 5.5:195)

That is, even when using terrellas of different coitional (attractive) strengths, the degree of dip will be the same for all of them at the same latitude. Moreover, he maintains that even on the same terrella the degree of dip is determined by latitude alone and is not simultaneously dependent on distance from the terrella's center. Gilbert's book reads as if these matters were actually observed, but it is dubious whether any experiment was performed.

In any event, on this basis Gilbert points out that the dip indicated in experiments using a versorium and a terrella should also quantitatively match that measured on the Earth's surface:

> The magnetick dip (which is the wonderful turning of magnetick things to the body of the terrella) in systematick course, is seen … to be the same thing upon the earth. (OM 6.1:212)

Gilbert thus simplifies the question of dip to determining the relationship between it and latitude, and in chapter 8 of book 5 he even includes a diagram for ascertaining "the latitude of a place anywhere in the world … without the help of the celestial bodies, sun, planets, or fixed stars, in fog and darkness" (OM 5.8:200). In fact, however, he bases this method on rather arbitrary reasoning and does not verify it by observation. On the actual Earth, the magnetic and geographical poles do not correspond and the degree of dip at any given point varies, rendering any such diagram meaningless. Edward Wright, who urged Gilbert to publish *On the Magnet* and wrote a preface to it, lauded Gilbert's findings regarding dip, but in fact it is unlikely that they were of any use at all.

Gilbert premises his experiments on "variation" (declination) and "declination" (dip) on the proposition that

> in all cases all magneticks conform themselves to the globe of the earth in the same ways and by the same laws by which another loadstone or any magneticks do to a terrella. (OM 3.1:119)

Because of this *a priori* assumption that the Earth is a large terrella, he develops rather simplistic and rash arguments in his interpretations of terrestrial magnetic phenomena.

## Section 7—The True Nature of Magnetic Force and the Form of the Sphere

Before examining how Gilbert viewed magnetic force, let us first consider what he called the "orbe of virtue" (Figure 17.2).

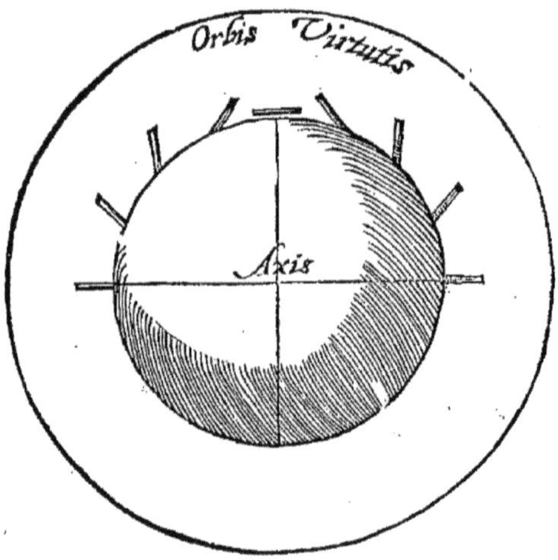


Figure 17.2: The *orbis virtutis* (domain of force). The poles are at either end of the horizontal axis. *Source*: Gilbert, *De magnete* (1600/1967).

According to the glossary at the beginning of *On the Magnet*, the "Orbe of Virtue"

> is all that space through which the Virtue of any loadstone extends. (OM, p. *vj)

And in chapter 7 of book 2 Gilbert asserts that

> from about a magnetical body the virtue magnetical is poured out on every side around in an orbe. (OM 2.7:76)

This concept is clearly borrowed from Della Porta. To modern commentators, it is tempting to interpret Gilbert as suggesting that the space within the orb is physically altered, and therefore to credit this idea as the forerunner of the concept of magnetic field. Among others, A. Crichton Mitchell, who wrote detailed histories on geomagnetism, science historian George Sarton, and electrical engineer J. Ambrose Fleming all understood "orbe of virtue" to mean "field of force."[23] But that is too modern an interpretation. Gilbert himself does not give the term that connotation, as is evident in the same chapter just quoted from, where he continues as follows:

> Yet there exists in nature no orbe or permanent or essential virtue spread through the air, but a magnet only excites magneticks at a convenient distance from it. And as light comes in an instant ... so much more quickly is the magnetick vigour present within the limits of its strength. (OM 2.7:76–77)

For Gilbert, then, magnetic force, unlike electric force, is simply action at a distance conveyed instantaneously to a spacially separate magnetic object, and the "orbe of virtue" is nothing more than the domain within which that action takes place. In his words,

> the orbes are magnetick, and yet not real orbes nor existent by themselves. (OM 5.11:206)

This brings us to Gilbert's conception of the true nature of magnetic force. He refers to this in many passages, particularly in comparison to electric force, as in the following:

> In all bodies in the world two causes or principles have been laid down, from which the bodies themselves were produced, matter and form. Electrical motions become strong from matter [*materia*], but magnetick from form [*forma*] chiefly; and they differ widely from one another and turn out unlike. (OM 2.2:52)

When Gilbert says that electric force arises "from matter" he means from the effluvia emitted by electricks, and his use here of the Latin *materia* should be understood in the literal sense of "matter."

On the other hand, he claims that the cause of magnetic force is nonmaterial. He bases this conclusion primarily on the observation that, unlike electric force, magnetic force remains unaffected by shielding:

> A versorium is moved as far off by a loadstone when an obstacle is put in the way, as through air and an open medium. (OM 2.32:102)

Likewise,

> loadstone and iron show sympathy with a loadstone though solid bodies be interposed. (OM 2.16:86)

Reasoning that magnetic force therefore must be nonmaterial action at a distance, he rejects reductionist explanations.[24] Similarly, concerning the magnetizing of a piece of iron by a magnet, he states:

> The stronger fortifies the weaker, not as though aught were taken away from its own substance, or its proper vigour, nor because any corporeal substance is imparted, but the dormant virtue of the one is aroused by the other, without loss. (OM 1.16:38)

Accordingly, such magnetic bodies

> do not move together on account of material affinity, but they take their motion and inclination from their form. (OM 3.6:130)

The key term "form" (*forma*) derives from Aristotle, and when Gilbert refers to the action of form upon matter (*materia*) he does so in the same sense as in Aristotelian natural philosophy, that is, as potentiality made actuality. He describes the magnetization of iron, for example, as follows:

> The natural magnetick virtue, confused and dormant in the iron, is aroused by the loadstone, is linked to the loadstone, and rejoices with it in its primary form; then smelted iron becomes a perfect magnetick, as robust as the loadstone itself. (OM 2.26:95)

Similarly, he holds that when a magnet demagnetized by heat is remagnetized it represents "the restitution and reformation of a confused form," and that accordingly there is no transfer of any *matter* from the magnet into the iron or back to the iron in these processes (OM 2.4:67). Rather, in Gilbert's interpretation, iron—indeed all earth-substance, the Earth's primary constituent—inherently possesses magnetic potentiality, and in its magnetization through contact with a magnet the iron transforms from potential magnet into actual magnet by acquiring the magnet's *form*.

Consider, in particular, Gilbert's views on the natural magnetization of iron. In chapter 12 of book 3 and in chapter 1 of book 5 he notes that iron melted in a furnace can be magnetized without using a magnet by letting it cool in a rod-shaped mold aligned in a north-south direction, or by hammering it out into a north-south-oriented length; and also that iron is naturally magnetized when left for a long period in a fixed north-south position on the Earth.

He interprets these observed facts as patent evidence of the Earth's peculiar power as a magnet. For Gilbert, moreover, it is surely the "form of the earth" (OM 1.17:42) that gives it that power.

When Gilbert thus attributes magnetic motion (magnetic force) to the Earth's "form," he means "the form of the primary and chief spheres," which "is unique, and particular, not the formal cause of the Peripateticks" (OM 2.4:65). The following is from chapter 11 of book 5:

> While some assign occult and hidden virtues of substances, others a property of matter, as the causes of the wonderful magnetical effects; we have discovered the primary substantive form of globes. (OM 5.11:207)

So, it is the spherical "form of the earth" that causes the Earth's magnetism. But what exactly does that mean?

Gilbert's views on "form" are complicated and obscure, but two statements provide some idea of the concept's specific character in the context of *On the Magnet.* The first comes near the end of book 1:

> Magneticks are conformable to and are regulated by the earth, and are subject to the earth in all their motions. All its movements harmonize with, and strictly wait upon, the geometry and form of the earth, as we shall afterwards prove by most conclusive experiments and diagrams. (OM 1.17:42)

By the "geometry" of the Earth he refers to the fact that the Earth (and likewise a terrella) is not simply a sphere but possesses poles, an axis, meridians, and an equator as natural rather than merely mathematical attributes. The second statement is in book 2:

> Electrical motion is a motion of aggregation of matter; magnetical motion is one of disposition and conformation. (OM 2.2:60)

In short, this means that the motion of a magnetic body is a unique type of motion that orients the magnet's axis parallel to its meridians and makes magnetic dip horizontal at the equator, perpendicular at the poles, and dependent on latitude in between. Hence,

> all magnetical bodies are reduced to order by the dominant form of the earth. (OM 2.16:86)

Gilbert regards the concept of dip (what he calls "declination") as particularly important. In his view, the peculiarity of magnetic force is demonstrated by the fact that the degree of dip in the Earth's or a terrella's orb of virtue is determined by latitude alone, regardless of distance from the orb's center or the magnetic strength of the terrella. Figure 17.3 is Gilbert's diagram showing multiple spheres of influence around a terrella—or rather, the different orbs of virtue of terrellas of the same size but different magnetic strengths—and how the versorium would incline at certain positions on each sphere. He describes this behavior of versoriums and other magnetic bodies as follows:

> They do not look toward the same part of the terrella or the same point at any distance from the same (unless they be on the axis of the orbes and of the terrella); but they always tend to those points of their own orbes, which are distant by similar arcs from the common axis of the orbes. As, for example, in the following diagram, we show the body of a terrella, with its poles and aequator; and also a versorium on three other concentrick orbes around the terrella at some distance from it. In these orbes (as in all those which we may imagine without end) the

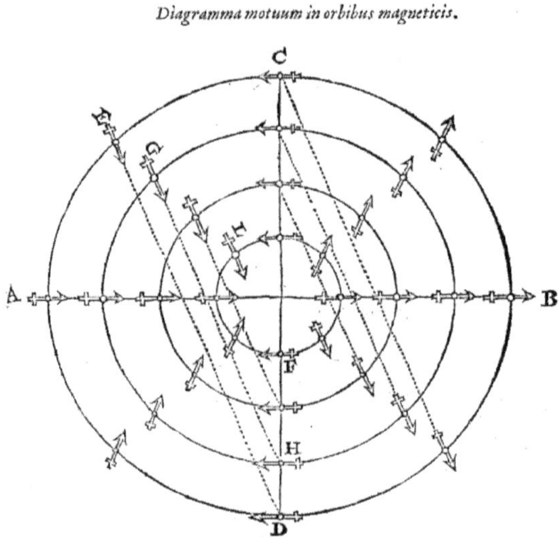


Figure 17.3: Multiple spheres of influence around a lodestone, and dipping needles at each sphere. *Source*: Gilbert, *De magnete* (1600/1967).

> magnetick body or versorium conforms to its own orbe in which it is located, and to its diameter and poles and aequator, not to those of the terrella. (OM 5.11:205)

In other words, at point L on the surface of the terrella (the innermost circle in the figure) the versorium points to F on the terrella's equator, but the versorium at E, the point on the outermost orb (ABCD) that corresponds to L, points not to F but to D, on the equator of the same orb, just as the one at G on the second outermost orb points to H. Thus, even though the radius of the sphere of influence varies according to the strength of the terrella's magnetic force, on any given orb the arrow representing the versorium direction points to the corresponding point on the same orb, "as if the orbes of virtue were solid and material loadstones" (OM 5.11:206). Consequently, all the versoriums at corresponding points of the concentric spheres align perfectly. Gilbert interprets this to mean that

> the power of magnetick globes ... is diffused and extended into orbes outside the body itself, the form being carried beyond the limits of the corporeal substance. (OM 5.11:205)

This also implies that the meridians, equator, poles, and so on of each orb are not merely conceptual but just as real as the physical Earth or terrella itself, the entities to which magnetic motion must conform. These features constitute the "primary form of the globes" (OM 5.12:208) and it is in this sense that the Earth's "form" can be thought to give the orbs the "geometry of the earth." This is what Gilbert refers to when he rather loftily congratulates himself for having

> by good fortune found out a new and admirable (beyond the marvels of all virtues magnetical) science of the orbes themselves. (OM 5.11:205)

Under this conceptual framework, the rest of the abovequoted passage about magnetic "form" is somewhat easier to understand:

> This form is ... the form of the primary and chief spheres [*globorum*] and of those parts of them which are homogeneous and not corrupted, a special entity and existence, which we may call a primary and radical and astral form; not the primary form [*forma prima*] of Aristotle, but that unique form, which preserves and disposes its own proper sphere. There is one such in each several globe, in the Sun, the moon, and the stars; one also in the earth, which is that true magnetick potency which we call primary vigour. (OM 2.4:65)

Although Gilbert uses the term "sphere" (*globus*) here, from his other expressions—such as "in each several globe [*in singulis globis*], in the Sun, the moon, and the stars," and, elsewhere, "all the globes [*omnes globos*], all the stars, and also the noble earth"

(OM 5.12:209)—it is clear that the term "sphere" refers generally to all globe-shaped celestial bodies and parts thereof that "are homogeneous and not corrupted." In other words, each complete spherical astronomical body is endowed with its own "form of the sphere" or "primary vigour." This is the natural-philosophical and metaphysical underpinning of the cosmos as Gilbert depicts it.

In the particular case of the Earth, its specific "sphere form" is its magnetism, which naturally gives it its poles, meridians, and equator, or in other words, its "geometry."

At the same time, Gilbert cautions that this

> magnetick nature peculiar to the earth … is neither derived nor produced from the whole heaven by sympathy or influence or more occult qualities, nor from any particular star. (OM 2.4:65)

He thus refutes the notion that magnets derive their force from the heavens, a concept common to Scholasticism, neo-Platonism, and magical thought from the late Middle Ages through the Renaissance and shared by everyone from Thomas Aquinas and Roger Bacon to Marsilio Ficino and Cornelius Agrippa. With this, in fact, Gilbert became the first in recorded history to assert, albeit in the language of Scholasticism, that magnets derive their power from the Earth, or more precisely, that magnetic force is inherent in the Earth.

Although Gilbert thereby dispels the magical understanding of magnetism as a "hidden force," he replaces it with a rather strange metaphysics of globes. As we will soon see, this metaphysics is grounded in Gilbert's peculiar hylozoistic view of the Earth.

## Section 8—The Motions of the Earth and Magnetic Philosophy

As we have seen, Gilbert discusses five types of magnetic motion, among which he attributes "direction," "variation" (declination), and "declination" (dip) to the effects of the Earth as a magnet. Today we understand this to mean that the spherical magnet of the Earth itself acts upon magnets and compass needles to turn them and give them direction, just as a terrella, a miniature of the Earth, affects a versorium placed upon it. In book 6's first chapter, titled "On the globe of the earth, the great magnet," Gilbert asserts:

> Those experiments which have been proved by means of the terrella, how magnetick things conform themselves to the terrella, are all or at least the principal and most important of them, displayed by means of the earth's Body. (OM 6.1:211)

This is the key premise upon which he develops his arguments.

In Gilbert's theory, however, the Earth-as-magnet does not merely act on magnets on its surface. Given his view that magnetic motions are natural motions of the earth-substances that are the Earth's true constituents, it is only natural to conclude, as he does, that

> the chief part of the visible earth is also magnetical, and has magnetick motions. (OM 1.17:42)

In other words, the Earth's magnetic nature also causes magnetic motions of the Earth itself.

In book 3, in which he discusses magnets' directionality, Gilbert notes that a magnet positioned so as to be able to rotate freely conforms its poles to those of Mother Earth. He surmises, furthermore, that

> if the earth were to deviate from its natural direction and its true position in the universe, or if its poles were to be drawn aside (if this were possible) toward the sun-rising or the sun-setting or toward any other points whatsoever in the visible firmament, they would return again to the north and south by magnetical motion, and would settle at the same points at which they are now fixed. (OM 3.1:117)

In short, this means that the Earth's magnetic force determines the Earth's own direction and is also what keeps its equatorial plane at an angle of 23.5 degrees from the ecliptic plane, so that the Earth's axis always points toward the vicinity of the North Star. In this understanding, the Earth's magnetic poles must correspond to its geographical poles.

Gilbert expounds his theory as follows:

> Directive force, which is also called by us verticity, is a virtue which spreads by an innate vigour from the aequator in both directions toward the poles. That power, inclining in both directions towards the termini, causes the motion of direction, and produces a constant and permanent position in Nature, not only in the earth itself but also in all magneticks. (OM 3.2:119)

Thus, in his view, it is the inherent action of magnet-Earth itself that gives direction to the Earth's axis, and the polarity of compass needles and magnets on the Earth's surface is, rather, an incidental effect of the same action. Although Gilbert expresses this idea very simply—"the natural movements of the whole and of the parts are alike" (OM 6.4:223)—in fact he attributes to magnetic force much greater effects on the Earth than is now known to be the case.

In Gilbert's theory, that is, those effects are not confined to the Earth's directionality (verticity). As indicated by the first words of the title of the third chapter of book 6—"On the magnetick diurnal revolution of the Earth's globe" (OM, p. 214)—Gilbert also considers diurnal rotation to be a natural motion of the Earth caused by its magnetic form. Here he applies Peregrinus's ideas on the diurnal motion of terrestrial magnets to the Earth itself as a magnet. In essence, Gilbert's magnetic philosophy holds that the Earth is an active and therefore noble entity that, because of its magnetic nature, possesses its own principle of self-motion.

This view undercuts a vital point of conventional Aristotelian cosmology—that the Earth is an ignoble, inert lump of clay at the center of the universe—and conflicts with the premises on which the entire geocentric theory rests. Indeed, Gilbert openly criticizes the static-Earth view upheld since Aristotle's time:

> Those who wrote about the world and about natural philosophy a century ago, especially those remarkable elementary philosophers, and all those who trace their knowledge and training to them down to our own times, those men, I say, who represented the earth as always at rest and, as it were, a useless weight, placed in the centre of the universe at an equal distance from the sky on every side, and its nature to be simple, imbued only with the qualities of dryness and cold, sought diligently for the causes of all things and of all effects in the heavens, the stars, the planets, in fire, air, waters and substances of mixed natures. Never indeed did they recognize that the terrestrial globe had, besides dryness and cold, some special, effective, and predominant properties, strengthening, directing, and moving the globe itself through its whole mass and its very deepest vitals; nor did they ever inquire whether there were any such. (OM 3.1:116)

At this juncture it is worth noting Gilbert's views on the Copernican theory. He refers to Copernicus in exalting tones as "the Restorer of Astronomy" (OM 6.9:240) and "a man most deserving of literary honour" among medieval men of learning (OM 6.3:214). In fact, Gilbert was one of the earliest English thinkers to adopt the heliocentric theory, at least concerning the diurnal motion of the Earth.

Gilbert begins his treatment of this topic by stating the two alternatives at issue:

> either the Earth herself must needs be set in motion with a diurnal movement from West to East, or the whole heaven and the rest of nature from the East to West. (OM 6.3:215)

He then immediately calls into question the until then widely accepted notion of stars fixed in a depthless celestial vault, asking "who has ever made out that the stars which we call fixed are in one and the same sphere?" And indeed, without further ado, he states outright that

> no one has ever proved this as a fact; nor is there a doubt but that just as the planets are at unequal distances from the earth, so are those vast and multitudinous lights separated from the Earth by varying and very remote altitudes. (OM 6.3:215)

He even suggests the possibility of a limitlessly vast universe:

> How immeasurable then must be the space which stretches to those remotest of fixed stars! (OM 6.3:215)

In fact, Gilbert did believe the universe to be unbounded. In a figure that appeared in his posthumous *De mundo*, he depicted

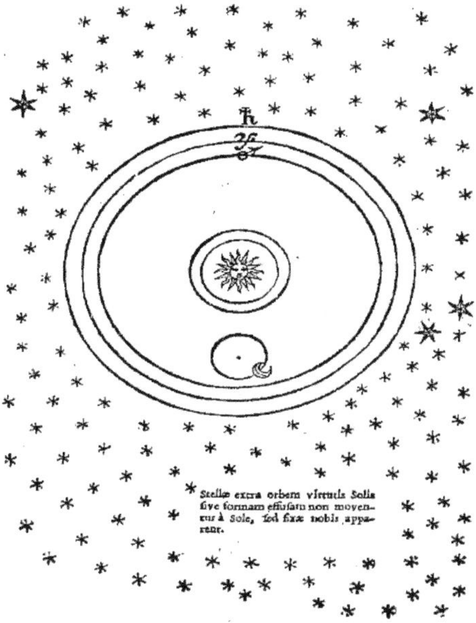


Figure 17.4: Gilbert's diagram of the universe, from a posthumous work. *Source*: Kay, *William Gilbert's Renaissance Philosophy of the Magnet.*

planetary orbits around the sun and, beyond the orbit of the outermost planet, numerous stars at irregular distances from the center and therefore not forming a single stellar sphere (Figure 17.4).

From such considerations, Gilbert reasons that

> there can be no movement of infinity and of an infinite body, and therefore no diurnal revolution of that vastest *Primus Mobile*. (OM 6.3:216)

Accordingly, he concludes that the idea of an enormous, three-dimensional fixed orb of stars revolving once every twenty-four hours is "a superstition and in the view of philosophy a fable" (OM 6.3:216).

Here we can clearly see the influence of Thomas Digges's 1576 text *A Perfit Description of the Caelestiall Orbes*, which introduced the heliocentric theory to England and went even further than Copernicus by proposing, for the first time, an infinite universe. Digges states that

> this orbe of starres fixed infinitely vp extendeth hit self in altitude sphericallye, and therfore immouable[25]

and he too provides a diagram depicting many stars even beyond the largest sphere of planetary orbit (Figure 17.5). Digges's influence is obvious from the similarity of his diagram to Gilbert's.

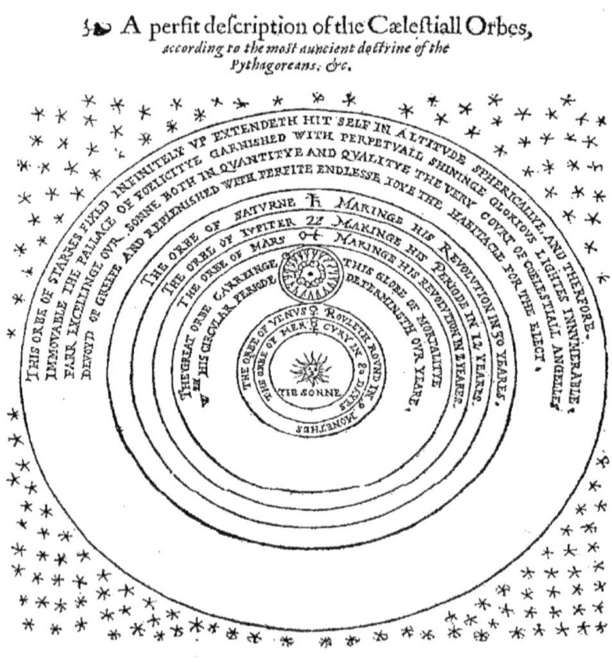


Figure 17.5: Digges's diagram of the universe. *Source*: Johnson and Larkey, "Thomas Digges, the Copernican System, and the Idea of the Infinity of the Universe in 1576."

Gilbert also responds to the common criticism that if the Earth revolved on its own axis all objects on its surface would be left behind by its onrush. He explains that, in the "effluvia" surrounding the Earth, "heavy bodies projected in any way by force, move on uniformly along with the Earth" (OM 6.5:229), or in other words that even airborne matter is fixed to the Earth by the atmosphere.

Gilbert's conclusion is that

> for these reasons, not only probable but manifest, does the diurnal rotation of the earth seem, since nature always acts through a few rather than through many; and it is more agreeable to reason that the Earth's one small body should make a diurnal rotation, than that the whole universe should be whirled around. (OM 6.3:219–20)

Thus, he is an explicit supporter of the Copernican theory, at least as far as diurnal rotation is concerned.

But there was more to the issue than that. The difference between the geocentric and heliocentric theories cuts to the heart of human understanding of nature. At the risk of redundancy, the matter can be summed up as follows. In Aristotelian cosmology, the sublunary (terrestrial) world and the celestial world were distinct and comprised of different kinds of matter. Unlike the eternally orbiting celestial bodies, the clod of terrestrial matter had no principle of self-movement other than the natural motion of falling toward the ground. Being made up of cold, inert earth-substance, the Earth was inherently incapable of rotating or orbiting without the aid of some external force. To Aristotle, the notion that the Earth was stationary was a logical corollary of his natural philosophy. Accordingly, the difference between the geocentric and heliocentric models was not a relative difference of observers' perspectives but an absolute difference in their understanding of nature and matter. For this reason, to win acceptance of the Copernican theory it was not enough merely to refute the popular criticism that objects on the Earth's surface are not left behind by its rotation; one had to grapple with Aristotelian natural philosophy head-on.

Indeed, Aristotle was not the only adversary. The idea that earth (earth-substance) was inert went back even further. The philosophers of ancient Miletus considered water, air, and fire to be alive and therefore animate, and yet they regarded earth as inanimate. Similarly, in Plato's four-element theory, atoms of water, fire, and air were active in that they could change their form one into the other, but earth atoms alone lacked that reciprocity and were therefore deemed inert. Such thinking persisted even after Copernicus. In 1588, Tycho Brahe, the greatest astronomical observer of the day, while acknowledging the supremacy of the Copernican system nonetheless presented an eclectic picture of the solar system, according to which all planets except the Earth orbited the sun, the Earth alone was stationary, and the sun, dragging all the other planets with it, orbited the Earth. This view was based on Brahe's conviction that the Earth was "large, sluggish and inapt for motion."[26]

Around the same time, Giordano Bruno, who supported the heliocentric theory, offered a different view. In *On the Infinite Universe and Worlds*, written in London in 1584, Bruno maintains that the Earth is

> impelled by her own intrinsic principle, soul and nature, taketh her course around the sun … and spinneth around her own center.[27]

In order to gain acceptance of heliocentric ideas, Bruno clearly felt impelled to endow the Earth with a soul as the principle underlying its movement. Remember that this was an age in which the principles of inertia and conservation of angular momentum were still undiscovered. To give credence to the idea that the Earth rotated on its axis and orbited the sun, it was necessary to explain what made it move.

By discovering the Earth's magnetic "form," Gilbert believed he had succeeded in identifying the intrinsic principle of its self-motion, the driving force that made it revolve and determined the orientation of its axes. Thus, he concludes, the Earth

> turns herself about by magnetick and primary virtue. (OM 6.4:224)

Or perhaps Gilbert's reasoning unfolded the other way around. He declares a couple of pages later that

> if the Earth revolve, that … must needs occur not by reason of the first sphaere, but of its innate forces (OM 6.5:226)

and this understanding may have come first in his thinking. Science historian Gad Freudenthal argues that Gilbert formed his hypothesis that the Earth was a magnet from this prior idea of innate forces, and indeed that seems likely.[28]

In any case, by providing a philosophical basis for a self-moving Earth, which was inconceivable in Aristotelian natural philosophy, Gilbert's magnetic philosophy removed a major hurdle to acceptance of the heliocentric theory.[29] This is probably what Christopher Wren meant when he wrote in 1657 that Galileo had "labour'd to prove the Motion of the Earth, negatively, by taking off Objections, but Gilbert positively."[30] It would appear, in fact, that from 1600 forward some English and other scholars, following Gilbert's lead, accepted the idea of the Earth's rotation. Flemish engineer and mathematician Simon Stevin wrote in his 1605 work *De Hemelloop* (The heavenly motions) that

> since this motion [the diurnal motion of the Earth] is thus simply described by him [Copernicus], without any natural argument or proof, this supposition long troubled me in my mind.... But thereafter there was published the book about the great terrestrial magnet, described by *Guilelmus Gilbertus* [William Gilbert], in which the natural cause of this motion in my opinion is hit off and revealed.[31]

And in 1613, English physician Marke Ridley wrote in his *A Short Treatise of Magneticall Bodies and Motions* that

> the Earth hath, at her first creation, bestowed vpon her a globous and starlike vigor, or intelligence, whereby she may, hauing her whole parts vnited together, by the vertue of grauity vnto the center, and her place made sure *Magnetically* by her poles vnmoueable; yet moue naturally, keeping her place circularly & diurnally out of the West into the East, to the end that all creatures should receiue the comfortable and liuely beames of the Sunne, and the influences of the rest of the Planets and Starres.[32]

By 1620, even Oxford theologian Nathanael Carpenter was among those asserting that the Earth revolved by virtue of its magnetic nature.[33]

## Section 9—The Magnet Earth and its Soul

A lifeless Earth that nonetheless moves by its own power—in an age when modern mechanics was still unknown, this idea needed explaining. Gilbert may have surpassed Aristotle in natural philosophy, but his logic and even his phraseology are Aristotelian. Moreover, the meaning he ascribes to diurnal rotation is teleological and not at all modern. That is, after declaring that diurnal rotation is the cause of the alternation of day and night, the even distribution of the sun's heat and the tides over the entire Earth, and therefore the sustenance of human and indeed all life, Gilbert then essentially reverses that logic by stating that "there is a diurnal movement on the part of the Earth for its own sake and for its advantage" (OM 6.5:227). Similarly, concerning axial tilt, he notes that if the Earth's axis did not incline toward the ecliptic pole there would be no change in the seasons, and that if it tilted at a greater angle regions in high latitudes would be wastelands. He therefore offers the teleological interpretation that

> Nature has taken care, through the Earth's soul or magnetick vigour ... that the points toward which the Earth's pole is directed should be 23 degrees and more from the poles of the Ecliptick. (OM 6.8:234)

In sum, Gilbert proposes that

> the Earth is moved by her own primary form and natural desire, for the conservation, perfection, and ordering of its parts, toward things more excellent. (OM 6.4:224)

Thus, in Gilbert's view, the Earth is inherently possessed of both the principle and purpose of self-generated motion. It was Aristotle who had defined the soul as "a thing's essence ... having *in itself* the power of setting itself in movement and arresting itself,"[34] and Gilbert concurs that "the power of moving itself seems to point to a soul" (OM 2.4:68).

In the title of the final chapter of book 5, the book dealing with "declination" (dip), Gilbert asserts that "magnetick force is animate, or imitates life." He begins the chapter with the following statement:

> A loadstone is a wonderful thing in very many experiments, and like a living creature. And one of its remarkable virtues is that which the ancients considered to be a living soul in the sky, in the globes and in the stars, in the sun and in the moon. (OM 5.12:208)

Shortly thereafter he presents his own conclusion on the matter as follows:

> We … find this life in globes only and in their homogenic parts; and though it is not the same in all globes (for it is much more eminent in the sun and in certain stars than in others of less nobility) yet in very many the lives of the globes agree in their powers. (OM 5.12:208)

To Gilbert, "globes"—that is, spherical celestial bodies—are special, superior entities that, due to their form, act in the universe as if they were alive. This metaphysics of globes is founded on his ascription to them of a kind of soul; or as he puts it,

> the bodies of the globes, as important parts of the universe, in order that they might be independent and that they might continue in that condition, had a need for souls to be united with them. (OM 5.12:209)

Aristotle had thought that only the sun, the moon, and the stars had life and that the elements comprising the sublunary world were lifeless, which logically entailed that the Earth was a dead, inert object. Furthermore, the Aristotelian order of being involved a hierarchy of value, with the heavens as noble and the Earth as ignoble, a concept that provided the ideological basis for the geocentric theory. Gilbert, however, cannot accept this. Wondering "why the globe of the earth alone ... is condemned by him [Aristotle] and his followers and cast into exile (as senseless and lifeless), and driven out of all the perfection of the excellent universe" (OM 5.12:208), Gilbert proposes instead that all astronomical bodies, including the Earth, are equally alive:

> We ... consider that the whole universe is animated, and that all the globes, all the stars, and also the noble earth have been governed since the beginning by their own appointed souls and have the motives of self-conservation. (OM 5.12:209)

This passage suggests a considerable influence from Bruno, who held that each star "is so large and comprehensive as to be worthy to be considered a world in itself; not one lacketh the efficient principle and power to preserve and maintain perpetual generation and life to innumerable and excellent individuals."[35] Whatever the influences behind it, Gilbert's view is that "the Earth is not so vile and abject a body as it is generally considered" (OM 6.5:227). Moreover, insofar as he deems the Earth to be self-moving by its own inherent principle and interprets this motion teleologically, this more or less necessitates his conclusion that the Earth is a vital entity with some sort of soul. Likewise he maintains that

> Thales, not without cause ... held that the loadstone was animate, being a part and a choice offspring of its animate mother the earth. (OM 5.12:210)

Thus, Gilbert's inquiry into magnetism leads him to his distinctive hylozoistic view of the Earth:

> By the wondrous wisdom then of the Builder forces, primarily animate, have been implanted in the Earth, that with determinate constancy the Earth may take its direction, and the poles have been placed truly opposite, that about them as the termini, as it were, of some axis, the motion of diurnal turning might be performed. But the constancy of the poles is regulated by the primary soul. (OM 6.4:221)

This understanding of the Earth as in some sense animate is the heart of Gilbert's entire magnetic philosophy, yet many science historians who praise him as the father of modern experimental science conveniently ignore this point. One commentator goes so far as to complain that "the last book of *De magnete* is a curious and fascinating one [in which the] reasoning is so utterly at variance with that of the earlier parts of the treatise that one finds it difficult to understand why this part was included at all."[36] And not a few lament that, as one puts it, the "'magnetical philosophy' of the earth was so fantastic that to a degree it vitiated the scientific usefulness of Gilbert's *De Magnete*."[37] But Gilbert's notion that the Earth is animate and has a soul cannot be divorced from his discovery that the Earth is a magnet. It is wrongheaded to praise the latter as pioneering while dismissing the former as backward. In his thinking, the two were inseparable elements comprising and characterizing what he envisioned as a soul-endowed Earth that revolves and determines its own direction by its own magnetic nature. This understanding is what enabled him to embrace heliocentrism.

Marie Boas maintains that, although *On the Magnet* is often taken to be "the first great work in modern experimental science," it would be more accurate to consider it "the last important work in natural magic," because "Gilbert was, after all, a natural magician, not a natural philosopher."[38] British science historian Hugh Kearney identifies three traditions within the scientific revolution in modern Europe—the organic tradition, going back to Aristotle; the magical tradition, springing from neo-Platonism; and the mechanistic tradition—and judges Gilbert's views to "fall into place as part of the magical tradition."[39] It is true that Gilbert invoked both Hermes Trismegistus and Zoroaster when he claimed that all astronomical bodies, including the Earth, were animate (OM 5.12). At the same time, however, he rejected magical interpretations of magnetism in terms of "occult power" or "sympathy and antipathy," as well as Platonic and Plutarchian mechanistic views of magnetism, presenting instead his own hylozoistic and organic picture of the Earth. It has also been pointed out that Gilbert's education was steeped in Aristotelianism[40], and indeed his theory of soul is essentially Aristotelian. Thus, although Gilbert espoused heliocentrism, his understanding of nature was fundamentally derived from the organic tradition, and it was from that standpoint that he attempted to create a foundation for heliocentrism.

* * *

Gilbert's achievements were significant. First, he distinguished the science of electricity from that of magnetics. In doing so, he laid the foundation for experimental study of electricity by conceiving the "electric versorium" namely electroscope, and with his model of electric force as action through a medium of effluvia he provided both the experimental and theoretical starting point for subsequent studies on static electricity. Second, by discerning that the Earth was a giant magnet, he founded the science of geomagnetism. And third, through a "magnetic philosophy" positing the Earth as an active entity endowed with magnetism as its primary form, he provided, albeit imperfectly, a natural-philosophical and metaphysical explanation for the Earth's self-motion as demanded by the heliocentric theory that had emerged half a century earlier. This last achievement has received scant recognition in science historiography because it is so alien to subsequent scientific thinking. But in the early seventeenth century, when modern physics and the modern view of the universe were just emerging, this was the point that had the greatest impact, because it also dispelled the ideological underpinnings of the geocentric model—specifically, the notion that the Earth, being inferior to other astronomical bodies, was inert and unable to move on its own. The influence of Gilbert's idea is clearly evident in Galileo's comment, made ten years later, that the Earth "is movable and surpasses the Moon in brightness, and … is not the dump heap of the filth and dregs of the universe."[41] Gilbert's magnetic philosophy was carried on by Kepler in Germany and John Wilkins and Robert Hooke in England and became a major driving force in the formulation of a new cosmology. It thus served as a vital bridge from the geometrical Copernican hypothesis developed in the mid-sixteenth century to the modern physics- and dynamics-based cosmology that emerged in the second half of the seventeenth century.

On the other hand, Gilbert's arguments themselves were far from modern in character. By this I mean more than simply that he engaged in metaphysics and slipped into a species of hylozoism. More importantly, whereas the laws of modern physics must be founded on quantitative measurement and expressed mathematically, Gilbert showed no inclination whatsoever for such an approach. Rather, in the interpretation of experiments his theory followed Aristotelian natural philosophy, a *qualitative* physics. Although this thinking proved useful on the qualitative side—such as in distinguishing electricks from non-electricks—it was completely devoid of aspiration toward precise quantitative measurement. There is also Gilbert's remark near the beginning of *On the Magnet* that "after the magnetick nature shall have been disclosed" by his treatise, "all darkness [i.e. in thought on the subject] will disappear" (OM 1.1:7), a claim that betrays a thoroughly Scholastic outlook. In short, the purpose of his inquiries was to exhaustively explore the essential nature of magnets, not to establish laws of magnetic force. Gilbert in fact appears not to have made any quantitative measurements; despite attaching great importance to magnetic dip, in *On the Magnet* he cites not a single measurement concerning that phenomenon, not even one of Norman's, nor, indeed, concerning anything.

Zilsel points out that, whereas Gilbert referred to many ancient and medieval scholars, he never mentioned Euclid, Archimedes, or Vitruvius, who were well known in Europe in Gilbert's time. "The three omissions," Zilsel concludes, "show that Gilbert was not concerned with the mathematical literature of the period, [and] that he was not interested in mechanics."[42] To this one can add that *On the Magnet* also fails to mention either Robert Recorde or John Dee, who were widely known in England and are indispensable to any discussion of the new mathematical technology and science. Accordingly, the nature of Gilbert's experiments and observations were fundamentally different from those conducted in the seventeenth century by, say, Galileo on inclined planes, or Robert Boyle and Robert Hooke on atmospheric pressure. While praising *On the Magnet* highly, Galileo also chided that "what I might have wished for in Gilbert would be a little more of the mathematician, and especially a thorough grounding in geometry."[43] In that sense, Gilbert can be viewed, as Kay puts it, "not as the anticipation of the seventeenth-century [sic] but as the culmination of the sixteenth."[44] It is also worth mentioning in this connection that some have credited Edward Wright as the true author of the most mathematical and technical portion of *On the Magnet*, chapter 12 of book 4.[45]

Nonetheless, it was under the tremendous influence of Gilbert's work that Kepler developed his new astronomy. Kepler's laws of planetary motion represented the very first laws of physics that were modern in the sense that they were expressed in terms of mathematically rigorous concepts and supported by quantitatively precise observations. Kepler announced the first and second laws in 1609, just nine years after the publication of *On the Magnet*, and the third law in 1619. From these laws on the one hand and Gilbert's magnetic philosophy on the other, Kepler derived the pivotal concept of gravity acting between astronomical bodies, and it was from this point that the modern view of the universe evolved. In other words, Gilbert's magnetic philosophy provided the foundation for the view of the Earth from which Kepler conceived his idea of gravity, as we will examine in the next chapter. Thus, Gilbert's surmise that the Earth moved of its own accord because it was a magnet became an important inspiration for the dramatic advances in physics that followed in the seventeenth century.

Chapter 18

# Johannes Kepler and the Magnetical Philosophy

## Section 1—Kepler's Starting Point

German mathematician and astronomer Johannes Kepler (1571–1630) was born in December 1571 to a Lutheran household in what was then the Free Imperial City of Weil, enclosed within the Duchy of Württemberg in southwest Germany. The following November, Danish astronomer Tycho Brahe was one of many who independently discovered that a new star (nova) had appeared in the constellation Cassiopeia. Tycho soon demonstrated that, since the nova did not seem to change its position among the fixed stars, it had to have come into existence above the moon. The nova continued to shine brightly until finally fading from view at the beginning of 1574; at its brightest, reportedly, it was visible to the naked eye even during daytime. Its appearance was a sensational event because it exploded the belief that the world above the moon's orbit was immutable, a dogma of Aristotelian cosmology that had been upheld for some two thousand years. Aristotelianism was rocked to its foundations not merely at the level of words and ideas but by a clearly visible real-world phenomenon. The Saint

Bartholomew's Day massacre, a spate of Catholic-instigated killings of Huguenots and their supporters, had erupted in France only three months earlier, and many people were terrified that the new star's appearance represented a "dire warning of God's anger and an omen of punishment."[1] This was the age of Reformation and Counter-Reformation turmoil and cruel witch trials, and with its mixed population of Catholics and Protestants, Germany, like other parts of Europe, was in a volatile political state. Such were the harsh and uncertain times that Kepler was born into.

Kepler's forebears had been important figures in the city, but by his father's day the family had fallen on hard times. However, the 1555 Peace of Augsburg had brought a basic settlement of the Reformation issue in Germany, with fortunate consequences for Kepler: needing not only capable public officials but also a corps of sharp-minded clergymen to contest the continuing religious debates, Protestant Württemberg had established quite a good education system by the time he was born. Despite his family's poverty, therefore, Kepler was able to study at a seminary and from 1589 in the theology department of the University of Tübingen. But although he had planned to become a minister of the Evangelical church, for one reason or another in 1594 he was persuaded to take the position of mathematics teacher at a Protestant seminary in Graz.

Not long after embarking on a career as a mathematician, the young Kepler leapt to fame throughout Europe as an astronomer with the 1596 publication of *The Secret of the Universe*,[2] in which he claimed to have unraveled the essential mystery of the solar system's motions. In this work, Kepler attributes his turn toward astronomy to the avid instruction in Copernican theory that he received from Michael Maestlin (1550–1631), professor of astronomy at the University of Tübingen. Kepler was fortunate that the University of Tübingen, an institution developed under the ideological influence of Martin Luther, Philipp Melanchthon, and other key Protestant Reformists who had rejected the notion of a heliocentric universe, nonetheless had teachers such as Maestlin on its staff. While pursuing his theological studies Kepler had continued to inquire diligently into Copernican theory, a task to which he applied himself even more earnestly upon assuming his teaching position at the seminary in Graz.

As is well known, Copernicus, by regarding the sun as being at the center of the universe and the Earth and the other planets as revolving around the sun, showed that the apparent anomalies of planetary motions, such as retrogression, could easily be explained as optical illusions perceived by observers on the moving Earth. Incidentally, the greatest advantage that his heliocentric theory had over the Ptolemaic geocentric theory is that the former enabled Copernicus to calculate the relative sizes of planetary orbits and thereby determine the true order of their arrangement, whereas the latter had merely treated the orbit of each individual planet independently of the others. In a word, with Copernicus's model the solar system was grasped for the first time as a single system. Copernicus himself declared that

> by long and intense study I finally found that if the motions of the other planets are correlated with the orbiting of the earth, and are computed for the revolution of each planet, not only do their phenomena follow therefrom but also the order and size of all the planets and spheres, and heaven itself is so linked together that in no portion of it can anything be shifted without disrupting the remaining parts and the universe as a whole.[3]

Copernicus discovered "a marvelous symmetry of the universe, and an established harmonious linkage between the motion of the spheres and their size, such as can be found in no other way."[4] Kepler starts from this point.

In *The Secret of the Universe*, Kepler is no longer concerned with whether or not heliocentric theory is correct; rather, on the premise that it is correct, he wishes to explore the underlying principle of the solar system as a whole. "There were," he explains, "three things in particular about which I persistently sought the reasons why they were such and not otherwise: the number, the size, and the motion of the circles [i.e., planetary orbits]" (SU, Preface, 63). The "number" of planetary orbits referred to here is six, one for each of the six planets from Mercury to Saturn (including Earth), the only planets known at the time; "size" is denoted by the radius of each orbit; and "motion" is expressed as each planet's orbital period. Up to that point, no one had even considered the question of the relative sizes (ratios) of the planets' orbital radii in heliocentric theory, much less inquired into what possible relationship there might be between the planets' orbital radii and orbital periods. It is also noteworthy that all of these questions that Kepler posed were quantitative ones.

But the reasoning behind these original ideas is not entirely modern. Kepler's explanation is that the stationary parts of the universe, namely, the sun, the fixed stars, and the space between them, correspond to the Father, Son, and Holy Spirit in the Christian doctrine of the Trinity; that some similar kind of order and harmony also obtains among the universe's moving parts, meaning the planets, including Earth; and that this order, too, shows the signs of God's creative hand at work. In Kepler's view, "by a most perfect Creator it was absolutely necessary that a most beautiful work should be produced" (SU 2:93).

What this means to Kepler, a Platonist, is that the universe was created in conformance to a geometric model—or in other words that, following Plato, we must acknowledge that "God is always a geometer" (SU 2:97). Accordingly, Kepler's aim in *The Secret of the Universe* is

> to show ... that the most great and good Creator, in the creation of this moving universe, and the arrangement of the heavens, looked to those five regular solids, which have been so celebrated from the time of Pythagoras and Plato down to our own, and that he fitted to the nature of those solids, the number of the heavens, their proportions, and the law of their motions. (SU, Preface, 63)

Kepler arrives at the following answer to his unique question concerning the size and number of planetary orbits:

> The Earth [i.e., its orbit; likewise for the other planets mentioned in this passage] is the circle which is the measure of all. Construct a dodecahedron round it. The circle surrounding that will be Mars. Round Mars construct a tetrahedron. The circle surrounding that will be Jupiter. Round Jupiter construct a cube. The circle surrounding that will be Saturn. Now construct an icosahedron inside the Earth. The circle inscribed within that will be Venus. Inside Venus inscribe an octahedron. The circle inscribed within that will be Mercury. (SU, Preface, 69)

He thus concludes that the relative sizes of the planets' orbits correspond to an arrangement of spheres either inscribed within or circumscribed around each of the five Platonic solids (regular polyhedra), one inside another in a particular order. The idea is presented graphically in Figure 18.1, an illustration that appears in *The Secret of the Universe.* In the figure, the outermost globe ($\alpha$) represents the sphere that includes Saturn's orbit, the one

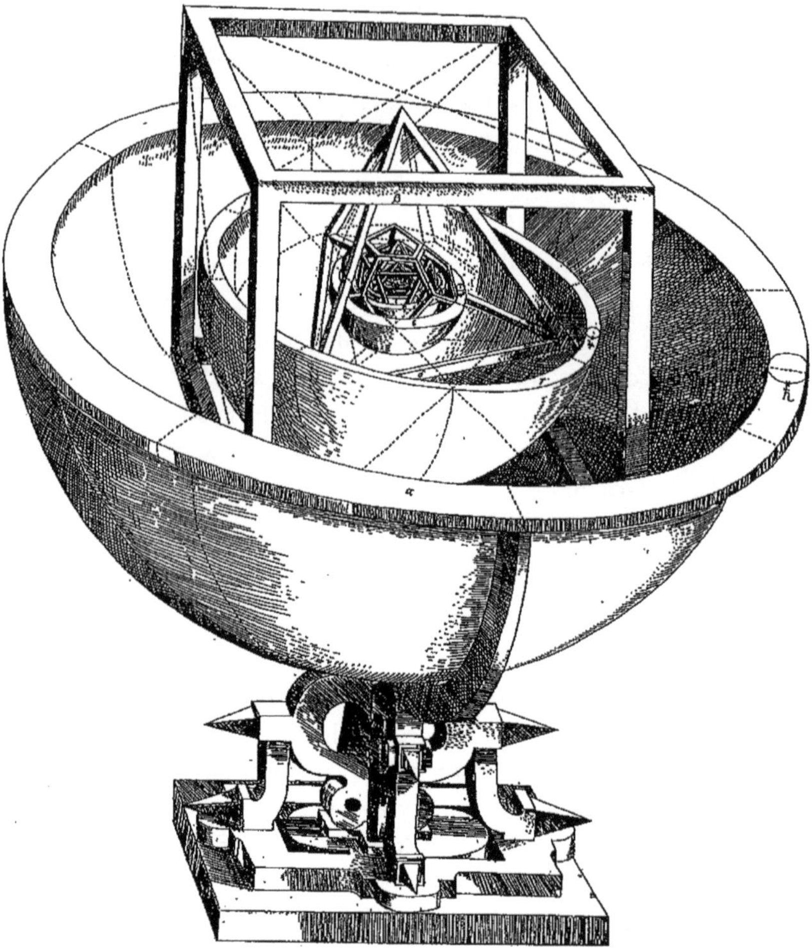

Figure 18.1: Kepler's solar system based on the five regular solids.
*Source*: Kepler, *Mysterium cosmographicum* (1596/1621).

immediately within that ($\gamma$) the sphere containing Jupiter's orbit, the next one ($\varepsilon$) the sphere containing Mar's orbit, and so on. Since there are only five Platonic solids, only six such inscribed and circumscribed globes could be produced by nesting them within one another, which meant, in Kepler's hypothesis, that the number of planets was limited to six. In a letter he wrote to Galileo on October 13, 1597, it is fitting that Kepler refers to "Plato and Pythagoras, our true masters";[5] his argument is precisely a faithful, almost fanatical application of both the Pythagorean belief that numbers are endowed with special significance and unique power and the Platonic idea that God configured the world geometrically. Given that the count of six planets was wrong to begin with, from our perspective today the problem as such is meaningless; but it does give a clear indication of the nature of Kepler's concerns at the time. That the ratios of orbital radii that Kepler calculated on the basis of his geometrical model happened to be fairly close to those arrived at by Copernican theory further reinforced his convictions.

Although grounded in Copernican theory, at this point Kepler's cast of mind seems still rather medieval and, by present-day standards, even mystical. Nonetheless, the fact that *The Secret of the Universe* enjoyed considerable popularity and high acclaim at the time implies that such ideas, far from seeming odd, struck a chord in the European society of his day. The book's success convinced Kepler that his true calling was to elucidate God's plan not as a minister but rather as an astronomer, a decision that set him on a path to the discovery of the three laws of planetary motion now named after him and to a major reformation of the science of astronomy.

## Section 2—Kepler's Reformation of Astronomy

Standard histories of science identify Nicolaus Copernicus's proposal of heliocentric theory as the starting point of modern astronomy and, by extension, of modern physics. From the viewpoint of physics, however, it is more accurate to regard Kepler as the one who effected modern science's true departure from what had gone before. Although Copernicus had correctly advocated a model of the solar system with a stationary sun, in terms of astronomy's character as a science his theory had yet to completely break free of the kind of thinking that had persisted since ancient times.

In contrast, Kepler's reformation of astronomy did more than simply place the sun at the center of the heavens and replace circular orbits with elliptical ones. Even more crucially, it introduced the idea that planetary motion was caused by a force exerted by the sun, a notion that transformed astronomy from a geometry of orbits into a dynamics of astronomical bodies and from a geography of the heavenly firmament into a physics of the celestial domain. Whereas Copernicus had advocated the heliocentric solar system mathematically, Kepler sought its physical or metaphysical grounds. Underlying this was a pivotal shift in thinking about what the objectives of the science of astronomy should be.

The admitted task and purpose of astronomy prior to Kepler is aptly summed up by sixth-century neo-Platonist Simplicius in the following description of Platonic astronomy that appears in his commentary on Aristotle:

> Plato lays down the principle that the heavenly bodies' motion is circular, uniform, and constantly regular. Thereupon he sets the mathematicians the following problem: What circular motions, uniform and perfectly regular, are to be admitted as hypotheses so that it might be possible to save the appearances presented by the planets?[6]

For one thing, then, there was an *a priori* assumption that planetary motions were a set of circular motions.[7] Even in Aristotelian natural philosophy, the view that celestial bodies, composed of a supposed fifth elemental substance, "move always in a circle" simply by their "nature" was taken to be a self-evident fact beyond dispute or doubt.[8]

We find the same thinking in the astronomy of Ptolemy, who declares that

> it is our purpose to demonstrate for the five planets, just as we did for the sun and moon, that all their apparent anomalies can be represented by uniform circular motions, since these are proper to the nature of divine beings, while disorder and nonuniformity are alien [to such beings].[9]

Here again circular motion is regarded as an attribute of celestial bodies that requires no further explanation. And almost 1,400 years later, on December 9, 1599, Tycho Brahe writes to Kepler that

> one must compose the revolutions of celestial objects definitely from circular motions; otherwise they could not come back on the same path eternally in equal manner, and an eternal duration would be impossible.[10]

Astronomy prior to Kepler thus consisted entirely in finding the right arrangement of orbital spheres, circular orbits, and so on to closely recreate observations of the planets' positions, and it was therefore considered sufficient to be able to calculate and predict the observed motions of the planets in the celestial sphere by some kind of mathematical or geometrical hypothesis (model). This is what was meant by the expression "to save the appearances." To that extent, it is fair to say that conventional astronomy had not proceeded from any ontological inquiry or natural-philosophical, causal questioning of whether the planets really did move in that manner and why. Prior to Kepler, then, astronomy had been a geometry of orbits that did not probe the physical causes of motion. Natural philosophy (physics), meanwhile, had been seen as a qualitative discipline essentially incompatible with the mathematical science of astronomy.

But then came Copernican heliocentrism. By treating the Earth as one of the planets and consequently abandoning the old sublunary-celestial dualism, Copernicus was confronted with the problem of creating a new natural philosophy. Thus, he acknowledges that the sun is not only at rest but also a special kind of entity, declaring that "indeed, as though seated on a royal throne, the sun governs the family of planets revolving around it." He also accepts that gravity exists in all heavenly bodies:

> For my part I believe that gravity is nothing but a certain natural desire, which the divine providence of the Creator of all things has implanted in parts, to gather as a unity and a whole by combining in the form of a globe. This impulse is present, we may suppose, also in the sun, the moon, and the other brilliant planets, so that through its operation they remain in that spherical shape which they display.[11]

Copernicus was thus already beyond regarding the fall of heavy objects on or near the surface of the Earth as a natural motion toward the sole center of the universe.

In Copernicus's view, however, gravity was simply a tendency, inherent in objects on or near an astronomical body that are fragments of that body, to spontaneously move toward it as the parent entity from which they arose; it was not that the celestial body drew the objects toward it. He regarded the cause rather as a kind of affinity or sympathy, a view which boiled down to the ancient argument of mutual attraction between things of the same kind. Copernicus therefore did not think that this force acted between different celestial bodies, much less that the planets orbited the sun by virtue of the sun's motive power or were held in the sun's environs by its attractive power.

These opinions were consequences of Copernicus's assumptions that "the motion appropriate to a sphere is rotation in a circle" and that, since periodic motion ought to be impossible unless it is circular motion, the planets' motions "are circular or compounded of several circles." In addition, Copernicus considers it an "axiom" of astronomy that "the heavenly bodies' motion is uniform and only apparently seems nonuniform."[12] Thus, although Copernican astronomy made the sun—or, more precisely, the fictional "mean sun" at the center of the Earth's orbit—the center of planetary orbits, this was still an orbital geometry adhering to the apriorism of uniform circular motion. By providing somewhat more simplicity than the Ptolemaic system, mathematically this was an improvement, but it nonetheless lacked an understanding of the physics and dynamics of celestial motion, and therefore failed even to pose the question of what impelled and controlled the planets.

In the foreword that he anonymously added to the first edition of Copernicus's *On the Revolutions of the Celestial Spheres*, Lutheran preacher Andreas Osiander inserted the proviso that "these hypotheses need not be true nor even probable. On the contrary, if they provide a calculus consistent with the observations, that alone is enough."[13] In light of the abovementioned shortcomings in Copernicus's thinking, this remark becomes more than just an apology directed at bigoted conservatives and church authorities; even if it did run counter to Copernicus's own intentions, it nonetheless expresses the objective limitations of the Copernican theory. Indeed, although Erasmus Reinhold's *Prutenic Tables*, a set of astronomical tables published in 1551, was based on Copernicus's mathematical models, it was constructed essentially without regard for the mobility of the Earth and was widely used by advocates of geocentrism. And in his 1602 work *The City of the Sun*, Tommaso Campanella reflects prevailing perceptions when, comparing Ptolemy and Copernicus, he writes that "the one does his counting with pebbles and the other with broad beans, and neither of them uses the very things being counted. Consequently, they pay the world off with play money instead of gold."[14] Neither theory was regarded as representing natural-philosophical or ontological truth.

In *The Secret of the Universe*, one thing in particular distinguishes Kepler from those earlier thinkers. Whereas "Copernicus's purpose was not to deal with cosmography, but with astronomy"—meaning that "he is … concerned … only with establishing from the observations the values which are best suited for deriving the motions of the planets and computing their positions" (SU 15:159)—Kepler repeatedly casts himself in the role of physicist (natural philosopher). Kepler contrasts "astronomy" and "mathematics" on the one hand with "physics" (natural-philosophy) and "metaphysics" on the other, and explicitly declares his difference from Copernicus in that respect. In his preface, for instance, in regard to reaching a certain hypothesis about planetary motions Kepler notes that, "where Copernicus did so through mathematical arguments, mine were physical, or rather metaphysical" (SU, Preface, 63). Thus, in his debut work, Kepler recorded the first step in what subsequently became his crucial revision of the science of astronomy.

In *A Defence of Tycho against Ursus*, a short treatise written around the end of 1600, Kepler asserts that, although "one who predicts as accurately as possible the movements and positions of the stars performs the task of the astronomer well," that is not enough: "For even though what he [Ursus, i.e., astronomer Nicolaus Reimers] mentions is the primary task of an astronomer, the astronomer ought not to be excluded from the community of philosophers who inquire into the nature of things."[15] Specifically, Kepler's approach dispelled what had until then been a clear distinction between "astronomy," which dealt solely with the geometry of orbits, and "physics," which inquired into the essential natures of things and the causes of phenomena. He regarded astronomy as part of physics, and his aim was to develop a dynamics and physics of celestial bodies. This entailed more than simply introducing efficient cause and other concepts of causality into astronomy; it also meant reforming natural philosophy itself, until then only a qualitative discipline, into the mathematical science of physics. Hence the subtitle of *New Astronomy*, the 1609 treatise in which Kepler first presented the first two of his now-famous laws of planetary motion: "Based upon Causes or Celestial Physics treated by means of commentaries on the Motions of the Star Mars from the observations of Tycho Brahe, Gent."[16] In this work's introduction, Kepler explicitly states that, as part of the process of reforming astronomical theory, he will "inquire into celestial physics and the natural causes of the motions" (NA, Introduction, 48).

At the beginning of book 1 of his magnum opus, *Epitome of Copernican Astronomy*, written much later (books 1–3 appeared in 1618, book 4 in 1620, and books 5–7 in 1621), Kepler poses the question "What is astronomy?" He answers that it is "a science setting out the causes of those things which appear to us on earth as we attend to the heavens and the stars" and that it is "a part of physics, because it seeks the causes of things and natural occurrences."[17]

In the same vein, book 4, in which Kepler for the first time presents a full-fledged cosmology based on heliocentrism and all three of his laws of planetary motion, carries a subtitle declaring that "Herein the natural and archetypal causes of Celestial Physics, that is, of all the magnitudes, movements, and proportions in the heavens, are explained." On the title page of the original Latin edition, furthermore, the words "Physica Coelestis" (Celestial Physics) are printed in large, bold type. It is also in book 4 that he explicitly points out that "the ancients, Eudoxus and Callippus, and their follower Ptolemy did not advance beyond circles, wherewith they were accustomed to demonstrate the phenomena—not worrying as to how the planets completed these circles" (ECA, 888). One can also point to Kepler's note in the second edition of *The Secret of the Universe*, published the following year (1621), where he refers to "the Celestial Physics [which] is established in the *Commentaries on Mars* [i.e., *New Astronomy*], and elaborated in Book 4 of the *Epitome of Astronomy* [sic]" (SU 20:203, author's note 3). It seems fair to say that by then Kepler was well aware of the significance of the revolution he had brought about.

## Section 3—Celestial Physics and "Moving Souls"

At the time of his debut work, *The Secret of the Universe*, the most immediate issue compelling Kepler toward reforming astronomy was that of how the planets were related. Taking the entire solar system to be a single harmonious order, he believed that some kind of significant relationship must therefore exist among the motions of the planets. Specifically, he sought a relation between the planets' orbital radii and their orbit periods, to determine whether "for the ratio of the motions to the distances a more accurate account can be obtained from Copernicus than from the customary hypotheses" (SU 20:197). Compared with his argument concerning the Platonic solids, which histories of science have often taken up when discussing *The Secret of the Universe*, this question has received relatively little attention, but in fact it held much greater significance for the subsequent development of Kepler's thought.

As he inquires into the problem, Kepler's argument suddenly veers into the province of dynamics. It is at this juncture that he begins to conceptualize an efficient cause applicable to the planets:

> But if ... we wish to make an even more exact approach to the truth, and to hope for any regularity in the ratios, one of two conclusions must be reached: either the moving souls [*anima motorix*] are weaker the further they are from the Sun; or, there is a single moving soul in the center of all the spheres, that is, in the Sun, and it impels each body more strongly in proportion to how near it is. In the more distant ones on account of their remoteness and the weakening of its power, it becomes faint, so to speak. Thus, just as the source of light is in the Sun, and the origin of the circle is at the position of the Sun, which is at the center, so in this case the life, the motion and the soul of the universe [*vita, motus et anima mundi*] are assigned to that same Sun; so that to the fixed stars belongs rest, to the planets the secondary impulses of motions, but to the Sun the primary impulse. (SU 20:199)

In a note appended to the second (1621) edition of *The Secrets of the Universe*, Kepler asserts as follows about the former hypothesis of "moving souls" inherent in the planets: "Of which I have proved there are none in the *Commentaries on Mars* [i.e., *New Astronomy*]" (SU 20:203, author's note 2). In the original text itself, moreover, he effectively ignores that option. This leaves him with just the latter hypothesis to consider—that there is a single moving soul in the sun. Here for the first time was put forward—albeit in hylozoistic language—an idea central to modern celestial mechanics: the sun exerts upon all planets in its system an influence that is the same though quantitatively different (weaker) the further away from it they are, and the motion of each planet is determined by that influence. For Kepler, heliocentrism meant more than a mere mathematical practice of placing a stationary sun at the origin of the coordinate system used to describe planetary motion; it also entailed the dynamical and hylozoistic understanding that the sun was the source of the activity of the entire solar system and that all the planets moved by virtue of the sun's physical effects and soul-like influence.

Furthermore, Kepler's account of the attenuation of the sun's influence on the planets as their distance from it increases is not just qualitative. From the analogy of the weakening of light the further it emanates from its source, he conjectures that the attenuation of the sun's power is likewise quantitatively determined, and in this way he seeks to identify a connection between each planet's orbital period and its orbital radius:

> Let us suppose, then, as is highly probable, that motion is dispensed by the Sun in the same proportion as light. Now the ratio in which light spreading out from a center is weakened is stated by the opticians. For the amount of light in a small circle is the same as the amount of light or of the solar rays in the great one. Hence, as it is more concentrated in the small circle, and more thinly spread in the great one, the measure of this thinning out must be sought in the actual ratio of the circles, both for light and for the moving power. Therefore in proportion as Venus is wider than Mercury, so Mercury's motion is stronger, or swifter, or brisker, or more vigorous than that of Venus … But in proportion as one orbit is wider than another, it also requires more time to go round it, although the force of the motion is equal in both cases. Hence it follows that one excess in the distance of a planet from the Sun acts twice over in increasing the period: and conversely, the increase in the period is double the difference in the distances. (SU 20:201)

This argument is expressed mathematically as follows—keeping in mind that both light and "moving power" are here conceptualized as extending not through three-dimensional space but rather in a two-dimensional plane that includes the sun and the planets' orbits. Let $I(r)$ be the light intensity (the energy passing through an arc of a given unit length in a given unit of time) at the circumference of the circle with radius $r$ around a point light source. The total energy passing the circumference and going out beyond the circle in a given unit of time is constant regardless of the circle's size; hence

$$I(r) \times 2\pi r = \text{const.}$$

and therefore

$$I(r) \propto \frac{1}{r}$$

In exactly the same way, if $F(r)$ is the magnitude of the "moving power of the sun" exerted on a circular orbit of radius $r$, then

$$F(r) \times 2\pi r = \text{const.} \quad \therefore \quad F(r) \propto \frac{1}{r} \tag{18.1}$$

Kepler discusses this relationship later, in *New Astronomy*, as the *lex effluxus*, the literal translation of which is "law of efflux" (GW 3:34). Had he formulated this argument for force extended in three dimensions, it would have led to Gauss's theorem [$F(r) \times 4\pi r^2 = \text{const.}$] and thus directly to the inverse-square law of force. As it happened, however, this step was not accomplished until half a century later, by Ismaël Boulliau (Ismaël Bullialdus).[18]

In any case, Kepler thought that the measure of tangential velocity $v(r)$ was proportional to this force. That is—as I will explain in more detail later—Keplerian dynamics posited that velocity was maintained by this momentary force and that therefore

$$v(r) \propto F(r) \tag{18.2}$$

Combined with the equation above, this yields

$$v(r) \propto \frac{1}{r} \tag{18.3}$$

Accordingly, the orbital period $T$ of a planet moving in a circular orbit of radius $a$ is

$$T = \frac{2\pi a}{v(a)} \propto a^2. \tag{18.4}$$

This is the meaning of the final clause of the passage from *The Secret of the Universe* just quoted, the proposition that "the increase in the period is double the difference in the distances."

On the basis of Tycho Brahe's observational data, Kepler later induced his famous third law of planetary motion, which states the relationship between the orbital period $T$ and the elliptical orbit's semimajor axis $a$ as

$$T \propto a^{3/2}, \tag{18.5}$$

which he announced in proposition 8, chapter 3, book 5 of *The Harmony of the World*, published in 1619. Although this meant that proposition 18.4 above was wrong, it is nonetheless noteworthy that, even at the time of writing *The Secret of the Universe*, Kepler already had an awareness of the issue that would later lead him to discover the third law expressing a single, quantitative relation among all components of the solar system.

Having discovered the third law, Kepler writes the following in a note in the second (1621) edition of *The Secret of the Universe*: "If what Ptolemy says about the motion of earthly bodies and the arrangement of the orbits is true, then there is no ratio of the motions or of the periodic times to the orbits which is permanent and constant for all the planets" (SU 20:207, author's note 8). In book 4 of the *Epitome*, written around the same time that the second edition of *Secret* appeared, Kepler asks himself "By what reasons are you led to make the sun the moving cause or the source of movement for the planets?" He cites as his first reason the validity of his third law, namely, that "in so far as any planet is more distant from the sun than the rest, it moves the more slowly—so that the ratio of the periodic times is the ratio of the 3/2 th powers of the distances from the sun" (ECA, 895).

Kepler felt that the establishment of the third law interrelating all the planets provided a physical and categorical corroboration of the heliocentric theory. Crucially, the third law put everything in physical terms, showing that, although the effect weakens with distance, the sun exerts qualitatively the same influence on all the planets. The important thing was that there existed a ratio applicable to all planets—a single relation tying all constituents of the solar system to the sun. Whether that ratio took the form of equation 18.4 or 18.5 above was only a secondary issue. The earlier formulation had already strongly implied the sun's central role in the solar system.

In the sense that he continued to express his ideas about the sun's influence in the hylozoistic terms of a "moving soul" (*anima motorix*), at that point Kepler still had one foot in the Middle Ages. On that aspect of his theory, however, he offers the following remarks in a note in the second (1621) edition of *The Secret of the Universe*:

> If for the word "soul" [*anima*] you substitute the word "force" [*vis*], you have the very same principle on which the Celestial

> Physics is established in the *Commentaries on Mars* [i.e., *New Astronomy*], and elaborated in Book IV of the *Epitome of Astronomy* [sic]. For once I believed that the cause which moves the planets was precisely a soul ... But when I pondered that this moving cause grows weaker with distance, and that the Sun's light also grows thinner with distance from the Sun, from that I concluded, that this force is something corporeal, that is, an emanation which a body emits, but an immaterial one. (SU 20:203, author's note 3)

Thus Kepler revised his concept of the sun's influence on the planets from his initial idea of a "moving soul" (*anima motorix*) to that of a dynamical "moving force" (*vis motorix*). What prompted this shift was the discovery of the fact that force weakens with distance. Here we witness the sublation of magical "effect" and animistic "influence" into a physical (natural-philosophical) concept of action at a distance. With this shift, Kepler accomplished a true reformation of the science of astronomy.

## Section 4—Gilbert's Theory of Gravity

Kepler succeeded in revising astronomy in line with the notion that the planets move by virtue of a force emanating from the sun, but the decisive impetus for this breakthrough in fact came from his encounter with the "magnetic philosophy" of William Gilbert. Kepler conceived his idea of a physical force at work among distant celestial bodies under the inspiration of Gilbert's theory that the Earth was a magnet. To understand Kepler's thought, it is therefore useful to backtrack slightly and briefly review Gilbert's ideas about gravity.

In his treatise *On the Magnet*, Gilbert observes that "the separated parts [of the Earth] tend toward their own origin [i.e., the Earth], but that tendency we call weight."[19] Gilbert's notion of gravity, referred to here as "weight" (*pondus*), does not, however, invoke Aristotle's idea of the natural movement of heavy bodies toward the Earth's center as the sole center of the universe. The following passages are from book 6 of *On the Magnet*:

> The whole Earth remains immovable in its place, united and compacted together by its own weight. That cohaesion of parts and aggregation of matter exist in the Sun, in the Moon, in the planets, in the fixed stars, in fine in all those round bodies whose parts cohaere together and tend each to their own centres.

> For that straight motion [i.e., of bodies toward the center] is only a tendency toward their own origin, not of the parts of the Earth only, but of those of the Sun also, of the Moon, and of the rest of the sphaeres.[20]

Thus, what Gilbert meant by "gravity" was, as with Copernicus, the tendency, common to all heavenly bodies, for objects on or near such a body (i.e., fragments of it) to move toward it as their parent entity—the sympathy of the part for the whole. It was not a quality exclusive to heavy bodies in the sublunary world.

In Gilbert's thinking, rather, this notion of gravity as a tendency toward like things (*inclinatio ad simile*) is expanded to include the relation between the Earth and the moon and, by extension, that between the Earth and the sun. The following is from book 6 of *On the Magnet*:

> It is more agreeable to the relation between the Moon and the Earth, that that harmony of motion should be due to the fact that they are bodies rather near together, and very like each other in nature and substance, and that the Moon has more

> evident effects upon the Earth than the rest of the stars, the Sun excepted; also because the Moon alone of all the planets conducts her revolutions, directly (however diverse even), with reference to the Earth's centre, and is especially akin to the Earth, and bound to it as with chains.[21]

Thus the moon is regarded no longer as a perfect sphere consisting of a fifth element but rather as made of the same kind of substance as the Earth. One of the examples Gilbert cites of the moon's effect upon the Earth is its influence on the tides. But the influence of the moon on the Earth was relatively easy to grasp. The real problem concerned the influence of the sun on the Earth's motion.

Before we proceed any further, it is worth considering at this point whether or not Gilbert was a supporter of the heliocentric theory. In *On the Magnet* he asserts that "neither the stars nor the wandering globes move about the Earth's centre"; and in *De mundo* that "the planets in their circular motion do not observe the Earth as a center of motion, but the greater Sun."[22] In an illustration that appears in *De mundo*, furthermore, he places the sun at the center of the orbits of all the planets (see Figure 17.4 on page 585).

According to most history books, however, although Gilbert acknowledged the Earth's rotation on its own axis (diurnal motion), he never publicly expressed any views about its revolution around the sun (annual motion). Notice, furthermore, that in the illustration just mentioned the Earth is the only planet without a circle representing its orbit around the sun. On the other hand there is the following rather ambiguously worded passage in book 6 of *On the Magnet*:

> The Earth, then, which by some great necessity, even by a virtue innate, evident, and conspicuous, is turned circularly about

> the Sun, revolves; ... The Sun (the chief agent in nature) as he forwards the courses of the Wanderers, so does he prompt this turning about of the Earth.[23]

In some interpretations, the original Latin expression, *telluris conversio*, translated here as "turning about of the Earth," is taken to mean the Earth's rotation on its axis. As we saw in the previous chapter, however, Gilbert's view of the Earth's diurnal rotation was that it was a spontaneous natural motion due to the Earth's own magnetic nature, not an externally caused effect of the sun. On that basis, the Latin expression in question cannot refer to the Earth's spin. Hence, in the Thompson translation used here, the relevant phrase is taken to denote orbit around the sun, by which interpretation the abovequoted passage becomes an assertion that all the planets, including the Earth, revolve around the sun by virtue of the sun's power. Other scholars and commentators also understand the original text in this way.[24]

But even leaving aside the question of whether or not Gilbert acknowledged the Earth's orbiting of the sun, from his description of the sun as "the agent and incitor of the universe in motion"[25] it is clear that he regarded it as the central agency controlling the motions of all the planets—a key premise on which the concept of gravitational force exerted by the sun on the planets was established.

Gilbert did not state outright that gravity was magnetic force. Nonetheless, given his far from clear-cut treatment of the subject, as well as such comments as the one in his posthumously published *De mundo* that "the Moon is drawn to the Earth by magnetism,"[26] it cannot be denied that he left ample room for being interpreted as regarding gravity in that way. In 1620 England, Francis Bacon described "Gilbert's opinion" as involving "the magnetic force of the earth to attract heavy objects."[27] Similarly, in Francis Godwin's novel *The Man in the Moone*, written around the same time, the traveler to the moon refers to the Earth's attraction exerted during his trip toward the moon as "the attractive Beames of that tyrranous Loadstone the earth."[28] And in 1666, natural philosopher Robert Hooke asserted that "Gilbert began to imagine [gravity] a magnetical attractive power, inherent in the parts of the terrestrial globe."[29]

Thus was Gilbert understood—one might say misunderstood, or at least broadly interpreted—throughout the seventeenth century. Indeed, this view persisted well into the twentieth century, with even Herbert Butterfield maintaining in his famous *The Origins of Modern Science* (1949) that "the force of the magnetic attraction was the real cause of gravity, said Gilbert."[30]

## Section 5—Gilbert's Influence on Kepler

After the publication of *The Secret of the Universe*, Kepler, interpreting Gilbert in precisely the way just described, began to conceptualize gravity as magnetic force. This is clear in the following passage from a letter Kepler wrote to astronomer David Fabricius on November 10, 1608, during the period when he was writing *New Astronomy*:

> How is it possible that a sphere, thrown vertically upward—while the earth rotates meanwhile—does return to the same place? The answer is that not only the earth, but together with the earth, the magnetic invisible chains [*catena magnetica*] rotate by which the stone is attached to the underlying and neighboring parts of the earth and by which it is retained to the earth by the shortest, that is, the vertical line.[31]

The first mention of Gilbert in Kepler's writings comes in the following passage in *A Defence of Tycho against Ursus*:

> William Gilbert the Englishman appears to have made good what was lacking in my arguments on Copernicus' behalf … in the study of magnets.[32]

Kepler is thought to have written the *Defence* between October 1600 and April 1601, just after Gilbert's *On the Magnet* was published in England. We cannot be sure to what extent Kepler may already have had feelers out in the academic community in search of precisely such new ideas and developments, but he was certainly very quick to respond to Gilbert's work. Until then, Kepler had posited the existence of the sun's "moving soul" acting upon the planets but had been unable to fully grasp exactly what that force was or how its existence might be demonstrated. But now Gilbert's argument in *On the Magnet* seemed to provide him with exactly what he had been looking for. In a letter to his friend Hervart von Hohenburg dated January 12, 1603, Kepler writes that he wishes he "had wings with which to travel to England to confer with [Gilbert]," adding: "I certainly think I can demonstrate all the motions of the planets with these same principles."[33] And in another letter, written to Englishman Christopher Heydon in October 1605, Kepler expresses his gratitude for Gilbert's discovery of magnetic philosophy.[34] Kepler thought that Gilbert's "magnetic philosophy" could provide the physical and causal underpinnings for Copernican theory, and it was from this point forward that Kepler's distinctive theory of magnetic gravitation began to take shape. The process of that theory's formation unfolded together with the process by which, on the basis of Tycho Brahe's observational data, Kepler discovered his first and second of the laws of planetary motion. That discovery is now universally acknowledged while his magnetic gravitation theory is ignored and forgotten, but to Kepler himself each integrally supported the other.

One of the motivations behind Kepler's decision to assist Tycho Brahe was that the Protestant seminary in Graz where Kepler taught was closed down amid the Austrian archiduke's suppression of Protestantism. A more compelling inducement, however, was Kepler's keen desire to obtain the abundance of precise observational data that Tycho had accumulated, a desire honed by the fact that the radii of the celestial spheres, including the planetary orbits, that Kepler had calculated using his Platonic solids theory differed slightly from Copernicus's numerical values. Because planetary orbits are not strictly circular, in the model of the solar system Kepler proposed in *The Secret of the Universe* each of the spheres containing or contained by the Platonic solids was a spherical shell with a finite thickness determined by the radius and orbital eccentricity. The problem, however, was that the numerical values Copernicus had used were not very precise.

Kepler was therefore eager to obtain accurate orbital-radius and eccentricity data, and the only person in the world who possessed such data at that time was Tycho. Tycho had devoted virtually his entire life to astronomical observation, and by improving and enlarging observational instruments he had obtained data of far greater precision than any previously available—so much greater that he is said to have attained the very limits of what is possible with the naked eye. Furthermore, having conducted his observations ceaselessly for many years, he had compiled statistically reliable data on the planets' motions that covered their entire orbits. This comprehensiveness is what made Tycho's observations incomparably superior, since previous data was not only inferior in precision but also largely limited to the relatively few observations made sporadically when planets were in special positions, such as at times of eclipse or conjunction.

Kepler moved to Prague to assist Tycho in September 1600, and was assigned by him the task of determining the orbit of Mars. Viewed in retrospect, this was an extremely fortuitous choice of assignments. Indeed, in *New Astronomy* Kepler reflects that, since the motions of Mars "provide the only possible access to the hidden secrets of astronomy, without which we would remain forever ignorant of those secrets," things must have happened as they did "by divine arrangement" (NA, 185). Of all the superior planets known at the time, Mars had the greatest eccentricity, its orbit departing markedly from the circular, and that is what made the discovery of elliptical orbits possible. The orbits of all the other planets except Mercury are so close to circular, while Mercury is so near the sun and observation of it so difficult, that had Kepler worked on any of those instead, he would probably not have been able to discover their elliptical nature, even with the superior precision of Tycho's data.

Tycho died in October 1601, just over a year after Kepler began assisting him. After various complications, Tycho's vast body of observational data was eventually entrusted to Kepler. At that time, Kepler was the only person who understood the true value of the data Tycho had compiled for over thirty years, or who had the ability to really make use of it. Thus, as Leibniz later remarked, Tycho's data appeared to drop into Kepler's hands by "divine providence."[35] Then, after several years of grappling with the data, particularly the almost twenty years' worth of information from positional observations of Mars, Kepler arrived at his first and second laws of planetary motion, namely, that planets move in ellipses with the sun at one focus, and that the radius vector (the vector from the sun to a planet) sweeps out equal areas during equal intervals of time. To qualify as a law of modern physics, a proposition must be based on precise quantitative observation and data of statistically assured accuracy and also be expressed in rigorously defined mathematical language. In that sense, Kepler's laws were the very first laws of modern physics in the field of astronomy.

Kepler's first law (elliptical orbit) was mentioned in the 1609 edition of *New Astronomy* only in regard to Mars, but in his subsequent works—*The Harmony of the World*, *Epitome of Copernican Astronomy*, and the 1627-published *Tabulae Rudolphinae* (Rudolphine tables)—it is applied to all planets. As for the second law (law of area), in *New Astronomy* Kepler confuses the correct expression $rv_{\perp} = \text{const.}$ (where $v_{\perp}$ is the component of the velocity vertical to the radius vector) with the approximation $rv = \text{const.}$, which is correct only at the aphelion and perihelion; but in *Epitome of Copernican Astronomy* he uses the correct expression $rv_{\perp} = \text{const.}$

Kepler's first and second laws dealt a deathblow to both the theory of circularity of planetary orbits and the notion of uniformity of planetary motion, and made the Platonic and Aristotelian views obsolete. If the planets had been found to move at constant speed in circular orbits, then such a system could be readily accepted by theological and aesthetic interpretations as reflecting the perfection of nature's creator, or else as indicating the natural motion of a perfect "fifth element." But such views could not explain a universe in which planets move elliptically and at varying speeds. The seemingly minor difference between the planets' orbits being circular and their being almost circular but slightly distorted—what we now call broken symmetry—is actually a crucial one. For Kepler, a Platonist convinced that nature "loves simplicity, she loves unity" and that "nothing ever exists in her which is useless or superfluous" (SU 1:77), it must have been an especially compelling problem. It immediately raised the question of the cause of such broken symmetry, and the explication of that cause was precisely a question of physics—above all, a question of force. In his subsequent *Epitome of Copernican Astronomy*, Kepler argues the point as follows:

> [The ancients] philosophized that of all movements which return into themselves the circular is the most simple and the most perfect and that something of straightness is mixed in with all the others, such as the oval and similar figures: accordingly this circular movement is most akin to the very simple nature of the bodies, to the motors, which are divine minds—for its beauty and perfection is somehow of the mind—and finally to the heavens, which have a spherical figure.
>
> *How must this be refuted?*
> To this I make answer as follows: Firstly, if the celestial movements were the work of mind, as the ancients believed, then the conclusion that the routes of the planets are perfectly circular would be plausible.... But the celestial movements are not the work of mind but of nature, that is, of the natural power of the bodies, or else a work of the soul acting uniformly in accordance with those bodily powers; and that is not proved by anything more validly than by the observation of the astronomers, who, after rightfully removing the deceptions of sight, find that the elliptical figure of revolution is left in the real and very true movement of the planet; and the ellipse bears witness to the natural bodily power and to the emanation and magnitude of its form. (ECA, 932)

Thus Kepler extricates himself from the spell of circularity and uniformity and thereby arrives at a conception of astronomy as a branch of physics built around the concept of force.

This shift is symbolically signified by the fact that, whereas Copernicus had located the center of each planet's orbit not in the sun itself but at the center of the Earth's orbit—the mean sun—Kepler locates it in the sun itself—the actual sun. With this approach, Kepler became the first to show that the planetary orbits move in fixed planes including the sun, a fact now sometimes referred to as Kepler's zeroth law.[36] And instead of ascertaining the sun's centrality *a posteriori* from observation, he makes it an *a priori* requirement. In his view, it is unthinkable, in terms of both physics and metaphysics, that the center of the universe could be a mere geometrical point in empty space like the supposed center of the Earth's orbit. Rather, he reasons, the center of the cosmos, as the physical center of motion, has to be a physical, material entity that is also the mother-source of light and power—in short, it has to be the sun. In that sense, it is fair to say that true heliocentrism began not with Copernicus but with Kepler. This step was also the beginning of celestial physics. Kepler himself states the point in his introduction to *New Astronomy*: "My first step in investigating the physical causes of the motions was to demonstrate that [the planes of] all the eccentrics intersect in no other place than the very centre of the solar body (not some nearby point), contrary to what Copernicus and Brahe thought" (NA, Introduction, 48; brackets in source).

The greatest impetus he received in fleshing out and realizing this theory came from Gilbert's discovery that the Earth is a magnet and exerts force—or rather, from Kepler's broad interpretation of that discovery. This can be seen in a number of works and letters Kepler wrote in the process leading up to *New Astronomy*.

In *Astronomiae pars optica* (The optical part of astronomy), which he wrote in 1604, Kepler states: "The sun is … a particular body, in it is this faculty of communicating itself to all things, which we call light; to which, on this account at least, is due the middle place in the whole world, and the center." And just before these words we find the following:

> [Corporeal bodies] were endowed with various powers, which, though they do have their nests in the bodies, nevertheless, being somewhat freer than the bodies themselves and lacking corporeal matter (though they do consist of their own kind of matter which is subject to geometrical dimensions), may proceed forth and might try to achieve an orb, as appears chiefly in the magnet, but appears plainly in many other instances.[37]

The conjecture that the powers "proceed forth and try to achieve an orb" refers to the law of efflux expressed above in equation 18.1. This represents force as emanating isotropically in a plane from a central body in the same way light does. The remark "as appears chiefly in the magnet" aptly evinces the fact that Kepler regarded magnetism as the prototype of such forces exerted by bodies.

Especially striking in this connection is the following passage from a letter Kepler wrote on February 10, 1605, by which time his studies on Mars had progressed to quite an advanced stage:

> I am much occupied with the investigation of the physical causes. My aim in this is to show that the celestial machine is to be likened not to a divine organism but rather to a clockwork ..., insofar as nearly all the manifold movements are carried out by means of a single, quite simple magnetic force, as in the case of a clockwork all motions [are caused] by a simple weight.[38]

Here Kepler momentarily anticipates the emergence of seventeenth-century mechanistic physics that, spearheaded by the likes of Rene

Descartes and Robert Boyle, posited a mechanical universe driven by motive force. But Kepler's position differed from that of the naive mechanists in one important respect: whereas mechanism regarded matter as inert and passive (as we will see in more detail in later), Kepler, as is clear from the abovequoted passage from *Astronomiae pars optica*, considered physical bodies, and particularly heavenly bodies, to be active entities with powers that they exerted on other bodies. And at that time, the only known example of such powers was magnetic force.

Soon thereafter, on March 5 that year, Kepler wrote in a letter to his former teacher Maestlin that "the body of the sun is a spherical magnet and rotates within its own space, sending forth into the sphere of its influence a power not attractive but driving"[39]. In his use of such terminology as "sphere of power" (*orbis virtutis*) we can clearly discern the influence of Della Porta and Gilbert. Gilbert's influence on Kepler was thus direct and pronounced.

In another letter, written to his friend Johann Brengger on November 30, 1607, Kepler is explicit about the magnitude of Gilbert's influence: "If I believe anything, you after reading my book [*New Astronomy*] will be persuaded that I have placed a celestial rooftop upon the magnetical philosophy of Gilbert, who himself has built the terrestrial foundation."[40] Even in the fourth book of *Epitome of Copernican Astronomy*, written years later, Kepler reflects: "I build my whole astronomy upon Copernicus' hypotheses concerning the world, upon the observations of Tycho Brahe, and lastly upon the Englishman, William Gilbert's philosophy of magnetism" (ECA, 850). Today the part played by the third inspiration mentioned here has been generally forgotten, or else disregarded, but to Kepler himself Gilbert's influence was not only strong but decisive.

In his 1657 inaugural lecture as professor of astronomy at Gresham College, Christopher Wren was not merely showing favoritism for a fellow Englishman when he referred to Gilbert as "the one I must reverence for giving Occasion to *Kepler* (as he himself confesses) of introducing Magneticks into the Motions of the Heavens, and consequently of building the elliptical Astronomy."[41] Kepler did indeed find the physical grounds for his theory of planetary motion in Gilbert's philosophy of magnetism.

## Section 6—Kepler's Dynamics

Gilbert's influence on Kepler is even more apparent in the main text of *New Astronomy*. This work opens with summaries of all of its chapters, and in the summary for chapter 32 Kepler provocatively exhorts "Physicists, prick up your ears!" because it is in that chapter that he embarks on his consideration of the physical causes of planetary motion. What he presents there is a unique dynamical model of the solar system in which, extrapolating from Gilbert's discovery about the Earth's being a spinning magnet, he asserts that the sun too is a giant magnet likewise rotating on its axis, and that the sun's magnetic power propels planets along their respective orbital paths.

Before considering that model, however, let us look at his dynamics, specifically what may be called "Kepler's equation of motion" and his related introduction of the concept of "inertia."

According to eighteenth-century mathematician Leonhard Euler, the term 'inertia'was coined by Kepler.[42] In his introduction to *New Astronomy* Kepler notes that "every corporeal substance, to the extent that it is corporeal, has been so made as to be suited to rest in every place in which it is put by itself, outside the sphere of influence of a kindred body" (NA, Introduction, 55); and an identical passage appears in a letter he wrote to Fabricius on October 11, 1605.[43] What Kepler calls "inertia" is thus limited to the inertia of bodies at rest, and to that extent he carries on a long-standing error. But the novelty in his theory is as follows: On the one hand, his conception of "every corporeal substance" endowed with this quality includes not only terrestrial bodies but also celestial bodies, that is, the planets. On the other hand, terrestrial heavy bodies high above the ground would be inclined to rest there if the Earth were removed. Whereas Aristotelian natural philosophy asserts that heavy bodies move spontaneously to the center of the world whether or not the Earth is at the center, Kepler's theory thus posits that a body will stay at rest in *any* place where it is unaffected by other bodies—that there is no special or original place for it to rest or to which it spontaneously strives to return. This view therefore inevitably entails the collapse of the Aristotelian theory of space that attributed special importance to the center of the universe. The planets are no longer viewed as perfect bodies composed of a mythical fifth element and moving by themselves in eternal orbits. Rather, "the nature of the planetary globes is material, from an inherent property, arising from the origin of things, to be inclined to rest or to the privation of motion" (NA 34:388); and "the body of a planet is inclined by nature to rest in every place where it is put by itself" (NA 39:407).

In book 4 of *Epitome of Copernican Astronomy*, published in 1620, Kepler maintains that "there is … a natural inertia in the planets themselves with respect to movement" (ECA, 899). And in a note in the second edition of *The Secret of the Universe*, published the following year, he remarks that the planets "possess the ability to resist a motion applied externally, in proportion to the bulk of the body and the density of its matter," and calls this ability "inertia of matter" (*inertia materiae*; SU 16:171, author's note 5). Thus, in addition to being the first to use the term "inertia," Kepler asserts that inertia is proportional to mass. To be sure, his concept of inertia was limited to bodies in a state of rest and was therefore erroneous by today's standards; but the fact that he introduced the concept and proposed that a body's inertia could be quantified in terms of its "mass" (i.e., density × volume) is nonetheless highly significant.

According to this theory, because astronomical bodies (planets) have inertia, some power must be at work upon them to make them move. Kepler thus posits the need for an unremitting force that sustains each planet's velocity. In part 2.3 of the fourth book of *Epitome of Copernican Astronomy*, he refers to this power that propels planets in their orbital directions as a "motor force" (*vis vectoria*) and "motor power" (*potentia vectoria*) and asserts that planetary motion is determined by competition between this motive force, which emanates from the sun, and the planets' inertial resistance. In the modern language of vectors we can represent this as $\boldsymbol{v} \propto \boldsymbol{F}/m$, which can therefore be expressed as:

$$m\boldsymbol{v} \propto \boldsymbol{F} \tag{18.6}$$

This is precisely Kepler's equation of motion, mentioned above. Kepler thus postulates that velocity is sustained by a force and is always oriented in the direction of that force. The correct equation is now known to be $m\boldsymbol{a} = \boldsymbol{F}$, where $\boldsymbol{a}$ denotes acceleration and is defined as $\boldsymbol{a} = d\boldsymbol{v}/dt$, implying that the planets move in a tangential direction by their inertia while being deviated toward the sun by the sun's gravitational force. But clarification of that would have to wait for Robert Hooke and Isaac Newton. In Kepler's thinking, his misunderstanding of inertia and the flaws in his laws of motion support one another. He adhered to these mistaken principles of dynamics for the rest of his life.

In chapter 32 of *New Astronomy*, Kepler shows that, at both the aphelion and perihelion of its orbit, a planet's velocity ($v$) is in inverse proportion to its distance ($r$) from the sun ($v \propto 1/r$). These were special instances of the law of area (his second law of planetary motion) in that it is only at the aphelion and perihelion that the expression is strictly correct. Supposing, however, that the expression is always valid, and applying it to equation 18.6 above, Kepler finds, at the beginning of chapter 33, that $F$, the force exerted on a planet, is indeed in inverse proportion to its distance from the sun. To Kepler, this means that his "law of efflux" (equation 18.1) has been directly confirmed from observational data—though of course his reasoning is premised on his erroneous equation of motion.

Furthermore, Kepler is certain, in terms of both physics and metaphysics, that a mere geometric point in empty space cannot be a source of force. For him, the source therefore has to be at one or the other extremity of the distance ($r$), or in others words in either the planet or the sun; and by way of an abstruse argument he concludes that the source of force is indeed in the sun. But while this is an *a posteriori* argument from observation, he notes that the same conclusion can be arrived at with an *a priori* argument from conventional notions of the sun's "worthiness and eminence." Presumably, Kepler in fact believed from the outset that the source of force lay in the sun, and presented his tedious, roundabout proof simply to convince readers to share his conviction. In any case, in chapter 33 he thus identifies the sun as the origin of the activity of the whole solar system: "The source of the world's life (which is visible in the motion of the heavens) is the same as the source of the light which forms the adornment of the entire machine, and which is also the source of the heat by which everything grows" (NA 33:379). He goes on to conclude as follows:

> Just as light, which lights the whole earth, is an immaterial *species* of that fire which is in the body of the sun, so this power which enfolds and bears the bodies of the planets, is an immaterial *species* residing in the sun itself, which is of inestimable strength, seeing that it is the primary agent of every motion in the universe. (NA 33:381–82)

He refers here to an "immaterial species" (*species immateriata*) and in chapter 34 to a "motive species" (*species motrix*) and "motive power" (*vis motrix*)—expressions reminiscent of Roger Bacon's terminology. In any case, by thus proposing that the planets are propelled by this force propagated from the sun, Kepler transfers the role of the cosmos's "first mover" from an invisible divine being existing outside the universe to the hotly radiant corporeal sun at the center of it. This step marked the beginning of true heliocentrism.

In Kepler's equation of motion, because velocity is considered always to be oriented in the direction of force, the "power which enfolds and bears the bodies of the planets" and which the sun exerts upon them must be oriented tangentially to their orbits. This is what he means by the phrase "a power not attractive but driving" in the abovequoted letter to Maestlin of March 5, 1605. Accordingly, if a planet's orbit were circular, this force would be perpendicular to its radius vector. This perhaps explains Kepler's strange choice of words in chapter 33 of *New Astronomy* where he remarks that "the planets are moved rather in the manner of the steelyard or lever" (NA 33:378). The immaterial *species* (motive power) propagated from the sun is like a rod extending from the sun to the planet and acting like a lever to propel the planet along its orbit (in a direction perpendicular to the rod). But if the planets move in their orbits in that way, then these rods of force must also rotate, which means that the sun itself must spin on its own axis. Reasoning in this way, in chapter 34 Kepler draws precisely that inference:

> Since the *species* is moved in a circular course, in order thereby to confer motion upon the planets, the body of the sun, or source, must move with it, not, of course, from space to space in the world—for I have said, with Copernicus, that the body of the sun remains in the centre of the world—but upon its centre or axis, both immobile, its parts moving from place to place, while the whole body remains in the same place. (NA 34:386)

With these remarks in *New Astronomy*, Kepler anticipated the discovery of the sun's rotation years before Galileo confirmed it by observing sun spots.

In his introduction to *New Astronomy*, Kepler gives the following striking description of the mechanism of the propulsion of planets based on this inference of a spinning sun:

> The sun, although it stays in one place, rotates as if on a lathe, and out of itself sends into the space of the world an immaterial *species* of its body, analogous to the immaterial *species* of its light. This *species* itself, as a consequence of the rotation of the solar body, also rotates like a very rapid whirlpool throughout the whole breadth of the world, and carries the bodies of the planets along with itself in a gyre, its grasp stronger or weaker according to the greater density or rarity it acquires through the law governing its diffusion [i.e., the law of efflux]. (NA, Introduction, 67)

Leibniz believed that Descartes's vortex theory for the motions of the solar system was plagiarized from this "whirlpool" idea of Kepler's.[44] It is not inconceivable that Descartes stole the idea, but more important for our present purposes is how this model ties in with Kepler's own understanding of the solar system.

## Section 7—Astronomical Bodies as Magnets

From the hypothesis of a spinning sun, in the latter half of chapter 34 of *New Astronomy* Kepler argues further that, like the Earth, the sun too is a magnet. He frames the question by supposing that "one might inquire of me, what sort of body I consider the sun to be, from which this motive species descends" (NA 34:390), then proceeds to answer it by pointing out three aspects of the magnet to which he suggests the sun is similar.

First, the magnet's power inheres in its entire body and increases as the magnet's mass increases. Second, the magnet's power to attract iron spreads out in an orb from the magnet, and iron within that sphere is attracted more strongly the nearer it is to the magnet. And third, rather than attracting iron with all of its parts, the magnet has along its length "filaments" or "fibers" by which, if a small piece of iron is placed midway between its poles, instead of attracting the iron the magnet only turns it parallel to its own "fibers." It is not difficult to see Gilbert's influence in this three-point description of the characteristics of magnets.

To each of these he finds a corresponding characteristic in the sun. First, "in the sun the moving power appears so much stronger that it seems likely that its body is of all the most dense." Second, "the power moving the planets is propagated from the sun in an orb, and is weaker in the more remote parts of the orb." And third, “there is in the sun no force whatever attracting the planets, as there is in the magnet … but only a directing force” (NA 34:390). The gist of Kepler’s argument here is that the similarities between the peculiar properties of magnetic force and the sun’s motive power indicate that the sun itself is a gigantic magnet.

In Kepler’s thinking, this conclusion also gains strong support from the fact that the Earth—which he views as causing the moon to revolve around it just as he imagines the sun propels the planets in their orbits—is itself a magnet:

> By the demonstration of the Englishman William Gilbert, the earth itself is a big magnet, and is said by the same author, a defender of Copernicus, to rotate once a day, just as I conjecture about the sun. (NA 34:390–91)

He concludes that

> it is therefore plausible, since the earth moves the moon through its *species* and is a magnetic body, while the sun moves the planets similarly through an emitted *species*, that the sun is likewise a magnetic body. (NA 34:391)

This means that the Earth and the moon comprise a kind of miniature solar system, and that the relationship between them is analogous to that between the sun and the planets. Such is Kepler’s argument in chapter 34 of *New Astronomy*, where he applies the following calculations. According to him, whereas

$$\text{Earth's radius} \cong \text{moon's orbital radius} \div 60,$$

“the moon is carried along by the rotation of the earth and the motion of its magnetic power, only thirty times slower” (NA 34:391).

That is, the time the moon requires to orbit once around the Earth (approximately thirty days) is roughly thirty times the duration of one rotation of the Earth (one day). Hence,

$$\text{Earth's rotation period} \cong \text{moon's orbital period} \div 30.$$

Meanwhile, assuming the corresponding relation between the sun and the first planet, Mercury, is similarly

$$\text{sun's radius} \cong \text{Mercury's orbital radius} \div 60,$$

then, just as for the Earth and the moon, it should be the case that

$$\begin{aligned}\text{sun's rotation period} &\cong \text{Mercury's orbital period} \div 30\\ &= 88 \text{ days} \div 30 \cong 3 \text{ days}.\end{aligned}$$

Kepler thus reckons the rotation period of the sun to be roughly three days. [Note that we now know the actual ratio of Mercury's average orbital radius ($579 \times 10^5$ km) to the sun's radius ($6.96 \times 10^5$ km) to be $579 \div 6.96 \cong 83$.]

The basis for these odd calculations is his assumption that, since the planets and the moon are driven along in their orbits by the same mechanism of rotation of the central body propelling them (i.e., the sun and the Earth, respectively), for all of them the relationship between orbital period and the rotation period of either the sun or the Earth (as the case may be) must be the same. This is presumably why Kepler was especially delighted by Galileo's telescopic discoveries soon thereafter of the sun's rotation and of the moons of Jupiter. In particular, upon finding that his third law of planetary motion applied equally to the four Jovial moons (satellites) that Galileo discovered, Kepler interpreted their discovery as unequivocal proof that his model of planetary revolution due to the sun's rotation could be applied without alteration to planet-satellite relationships as well. In book 4 of *Epitome of Copernican Astronomy* he notes that his third law, in the form expressed in equation 18.5 above, holds as well for the four satellites of Jupiter, and from this he draws the following general conclusion:

> Since the agreement [i.e., in terms of Kepler's third law] of the Jovial satellites with the six primary planets is so exact, not only did we rightly infer above from this that the body of Jupiter turns around its axis like the sun, so that the proportion holds for all its members; but here already, over and above that, we are confirming not improperly the general statement that this rotation of the primary bodies around their axes is the cause of the circuit of the secondary bodies around the primary bodies. (ECA, 919)

Today's understanding of the dynamics of lunar and planetary orbital motion is of course different from Kepler's, and admits no dynamical correlation between the sun's rotation and the planets' revolutions. And needless to say, Kepler's reckoning of the sun's rotation period to be three days is meaningless. But even though his argument about the rotation of primary bodies and the revolutions of other celestial bodies around them is obsolete, it nonetheless clearly pointed to the dynamical similarity between the sun-planet relationship and the planet-satellite relationship. It therefore lent strong impetus to the idea that the force exerted by the sun upon the six known planets and that exerted by the Earth on the moon and by the other planets on their satellites were all of a kind. Indeed, this line of thinking was crucially important to the establishment of Newton's theory of universal gravitation half a century later.

Assuming, as Kepler argues, that the sun is a magnet, then if the sun exerts its magnetic power on the planets, it follows that the planets must also be magnets. In chapter 57 of *New Astronomy*, Kepler himself posits that the planets are magnets as he attempts to explain their variations in distance from the sun:

> What if all the bodies of the planets are enormous round magnets? Of the earth (one of the planets, for Copernicus) there is no doubt. William Gilbert has proved it. (NA 57:550)

It is from this point that Kepler develops his unique theories of magnetic gravitation and planetary magnetism.

Note that the driving force ("moving power") Kepler describes as exerted by the sun upon the planets is a force that propels them tangentially to their orbits; he himself grants that it is not a force that pulls the planets toward the sun. In fact, however, each planet gets further from the sun as it moves from perihelion to aphelion and closer to the sun on its return journey—which naturally raises the question of the physical cause of this variation in distance between the planet and the sun. Kepler attempts to explain this phenomenon by the fact that each planet is itself a magnet with two poles. His idea is represented in Figure 18.2, an illustration that appeared in book 4 of *Epitome of Copernican Astronomy*, published years after *New Astronomy*. According to this model, the planet-magnet has two poles, one of which (at the arrow's tip) is subject to the sun's attractive force and the other of which is subject to the sun's repulsive force. In the figure, the planet moves in a counterclockwise orbit. At its aphelion (*A*), the poles are equidistant from the sun, but between there and *B* the planet's sun-attracted pole starts to face toward the sun, with the result that

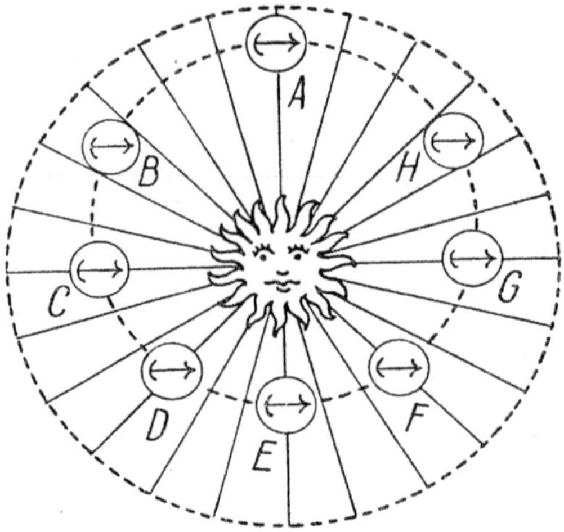


Figure 18.2: Kepler's conception of motion around the sun, controlled by magnetic interaction. *Source*: Kepler, *Epitome astronomiae Copernicanae.*

the planet draws nearer the sun as it travels from *B* to *C* to *D*. In that way it reaches its perihelion (*E*), where its poles once again become equidistant from the sun. From that point forward the sun-repelled pole faces toward the sun, so now the planet moves away from the sun through points *F*, *G*, and *H*, until returning to its aphelion at *A*.

In this model, the sun would have to be a magnetic monopole, because it constantly attracts one pole of the planet-magnet while constantly repelling the other. Kepler appears to have noticed this difficulty, and by way of explanation he postulates in the *Epitome* that

> the centre of the solar body corresponds to one extremity or region of a loadstone, but the whole surface to the other region of the loadstone. (ECA, 898)

No such magnet exists, however, and the argument is simply too ad hoc. The theory cannot apply to the Earth's case, for a start, because the Earth-magnet's direction (the direction of the arrows in the figure) is not in line with its orbital plane. But although Kepler's argument was thus rather far-fetched, his entire magnetic model was only hypothetical in the first place, so there is little point in singling out that flaw. For our present purposes, it is more important simply to understand the great impact on Kepler's thinking of Gilbert's magnetic philosophy, principally the claims that the Earth is a magnet and that its magnetism is the cause of its motion. It is worth noting, incidentally, that in the latter half of the seventeenth century the English natural philosopher Robert Boyle and the English astronomer John Flamsteed also came to regard the sun as a magnet.

## Section 8—Kepler's Theory of Gravity

Even at the time of *The Secret of the Universe* (1596), Kepler had already supposed that the sun's "moving soul" or motive power decreased with distance, and he seems to have formulated a notion close to universal gravitation at quite an early stage. In a letter dated March 28, 1605, when his long study of Mars was drawing to an end, he postulates that

> if we were to place the earth at rest in some place and bring near to it a larger earth, the first would become a heavy body [*gravis*] in relation to the second and would be attracted by the latter.[45]

From his use of the passive voice in the verb *attraheretur* ("be attracted") and his comment later in the letter that "weight is a passive action," Kepler clearly indicates his understanding that heavy bodies do not move toward the center of attraction of their own accord but rather are pulled by the central body. We also find the following in his letter to Fabricius of October 11 the same year:

> Suppose that a stone of a magnitude having some perceptible ratio to the size of the earth were placed behind the earth. Let it be the case that both are exempt from all other motions. Then I say that not only will the stone move toward the earth, but also the earth will move toward the stone. They will divide the intervening space in the inverse ratio of their weights. Then $C$ being the place where they will meet, as $A$ [the earth] is to $B$ [the stone] in size, so [the distance] $\overline{BC}$ will be to [the distance] $\overline{AC}$, in exactly the same ratio as is utilized in a balance with unequal arms.[46]

This passage is the earliest known expression of the fact that gravity is reciprocal—that in addition to the stone's being attracted by the Earth, the Earth would likewise be attracted by the stone. That is, it is the first clear statement of the law of action and reaction concerning gravity as a force acting at a distance. If we ignore the fact that Kepler confuses mass with weight, this also qualifies as the first explicit expression of the fact that displacement due to the effect of force is not only proportional to force but also inversely proportional to mass.

In somewhat more modern terms the idea can be expressed as follows. If we represent infinitesimal displacement during the infinitesimal time ($\Delta t$) as $v\Delta t = \Delta s$, then from equation 18.6 we get

$$m\Delta s = mv\Delta t \propto F\Delta t.^{\ddagger} \qquad (18.7)$$

‡ The correct equation of motion, $ma = F$, yields $m\Delta s \propto F(\Delta t)^2$, but since variation $\Delta s$ is proportional to $F$ while inversely proportional to mass $m$, the remainder of argument holds in the same way.

If we then express the total displacement of $A$ (mass $m_A$) and $B$ (mass $m_B$) respectively as $\overline{AC} = \sum \Delta s_A = \sum v_A \Delta t$ and $\overline{BC} = \sum \Delta s_B = \sum v_B \Delta t$, where the summation denoted by $\Sigma$ means the sum over all the time intervals, then, since the magnitudes of the force of mutual attraction between are equal due to the law of action and reaction, we get and hence

$$\overline{AC} : \overline{BC} = \sum \Delta s_A : \sum \Delta s_B = m_B : m_A. \tag{18.8}$$

This is the meaning of the final sentence of the passage from Kepler's letter just quoted.

Kepler elaborated his theory of gravity in *New Astronomy*, in *Epitome of Copernican Astronomy*, and in the posthumously published *Somnium* (Dream). Let us look first at how he treats the topic in his introduction to *New Astronomy*:

> Gravity is a mutual corporeal disposition among kindred bodies to unite or join together; thus, the earth attracts a stone much more than the stone seeks the earth....
>
> Heavy bodies (most of all if we establish the earth in the centre of the world) are not drawn towards the centre of the world *qua* centre of the world, but *qua* centre of a kindred spherical body, namely, the earth. Consequently, wherever the earth be established, or whithersoever it be carried by its animate faculty, heavy bodies are drawn towards it.
>
> If the earth were not round, heavy bodies would not everywhere be drawn in straight lines towards the middle point of the earth, but would be drawn towards different points from different sides.

> If two stones were set near one another in some place in the world outside the sphere of influence of a third kindred body, these stones, like two magnetic bodies, would come together in an intermediate place, each approaching the other by an interval proportional to the bulk [*moles*] of the other.
>
> If the moon and the earth were not each held back in its own circuit by an animate force or something else equivalent to it, the earth would ascend towards the moon by one fifty-fourth part of the interval, and the moon would descend towards the earth about fifty-three parts of the interval, and there they would be joined together; provided, that is, that the substance of each is of the same density. (NA, Introduction, 55; brackets in source)

While essentially identical to what he wrote in the abovementioned letter to Fabricius, this is the first mention of such mutual attraction not only between the Earth and a stone but also between the Earth and the moon and even between two stones. It clearly anticipates Newton's representation of gravity as "universal gravitation."

A couple of pages later Kepler adds the following comments on "levity":

> Nothing that consists of corporeal material is absolutely light. It is only comparatively lighter, because it is less dense, either by its own nature or through an influx of heat. By "less dense" I do not just mean that which is porous and divided into many cavities, but in general that which, while occupying a place of the same magnitude as that occupied by some heavier body, contains a lesser quantity of corporeal material.
>
> The motion of light things also follows from their definition. For it should not be thought that they flee all the way to the surface

> of the world when they are carried upwards, or that they are not attracted by the earth. Rather, they are less attracted than heavy bodies, and are thus displaced by heavy bodies, whereupon they come to rest and are kept in their place by the earth. (NA, Introduction, 57–58)

This view marked a major milestone in discourse on gravity. Kepler's assertion that the difference between "gravity" and "levity" was nothing more than a relative difference in the strength of the Earth's attractive force completely dispelled the Aristotelian theory that light bodies move away from the center of the universe by a natural motion. At the same time, it also exploded the Aristotelian theory of space in which the distinction between the universe's center and its periphery was regarded as an ontological absolute. The implication is that Kepler succeeded in transcending Aristotelian qualitative physics, which attributed to "heavy" and "light" bodies opposing natures that precluded any gradual transition from one to the other, and in recasting the issue according to concepts of modern quantitative physics that unified "gravity" and "levity" in terms of greater or lesser "weight."

*Somnium* is a peculiar work, part fantasy about a visit to the moon and part scientific description of what the universe would look like from the moon. Appended at the back of the book are detailed notes by Kepler himself, written over a span of more than thirty years and holding great scientific interest for their elucidation of his astronomical and physical thinking.[47] In one he likens gravity to magnetic force:

> I define "gravity" as a force of mutual attraction, similar to magnetic attraction. But the power of this attraction in bodies near to each other is greater than it is in bodies far away from each other. (N. 66)

Moreover, he states, because "the theory of magnetism" holds that "the moon is a body akin to the earth" (n. 62)—that is, both are magnets—a person en route from the Earth to the moon would be subject to the magnetic gravitation of both. Indeed, Kepler uses the following expressions in describing the varying states of a resultant force: "when the magnetic forces of earth and moon are canceled out by their opposing attractions, it is as though nothing pulled the [intervening] body in any direction" (n. 75); "when the body has been carried so far beyond the region of the earth's magnetic force that the lunar globe's magnetic force is already preponderant" (n. 74); and "the region of the moon's magnetic force is nearby and therefore predominant" (n. 77). Thus he already regards gravity and magnetic force in entirely the same light.

Particularly important to the theory of gravity itself is the following:

> A body located between the two globes [i.e., the moon and the Earth] at the point where its distance from each is in the same ratio as are the bodies to each other will remain motionless, since the pulls in opposite directions eliminate each other. This will happen if the body's distance from the earth is 58 1/59 earth-radii, and from the moon 58/59 earth-radii. But when the body moves a little closer to the moon, it will be subject to the moon's attraction, since the moon's force is dominant on account of its nearness. (N. 77)[§]

This is the first clear statement of the principle that gravity is not only inversely proportional to distance but also proportional to

---

[§] The grounds for the numerical ratio 58 1/59 : 58/59 can be surmised as follows. Kepler posits the distance between the moon and the Earth as $59R$, where $R$ is the Earth's radius. Since the apparent diameter of the moon is 0.5°, the moon's radius is derived as $R' = 59\text{R} \times (0.5/2) \times \pi/180 = 0.258R$. Supposing the density of the moon and that of the Earth to be the same, the ratio of their mass is $m' : m = R'^3 : R^3 \cong 1 : 58$.

the mass of the attracting bodies. Thus, in conjunction with equation 18.1, we can represent the force of gravity between a body with mass $m_1$ and another with mass $m_2$ at distance $r$ as

$$F \propto \frac{m_1 m_2}{r} \tag{18.9}$$

Strictly speaking, gravitational force should be described not as inversely proportional to distance but as inversely proportional to the square of the distance ($F \propto m_1 m_2 / r^2$); but such inaccuracy is a rather trivial defect compared with the basic discovery that gravitational force between two bodies can thus be represented in a mathematical function as proportional to mass and as decreasing with distance. Though the outcome was flawed, the idea itself was ahead of its time. Indeed, Kepler's presentiment clearly anticipated the thinking behind Newtonian and later physics that came to express force as a mathematical function.

Also notable is the fact that in his account of gravity Kepler has by this point abandoned the proviso "between kindred bodies" (*inter cognata corporis*), which had appeared in *New Astronomy* and *Epitome of Copernican Astronomy.* What he refers to in *Somnium* is the attractive force not between the moon and the Earth but rather between either of them and a *person.* Furthermore, whereas in the introduction to *New Astronomy* he had emphasized the moon's influence on the tides—"The sphere of influence of the attractive power in the moon is extended all the way to the earth, and in the torrid zone calls the waters forth" (NA, Introduction, 56)—in his notes to *Somnium* he takes the idea further:

> The causes of the ocean tides seem to be the bodies of the sun and moon attracting the ocean waters by a certain force similar to magnetism. (N. 202)

Kepler thus supposed that not only the moon but the sun also exerted a force upon terrestrial bodies (including ocean waters), and this line of thinking brought him one step closer to the concept of universal gravitation.

With this we arrive at the threshold of the dynamical explication of the solar system based on the laws of motion advanced by Galileo and Descartes and on the Newtonian theory of universal gravitation to which those laws gave rise.

* * *

Although Kepler sowed the seeds of the concept of universal gravitation, the premises behind his thinking were not entirely modern. In book 4 of *Epitome of Copernican Astronomy* he offers the following account of the sun's role as a source of power:

> The function of the sun in the world seems to persuade us of nothing else except that just as it has to illuminate all things, so it is possessed of light in its body; and as it has to make all things warm, it is possessed of heat; as it has to make all things live, of a bodily life; and as it has to move all things, it itself is the beginning of the movement; and so it has a soul in itself. (ECA, 897)

It must be admitted that, even while on the one hand attaining a quite modern view of force and motion, on the other Kepler failed to make a complete break from the hylozoistic view of the sun and other celestial bodies.

It is also known that Kepler practiced astrology. This was at least partly to make a living, but it is still not clear what he really thought of the practice. While on the one hand critical of the kind of irresponsible astrology that was popular around town, on the other he warned his fellow intellectuals that "while justly rejecting the stargazers' superstitions, they should not throw out the child with the bathwater."[48] Ultimately, it seems, he accepted some part of astrology, with certain reservations. In book 4 of *The Harmony of the World* he rejects superstitious astrology as dubious but affirms what he regards as astrology proven empirically. The latter kind refers to his "theory of aspects," an "aspect" being "an angle formed at the earth by the luminous rays of two planets which is effective in stimulating a sublunar being." At the very least, Kepler allowed that the disposition of celestial bodies exerted some kind of influence on terrestrial bodies, and he maintained that stance throughout his entire life.[49]

In *The Mathematicall Praeface*, John Dee describes "Astronomie" as "an Arte Mathematicall, which demonstrateth the distance, magnitudes, and all naturall motions, apparences, and passions propre to the Planets and fixed Sterres: for any time past, present and to come"; in other words, a merely descriptive science of the heavens that does not inquire into the physical causes of the motions of celestial bodies. He contrasts this with "Astrologie," which he describes as "an Arte Mathematicall, which reasonable demonstrateth the operations and effectes, of the naturall beames, of light, and secrete influence: of the Sterres and Planets: in euery element and elementall body: at all times."[50] Correlating this distinction to Kepler's distinction between astronomy prior to Copernicus and his own "celestial physics," we can see that his "celestial physics" was actually closer to Dee's idea of astrology, and that in that sense the astrological view of nature was one of the sources of the concept of gravitational force between astronomical bodies.[51]

Among examples of celestial influence, along with the moon's effect on sea and river tides Dee also cites "Force, as they see in a litle peece of a *Magnes stone:* which, at great distance, sheweth his operation."[52] Here too, then, magnetism provides the model for action at a distance between celestial bodies. Moreover, if instead of "secrete influence" we call that action "gravity as action at a distance," we see how close Dee's "astrology" came to Newton's theory of tides.[53] A further break with previous thought was the move to quantify gravity and comprehend it by mathematical concepts.

For Kepler, too, who was already under the influence of Gilbert's magnetic philosophy, magnetism was the model for force acting over distance between planets. Modern mechanics and modern astronomy began when Kepler conceived of the force exerted between celestial bodies (gravity) as something akin to magnetism. Unlike in previous astronomy, which had sought to determine the orbits and positions of the individual planets separately, his starting point was to regard the entire solar system not as a miscellany of planets but as a single, harmonious dynamical order and to elucidate the underpinnings of that order—in other words, to interpret in physical terms the mathematical, heliocentric solar system of Copernicus. He sustained that approach consistently from the Platonic solids theory of his debut work, *The Secret of the Universe*, right up to the establishment of his third law of planetary motion in *The Harmony of the World*, written toward the end of his life. The result was the idea of a force that acts between astronomical bodies and can be expressed as a single mathematical function. This was something that could not be found in the philosophies of either Aristotle or Plato. If one had to identify its precursors, they would be the "hidden powers" speculated upon in natural magic and astrology and epitomized by magnetism.

I have noted that by inquiring into the physical causes of the motions of heavenly bodies Kepler revolutionized astronomy, but that also implied a revolution in physics itself. When he remarks in his introduction to *New Astronomy* that "I have mingled celestial physics with astronomy in this work" (NA, Introduction, 47), we can infer that until then physics (natural philosophy) and astronomy were regarded as separate. Previously, astronomy, though geometrical and quantitative, did not involve ontology and therefore did not inquire into causes; while natural philosophy, which did seek causal explanations, was nonetheless a qualitative discipline that eschewed quantitative understanding and mathematical representation.[54] Kepler made it possible to combine astronomy and physics, on the one hand by infusing astronomy with an efficient cause (the concept of force) and causal understanding, and on the other by understanding gravity—something previously bundled together with "hidden powers"—as a mathematical function and thus preparing the ground for physics (natural philosophy), too, to become a quantitative discipline.

In terms of Dee's distinction, the essence of the pivotal shift Kepler made was his reinterpretation of "Astronomie" and "Astrologie" by incorporating both. If so, then rather than considering astronomy to be the child of astrology, it is perhaps more accurate to say that astrology's "secrete influences" and magic's "hidden powers" were reincarnated in the form of physical force represented as a mathematical function; and that what until then had been astronomy as geometry and astrology as a study of force (influence) were both integrated and sublated into a dynamical astronomy (celestial mechanics). This achievement then gave rise, via refinement in Newton's more correct laws of motion, to modern physics. In that sense, it was the point where the history of the physics of magnetism and gravity finally began.

Chapter 19

# Seventeenth-century Mechanism and Notions of Force

## Section 1—The Features of Mechanism

From the late sixteenth to the early seventeenth century, as technical advances and the global expansion of European activity brought to light more and more previously unknown aspects of the natural world, the Aristotelian and Ptolemaic conceptions of the universe started to come apart at the seams in places. Their failure and insufficiency were confirmed by new phenomena such as the appearance of a series of comets and nova (one nova in 1572 and another in 1604). With the broadening of knowledge about the Earth, numerous facts were ascertained about magnets and the Earth's magnetic field; and, as exemplified by Galileo's telescopic discovery of Jupiter's satellites and observation of the lunar surface early in the seventeenth century, newly developed observational equipment opened up to human scrutiny previously inaccessible aspects of nature. The fact that Tycho Brahe's precise astronomical observations undermined existing theories also held great significance.

The mounting impetus of those various findings forced a radical overhaul of conventional views of and research approaches toward the natural world. It was the pressure of those newfound realities that engendered such developments as Gilbert's magnetic philosophy and Kepler's theory of planetary motion. But theory failed to keep pace with factual discoveries. Gilbert was basically still under the thrall of Aristotelianism, Kepler was still working from a neo-Platonic basis, and both still carried the baggage of hylozoism. This is why they were unable to fully express their own discoveries in a language truly appropriate to those discoveries.

It was under such circumstances that a movement arose throughout seventeenth-century Europe to create a new philosophy and new paradigm of learning that would supersede the overwrought, unrealistic "proofs" of Scholasticism, the mystical teachings of neo-Platonism, and the subjective, expedient arguments of magical thought. In that century, natural philosophy was forced to undergo major renovation not only in terms of its specific content but also in the methods and fundamental thinking behind it.

The two most celebrated figures in this connection are usually said to be Galileo Galileo (1564–1642) of Italy and Rene Descartes (1596–1650) of France. Galileo has been called "the world's first scientist in the modern sense" and Descartes "the first modern critical thinker."[1] I take these comments from books that just happen to be at hand, but they are typical of the evaluations of Galileo and Descartes given in most histories of science or philosophy. But while it is true that in the field of mechanics both men played major parts in founding the laws of motion (the principles of mechanics), when it came to cosmology and astronomy neither was able to grasp the significance of Kepler's three laws of planetary motion, and both thus let slip the chance to discover the key to modern cosmology, universal gravitation. Galileo had far too rationalistic a spirit to accept the idea of force acting at a distance in the form of gravity; and Descartes, perhaps because he likewise placed too high a value on human reason, viewed the natural world too simplistically.

The views of nature that these thinkers expounded are regarded as types of mechanism. Mechanism is a form of reductionism that regards material bodies solely in terms of objective aspects, such as their geometric shape and size and their motion, arrangement, and number, and that regards the various sense-perceptible attributes that bodies exhibit as effects to be explained in those more fundamental, objective terms. In his 1623 treatise *The Assayer*, Galileo asserts that "to excite in us tastes, odors, and sounds I believe that nothing is required in external bodies except shapes, numbers, and slow or rapid movements [of their constituent parts]."[2] In much the same manner, Descartes maintains in *Principles of Philosophy* that whereas "diverse sensations, namely those which we call the sensations of taste, of odor, of sound, of heat, of cold, of light, of colors, and of similar things … represent nothing situated outside thought," "sizes, figures, motions, and such" are presented to the mind as "certain things or modes of things, existing or at least capable of existing, outside thought." This line of thinking leads Descartes to conclude that "the nature of matter, or of body considered in general, does not consist in the fact that it is hard, heavy, colored, or affects the senses in any other way; but only in the fact that it is a thing possessing extension in length, breadth, and depth."[3]

Mechanism thus holds that not only the tactile sensations associated with a material body, such as texture, heat, and cold, but even such qualities as its luster, color, odor, and taste are nothing more than products of the stimulation of human sense organs by the shapes or motions of the perceived body's constituents, which do not themselves possess any such attributes; and that therefore such qualities should be accounted for solely in terms of those shapes and motions. Atomism differs from Descartes' theory in that the former grants the existence of a vacuum, but in this understanding of the attributes of material bodies the two views are identical. This is evident in atomist Pierre Gassendi's observation that "taste does not seem to consist in anything else than in corpuscles of such a configuration that by penetrating the tongue or palate they affect the contexture of this organ and set it in motion in a manner that gives rise to the sensation we call taste."[4]

At the heart of such thinking is the view that physical matter is essentially inert and passive. Hence, according to mechanism, physical bodies cannot act upon other physical bodies except by the direct contact of collision or pressure. One recent book on the history of atomism characterizes mechanism as a position involving a threefold denial, namely, of (a) action at a distance, (b) the spontaneous initiation of motion, and (c) incorporeal causal agents capable of moving bodies.[5]

Mechanism was thus fundamentally incompatible with the whole range of views of nature proposed up to that point, from Scholasticism, which understood the distinctive attributes and effects of matter in terms of "substantial form," to magical thought, which posited the existence of specific relations of "sympathy" and "antipathy" between natural bodies and explained away such imperceptible qualities as magnetism with the notion of "occult properties." Mechanism's stance in regard to those established views of nature, and its sense of superiority over them, are aptly expressed in Galileo's scoffing remark that "'sympathy,' 'antipathy,' 'occult properties,' 'influences,' and other terms [are] employed … as a cloak for the correct reply, which would be: 'I do not know'";[6] and also in Descartes' assertion that "there are, in rocks or plants, no forces so secret, no marvels of sympathy or antipathy so astounding."[7]

The same arguments were extended to the concept of gravity. From the standpoint of mechanism, describing the weight of terrestrial bodies in terms of the part's "sympathy" for the whole, or accounting for the gravity at work between astronomical bodies by saying that all matter has the property of acting across intervening space, were typical cases of explaining something one does not understand with something else one does not understand, and thus amounted to abandoning the very attempt at explication.

## Section 2—Galileo and Gravity

Galileo's perspective on gravity between astronomical bodies is most apparent in his discussion of tides. As we saw in the previous chapter, Kepler sought the cause of tides in the force of gravity exerted upon the ocean's surface by the moon and the sun. As is clear from its mention in Pliny's *Natural History*, the correlation between the tides and the moon had been well known since antiquity. In more modern times, the English thinkers William Gilbert and Francis Bacon both addressed the topic of the tides similarly; Italy's Giambattista della Porta asserted in *Natural Magick* that the only conceivable cause of tidal ebb and flow was the moon;[8] and in 1608 the Flemish engineer Simon Stevin developed a theory of tides caused by the moon's "attraction," declaring that "it is found by daily continual experiences that ebb and flow are governed by the Moon."[9] Nor was the idea limited to scholars. Dante's Divine Comedy, written in the early fourteenth century, and John Gower's late-fourteenth-century poem *Confessio amantis* (Lover's confession) both contain lines attributing the tide's ebb and flow to the moon.[10] And in *The Winter's Tale*, Galileo's contemporary Shakespeare has one of his characters express the idea of futile effort by saying that one might as well "Forbid the sea for to obey the moon."[11] By the early seventeenth century, the moon's influence over the tides was widely acknowledged as an indubitable fact.

Galileo, however, dared to reject that common view. In the "Fourth Day" discussion, the last part of his famous *Dialogue concerning the Two Chief World Systems: Ptolemaic and Copernican* (the *Dialogue*), Galileo's persona states that "the natural cause of the tides is the motion of the earth."[12] Galileo in fact advanced a unique theory of the tides as the combined effects of the Earth's diurnal axial rotation and its annual orbital motion. According to this theory, the rotational velocity of the Earth's surface varies depending on whether the place in question is moving in the same direction as the Earth's orbital motion or in the opposite direction, and the ocean's surface rises and falls because the ocean cannot keep up with these changes in the Earth's surface speed. In other words, the tides are the sloshing up and down of the ocean's waters as their vast container rocks to and fro. Galileo took this phenomenon to be a direct indication of the existence of the Earth's axial rotation and orbital motion.

As indicated in its full title, the *Dialogue*, which appeared in 1632, is a book of disputation aimed at determining the relative merits of the geocentric and heliocentric models of the universe, and in it the phenomenon of tidal motion occupies a pivotal place. Indeed,

Galileo at first referred to the work as "my dialogues on the tides."[13] In his view, whereas the various phenomena previously cited as supporting heliocentrism could actually be explained by either a heliocentric or a geocentric model, and in that sense suggested that the former was only probably true, the tides could be explained by the heliocentric theory alone, and therefore showed that the heliocentric model had to be true. He saved this argument for the last day of the book's four-day discussion because he regarded it as the clincher for the heliocentric position. In fact, however, Galileo's model had the fatal flaw of not being able to account for the semidiurnal period of the tidal cycle. Indeed, it was by then widely known that, as Francis Bacon put it around the same time, "the ebb and flow of the sea [repeats] twice a day, six hours for each incoming and outgoing tide."[14]

That Galileo clings to such a theory in the face of empirical facts is because he fundamentally rejects the idea of gravity as a force acting at a distance. In this connection he derides Kepler by name, finding it astounding that Kepler could have "lent his ear and his assent to the moon's dominion over the waters, to occult properties, and to such puerilities," and declaring that such things are "far from being actual or possible causes of the tides."[15] From today's perspective, this attitude of Galileo's toward Kepler, whom in the intellectual climate of that time he ought to have regarded as one of his strongest allies, seems absurdly arrogant and presumptuous. But it can also be said that, unlike Kepler, who still had one foot in the Middle Ages, Galileo was in fact a more "modern" thinker for whom notions such as that the moon exerts a force or influence upon distant terrestrial bodies (including oceans) were no more than the delusions of magical thought or the nonsense of astrology and could not possibly be allowed.

Indeed, Galileo's judgment in this respect is far from absurd given the kinds of nonmechanist explanations he wished to avoid. Pliny had described the tides as the result of "sympathy" between the moon and water, and since ancient times the tides had often been explained in terms of the astrological influence of celestial bodies.[16] Even as late as the fifteenth century, Marsilio Ficino had avowed that "the moon attracts the seas" in an illustration of the power of "enchantment";[17] and in the sixteenth century, John Dee had considered the tides to be evidence of the astrological influence of the stars.[18]

Having rejected such notions of lunar influence on terrestrial matters as irrational, Galileo likewise rejects the idea of the sun's attraction as the cause of planetary motion. This is because, while denying the Aristotelian duality of separate sublunary and celestial realms and supposing a sun-centered solar system, he nonetheless carries on the Aristotelian idea that the natural motions of astronomical bodies are circular. In the *Dialogue*, he asserts that "straight motion being by nature infinite … it is impossible that anything should have by nature the principle of moving in a straight line," and that "circular motion is natural for the whole and for the parts." He thus arrives at the conclusion that "if all integral bodies in the world are by nature movable, it is impossible that their motions should be straight, or anything else but circular."[19] Thus Galileo rejects from the outset the Keplerian law of elliptical planetary motion.[20] In this connection, the inertial law for terrestrial bodies that Galileo discusses concerns the inertia of circular motion parallel to the Earth's surface. Preserving the Aristotelian distinction between "natural motion" and "compulsory motion," he denies the naturalness of straight motion and elevates circular motion to the position of the only possible kind of "natural motion." Accordingly, in Galileo's view, the planets' orbiting of the sun is a spontaneously occurring "natural motion" the maintenance of which requires no exertion of force whatsoever by the sun upon either the Earth or any of the other planets. Here again, then, no room is left for the notion of gravity between astronomical bodies. Ernst Mach sums up Galileo's position thus: "The planetary system had not yet taken the shape of a genuine physical problem for him."[21]

Galileo also dismisses the question of the Earth's gravity acting on terrestrial bodies, seeing it as something requiring no explanation. In his 1590 treatise *On Motion*, the young Galileo, still following Aristotle, treats the falling of objects as the natural motion of heavy bodies, and in his 1613 report *Letters on Sunspots*, he writes:

> I seem to have observed that physical bodies have physical inclination to some motion (as heavy bodies downward), which motion is exercised by them through an intrinsic property and without need of a particular external mover, whenever they are not impeded by some obstacle.[22]

In the *Dialogue*, when the Aristotelian Simplicio asserts about the downward motion of heavy bodies that "the cause of this effect is well known; everybody is aware that it is gravity," Galileo's persona, Salviati, retorts: "You are wrong, Simplicio; what you ought to say is that everyone knows that it is called 'gravity.'"[23] What he means is that to attribute the fall of bodies to gravity is simply to give the cause a name; it does not explain it. But then nor does Galileo provide an explanation of gravity. Speaking of the component parts of the Earth, the moon, or any astronomical body, Salviati remarks: "If at any time one of these parts were forcibly separated from the whole, is it not reasonable to believe that it would return spontaneously and by natural tendency?"[24] Thus, while transcending celestial-terrestrial dualism, Galileo's argument about gravity remains at the Copernican level.

The position Galileo ultimately arrived at on the question of the fall of terrestrial bodies is summed up in a book published in 1638, near the end of his life, his *Two New Sciences*. In that work, he refines his view of falling motion as uniformly accelerated motion, asserting that

> a heavy body has from nature an intrinsic principle of moving toward the common center of heavy objects (that is, of our terrestrial globe) with a continually accelerated movement, and always equally accelerated, so that in equal times there are added equal new momenta and degrees of speed.[25]

But even in that work, he does not regard such acceleration as a result of the Earth's gravity. Rather, he explicitly sidesteps the issue of cause:

> The present does not seem to me to be an opportune time to enter into the investigation of the cause of the acceleration of natural motion ... For the present, it suffices our Author [i.e., Galileo] that we understand him to want us to investigate and demonstrate some attributes of a motion so accelerated (whatever be the cause of its acceleration).[26]

In short, whereas previous natural philosophy had asked *why* bodies fall, Galileo rejected the question itself and instead limited the scope of natural science to the matter of *how* they fall.

Galileo adopted this attitude toward all natural phenomena. The following is from his *Letters on Sunspots* of 1613:

> In our speculating we either seek to penetrate the true and internal essence of natural substances, or content ourselves with a knowledge of some of their properties. The former I hold to be as impossible an undertaking with regard to the closest elemental substances as with more remote celestial things. The substances composing the earth and the moon seem to me to be equally unknown, as do those of our elemental clouds and of sunspots. I do not see that in comprehending substances near at hand we have any advantage except copious detail....
>
> But if what we wish to fix in our minds is the apprehension of some properties of things, then it seems to me that we need not despair of our ability to acquire this respecting distant bodies just as well as those close at hand ... Hence I should infer that although it may be vain to seek to determine the true substance of the sunspots, still it does not follow that we cannot know some properties of them, such as their location, motion, shape, size, opacity, mutability, generation, and dissolution.[27]

Galileo's mechanistic view of nature thus takes shape by abandoning the question of what a certain thing is in terms of its true essence or substance and instead limiting the objects of study to phenomena.

To Galileo, furthermore, those phenomena are prescribed and "read" by means of geometrical and mathematical concepts. A decade after *Letters on Sunspots*, in *The Assayer*, he writes:

> Philosophy is written in this grand book, the universe, which stands continually open to our gaze.... It is written in the language of mathematics, and its characters are triangles, circles, and other geometric figures without which it is humanly impossible to understand a single word of it; without these, one wanders about in a dark labyrinth.[28]

Ultimately, Galileo's natural science boils down to regarding things and phenomena purely as the aggregates of their various attributes that can be expressed and understood mathematically and geometrically, and to deciphering the mathematical laws that hold for those attributes. It does not pursue such further questions as what constitutes the essence or substance of a thing or wherein lies the true cause of acceleration.

Accordingly, Galileo sees the study of the motion of terrestrial bodies as limited to three basic steps. First, one hypothesizes that all terrestrial bodies would fall at a uniform rate of acceleration if all external impediments, air resistance, and so on were eliminated. Second, on the basis of that hypothesis one deduces, purely mathematically, the law of falling bodies—that is, the proposition that in uniformly accelerating descent the velocity of a body falling from a stationary state is proportional to the fall time, while the fall distance is proportional to the square of the fall time. And third, one verifies this mathematical proof with experiments performed under optimized conditions and with measurements made using devices specially conceived for that purpose. This three-step procedure exhausts all that needs to be done. If it proves the original hypothesis about the uniform rate of acceleration of falling terrestrial bodies, then that is as far as the investigation goes; it does not extend to probing the origin or cause of acceleration. In this manner, Galileo conceived the modern-scientific methodology of hypothesis, proof, and experiment.

Galileo's approach, however, was completely devoid of the concept of force. Consisting of only mathematically expressed values for velocity, acceleration, and so on, his mechanics thus fell short of becoming a theory of dynamics, remaining instead only a mathematical kinematics. Particularly with respect to the order of the solar system, the arguments he advanced and the discoveries he made using telescopes had great potency in terms of refuting objections to heliocentrism; but concerning planetary motion he offered nothing that could reasonably be called a theory. Galileo is often characterized as a champion of the new vision of the universe and as a martyr to heliocentrism, and it is true that he did much to promote awareness and widespread acceptance of the Copernican theory; but he was nonetheless completely ignorant when it came to the dynamism of planetary motion. Max Casper, supervising editor of Kepler's complete works, puts it thus:

> While fully appreciating Galileo's accomplishments in the domain of mechanics, it must still be emphatically pointed out that he completely failed to comprehend the idea of a celestial mechanics.[29]

## Section 3—Cartesian Mechanics and Gravity

Descartes' view of nature is also called mechanistic and shares with Galileo's the premise that matter is inert, passive, and devoid of qualitative properties. But in terms of the meaning and methods of natural philosophy, Descartes' position is quite different from Galileo's. We noted earlier that in the *Principles of Philosophy* (the *Principles*) Descartes asserts that the nature of matter consists only in the fact that it possesses extension in length, breadth, and depth. This was published (originally in Latin) in 1644, but in a French-language edition revised and annotated by Descartes himself and published in 1647 he writes:

> I first considered generally all the most clear and distinct notions of material things to be found in our understanding, and that finding none but those we possess of figures,

> magnitudes and motions, and the rules whereby these three things can be diversified by one another, which rules are the principles of Geometry and Mechanics, I judged that all the knowledge man can have of nature must be derived from this source alone, because all the notions that we have of sensible things, being confused and obscure, cannot serve to give us any acquaintanceship with anything outside ourselves, but may on the other hand serve to impede it.[30]

In Descartes' view, then, the first concepts serving as the starting point toward knowable truth are discerned only in the understanding, and this is also the basis for his mechanism, which regards matter as mere extension. Taking this mechanistic view of matter as its "clear and distinct" point of departure, his brand of natural philosophy is an *a priori*, top-down system whereby all the attributes and behaviors of things are deduced from those first principles by means of rigorous, airtight reasoning. In addition, Descartes regards sensory perception as something that may impede understanding, and thus argues that "even if experience were to appear to show us the opposite, we would nevertheless be obliged to place more trust in our reason than in our senses" (PP 2.52, footnote). This is why his natural philosophy almost totally disregards experimental verification. This illustrates how experimental verification as opposed to deductive argumentation is not always what characterizes modern thought or distinguishes it methodologically from medieval thought. In the Cartesian view, the trueness of any discipline of learning ultimately consists entirely in the trueness of its fundamental ideas (first principles) and of the deductive reasoning (chain of proofs) that flows from them.

Descartes' first principles are, furthermore, quite different from what we today might suppose them to be. In *Discourse on the Method* (the *Discourse*), first published in 1637, he notes the following about the process by which his theory of the universe took shape:

> I have first tried to discover generally the principles or first causes of everything that is or that can be in the world, without considering anything that might accomplish this end but God Himself who has created the world, or deriving them from any source excepting from certain germs of truth which are naturally existent in our souls.[31]

And in the *Principles*, the proposition that he takes as the starting point of mechanics is that "God is the primary cause of motion [and] He always maintains an equal quantity of it in the universe" (PP 2.36). From this *a priori* premise, Descartes derives the following three laws—the principles of Cartesian mechanics—as the foundations of natural philosophy:

> The first law of nature: that each thing ... always remains in the same state; and that consequently, when it is once moved, it always continues to move. (PP 2.37)

> The second law of nature: that all movement is, of itself, along straight lines; and consequently, bodies which are moving in a circle always tend to move away from the center of the circle which they are describing. (PP 2.39)

> The third law: that a body, upon coming in contact with a stronger one, loses none of its motion; but that, upon coming in contact with a weaker one, it loses as much as it transfers to that weaker body. (PP 2.40)

With the first two of these propositions Descartes correctly formulated a law of inertia, something which Galileo had expressed in an incomplete form and which Gassendi had conceived. And with the third, the principle of motion transfer in collisions between bodies, Descartes proposed a nascent form of what would become the law of conservation of momentum.**

But although these were indeed major contributions to the development of early mechanism, from this standpoint the only kind of action possible between bodies is a transfer of motion through direct contact—either impulse or pressure—between property-less bodies having a certain extension and characterized solely in geometrical terms. Hence, "the force of each body to act against another or to resist the action of that other consists ... in the single fact that each thing strives, as far as is in its power, to remain in the same state" (PP 2.43). In essence, then, Cartesian mechanism is no more than a theory of collision. What appears to be action at a distance between two bodies must therefore be explained as the result of the transmission of effects via the pressure and impulse of an imperceptible material medium—"minute matter"—that entirely fills the intervening space.

Thus the premises of mechanism logically lead to the existence of "minute matter." In Descartes' view, that is, because the nature of matter consists solely in a thing's having extension, extension (as simple space) and extended substance are in effect the same thing, even though they differ in the mode in which people perceive them (PP 2.8, 2.10). Hence, without any kind of experimentation, the existence of a vacuum is logically precluded (PP 2.16). This implies

** In this third law, on the conservation of momentum, the part about contact with a "stronger" body is incorrect, and Descartes restates it correctly in the seventh chapter of *The World* as that "when one body impels another, it cannot impart to it any motion without at the same time losing so much of its own, nor take from it but so much as its own is thereby increased" (*The World*, 325). This is not to say, however, that Descartes precisely understood the concept of momentum as a vector quantity.

further that there must be "minute matter"—the "first element"—that fills all the apparent spaces between bodies, no matter how small (PP 3.52).

Descartes supposes that outer space is likewise filled with this minute matter and that such matter forms an enormous vortex around each astronomical body. This is his famous vortex theory, according to which a colossal eddy of this minute heavenly matter swirls around the sun carrying the planets along in their orbits, and a similar vortex exists around the Earth and propels the moon in its orbit. He also claims that this supposed whirl of such fantastic matter accounts for the effect of gravity (weight) in heavy bodies on the surface of the Earth. In the *Discourse*, he maintains that "God had not placed any weight in the matter of which [the Earth] is composed";[32] and in the *Principles*, he likewise asserts that "there is no weight in any body considered in isolation, but only insofar as that body depends on the situation and movement of other bodies, and relates to them" (PP 4.202). It is clear from such passages that Descartes regards gravity (weight) as an effect arising from the interplay of motion between bodies and minute matter. Following the same reasoning, in the eleventh chapter of *The World* he describes the concrete mechanism of gravity as follows:

> I desire now that you consider what the gravity of this earth is, that is to say, the force which unites all its parts, and which makes them all tend toward its center, everyone more or less according as they are more or less large and solid; which is nothing else and consists only in this, that the parts of the small heaven which surrounds it, turning much more swiftly than its own do around its center, tend also with much more force to withdraw themselves from it, and consequently push them back there.[33]

What he proposes here is that the acceleration of terrestrial bodies vertically downward (i.e., falling) occurs because the cosmic minute matter swirling around the Earth tends, by the centrifugal force of its rotation, to move away from the Earth and consequently usurps the position of any terrestrial heavy body in its way. This is, however, not a quantitative but rather an entirely qualitative explanation. Moreover, it has a fundamental drawback in that, whereas gravity on the Earth's surface is spherically symmetrical, the vortex is axisymmetrical around its axis of rotation, which would mean that gravity caused by the vortex would be two-dimensional and at right angles to that axis. Since the Earth spins on its own axis due to the motion of the vortex, the vortex's rotational axis and the Earth's rotational axis would coincide. This would entail that the force of gravity engendered by the vortex, if such did exist, would operate vertically downward at the equator but gradually shift away from the vertical as latitude increased, until it became horizontal at either pole. It seems rather odd that Descartes himself failed to notice this discrepancy in his theory. As we will see further on in our story, this defect was pointed out later by Robert Hooke.

Regarding the phenomenon of tides, about which Galileo was thoroughly mistaken, Descartes, in articles 49 through 52 of part 4 of the *Principles* and in the twelfth chapter of *The World*, offers the following explanation based on his vortex theory. Figure 19.1 shows the ocean surface enveloping the Earth as the circle 1234, the surface of the air around the ocean as the circle 5678, and the moon as the circle close to B. If the moon did not exist, the center of the Earth (T) would coincide with the center of the vortex around the Earth (M); but because the moon is near B, T shifts slightly to a point between M and D. The reason for this is that, were T and M to coincide at the midpoint between B and D, the distance between

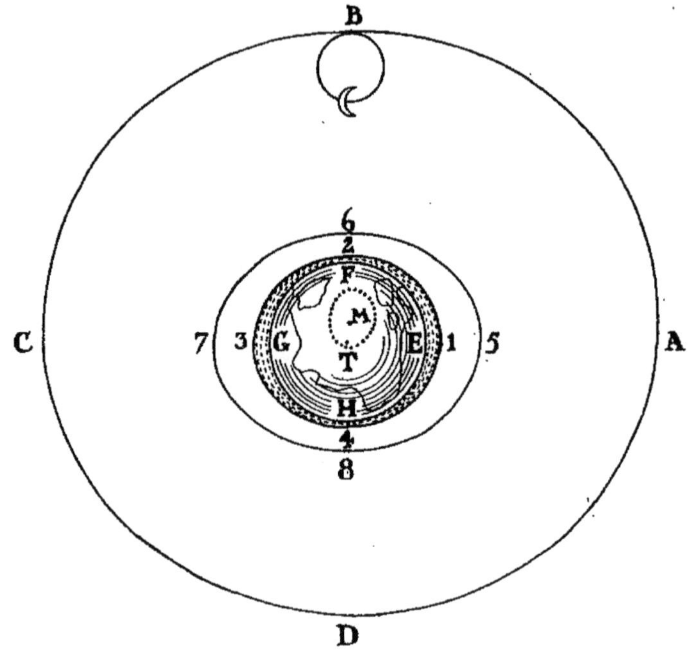


Figure 19.1: Descartes' conception of the phenomenon of tides.
*Source*: Descartes, *Principles of Philosophy*.

the Earth and the moon would be less than the distance between the Earth and D, which would make it more difficult for the "heavenly matter" whirling around the Earth to pass through the space between the Earth and the moon. In other words, if T and M were in the same place, the pressure of the vortex of celestial matter would be stronger at B than at D, and consequently the Earth would shift toward D. In this manner, the position of the Earth is determined so that the pressure exerted by the heavenly matter on the side facing the moon is equal to that on the opposite side. Hence:

> It follows that this heavenly matter flows more rapidly in those spaces and therefore presses more upon the surface of the air at

> 6 and 8, and upon the surface of the water at 2 and 4, than it would if the Moon were not on diameter BD of the vortex. And since the bodies of air and water are fluid and easily obey this pressure, these bodies must be less deep above parts F and H of the Earth than if the Moon were not on diameter BD; and, on the contrary, [these bodies] must be deeper at G and E, so that the surfaces of the water 1, 3, and of the air 5, 7, swell there. (PP 4.49; brackets in source)

Although this explanation accounts for the semidiurnal cycle of the tides, it posits an ebb tide for the ocean surface facing toward the moon, which is clearly at odds with reality.

Concerning planetary motion, Descartes postulates that the sun, too, is surrounded by an enormous vortex carrying the Earth and the other planets along their orbits. The fatal flaw in this model, however, is that it cannot be used to derive the planets' movements quantitatively, and thus fails to explain Kepler's three laws, which rest on precise observations, with anywhere near the same degree of rigor. When we take all these shortcomings into account, Descartes' tone of self-congratulation rings rather hollow when he declares in the *Discourse* that "I have not ... observed anything which I could not easily explain by the principles which I had discovered."[34] Whereas Tycho's observations and Kepler's theories had already elevated the astronomy of the solar system, at least as descriptive astronomy, to the plane of modern mathematical science, Descartes' theory of planetary motion remained at a far more primitive level.

We can see from the foregoing discussion that both Galileo and Descartes failed in every sense to grasp the significance of Kepler's breakthrough discoveries, namely, his three laws of planetary motion and his concept of gravity. In particular, they not only misunderstood but actively rejected Kepler's notion of gravity acting between astronomical bodies. Early-seventeenth-century mechanical philosophy had emerged with vigorous promises of superseding Scholasticism, by then a clearly untenable system, with a more rational explanation of the natural world, but in reality mechanism was still at a loss to account for even such everyday occurrences as the tides and the falling of bodies. Galileo and Descartes both occupy central places in any history of physics that describes the development of the principles of mechanics, but in a historical description of the evolution of the concept of force they must be relegated to a short-lived subsidiary branch.

## Section 4—Cartesian Mechanics and Magnetism

The following passage appears in *De sensu rerum et magia* (On the sense of things and magic), a work written by Italian philosopher Tommaso Campanella in 1604, during his imprisonment for heresy and rebellion, and published in 1620:

> Everything that scientists do in imitating nature or helping it with unknown art is called magical work, not only by the vulgar crowd but by the community of men.... It was by magic that Archytas made a dove that flew like a natural one; and at the time of Emperor Ferdinand of Germany, a German made an artificial eagle and a fly that flew by themselves. For technology is always called magic until it is understood, but after a while it becomes ordinary science. The invention of gunpowder, the printing press and the use of the magnet were once regarded as magic, but now that everyone understands the art, it is common knowledge. Likewise clockmaking and other mechanical arts just as easily lose their reverence when their workings become obvious to ordinary people. But physical, astrological and religious things are rarely divulged, and it was from these that the ancients drew the art of magic.[35]

What Campanella points to here is the demystification of magic. A mechanized puppet may be a marvel at first, but once the mechanism of its movement is understood, it is just another plaything. Indeed, all devices, whether a matchlock gun, a mechanical clock, or even Della Porta's camera obscura, lose their mystique when we learn the principle behind their operation.

Just a few decades later, in 1648, English clergyman John Wilkins published *Mathematicall Magick*, a book primarily about effective applications of force using levers, pulleys, sloping surfaces, and so on. Among the devices described was a machine that, as depicted in Figure 19.2, purportedly could uproot a large tree with just the force of a person's breath transmitted and amplified several times over by means of gears and pulleys.[36] Due to such factors as metal strength and friction on the gears, such a machine would not actually operate as intended, but in principle it is feasible; at a smaller scale, it would be possible to actually construct a devise much like it. And if such a machine had actually been produced at the time, it would surely have been widely regarded as a marvel. Such mechanical devices are what Wilkins means by "mathematical magic," with "magic" in this context denoting nothing more than the hidden structures or machinery producing the marvelous effect.

By exposing to scrutiny such normally hidden mechanisms, seventeenth-century mechanical philosophy sought to strip away the veil of mystery surrounding magic and deconstruct its apparent marvels. In his *Principles*, Descartes states that "I have described this Earth, and indeed this whole visible world, as a machine, considering nothing in it except figures and motions" (PP 4.188), and explains the methodology of his natural philosophy as follows:

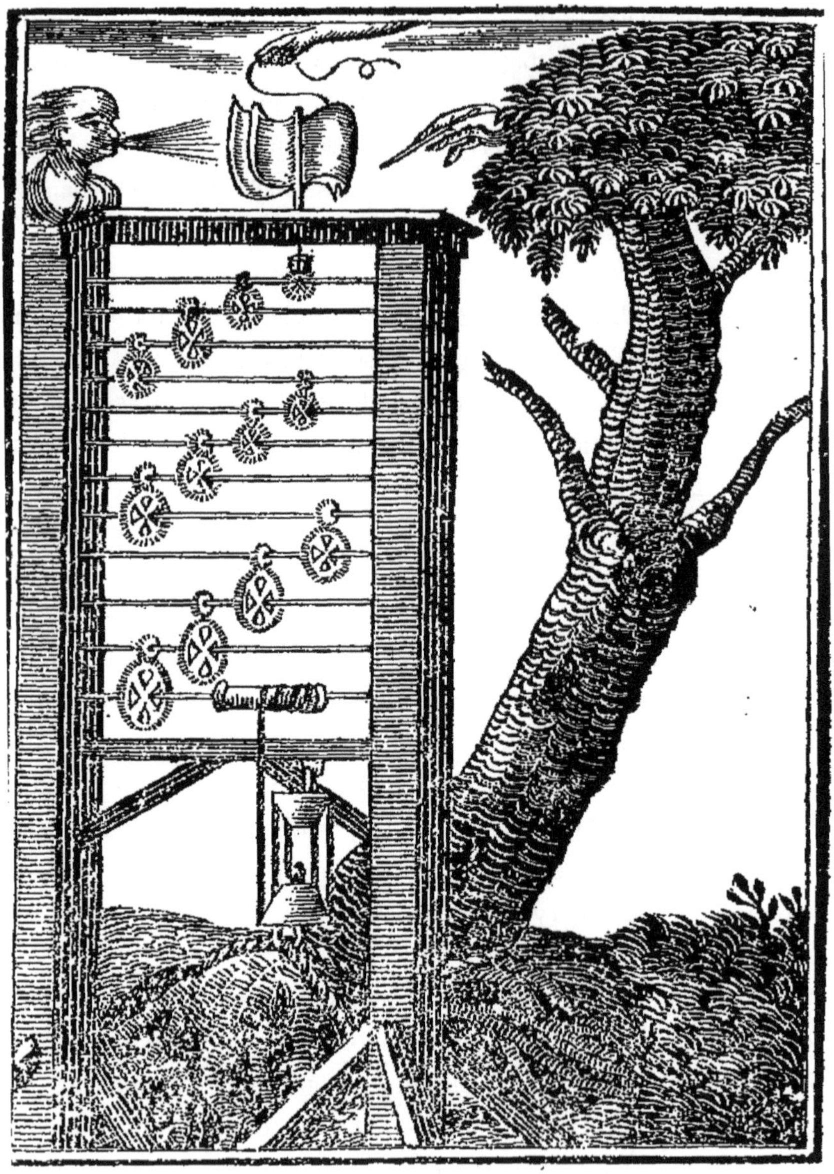

Figure 19.2: A machine driven by the force of a person's breath.
*Source*: Wilkins, *Mathematicall Magick*.

> Just as when those who are accustomed to considering automata know the use of some machine and see some of its parts, they easily conjecture from this how the other parts which they do not see are made: so, from the perceptible effects and parts of natural bodies, I have attempted to investigate the nature of their causes and of their imperceptible parts. (PP 4.203)

One figure who played a major role in disseminating and popularizing this Cartesian theory was the French thinker Bernard le Bovier de Fontenelle, who lived for mere weeks short of a hundred years, from the mid-seventeenth to the mid-eighteenth century. His book *A Conversation on the Plurality of Worlds* was published to high acclaim in 1686, thirty-nine years after Descartes' *Principles* and a year before Newton's *Mathematical Principles of Natural Philosophy* (*Mathematical Principles*). In this work Fontenelle avers that people "will have the World to be in Large, what a Watch is in Small, that is very regular, and depends only upon the just Disposition of the several Parts of the Movement." He also makes the following remark on the emerging mechanistic view of nature: "I fancy to myself, that *Nature* very much resembles an *Opera*; where you stand, you do not see the *Stage* as it really is, but as it is placed with Advantage, and all the Wheels and Movements hid, to make the Representation the more agreeable."[37]

According to this new understanding, the world is an automated machine, and just as we understand the movements of props on a stage as the effects of mechanical devices behind it, or consider the motions of a clock's hands in terms of the gears, springs, and weights under its face, so the explication of nature, and therefore the proper vocation of natural philosophy, consists in explaining directly observable natural phenomena in terms of the hidden mechanical operations underlying them. In this manner, wonders previously "explained" in such vague magical terms as "sympathy" and "occult power" could now be properly accounted for in terms of actual causes and thereby reduced to the level of ordinary occurrences.

The problem, however, was what to make of magnetism. Magnetic force was a clear case of force acting at a distance and a conspicuous example—indeed, virtually the only cited actual example—of sympathy, antipathy, and occult power, the key concepts of magical theory. As magical theory and mechanical philosophy vied with one another in the first half of the seventeenth century, the outcome of their rivalry thus hinged upon whether or not mechanical philosophy could uncover and identify the "mechanism" of magnetic force.

Descartes' discussion of magnets and magnetism appears in articles 133 through 183 of part 4 of the *Principles*. Given that this is the longest and most detailed portion of the *Principles* that deals with any single substance, it is clear that Descartes devoted particular attention to the topic. In article 145, he enumerates thirty-four properties of magnets, including the fact, discovered in 1634, that magnetic declination changes over many years. This is, however, nothing more than a listing of facts already known by that time; Descartes provides no new knowledge based on his own experiments or observations. This is because the aim of his natural philosophy was not to increase knowledge through experiment but rather to explain matters that were already known. Hence, the "new" aspects he offered consisted entirely in his interpretations. Although those interpretations are no longer significant today, Descartes himself was confident that his account of magnetism represented one of the most successful applications of his overall philosophy. This is evident when he remarks near the end of the *Principles* that

> those who notice how many things concerning the magnet, fire, and the fabric of the entire World have been deduced here from so few principles ... will perhaps still know that it could scarcely have occurred that so many things should be consistent with one another, if they were false. (PP 4.205)

Since his account of magnetism is a classic example of mechanistic argument, and one that exerted a long-lasting influence, let's take a brief look at what it involves.

In article 133, Descartes offers the following explanation of the Earth's magnetic field:

> We shall think that there are many pores in the Earth's intermediate region which are parallel to its axis, and through which the grooved particles coming from one pole freely proceed to the other. And these pores have been hollowed out to the measurement of these particles in such a way that those which accept the grooved particles coming from the South pole can in no way admit those which come from the North pole; conversely, those which accept the Northern particles do not admit the Southern ones: because of course these particles are twisted like the thread of a screw; some in one direction and the others in the opposite direction. Furthermore,... the same particles can enter through only one end of these pores, and cannot return through the opposite one ... (PP 4.133)

Essentially, then, he posits that celestial minute matter comprises two kinds of grooved particles (right-winding and left-winding); that the Earth has numerous pores all running parallel to its polar axis and having spiralling grooves in either of two types, each type corresponding to one kind of screwlike particle; and that each type of pore accordingly allows the passage of only one kind of particle and only in one direction. A grooved particle entering at the South

Pole (let's call it an *s* particle) is able to pass through the Earth only via a pore with a compatible thread (an *s* pore) and in only one direction, from the South Pole to the North Pole. Conversely, an *n* grooved particle, which enters at the North Pole, can pass through only *n* pores and only in the north-to-south direction. After the particles exit the Earth at one pole, they move through the atmosphere toward the opposite pole, where they once again enter a compatible pore. This circulation of particles inside and outside the Earth generates the geomagnetic field to whose influence terrestrial magnets are subject.

In Figure 19.3, an illustration from the *Principles*, the larger circle ABCD at the center of the upper diagram represents the Earth, and each of the smaller circles around it (I, K, L, M, and N) represents a magnet. Each magnet has pores like those in the Earth, with the end where *n* particles enter *n* pores being the magnet's "North pole" and the opposite end being its "South pole" (PP 4.149). An *s* particle that exits the Earth's North Pole at A will then enter an *s* pore in magnet I at b and exit the magnet at a, while an *n* particle that leaves the Earth's South Pole at B will enter an *n* pore in magnet N at a and exit it at b. When, as in the cases of magnets K, L, and M, a magnet's poles are not oriented in a north-south direction,

> these grooved particles rush obliquely into the pores of the magnet and drive it with that force which they have to continue their movement along straight lines, until they have brought it back to its natural situation. (PP 4.150)

Thus the entrances of a magnet's *s* pores and the exits of its *n* pores—or in other words the magnet's "South pole" b—will tend to turn toward the Earth's North Pole A. This, claims Descartes, is the cause of the directionality of terrestrial magnets.

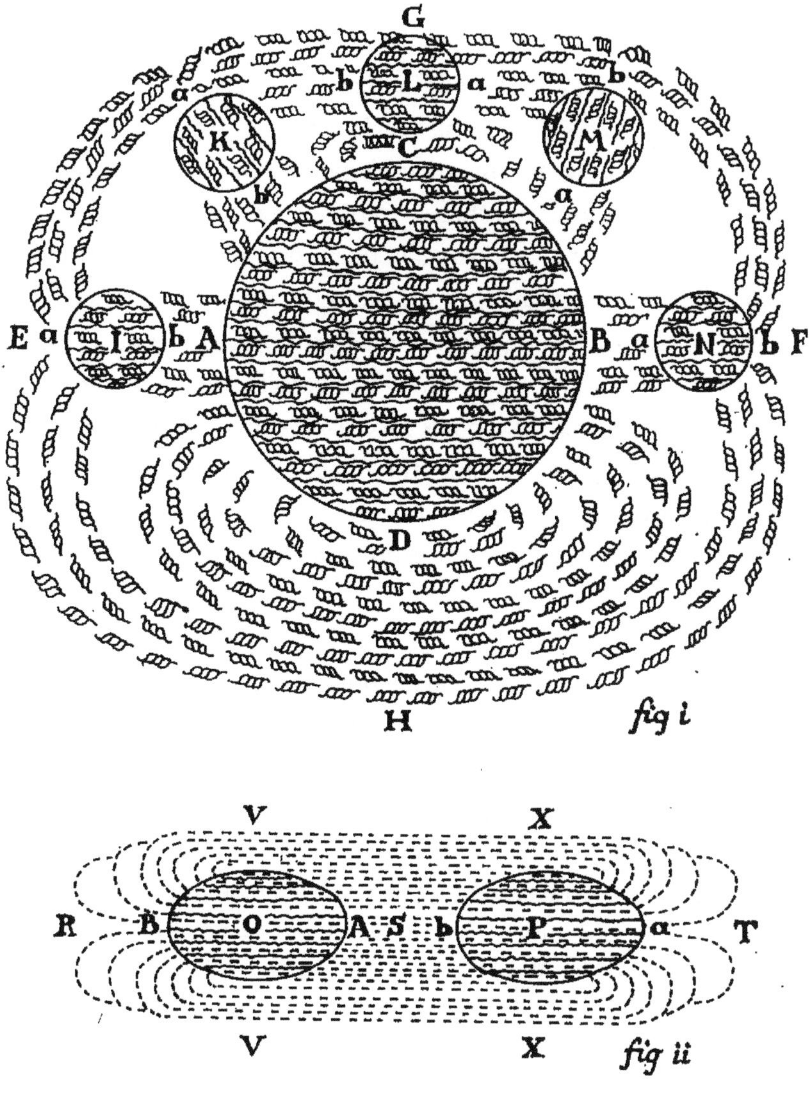


Figure 19.3: Descartes' conception of the phenomenon of magnetism.
*Source*: Descartes, *Principles of Philosophy.*

Now let's look at Descartes' explanation in articles 153 and 154 for the forces of attraction and repulsion between magnets. Consider two magnets, O and P, placed so that their respective spheres of influence overlap, as shown in the lower diagram of Figure 19.3. When, as in the diagram, magnet O's north pole A and magnet P's south pole b face one another, *s* particles exiting A pass directly into b while *n* particles exiting b likewise enter A. As this flow of particles pushes away the air between the two magnets (i.e., between A and b), the rest of the air surrounding O and P causes them to move toward each other. Conversely, when the magnets are positioned so that like poles face each other, air is forced into the intervening space, causing the magnets to move away from each other. This, according to Descartes, is the cause of attraction between opposite poles and repulsion between like poles of magnets.

Descartes explains attraction between magnets and iron in a similar fashion. He claims that iron has pores running in all directions, so that grooved particles from the magnet enter whichever pores of the iron which are available to them. Hence,

> the magnet attracts iron, or rather, a magnet and a piece of iron approach each other; for in fact there is no attraction there: rather, as soon as the iron is within the sphere of activity of the magnet, it borrows force from the magnet, and the grooved particles which emerge from both the magnet and the piece of iron expel the air between the two bodies: as a result, the two approach each other in the same way as two magnets do. (PP 4.171)

Meanwhile, Descartes' explanation for the attractive force seen with such materials as amber and glass—what we now call static electricity—is that when those substances are subjected to friction, certain thin, wide, long ribbons of matter (*fasciola*) emerge from tiny fissures in them, extend into the surrounding air and, finding no paths to easily pass into, return to the main body of amber or glass, drawing any nearby light bodies along with them as they do (PP 4.185, 4.186).

From these ideas Descartes concludes as follows:

> I wish it to be observed here that these particles, composed of the matter of the first element in the pores of terrestrial bodies, can not only be the causes of various attractions, such as are in amber and in the magnet, but also of innumerable other admirable effects.... [T]here are ... no effects in all of nature which are properly attributed to purely physical causes or causes lacking in mind or thought; the reasons for which cannot be deduced from these same principles. (PP 4.187)

In the same manner, toward the end of the *Principles* Descartes proclaims rather conceitedly that "no phenomena of nature have been omitted in this treatise" (PP 4.199) and that "it will scarcely seem possible for all the rest, at least the more general things which I have written about the World and the Earth, to be understood otherwise than as I have explained them" (PP 4.206). In his day, as many people sensed that a new age of thought was dawning, those who struck a bold attitude and made grand statements were the ones likely to succeed in that emerging milieu. Accordingly, from the latter half of the seventeenth century through the first third of the eighteenth century, Cartesian mechanism was lauded on the Continent as the champion of new science. Thereafter, however, it was just as quickly abandoned.

Galileo steered the mechanistic view of nature toward mathematical phenomenalism, focusing on quantifiable properties and mathematically manageable phenomena and probing the correlations between them. For the time being, he banished from the domain of science such issues as the cause of acceleration and the true nature of matter, and thus limited science's mission to elucidating the mathematical laws of motion. Moreover, by regarding those laws as hypotheses requiring experimental verification, Galileo correctly pointed the way toward modern physics as not only a mathematical but also an empirical discipline. On the subject of magnetism, similarly, he criticized Gilbert for not being mathematical enough, but unlike Descartes he did not concern himself with questions about the cause of magnetic force or the structure of its transmission. Rather, Galileo's brand of mechanism was built on abandoning ontology altogether, and it is fair to say that he never strayed far from that well-defined philosophical path.

Though also advocating mechanism, the overconfident Descartes, on the other hand, did so from within a system of top-down, *a priori* idealism; and by seeking to explain all causes and essences by means of facile analogies to machinery, he ended up falling into hollow speculation. Indeed, Descartes' approach, limited as it was by his age's poverty of knowledge about the actual natural world, could hardly but lead him to views quite out of touch with reality. His various conjectures—whether about two types of grooved particles, two kinds of pores in magnets and iron, or invisible ribbons of matter—had no experimental support whatsoever. For the most part, therefore, his explanations were not elucidations based on experiments on actual mechanisms but essentially specious fabrications concocted from arbitrary argumentation. They were, in short, the products of fancy.

It has even been asserted that early mechanism in fact held progress back—that

> the demand for mechanical explanations stood in the way of the other fundamental current of 17th century science, the Pythagorean conviction that nature can be described in exact mathematical terms. Despite its rejection of a qualitative philosophy of nature, the mechanical philosophy in its original form was an obstacle to the full mathematization of nature.[38]

In the *Principles*, Descartes boasts, "I do not accept or desire in Physics any other principles than in Geometry or abstract Mathematics" (PP 2.64); but what he actually does is not in fact mathematical. His natural philosophy, especially his cosmology, is utterly devoid of the fundamental requirement of modern physics and modern astronomy: verification of the results of mathematical deduction by precise measurement. He completely disregards Kepler's mathematical laws of planetary motion. On the subject of magnetic force, the idea of quantitatively measuring it is nowhere in his thinking; his theory of magnetism amounts to devising a model to represent the qualitative behavior of magnets in terms of a hypothetical substance. In that sense, it must be admitted that as natural philosophy Cartesian mechanism was an immature and therefore highly deficient antithesis to magical thought. Finally, it is also worth noting that, although Descartes did achieve a certain degree of success as a man of physics in his theory of light refraction, rainbows, and so on, he did so by formulating mathematical laws, not on the strength of the mechanistic model he superimposed on those laws.

## Section 5—Walter Charleton

The first work to introduce the atomism of Epicurus and Gassendi in Great Britain was *Physiologia, Epicuro-Gassendo-Charletoniana* (*Physiologia*), written by Walter Charleton (c. 1620–c. 1707) and published in 1654.[39] Although almost forgotten today, at the time this was an important introduction to mechanism and atomism on the British intellectual scene. It was, indeed, "the first systematic presentation of Gassendi's views in the vernacular,"[40] and was by some accounts the book that occasioned Newton's discovery of Gassendi's philosophy in his student days.[41]

Charleton was a physician who had studied medicine and philosophy at Oxford. His talents were recognized at a young age, and in 1643, immediately after receiving his doctoral degree in medicine, he was appointed physician to Charles I. He remained a Royalist until his death. After learning of Gassendi in the 1650s Charleton embraced atomism, persuaded that Gassendi's theory, which held that the motion of atoms was initiated by God and abided by laws set by God, had no connection with heresy or atheism. Charleton recounts this in *Physiologia*, and though it is beyond the scope of my present purposes to explore the point in detail, suffice it to note that his aims in writing the book included refuting the idea, held since the time of Lucretius, that atomism entailed atheism.

Charleton's view of nature is perhaps best seen in his discussion of "occult qualities." Previous thinkers, he notes, distinguished between "occult" (or "immanifest") qualities that are not available to our senses or whose causes are unknown, and "manifest" qualities that are available to the senses and whose causes are known. But since "to Ourselves all the Operations of Nature are meer Secrets," he finds this distinction ultimately meaningless. In reality, because

> many things, which are most obvious and open to the Sense, as to their *Effects*, may yet be remote and in the dark to the Understanding, as to their *Causes:* so on the Contrary, doth not

> the *Insensibility* of a thing necessitate, nay, nor aggravate the *Unintelligibility* thereof, but that many things, which are above the sphere of the Senses, may yet be as much within the reach of our Reason, as the most sensible whatever. (PEGC, 341–42; italics in source)

In other words, though we perceive something to be red when we see it, that alone hardly constitutes an explanation of its redness; and though we feel on our skin that something is hot, that does not mean we have understood the essence of heat. Conversely, even though magnetic force cannot be perceived by human senses, that does not mean it is entirely beyond our comprehension.

What this boils down to is that all qualities can be regarded as occult and therefore as requiring explanation by more fundamental principles. When Charleton calls the notion of occult qualities a "Sanctuary of Ignorance" (PEGC, 343), he is not denying the existence of such qualities but rather criticizing the attitude that, having designated certain things as "occult qualities" or in terms of "sympathy and antipathy," ceases to inquire into them any further. It is at this point that modern natural science splits away from the Aristotelian natural philosophy and magical thought that had prevailed until then. In practice, however, the more fundamental principles for explaining phenomena can be considered at various levels, and this is where the crucial difference arises in the methods, and by extension the aims, of science. This split in thought grows even more pronounced on the question of how to understand and warrant the fact of action at a distance.

In Charleton's case, those principles for explaining phenomena are expressed as three general laws of nature, namely, that "every Effect must have its Cause," that "no Cause can act but by Motion," and that "Nothing can act upon a Distant subject" (PEGC, 343). He elaborates as follows:[42]

> This small Difference only allowed, that in Gross and *Mechanique* operations, the Attraction, or Repulsion is performed by *Sensible* Instruments: but, in those finer performances of Nature, called *Sympathies* and *Antipathies*, the Attraction or Repulsion is made by Subtle and *Insensible*. The means used in every common and Sensible Attraction and Complection of one Bodie by another, every man observes to be Hooks, Lines, or some such intermediate Instrument continued from the Attrahent to the Attracted; and in every Repulsion or Disjunction of one Bodie from another, there is used some Pole, Lever, or other Organ intercedent, or somewhat exploded or discharged from the Impellent to the Impulsed. Why therefore should we not conceive, that in every Curious and Insensible Attraction of one bodie by another, Nature makes use of certain slender Hooks, Lines, Chains, or the like intercedent Instruments, continued from the Attrahent to the Attracted, and likewise that in every Secret Repulsion or Sejunction, she useth certain small Goads, Poles, Levers, or the like protruding Instruments, continued from the Repellent to the Repulsed bodie. Because, albeit those Her Instruments be invisible and imperceptible; yet are we not therefore to conclude, that there are none such at all. (PEGC, 344; italics in source)

Perhaps what Charleton had in mind were mechanisms such as those depicted in the illustrations in Agricola's *De re metallica* a hundred years earlier. Recall, for example, figure 13.5 (see chapter 13), which shows underground mining work using machines driven by aboveground horse power. We see constructed and installed in that scene a complex but uninterrupted series of devices for transmitting motive power from above the ground to the underground areas out of sight. Charleton's suggestion is that something much like this occurs on an invisibly minute scale in the microscopic world. The premise behind this is his *a priori* conviction that the microscopic world and the macroscopic world are fundamentally analogous: "the Instruments of *Natural* Attraction, Complectence, Repulsion, Sejunction, are *Corporeal*, and hold a neer Analogie to those of *Artificial*; only these *Gross* and *Perceptible*, those *Subtile* and *Imperceptible*" (PEGC, 345; italics in source).

In short, his view is that "nothing can act upon a distant subject but by some Instrument, either continued or transmitted" (PEGC, 404). Regarding the particular case of magnetism he argues that

> to affirm, that nothing Material is emitted from the Loadstone to Iron, which by continuity may Attract it; only because our sense doth deprehend nothing intercedent betwixt them: is an Argument of equal weight with that of the Blind man, who denied the Being of Light and Colours, because He could perceive none. (PEGC, 344)

Concerning the amber phenomenon, too, he similarly asserts that

> it is firme and indubitable, that though the Attraction of straws by Amber, be in some sort Admirable, yet it is not *Miraculous*, as is implied in that opinion, which would have it to be by some *Immaterial* (i.e. *Supernatural*) Virtue; and that it is effected by some *Corporeal*, though both impalpable and invisible Organs continued from the Attrahent to the Attracted. (PEGC, 346)

Thus, although Charleton acknowledges Gilbert's discovery of the Earth's being a great magnet, and even lauds him as "the Father of Magnetique Philosophy" (PEGC, 401), on the question of the origin of magnetic force he takes a thoroughly mechanistic position:

> the Virtue immitted into Iron, either from the Earth it self, or a Loadstone, is no simple, or immateriall Quality, as both *Gilbert* and *Grandamicus* earnestly contend; but a certain *Corporeal Efflux*, or Fluor, consisting of insensible bodies, or particles, which introduce upon the particles of Iron the same Disposition, and Local respect, as themselves have. (PEGC, 403–4)

Although Charleton's explanation of magnetic force in terms of "magnetical rayes," "magnetical atoms," and "magnetick effluxes" is actually somewhat different from Descartes' theory of grooved particles, from our present-day perspective their minor differences are of little consequence.

More important is the fact that Charleton accepts the theory of magnetic gravity:

> *All terrene bodies Descend, only because they are Attracted by the magnetique Virtue of the Earth.* (PEGC, 453; italics in source)

Here is where he parts ways with Gassendi. Concerning "the Essence of Gravity" Charleton pronounces that "*it is the meer Effect of the Magnetique Attraction of the Earth*" (PEGC, 288; italics in source), and gives his surprisingly simple reason for this view as follows:

> Because no other Attractive Force, which might begin and continue the Downward motion of a [falling] stone, can be imagined, unless it be that *Magnetique Virtue of the Earth*, whereby it Draws all Terrene Bodies to an Union with it self, in order to their, and its own better Conservation: we may lawfully Conclude, *that the Cause of the Downward motion of all Heavy Bodies, is the Magnetique Attraction of the Earth.* (PEGC, 450; italics in source)

On the question of gravity, Charleton's atomism completely disregards Descartes' vortex theory and unconditionally embraces the theory of magnetic gravity. As Charleton sees it, the reason people adhere to Aristotle's erroneous position about gravity (weight) is their "ignorance of the Earths [sic] being one Great Loadstone" (PEGC, 289). Here we see how, in the process of being transplanted to Britain, Cartesian mechanism and Gassendian atomism were influenced and significantly altered by the magnetic philosophy of Gilbert and Kepler.

For mechanism, gravity remained the greatest enigma and was ultimately its Achilles heel—and this is why, on the question of gravity, the magnetic philosophy of Gilbert and Kepler was preferred in Britain, even among mechanists. Indeed, mechanism's transformation in Britain under the influence of magnetic philosophy holds great significance for the ensuing phase of our account, as we shall see in the next chapter.

* * *

In its impatience to "explain" forces that Renaissance magical thinkers had previously discussed in such terms as "occult powers" and "sympathy and antipathy," early-seventeenth-century mechanism threw up one arbitrary model after another. But no matter how ingeniously one crafts a mechanical model of transmission of force using ideas about "minute matter" or "effluvia," until one can confirm the existence of the "grooved particles" or "magnetical atoms" or whatever one supposes, it is just speculation; it does not lead to the formulation of a new theory of physics. As the antithesis of magical thought, such naive mechanism tried to dismantle magic, to expose the hidden cogs and levers underlying seemingly magical phenomena; but insofar as it was preoccupied with that task it remained no more than magic inversed.

Nonetheless, to people fed up with previous notions—with "substantial form" and "sympathy and antipathy," with Scholasticism's fruitless inflation of terms, and with the vague, expedient explanations offered by magical thought—the simple, straightforward arguments of mechanism had the appeal of being fresh and clear-cut. Above all, mechanism was readily comprehensible to ordinary people. Indeed, Cartesian mechanism is not as mathematical as even Descartes himself, the founder of analytic geometry, imagined. Whereas the painstaking geometrical and mathematical proofs of Kepler's *New Astronomy* and Newton's *Mathematical Principles* were impenetrable to most people, Descartes' mechanism had many elements that appealed to the popular mind. Descartes' theory was even bandied about in salons, as depicted satirically in one Molière comedy in which characters make such remarks as "I much prefer subtle matter," "I quite agree with Descartes about magnetism," and "I like his vortices."[43] But the theory was specious and shallow nonetheless, and for all the attention it attracted its useful life turned out to be quite short.

Born roughly at the midpoint of the period between Descartes' birth and Fontenelle's, the Dutch scientist Christiaan Huygens (1629–1695) was, like many others, fervently taken with Cartesianism as a young man. Near the end of his life, however, he reflected as follows on how his attitude changed:

> When I read the book of Principles [i.e., Descartes' *Principles*] the first time, it seemed to me that everything proceeded perfectly; and when I found some difficulty, I believed it was my fault in not fully understanding his thought. I was only fifteen or sixteen years old. But since then, having discovered in it from time to time things that are obviously false and others that are very improbable, I have rid myself entirely of the prepossession

> I had conceived, and I now find almost nothing in all his physics that I can accept as true, nor in his metaphysics and his meteorology.[44]

Magnetism had been regarded since ancient times as a mysterious and sometimes as even a magical power. Then the idea of gravity acting between astronomical bodies as a force exerted at a distance emerged from geomagnetic theory. Galileo's mathematical phenomenalism, Descartes' mechanism, and Charleton's atomism all failed to recognize these notions' entitlement as worthy new scientific concepts, failed to determine the laws by which they operated, and failed to accord them their proper place in scientific theory. They neither paid attention to the laws that Kepler had arrived at nor nurtured the nascent concept of gravitational force that Kepler had conceived.

"The mechanical world view," observes one scholar, "rested on a single, fundamental assumption: *matter is passive*."[45] As long as it retained that premise, naive mechanism was inherently incapable of dealing with such phenomena as magnetism and astronomical gravity—phenomena, that is, of force exerted over distance by one body on another. Instead, ideas about magnetic force and gravitational force were gradually being cultivated elsewhere, as part of a view of nature that mechanism specifically sought to repudiate.

Chapter 20

# Robert Boyle and the Transformation of Mechanism in Britain

## Section 1—Francis Bacon

As one of the earliest proponents of a new philosophy for a new age, in Britain Francis Bacon (1561–1626) made his mark even before Galileo and Descartes did. Bacon's key works in this connection were *The Advancement of Learning* and *The New Organon.*[1] In the former, published in 1605, he begins his iconoclasm by commenting on "the schoolmen" (Scholastics) that "their wits [are] shut up in the cells of a few authors (chiefly Aristotle their dictator)" and their minds produce "cobwebs of learning, admirable for the fineness of thread and work, but of no substance or profit" (AL 1.4.5); and in the latter, published in 1620, he dismisses the philosophers of ancient Greece, remarking that "their wisdom seems wordy and barren of works [i.e., concrete results]" (NO 1.71). Purposely breaking from the past, Bacon sought, or rather agitated for, a complete overhaul of the aims and methods of philosophy. Such tall talk was fueled, as we saw earlier, by the new awareness and confidence that modern Europeans gained with their discoveries of entire continents that had been unknown to the ancients. "We should ... take into account," notes Bacon, "that many things in nature have come to light and been discovered as a result of long voyages and travels (which have been more frequent in our time), and they are capable of shedding new light in philosophy" (NO 1.84).

Up to Bacon's time, mainstream philosophy had been a single, closed system in which the answers to all questions were derived deductively by airtight logic from "first principles" and all phenomena were rigorously accounted for by strict reasoning. Moreover, once that system had been instituted by its first exponents, it had reigned supreme as the highest authority in intellectual matters and as a complete theoretical framework within which the only margin for difference or debate was in interpretation.

Bacon counterposed to this a different kind of learning characterized as developing inductively, "at first crude, then adequate, later refined, and always progressing" (NO 1.74)—or in other words, evolving and improving with the accumulation and expansion of experience. For a model of this new kind of knowledge-building he looked to the way in which technology developed through the collaboration of diverse craftsmen and engineers. The following passage from *The Advancement of Learning* aptly expresses what he meant:

> In arts mechanical the first deviser comes shortest, and time addeth and perfecteth; but in sciences the first author goeth furthest, and time leeseth and corrupteth. So we see, artillery, sailing, printing, and the like, were grossly managed at the first, and by time accommodated and refined: but contrariwise, the philosophies and sciences of Aristotle, Plato, Democritus, Hippocrates, Euclides, Archimedes, of most vigour at the first and by time degenerate and imbased. (AL 1.4.12)

Bacon thus sought a new ideal of knowledge that, like technology, which improves and evolves day by day on the basis of many people's actual application and experience, would be based on cumulative, flexible, ever-expanding, and open-ended theory, continually adjusted and augmented as our interaction with and experience of the natural world increased, and constantly refined through the cooperation of numerous people to make it ever more complete and comprehensive. This is why, in describing this proposal for a "renewal of learning," he points out that "it does not expect that the thing can be completely finished in the course of one lifetime, but provides for successors."[2]

Another reason Bacon has for rejecting the philosophies of the past is that, no matter how exquisitely constructed, such contemplative philosophies, being mostly just word play with no connection to immediate reality, are of little use as guidelines on how to make nature serve humanity. Such speculation does not, he says, constitute "a fruitful womb for the use and benefit of man's life" (AL 1.4.6). Bacon originally conceived *The New Organon* as part of a larger work titled *The Great Instauration*, though the only other parts of the latter that he actually ever wrote were the preface and an outline plan. In the preface he states plainly that we should pursue "the true ends of knowledge ... for the uses and benefits of life."[3] What he sought, in short, was firstly a science that served man's practical interests and secondly a system of knowledge that effectively incorporated feedback from man's ever-increasing experience.

*The New Organon* was written in response to the *Organon*, the collection of Aristotle's six works dealing with logic (*Categories*, *On Interpretation*, *Prior Analytics*, *Posterior Analytics*, *Topics*, *Sophistical Refutations*). As indicated by its subtitle, *True Directions for the Interpretation of Nature*, Bacon's work is a proposal for a unique method of induction as a new epistemological methodology, and an exhortation for experimental science based on that method.

In *Posterior Analytics*, Aristotle too argues for the necessity of inductive understanding, admitting that, although "scientific knowledge through demonstration is impossible unless a man knows the primary immediate premisses," nonetheless "we must get to know the primary premisses by induction." On the other hand, Aristotle's position is that "scientific knowledge is not possible through the act of perception" because, whereas perception is perception of particulars, "scientific knowledge involves the recognition of the commensurate universal."[4] Such knowledge is acquired through the intuition that what is apparent in various particular instances is true for all such instances. This entails that, at some level, it is necessary to make a logically discontinuous leap from the particular to the universal.

But to do so, counters Bacon, is "to leap and fly from particulars to remote and highly general axioms (such as the so-called *principles* of arts and things), and on the basis of *their* unshakable truth, demonstrate and explicate the intermediate axioms" (NO 1.104). This, he asserts, is how Aristotle "utterly enslaved his natural philosophy to his logic" (NO 1.54). Bacon criticizes even Gilbert for falling into the same error, saying he hastily "concocted a philosophy" from limited experiments on magnets (NO 1.54). Bacon's own position is that correct knowledge can be attained only "when the ascent [to it] is made on a genuine ladder, by regular steps, without gaps or breaks, from particulars to lesser axioms and then to intermediate axioms, one above the other, and only at the end to the most general" (NO 1.104).

On that point, however, the difference between Bacon and Aristotle is only a matter of emphasis. Their true difference lies in Bacon's assertion that to reach true knowledge researchers must carefully systematize their empirical experience and that in all such experience they must abandon all preconceptions, which he classifies into four "idols" or misleading dogmas in human thinking.

In Bacon's view, the inductive method must be based not merely on particular, contingent experience but rather on deliberate, systematically conducted experiment. Nature and things must be judged by "suitable and relevant experiments" (NO 1.50), by which he means "experiments which have been devised and applied specifically for the question under investigation with skill and good technique."[5] In speaking of experiments "devised with skill and good technique," Bacon refers to a peculiar kind of experimental design based on his distinctive inductive method and according to which, by exhaustively enumerating and carefully classifying and comparing all instances of things that can be regarded as having and not having the attributes under inquiry, one inevitably distills out the essential nature or "form" of the thing in question.[††]

Note, however, that the Baconian method differs distinctly from

---

[††] By "form" Bacon means the thing's essential characteristic. The concept is close to that of "substantial form" in Scholasticism. As John Losee puts it: "Bacon's Forms are neither Platonic form nor Aristotelian formal causes. Rather, Forms supposedly express those relations among physical properties that have the power to produce effects. In Aristotelian terms, Bacon's Forms refer to the material and efficient aspects of causation, as well as to the merely formal aspect." (Losee, *A Historical Introduction to the Philosophy of Science*, 66.) And as Fulton H. Anderson notes, to Bacon, therefore, "The form of a nature is such that, given the form, the nature infallibly follows. The form is always present when the nature is present." (Anderson, *The Philosophy of Francis Bacon*, 160.)

modern science in one crucial respect. In book 2 of *The New Organon*, Bacon sets out an inquiry into the nature of heat as a model case to illustrate the application of his new method of natural philosophy. His idea is that, by drawing up a "table of degrees or comparison on heat" that covers all "instances meeting in the nature of heat" and all "instances which are devoid of the nature of heat," then identifying among them "example[s] of exclusion or rejection of natures from the form of heat" to eliminate any misconceptions, one can isolate the "form of heat" and thus inductively ascertain its essential nature (NO 2.11–18). For example, since a flame produces both heat and light while the moon produces only light, it is established that the "form of heat" excludes the attribute of light.

In a similar fashion, Bacon comments as follows on phenomena that seem to involve action at a distance:

> As for combination [of the parts of bodies] at a distance, it is uncommon and rare, but occurs in more cases than is recognised. Here are some likenesses of it: when one bubble dissolves another; when medicines draw out humours because their substance is similar; when a string in one stringed instrument causes a string in another to make the same sound, and so on.... It is certainly evident in the magnet and in magnetised iron. (NO 2.48)

With the hindsight of present-day science, we know that the principle behind each of the examples Bacon cites here is different from that of all the others, and magnetism is the only one deemed to be action at a distance in a physical sense. No better understanding of gravity or magnetism can be gained by randomly enumerating various heterogeneous phenomena simply because they appear to entail some kind of action over distance.

Nonetheless, in this investigation of the "nature of heat," performed to illustrate the application of his inductive method, Bacon, at the end of a long-drawn-out discussion, arrives at the conclusion that "heat is a motion which is not uniformly expansive throughout the whole of a body, but expansive through its smaller particles, and is at once checked and repelled and bounced about" (NO 2.20). Through its subsequent more accurate revision by Robert Boyle, this conclusion would eventually prove to have anticipated the nineteenth-century kinetic theory of heat. As the example about action at a distance shows, however, Bacon's procedure can hardly be said to be equally valid in all cases.

Bacon failed to understand that effective observations required some kind of theoretical framework and that in conducting an experiment one first had to posit some kind of hypothesis as the proposition to be verified. He thought that in learning from experience one should rather *discard* all preconceptions and confront the facts with a completely blank and open mind. Thus, his brand of experimentation was not the hypothesis-testing kind whereby one supposes in advance a proposition to be tested, and for this reason his method was not aimed at the discovery of scientific laws in the modern sense.

Moreover, nature as Bacon saw it was entirely qualitative; his method was utterly devoid of anything conducive to precise measurement or quantitative understanding. In his thinking, heat and cold were mutually opposing qualities that did not admit any gradual transition from one to the other, a view closed off to the concept of quantitative temperature. And even though he recognized that "heat is a motion," he made no move toward probing the interchangeability of heat and motion or their quantitative relation. In this respect, Bacon remained on the same plane as Aristotle. "For, in effect," comments one historian, "these [Aristotelian] doctrines [that presumably influenced Bacon] said to the physicist *classify* when they should have said *measure*."[6] In short, the aim of Bacon's inquiry into nature was to acquire knowledge of the "essences" and "forms" of things, not to establish laws in the modern sense. This was the essential defect of his natural philosophy.

Indeed, even within studies in the theory of heat, the field's subsequent development as a modern science was made possible not by any supposed elucidation of the "essence of heat" but rather by the establishment of methods for quantitative measurement of heat and temperature and by the confirmation of the equivalence of heat energy and mechanical work (the first law of thermodynamics). For similar reasons, no amount of enumeration or classification of phenomena involving action at a distance can give rise to a science of magnetism. Magnetic force and electrical force likewise took their places in modern physics only after their magnitudes had been precisely and quantitatively measured and expressed as mathematical functions in the eighteenth century.

In the end, then, Bacon's philosophy had no direct link with the subsequent development of the mathematical natural sciences, and thus did not directly anticipate the new science of physics.

## Section 2—Thomas Browne

Baconianism optimistically advocated that with accumulated observation and experiment, free from preconceptions or prejudice, the essences or truths of nature would emerge of their own accord. Cartesian rationalism, meanwhile, complacently proposed that all the facts of nature could be deduced solely by rigorous reasoning from clear and distinct first principles. In fact, however, neither philosophy succeeded in engendering a new science.

The significance of Bacon's method, nonetheless, is that he tacitly based his classifications on modes of motion, as we saw in his analysis of heat. His thought thus had aspects consistent with mechanism. Indeed, Baruch Spinoza identified Bacon's position as predating Cartesian mechanism, commenting that whereas "Mr. Boyle is trying to show that all the tactile qualities depend only on motion, shape, and the remaining mechanical states … this has already been proved … more than sufficiently by Verulam [i.e., Bacon] and afterwards by Descartes."[7] This view seems to reflect how Bacon was generally understood in the seventeenth century. In *The New Organon*, Bacon does indeed state that "every natural action is transacted by means of the smallest particles, or at least by things too small to make an impression on the senses" (NO 2.6), and moreover asserts that "the school of Democritus [i.e., atomism] … penetrated more deeply into nature than the others" (NO 1.51). In some sense at least, Bacon had embraced atomism.[8]

Thus the mechanist philosophy of the Continent was received relatively favorably in Britain even among the generation of thinkers and intellectuals raised more or less under Bacon's influence. Nonetheless, mechanism underwent considerable change in the process of its integration with Baconianism. Baconian philosophy, especially its empiricism, set the direction for new developments in British scientific thought from the Puritan Revolution (English Civil War) through the Restoration to the establishment of the Royal Society in 1660.

Two figures in particular stand out as embodying the peculiar changes Cartesianism underwent under Baconian influence as it was imported into Britain: Henry Power (1623–1668) and his mentor Thomas Browne (1605–1682).

Browne, a physician educated at Oxford and at universities on the Continent, wrote several books on the science of his day; but with no original scientific achievements of his own, and unable to accept Copernican theory, he cannot be said to have been at the vanguard of scientific thinking. Indeed, mention of Browne is absent not only from comprehensive science histories, like those by J. D. Bernal and Friedrich Dannemann, but even from works focusing on the science of Browne's time, such as Richard S. Westfall's *The Construction of Modern Science* and Herbert Butterfield's *The Origins of Modern Science*. Browne is remembered in science history only as the first person to use the word "electricity."[9] But although he is no longer regarded as a scientist, according to one scholar "Browne's contemporaries thought of him as a serious scientist."[10] What this signifies is that Browne, rather than anticipating the future during a transitional phase in scientific thinking, typified that transition itself. "Perhaps no writer," remarks one scholar, "is more truly representative of the double-faced age in which he [Browne] lived, an age half scientific and half magical, half sceptical and half credulous, looking back in one direction to Maundeville, and forward to Newton."[11]

In *Religio Medici* (The religion of a physician), published in 1643, Browne remarks, "I think at first a great part of Philosophy was Witchcraft; which … was indeed no more but the honest effects of Nature: what, invented by us, is Philosophy, learned from [the Devil], is Magick."[12] Magic and natural philosophy, he suggests, are alike in seeking to uncover the secrets of nature, and differ only in the route by which that knowledge is derived. Thus Browne does not from the outset reject medieval notions about miracles and marvels. On the other hand, as a Baconian he proclaims the importance of experiment and fiercely rebukes blind deference to authority. In *Pseudodoxia epidemica* (which carries the subtitle *Or: Enquiries into Very Many Received Tenents and Commonly Presumed Truths*), published in 1646, he deplores Dioscorides, Pliny, Albert the Great, and others for their uncritical acceptance of authority and traditional belief, and through specific experiments thoroughly lays to rest a number of strange old myths, such as those about the effects of garlic or diamonds on magnets. Note, however, that he seems to have conducted those actual tests precisely because he thought no purported marvels could be categorically ruled out. In this way, Browne brought almost to completion the mop-up work—testing, verifying, and dispelling folk beliefs and superstitions about magnets—begun much earlier by Roger Bacon and carried further by Della Porta and Gilbert. These experiments by Browne are recorded in chapter 3 of book 2 of *Pseudodoxia epidemica*, and need not be recounted in detail here anew.

With respect to Cartesianism, however, the key point of Browne's thinking is his assertion in *Pseudodoxia epidemica* that theories of magnetism as action mediated through some hypothetical matter, such as Descartes's "grooved particles," "are conceits of eminent use to salve Magnetical Phenomena's."[13] What he means by "salve" (salvage) the phenomena is the same as in ancient astronomy: explaining them, by means of an artificial model, in such a way that at least they need not be dismissed as illusory or unaccountable. In other words, while granting that Cartesian theory provides, provisionally, a useful and plausible hypothesis, Browne does not regard it as the ultimate and fundamental solution. On the one hand, unlike Bacon, he acknowledges the utility and necessity of theoretically framed hypotheses, but on the other he rejects the *a priori* argumentation of Cartesian rationalism, realizing that it supplies only working hypotheses to be verified experimentally. Browne's approach is founded on a firm conviction that the truth or falseness of any proposition, whether magical or folkloric or Cartesian, would be distinguished as a matter of course when the proposition is tested in the cauldron of experiment. Far from being unique to Browne, this view became common among subsequent British natural philosophers in general. Robert Hooke eloquently expresses this new intellectual mood when he remarks in a 1664 work that "if the most Acute *Des Cartes* had applied himself experimentally to have examined what substance it was that caused that shining of the falling Sparks struck from a Flint and a Steel, he would certainly have a little altered his *Hypothesis*."[14] Thus was idealist Cartesian natural philosophy transplanted to the British soil of Baconian experimental science.

## Section 3—Henry Power and "Experimental Philosophy"

Henry Power, whose thinking was strongly influenced by Browne, was one of Britain's earliest adherents to Cartesian mechanism and the first to discourse on magnetism from that perspective. Power was a science enthusiast who studied medicine at Christ's College, Cambridge—the center of Cartesian thought in Britain at that time—before establishing a medical practice in Halifax.

At Cambridge in 1653, Power conducted experiments on air pressure. From 1660 to 1661, with the help of mathematician-astronomer Richard Towneley, he conducted further experiments with mercury in a Torricelli tube set up so that air remained in the top of the tube, above the mercury. By moving the tube to different elevations, Power and Towneley demonstrated (albeit only when the pressure is less than atmospheric pressure) that at a constant temperature the pressure and volume of a gas are inversely proportional. Soon thereafter, Robert Boyle, assisted by Robert Hooke, made similar measurements using a U-shaped tube and showed that the same rule applied at pressures higher than atmospheric pressure as well. Because Boyle's results were published first (in 1660 and 1662), the simple relationship between the pressure and volume of a gas in an isothermal change is now widely known as Boyle's law; in fact, however, Power and Towneley made their measurements prior to Boyle's and Hooke's research.[15]

Around the same time, Power became one of the first scholars in Britain to make observations of nature using a microscope, the results of which appeared in his book *Experimental Philosophy*, published in 1664.[16] This work attracts little attention today, primarily because it was eclipsed almost immediately by the 1665 publication of Hooke's *Micrographia*, in which Hooke's microscopic observations are presented with striking copperplate engravings. Nonetheless, Power's work remains "the first English book on microscopy, and the best description of microscopical observations of its time."[17]

On the one hand, Power admits the existence of "atoms" in the sense of "corpuscles" that constitute matter, as can be surmised from *Experimental Philosophy*'s subtitle: *Containing New Experiments Microscopical, Mercurial, Magnetical, With some Deductions, and Probable Hypotheses, raised from them, in Avouchment and Illustration of the now famous Atomical Hypothesis.* On the other hand, however, he follows Descartes in denying that a vacuum can exist. In Power's view, air is comprised of atoms floating in an all-permeating "aether" that has a "Subtil and Penetrative Nature" and "shoots it self through all Bodies whatsoever" (EP, 103). Indeed, he rejects the view that considers the void above the mercury in a Torricelli tube to be "an absolute Vacuity" as "not only non-philosophical, but very ridiculous." The first reason he gives for this is that "the Space deserted hath both Longitude, Latitude, and Profundity, therefore a Body; for the very nature of a Body consists onely in extension, which is the essential and unseparable property of all Bodies whatsoever"—a rationale in which the influence of Descartes is conspicuous. Power's second reason for rejecting the idea of an absolute vacuum is that light passes through such a space, and the third is that "the Magnetical Efluxions of the Earth are diffused through that seeming Vacuity, as *per Experiment*" (EP, 95). The mechanistic view was that both light and magnetic "effluvia" must either be kinds of matter themselves or else be conveyed by the mediation of some kind of matter. In *Treatise on Light* (1690), Christiaan Huygens, another seventeenth-century natural philosopher influenced by Descartes, likewise declares that, since the space in a Torricelli tube that has been voided of air nonetheless continues to allow light to pass through it, "this proves that a matter different from air exists in this tube."[18]

Power nonetheless makes a number of important changes to Cartesian mechanism. His first addition to Cartesianism within the framework of mechanical philosophy is his proposition that "Motion … may be indefinitely swift or slow, and yet never come to a quiescency," or in other words that "there is no such thing in the World as an absolute quiescence" (EP, b3v–b4r). This view, subsequently adopted by Boyle and Hooke, became a characteristic of British mechanism, which gave precedence to motion over figure.

The second change Power makes is to posit the existence of "subtle spirits" as the fundamental substance behind all motion apparent in the natural world and as the physical basis of matter's activity:

> First, then, we have not those narrow conceptions of these subtle Spirits to think that they are onely included within the Bodies of Animals, or generated … there, but we doe believe that they are universally diffused throughout all Bodies in the World, and that Nature at first created this aetherial substance or subtle particles, and diffused them throughout the Universe, to give fermentation‡‡ and concretion to Minerals; vegetation and maturation to Plants; life, sense, and motion to Animals; And indeed, to be the main (though invisible) Agent in all Natures three Kingdoms Mineral, Vegetal, and Animal. (EP, 61)

Descartes had regarded matter as inert and passive and all the motion it exhibited as originating in an initial impetus imparted by God. Power, however, asserted that lifelike "subtle spirits" pervaded the natural world as a specific kind of being and the basic substance of matter's activity. In doing so, he introduced into Cartesian mechanism an element essentially alien to it. Such thinking was carried on by Boyle as well, and eventually paved the way for the relatively easy acceptance in Britain of Isaac Newton's notion of "active principles," which the mechanistic natural philosophy could not accommodate.

But what sets Power categorically apart from Descartes is the former's firm conviction that, with the advancement of observational technology, all theories of natural philosophy, whether about the existence of atoms or any other topic, would eventually become experimentally testable and furthermore must be so tested. In this connection Lynn Thorndike aptly notes that "a sign of the growing popularity of experimentation and the experimental method [in that period] is seen in the *Experimental Philosophy* of Henry Power"[19]. After all, the very title *Experimental*

---

‡‡ At that time, the term "fermentation" was used to denote chemical reactions in general.

*Philosophy* plainly reflects the influence of Baconian empiricism as a challenge to Cartesian rationalism.

Furthermore, Power favors "experimental philosophy" as a methodological principle over "mechanical philosophy" as a working theory. He refers to the "Patriark of Experimental Philosophy, the Learned Lord Bacon" (EP, 82); on the title page of *Experimental Philosophy* he quotes the Latin text of Bacon's remark in *The New Organon* that "if Democritus had seen a magnifying glass, he would perhaps have jumped for joy, thinking a means of viewing the atom … had been invented" (NO 2.39).

As attested by the quotation just mentioned, particularly prominent in Power's thinking are his boundlessly optimistic expectations for and faith in the microscope, the telescope, and other newly developed observational equipment that greatly expanded the range of human sensory capabilities. In the preface to *Experimental Philosophy*, which opens with the words "Dioptrical Glasses are but a Modern Invention," Power notes that

> howsoever though the faculties of the soul of our Primitive father *Adam* might be more quick & perspicacious in Apprehension, than those of our lapsed selves; yet certainly the Constitution of *Adam's* Organs was not divers from ours, nor different from those of his Fallen Self, so that he could never discern those distant, or minute objects by Natural Vision, as we do by the Artificial advantages of the *Telescope* and *Microscope*. (EP, a4r)

Power believes that, by means of the "Modern Engine" of the microscope, people could now hope to actually see "what the illustrious wits of the Atomical and Corpuscularian Philosophers durst but imagine" (EP, b2r), namely,

> the Magnetical Effluviums of the Loadstone, the Solary Atoms of light (or *globuli aetherei* of the renowned *Des-Cartes*), the springy particles of Air, the constant and tumultuary motion of the Atoms of all fluid Bodies, and those infinite, insensible Corpuscles (which daily produce those prodigious … effects amongst us:). (EP, c2v)

By revealing the previously unknown structures of the microscopic world in their true complexity and intricacy, observation through microscopes infused the mechanists and atomists with greater confidence.

The new inventions of that time—Galileo's telescopes, the microscopes used by Power and Hooke, the vacuum pump developed by Boyle and Hooke, and so on—were indeed engines of civilization on a par with today's giant radio telescopes for observing distant nebulas and enormous particle accelerators for probing the very limits of matter. The early-modern scientific instruments also required considerable cost and labor to manufacturer, and their application by the scientists of that time effectively transformed natural-scientific enquiry from haphazard, hit-and-miss observation into purposive, systematic experiment and measurement.

Nor can we overlook the role played by such devices as the barometer and thermometer, also developed around that time, in showing the importance of quantitative measurement. In book 2 of *Experimental Philosophy* we find numerical values for measured pressure minutely recorded in terms of the height of a mercury column—measurements the like of which are absent from the works of even Bacon, Descartes, and Gilbert. In any case, such inventions clearly elevated Bacon's "experimental philosophy" to a whole new level, and for Power this enhancement pointed to the potential for experimentally verifying the previously only imagined notions proposed by mechanism and atomism.

Power devotes book 1 of *Experimental Philosophy* to "microscopical observations"; book 2 to "mercurial experiments," meaning experiments dealing with air pressure and vacuums; and book 3 to "magnetical experiments." Book 3 was written in 1663 as a refutation of the 1645 treatise *Nova demonstratio immobilitatis terrae* (New demonstration of the Earth's immobility) by Jacques Grandami, a French Aristotelian and Jesuit priest who continued to deny that the Earth rotated on its own axis.[20] Power's work is rarely mentioned in histories of electromagnetic science, perhaps because its subject was thus no longer novel even then, and also because it has been regarded as containing no especially remarkable observations or new knowledge about magnets. Even in Paul Fleury Mottelay's extensive *Bibliographical History of Electricity and Magnetism*, it appears only as a title in the chronological table at the back of the book; it is not mentioned in the main text. Nonetheless, book 3 of *Experimental Philosophy* is a typical expression of the now largely disregarded mechanistic understanding of magnetism that was current in Power's day, and as such it deserves a brief exposition at this point in our account.

In the first chapter of book 3, Power holds Grandami to be in error in his claims "that the virtue of the Magnet, and all Magnetick Bodies, is purely immaterial, and a bare simple Quality" and "that it proceeds intrinsecally from the proper form of the Loadstone" (EP, 154). Following this up in chapter 2, Power insists instead that "electrical bodies" and "magnetical bodies" "both work by Corporeal Effluviums." He then concludes that electrical effect and magnetic effect differ from each other in two respects. Firstly,

> the Effluviums of the one [i.e., electrical bodies], (as being more Gross and Corporeal) are intercepted by any medium; but Magnetical Effluviums are hindered (because of their exceeding tenuity) by the interposition of no Body whatsoever. Secondly, Whereas Electrical fluors do presently recoyl by short streight lines to their Bodies again, Magnetical Atoms do not so; but do wheel about, and, by a Vortical motion, do make their return unto the Loadstone again, as *Des-Cartes* hath excellently declared. (EP, 156)

It was the absence of a shielding effect in magnetic force that had prompted Gilbert to regard it as soul-like action at a distance and to distinguish it from electrostatic force as physically mediated action. Power, however, simply by taking "magnetical effluvia" to be tenuous and permeable, is able to readily accept Descartes's vortex theory as more or less self-evident without concerning himself too much about the differences between the two kinds of force.

Furthermore, Power goes into considerable detail in describing the motions of "magnetical atoms" and "magnetical effluvia" as the agents of those vortices:

> The Magnetical Effluviums do not proceed intrinsecally from the Stone [i.e., lodestone], but are certain extrinsecal particles, which approching to the Stone, and finding congruous pores and inlets therein, are channel'd through it; and having acquired a Motion thereby, do continue their Current so far, till being repulsed by the ambient Ayr, they recoyl again, and return in a Vortical Motion, and so continue their revolution for ever, through the Body of the Magnet. (EP, 157)

Facts cited as empirical evidence for this model included that, although a magnet loses magnetic power at red heat, it can, depending on how it is positioned, reacquire that power when cooled; and that its polarity can even be reversed by that process. Although Gilbert had already noted these facts themselves, his interpretation was that the return of magnetic power was a restoration of the "magnetic form" that had been disrupted by the heat. The new interpretation by Power, however, was thoroughly mechanistic, based on the idea that the magnet's pores are destroyed by the heat but then remade by the dynamical impact of the "magnetical effluvia" circulating through and around the Earth.

Regarding the magnet's directionality as well, Power similarly supposes that

> the stream of Atoms from without, by beating upon the Stone, do turn it to and fro, till they have laid it in such a Position as is fittest for them to run through it, as a stream of water turns a hollow trunk of wood, or a long stick, till it come to lye parallel to its current. (EP, 158)

This explanation, too, is typical of traditional mechanical philosophy. Of course, it requires that there actually be such a flow of "magnetical effluvia" through the Earth's poles, but in Power's thinking it is in fact the effluvia's existence, conversely, that is inferred from the directionality of magnets:

> Since a constant, steddy, and polary direction of parts is onely observable in Bodies Magnetical, we have reason to think and believe, that these Magnetical *Effluvia* (which are the cause of this peculiar direction) are not only transmitted and channel'd through the Earth, but through many other Coelestial Bodies also, as [the Sun, the moon, Mercury, Venus], and, perchance, the rest of the Planets yea and Fixed Stars too. (EP, 160)

It cannot be denied that the foregoing thesis cites empirical observations about magnetization in support of the theoretical model of magnets as having pores. Nonetheless, the argument shows only that such an interpretation is possible; it does not present anything that directly and independently proves the existence of the crucial "magnetical effluvia" themselves. Ultimately, this "explanation" rests on commuting the whole mystery of magnetic force to the existence of special kinds of matter called "magnetical atoms" and "magnetical effluvia." This point becomes even clearer when we consider the contributions of Robert Boyle to the debate.

## Section 4—The "Corpuscular Philosophy" of Robert Boyle

Robert Boyle (1627–1691) was born as the son of Ireland's first Earl of Cork four years after the birth of Henry Power and the year following Francis Bacon's death. After receiving his primary education at Eton College, at the age of eleven Boyle, accompanied by his tutor, embarked on a study tour of the Continent, traveling to Switzerland, France, and Italy, among other countries, and it was during this trip, reportedly, that his eyes were opened to the new developments in science. He was heavily influenced by Francis Bacon and Rene Descartes, especially by the former's *New Organon* and the latter's *Principles of Philosophy*. Boyle returned to Britain in 1644, the year the English Civil War broke out, and from his studies in medicine, particularly pharmacology, he subsequently broadened his intellectual interests to chemistry and then to physics.

Boyle's renown in the scientific community derives from a series of experiments he conducted on air and vacuums using the vacuum pump he developed in collaboration with Robert Hooke, and in particular from his announcement of the fact now known as Boyle's law. The discovery of Boyle's law was an outstanding example of the praxis of "new science" in the early seventeenth century, which sought to establish laws based on purposive experiment and elaborate measurement—that is, systematic experiment and quantitative measurement using equipment devised for the specific task at hand—and expressible in the language of mathematics. The results of Boyle's seminal research on air were published in his 1660 treatise *New Experiments Physico-mechanicall, Touching the Spring of the Air, and its Effects* and in the revised edition of the same, published two years later. In addition to thrusting Boyle to the very forefront of the science of his day, this research formed the basis of his own subsequent inquiries and the intellectual starting point to which he often returned.

Boyle called his view of matter "corpuscular philosophy." Heavily influenced in his youth not only by Descartes but also by Gassendi, Boyle developed his own thinking along lines characterized firstly by the idea that the natural world operated like an automaton and secondly by an understanding of matter according to which all the qualities it exhibited should be explained from that conception of nature. In short, his was a thoroughgoing mechanical philosophy. This meant that for Boyle the difference between atomism, which admitted vacuums, and Cartesianism, which did not, was not an especially important issue; rather, he focused on what the two positions had in common: mechanism. Accordingly, although chemistry was the main arena of his work, "the development of a satisfactory chemical theory as such was not Boyle's goal. Chemistry represented to him a means to demonstrate the validity of the mechanical philosophy of nature."[21]

Descartes, too, had expressed the view that the world was an automaton, but Boyle referred to the idea even more frequently. In the 1666 work *The Origin of Forms and Qualities according to the Corpuscular Philosophy*,[22] in which Boyle put forward his scientific thought ("corpuscular philosophy") in an almost complete form, we find the following:

> According to our doctrine, the world we live in is not a moveless or indigested mass of matter, but an *Ἀυτοματον* [automaton], or *self-moving engine*, wherein the greatest part of the common matter of all bodies is always (though not still the same parts of it) in motion. (*Works* 3:34; see also 3:48)

Similarly in *The Excellency of Theology compar'd with Natural Philosophy*, published in 1674, Boyle writes that "the world being but, as it were, a great piece of clock-work, the naturalist, as such, is but a mechanician" (*Works* 4:49); and the same point is made in one of his earliest books, *The Sceptical Chymist* (1661).[23] Hylozoism, which since ancient times had attributed all self-motion to some kind of soul, was completely abandoned. In Boyle's case, however, this shift had nothing to do with atheism; he was, on the contrary, a committed Christian. To him, the fact that the world was a purposely constructed machine exquisitely crafted down to the minutest detail meant that it could not be the product of mere chance; there had to be a superior intelligence—God—who existed outside of nature and who had planned, created, and set in motion the world.[24] Boyle maintained an unwavering conviction in this view of nature throughout his life.

But the crucial point is that, unlike Descartes, who had elaborated a grand yet simple cosmology contained entirely within the sphere of language, Boyle, a chemist, regarded that same mechanical philosophy as something whose claims must be individually verified by experiments dealing with the complex and diverse physical and chemical properties of terrestrial matter. He states this intention clearly in his preface to *The Origin of Forms and Qualities*:

> That then which I chiefly aim at, is to make it probable to you by experiments (which I think hath not yet been done) that almost all sorts of qualities, most of which have been by the schools either left unexplicated, or generally referred to I know not what incomprehensible substantial forms, may be produced mechanically; I mean by such corporeal agents, as do not appear either to work otherwise than by virtue of the motion, size, figure, and contrivance of their own parts (which attributes I call the mechanical affections of matter, because to them men willingly refer the various operations of mechanical engines). (*Works* 3:13)

The significance mechanism held for Boyle is aptly expressed by one historian when he points out that for Boyle "the atomic and Cartesian philosophies were working hypotheses, heuristic instruments in his experimental researches."[25] Following the tradition of British mechanism, Boyle sought not to provide a reasoned "proof" of mechanism but rather to test it.

The intellectual systems Boyle confronted and struggled against were, on the one hand, the Aristotelian world view based on the four elements of earth, water, air, and fire, together with the medieval Scholastic philosophy that attributed the peculiar qualities of each kind of substance to its having a specific "substantial form"; and, on the other, the position adopted by the Paracelsian chemists, who considered the three primary principles of "salt," "sulfur," and "mercury" to be the bases of all matter. In *The Sceptical Chymist*, Boyle elaborates a strategy on both of those fronts, reasoning that "the principles of the world, as it now is, [are] three, *matter*, *motion*, and *rest*," and that from these three can be deduced "colours, odours, tastes, fluidness and solidity, and those other qualities that diversifie and denominate bodies." Indeed, he goes even further, declaring that even the magnitude and figure of bodies should be deducible from matter and motion, because "the latter of these [i.e., motion] variously agitating, and, as it were, distracting the former, must needs disjoyne its parts; which being actually separated must each of them necessarily both be of some size, and obtain some shape or other."[26] In Boyle's view, then, both the Scholastics' "four elements" and the Paracelsians' "three principles" would be better explained in terms of the states of combination and states of motion of homogeneous constituent atoms.

In *The Origin of Forms and Qualities*, which has been described as presenting "a really complete exposition of Boyle's own views on the underlying structure of matter, and also an enunciation of the 'corpuscular philosophy,'"[27] he requires, furthermore, the existence of "one catholick or universal matter common to all bodies" that is also "a substance extended, divisible, and impenetrable." He reasons as follows:

> But because this matter being in its own nature but one, the diversity we see in bodies must necessarily arise from somewhat else than the matter they consist of. And since we see not how there could be any change in matter, if all its (actual or designable) parts were perpetually at rest among themselves, it will follow, that to discriminate the catholick matter into variety of natural bodies, it must have motion in some or all its designable parts: and that motion must have various tendencies, that which is in this part of the matter tending one way, and that which is in that part tending another. (*Works* 3:15)

Thus Boyle argues that matter and motion are the "universal

principles" of bodies. In *About the Excellency and Grounds of the Mechanical Hypothesis*,[28] written in 1674, he puts it plainly and succinctly: "Of the principles of things corporeal, none can be more few, without being insufficient, or more primary, than matter and motion" (*Works* 4:77; see also 4:70). By giving priority to motion over figure, Boyle's position surpasses both Cartesian mechanism, which placed figure, magnitude, and motion on the same plane, and earlier atomism, which tended to give figure priority and reduce the qualities of atoms primarily to figure.

This aspect of Boyle's thinking is most clearly evident in his conception of heat and cold. *Experiments, Notes, &c., about the Mechanical Origin or Production of Divers Particular Qualities* (hereafter *Experiments and Notes*),[29] published in 1675, contains a series of studies in which Boyle sought to account for all qualities in terms of matter and motion. A compilation of a number of his earlier treatises and a comprehensive statement of Boyle's mechanistic view of matter (his corpuscular philosophy), this work sets out his experiments and thinking concerning almost all the known qualities and effects exhibited by matter, including heat and cold, taste, odor, chemical properties, volatility and nonvolatility, corrosiveness, magnetism, and electricity. In Boyle's classification, heat and cold are among the "first qualities," as are the "sensible qualities" of tastes and odors; volatility, fixity, and corrosiveness are among the "second qualities"; and magnetism and electricity belong to the "occult qualities" (*Works* 4:235). It is noteworthy that he gives precedence to heat and cold, since his treatment of these qualities is in fact the most successful part of his corpuscular philosophy.

In Aristotelian theory, each quality in such pairs as heat/cold and wetness/dryness was an independent, irreducible opposite of the other; cold was not simply a lack of heat, nor dryness merely a lack of moisture. Meanwhile, atomism since Plato had sought to reduce such sensible qualities to geometrical shapes. In Gassendi's atomism, for example, a body's coldness is attributed to its containing "frigorific atoms" that have sharp protuberances or teeth, the sensation of cold being the stimulation of our sense of touch by those barbs. Here again, cold is not simply a deficiency of heat. Thus, although Aristotelianism and early atomism constituted quite different viewpoints, they were alike in maintaining that there could be no gradual progression between heat and cold and that therefore heat and cold were mutually opposing qualities that could not be quantitatively unified.

In response to this, Boyle, taking his cue from Francis Bacon, attributes heat—or rather the true nature of a body's being hot—to the intensity of motion of the minute particles of which matter is composed. In "Of the Mechanical Origin of Heat and Cold" he supposes that "the nature of heat consists either only or chiefly in the local motion of the small parts of a body" (*Works* 4:244); and as a condition of such "local motion" he then posits that (to translate into somewhat more modern terms than his) this is a motion of microscopic particles that is itself insensible, chaotic, and vigorous. Boyle's understanding here is thus very close to the present-day conception of thermal motion.

One crucial point lies in Boyle's reference in the same treatise to the possibility of mechanical motion's being converted into heat:

> If some attention be employed, in considering the formerly proposed notion of the nature of heat, it may not be difficult to discern, that the mechanical production of it may be divers ways effected. For, excepting in some few anomalous cases, … by whatever ways the insensible parts of a body are put into a very

> confused and vehement agitation, by the same ways heat may be introduced into that body. (*Works* 4:245–46)

Furthermore, in the same passage in which he asserts that heat is motion, Boyle remarks of "cold" that "when the minute parts of a body are in, or arrive at such a state, that they are more slowly or faintly agitated than those of our fingers, or other organs of feeling, we judge them cold"; and that therefore "a privation or negation of that local motion, that is requisite to constitute heat, may suffice for the denominating a body cold" (*Works* 4:244).

Although the extensive concept of "heat value" and the intensive concept of "temperature" had yet to be clearly differentiated, nonetheless Boyle thus unified "hot" and "cold" as referring not to a qualitative and absolute opposition but rather to a quantitative and relative difference of degree, specifically, in the vigor of thermal motion. This was the first step in the transition from a qualitative conception of nonmeasurable "hot" and "cold" to a quantitative understanding of measurable degrees of temperature ("high" and "low"). Along with Galileo's invention of the thermometer, this laid one of cornerstones for the evolution of thermal studies into a mathematical, modern physical science.

## Section 5—Mechanism and "Magnetical Effluvia"

Let us now consider Boyle's view of the central topic of this book: force. Boyle rejected the notion of "attraction," preferring to regard what appeared to be attraction operating at a distance as actually the result of some impulse or pressure caused by the direct contact of a material medium between the bodies acted upon. To that extent his position was in keeping with mechanism in general, but Boyle had especially compelling motives and grounds for holding that view.

As we saw earlier, the starting point for Boyle's physical research was his series of experiments on vacuums. Until that time, the fundamental understanding of the operation of an ascending water pump was based on the widespread hypothesis—or rather, preconception—that "nature abhors a vacuum." In his 1674 work *Of the Cause of Attraction by Suction*,[30] Boyle challenges this Aristotelian theory of nature's "abhorrence of a vacuum" with empirical fact, pointing out that, whereas "according to their hypothesis, water and other liquors should ascend upon suction to any height to prevent a vacuum, [this] is not agreeable to experience" (*Works* 4:132). He shows that when a Torricelli tube is placed in the chamber of a vacuum pump and the air is exhausted from the chamber, the column of mercury in the tube descends as the air is exhausted, a fact that exposes the error of explanations based on "abhorrence of a vacuum." In Boyle's view, it is not that the vacuum in the Torricelli tube sucks the column of mercury up but rather that the air in contact with the surface of the mercury outside the tube pushes it down, thereby pushing the mercury inside the tube up. This fact seems to have convinced him that other phenomena widely described as "attraction" should likewise be understood as resulting from direct contact force, that is, pressure. Indeed, from there he leaps to the generalization that "the examples of suction are not the only noted ones of attraction, that may be reduced to pulsion" (*Works* 4:129).

A corollary of this line of thinking is that magnetic force, normally considered the archetype of action at a distance, should also be explained as operating by mediation. Boyle does not, however, refer to magnetic force itself; his treatment of the topic is entirely limited to the magnetization (he calls it "excitement") of iron. His "Experiments and Notes about the Mechanical Production of Magnetism," which forms part of *Experiments and Notes*, is aimed at showing that magnetism is not what the Scholastics called a "substantial form" and that magnetization (magnetic induction) is a "mechanical operation"—or in other words, that "even magnetic qualities may be mechanically produced or altered." He notes, for example, that if an iron bar is left standing vertically it is "in tract of time, by the continual action of the magnetical effluvia of the terraqueous globe, turned into a kind of magnet" (*Works* 4:342). Likewise, a length of iron heated to red heat and then set to cool in a vertical position will also be magnetized. Boyle's proposed explanation of such facts is that

> the great commotion of the parts, made by the vehement heat of the fire, disposed the iron, whilst it was yet soft, and had its pores more lax, and parts more pliable, disposed it, I say, to receive much quicker impressions from the magnetical effluvia of the earth, than it would have done, if it had been still cold. (*Works* 4:343)

This is a typically mechanistic account of the phenomenon of magnetization (magnetic induction). In a somewhat peculiar treatise of 1671 titled *Of the Systematical or Cosmical Qualities of Things* Boyle elaborates this argument further but without any essential differences.[31]

In "Experiments and Notes about the Mechanical Production of Magnetism," Boyle thus arrives at the conclusion that

> the change of the magnetism, communicated to iron, may be produced, at least, in good part, by mechanical operations, procuring some change of texture in the iron. (*Works* 4:345)

It is clear from the above that Boyle, like Power, accepts without alteration the Cartesian model according to which magnets and iron have pores; and that he believes, again following Power, that what flows through those pores is “the magnetical effluvia of the earth.” Like Descartes’s “grooved particles,” in this theory the “magnetical effluvia” circulate through and around the Earth via its poles, and their impact opens up the pores in red-hot iron and gives directionality to magnets. Thus, once one accepts the existence of things like “magnetical effluvia” and “pores in iron,” it seems plausible to attribute magnetization to mechanical alteration of the iron’s internal structure, without having to describe magnetism in terms of either the “substantial forms” of Scholasticism or the “hidden virtues” of Renaissance magic.

The same can be said of Boyle’s explanation of electric force. “Experiments and Notes about the Mechanical Origin or Production of Electricity,” another component of *Experiments and Notes*, begins as follows:

> That it is not necessary to believe electrical attraction (which, you know, is generally listed among occult qualities) to be the effect of a naked and solitary quality, flowing immediately from a substantial form; but that it may rather be the effect of a material effluvium, issuing from, and returning to the electrical body (and perhaps in some cases assisted in its operation by the external air) seems agreeable to divers things, that may be observed in such bodies and their manner of acting. (*Works* 4:345–46)

But although they share the mechanical philosophy, in fact Boyle’s conception of “magnetical effluvia” and Descartes’s of “grooved particles” are quite different from each other in character. To be sure, they are alike in describing the magnetization of iron as occurring by the mechanical operation (direct impact) of the supposed effluvia or particles. But whereas Descartes’s “grooved particles" are distinguished from other matter solely by their screwlike figure and gyrating motion, and in that sense are in accordance with mechanism, Boyle makes no mention of such winding or rotating action. Rather, Boyle's "magnetical effluvia" and "effluvia from an electrical body" are themselves each a specific kind of substance with particular properties and operations, distinct from other kinds of matter. In Boyle's view, the fact that magnetical effluvia have a distinctive action is due to the distinctive essential nature of magnetical effluvia themselves.[32]

In this respect Boyle's "effluvia" are like Power's "subtle spirits," each type of effluvium or spirit having its own specific operational capacities and thus contradicting the mechanistic idea of matter as fundamentally passive and devoid of quality. This view is clear in Boyle's conception of air, as we shall see in the following section.

## Section 6—The Admission of Specific Operational Capacities

What Boyle essentially did was to introduce a specific type of "effluvium" for each particular type of force, each "occult quality." This approach inevitably invites the charge that it is inconsistent with Boyle's own basic stance of recognizing only "one catholick or universal matter," as previously noted. It seems, however, that Boyle's understanding of "universal matter" as "impenetrable" applied to solids and liquids but not to things such as air and effluvia. Indeed, in *The Origin of Forms and Qualities*, whereas metals, earth, and water are regarded as consisting of "universal matter," gases are not discussed in that way.

The same distinction can be inferred from the passage in *Of the Systematical or Cosmical Qualities of Things* where Boyle states that

> besides those more numerous and uniform sorts of minute particles that are by some of the new philosophers thought to compose the aether …, there may possibly be some other kind of corpuscles fitted to have considerable operations, when they find congruous bodies to be wrought on by them. (*Works* 3:316)

Boyle considered cosmic space to include many kinds of effluvia each with its own special properties. In this respect his mechanism departed widely from Cartesian mechanism. In any case, this departure opened up the possibilities that led to Boyle's subsequent development of chemical theory, particularly regarding the chemistry of gases.

Note, however, that for Boyle, who placed great emphasis on experiment, there was no *a priori* requirement for "magnetical effluvia" or "electrical effluvia" to exist. Their introduction into his theory was based on the peculiar view of air that he developed from his own vacuum experiments, the touchstone of his scientific thinking. He observed that when an animal or flame was placed inside a container and the air then evacuated from the container using a vacuum pump, then animal died or the flame went out. Boyle ascertained from this that air was essential for sustaining life and fire, and was convinced by this that air contained some special component indispensable to life and combustion. This was the genesis of research that led to the discovery that gases were not single-component substances, and hence was the starting point for the discovery of oxygen by Antoine Lavoisier and others.

Boyle's findings sent his imagination soaring into the heavens and delving into the depths of the Earth. Reflecting on those vacuum experiments in his 1674 work *Suspicions about Some Hidden Qualities in the Air*,[33] he relates that

> the difficulty we find of keeping flame and fire alive, though but for a little time, without air, makes me sometimes prone to suspect, that there may be dispersed through the rest of the atmosphere some odd substance, either of a solar, or astral, or some other exotic nature, on whose account the air is so necessary to the subsistence of flame

and concludes that "there may be latent qualities in the air." (*Works* 4:90).

Thus air is no longer regarded as consisting of the quality-less, passive, homogeneous "minute matter" supposed by Descartes, nor, of course, as the simple element that many following Aristotle believed it to be. Boyle's conception of air is that it "is not, as many imagine, a simple and elementary body, but a confused aggregate of effluviums from ... differing bodies" (*Works* 4:85).

At the same time, Boyle conjectures that "as divers wonderful and peculiar operations of the loadstone ... were not discovered till of later ages, ... so there may be other subterraneous bodies, that are endowed with considerable powers"; and that "some, if not many, of those bodies and their effluxions, may be of a nature quite differing from those we take notice of here about us, and consequently may operate after a very differing and peculiar manner." He also supposes that, in the air, "besides those vapours and exhalations, which by the heat of the sun are elevated into the air ... , there are, at least in some times and in some places, store of effluviums emitted from the subterraneal parts of the terrestrial globe." His further supposition that "the sun and planets (to say nothing of the fixed stars) may have influences here below distinct from their heat and light," prompts him to state: "it seems not absurd to me to suspect, that the subtil, but corporeal, emanations even of these bodies may (sometimes at least) reach to our air, and mingle with those of our globe" (*Works* 4:85–86).

Following this line of thought, Boyle maintains that

> it may not seem altogether improbable, that some bodies, we are conversant with, may have a peculiar disposition and fitness to be wrought on by, or to be associated with, some of those exotic effluvia, that are emitted by unknown bodies lodged under ground, or that proceed from this or that planet. (*Works* 4:95)

Even considering just the Earth, he proposes that

> there may be in those vast internal parts of the earth, whose thin crust only has been here and there dug into by men, considerable masses of matter, that may have periodical revolutions, or accensions, or eustations, or fermentations, or, in short, some other notable commotions, whose effluvia and effects may have operations, yet unobserved, on the atmosphere, and on some particular bodies exposed to it. (*Works* 4:98)

In his view, the Earth's magnetism is one of the few known examples of operations caused by such "effluvia" emanating from underground. To see this as anticipating the present-day understanding that the rotation of subterranean magma generates the Earth's magnetic field may be reading too much into Boyle's conjecture. Nonetheless, this was the unique and intriguing—though so far largely overlooked—picture of the Earth that he described. It was upon this vision of the planet and this conception of air, furthermore, that the existence of "magnetical effluvia," the crux of his magnetic theory, rested.

The key factor in Boyle's mechanism is that, while considering sensible qualities, occult qualities, and so on as the phenomena to be explained, it nonetheless eschewed the Cartesian doctrinaire approach of top-down, *a priori* deduction from first principles in favor of an empiricist stance emphasizing experiment. It therefore remained flexible enough to provisionally accept the possibility of certain things—occult qualities, for instance, or a primitive substance in the air that sustains life and combustion—even though they were not yet adequately explained. Indeed, in Boyle's mechanism the possible existence of any such quality had to be accepted as long as it was indicated experimentally.

Early in *Of the Systematical or Cosmical Qualities of Things*, Boyle forewarns the reader that

> lest you should think, that under the name of cosmical qualities I should introduce chimaeras into natural philosophy, I must betimes advertise you that you will meet with divers particles in the following discourse, fit to shew that these qualities are not merely fictitious qualities, but such whose existence I can manifest, not only by considerations not absurd, but also by real experiments and physical phaenomena. (*Works* 3:307)

We noted in chapter 4 that Boyle sought to experimentally investigate the supposed special operations and medicinal effects of precious stones. This signifies not that he simply accepted without question superstitions held since medieval times but rather that he allowed for the possibility that nature included many wondrous forces and effects that the science of his day could not explain. Whereas Descartes had regarded the natural world too simplistically, Boyle reminded man of nature's complexity.

What made Boyle's acceptance of those diverse operations and qualities theoretically possible was his introduction of various kinds of specific "effluvia." Science historian Robert E. Schofield depicts the development of eighteenth-century theory of matter as a struggle between mechanism and materialism. "The materialists believed," he states, "that the causes of phenomena inhere in unique substances, each possessing as an essential property the power to convey, in proportion to its quantity, some characteristic quality."[34] If so, then Boyle's proposition of diverse kinds of effluvia arguably represents the beginning of natural philosophy's transition from mechanism to materialism. Indeed, it can be regarded as the very source of the materialistic natural philosophy of the eighteenth century, which in addition to electrical and magnetic effluvia went on positing many other unique substances, such as phlogiston and caloric.

However, if one can suppose a specific substance as the agent or medium of each unique effect found in matter—be it magnetic force, electrostatic force, or even a property such as combustibility—then one can also suppose that matter itself possesses such peculiar properties, and indeed it is natural to consider the latter explanation the simpler. Such was the reasoning by which the naive mechanism and atomism expounded up to Boyle's time gradually transformed thereafter.

A prominent example of that change can be seen in the ideas of William Petty (1623–1687). Born in the same year as Henry Power, Petty developed a unique theory of matter that he presented to the Royal Society in 1674. In this discourse, he posits that the primary matter (*prima materia*) of the universe takes the form of "atoms," whose shape and magnitude do not change; and that through their motions atoms combine to form "corpuscles" or "concretions," whose shape and motion account for matter's sensible qualities.

Thus far, Petty's thinking is in keeping with previous atomism and with Boyle's corpuscular philosophy.

In Petty's theory, however, instead of being devoid of qualities, atoms themselves possess the qualities of tiny magnets: "[E] very Atom is like the Earths Globe or Magnet, wherein are three Points considerable, viz. two in the surface, called Poles, and one within the substance, called Center." Because the heavens, like the Earth, also have such poles and a center, the atoms themselves are "microcosms" and have the same motions that Copernicus assigned to the Earth—that is, they spin on their own axes, they orbit around other atoms, and their centers attract each other. Petty proposes, furthermore, that atoms are attracted to the Earth by gravity: "I suppose that all Atoms have, like a Magnet, two Motions, one of Gravity whereby it tendeth towards the Center of the Earth and the other of Verticity by which it tendeth towards the Earths Poles."[35] Thus, in Petty's view, atoms are not inert and passive but rather active entities possessed of gravity and magnetism as intrinsic qualities.

This theory illustrates not only the pronounced metamorphosis that mechanism and atomism underwent as they were transplanted to Britain but also the great impact that Gilbert's magnetic philosophy had in the process of that transformation. Let us therefore open a new chapter in our story to examine Gilbert's influence in seventeenth-century Britain.

* * *

Thomas Browne, who embodied the intellectual transition that marked the early seventeenth century, accepted Cartesian natural philosophy yet sought to transplant it to the soil of

Baconianism. Hence, on the one hand he did not adopt Bacon's stance of rejecting every hypothesis that was not supported by direct experiment, but on the other he considered what Descartes claimed to have "proved" by *a priori* deduction to be no more than working hypotheses the validity of which should be determined by experiment. This attitude toward Cartesianism was characteristic of British intellectuals in the seventeenth century. Under the influence of Bacon's empiricism, along with dramatic improvements in experimental and observational instruments, more and more thinkers accepted the view that the premises of mechanism must be tested experimentally.

Henry Power, one of the first in Britain to embrace mechanism, nonetheless broke with the Continental mechanist tradition by introducing the concept of "subtle spirits" as the basic substance underlying all motion in nature. This was the first step in the gradual transformation of mechanical philosophy.

Robert Boyle was also a mechanist who considered the world to be a vast automaton. At the same time, however, his own experiments, showing that air sustained fire and living things, formed the basis of his conviction that there exists in air a fundamental substance with unique qualities indispensable to life and combustion. From there he conjectured that air also contained a variety of other "effluvia" each with its own kind of active quality. This idea represented a clear departure from the principle, central to mechanistic natural philosophy, that matter itself is devoid of qualities. As a result, discourse about magnetism, electricity, and other so-called occult qualities was steered in a new direction according to which all manner of hypothetical effluvia were conceived in order to "explain" such qualities.

Thus, through the debates presented and provoked by Power, Boyle, and others, a materialistic approach to natural philosophy gradually emerged from within mechanism. This helps to explain why Newton's subsequent theory of universal gravitation, which was scorned on the Continent, was accepted relatively easily in Britain.[36]

Boyle's law about the pressure of air was initially tied to the mechanistic interpretation that air exhibited elasticity because its corpuscles had spring-shaped appendages. Although this interpretation was soon rejected, Boyle's law itself, which expresses a quantitative relation on the basis of precise measurement, survived independently of such models. In a similar way, for magnetic, electrostatic, gravitational, and other forces to become the proper subjects of physics as a modern, mathematical science, it was necessary not to concoct mechanisms by which such forces were conveyed or conjecture about the qualities of special effluvia for each one but rather to accept, provisionally, that such forces were indeed cases of action at a distance and then accurately and quantitatively *measure* the variations of their strength. Science had to wait until after Boyle for the dawn of this decisive new phase of its history.

In the will executed after his death in 1691, Boyle declares: "I give to Mr Robert Hooke now Professor of Mathematicks in Gresham Colledge my best Microscope and my best Load-Stone wch I shall have att the time of my death."[37] It was Hooke who, under the influence of both the mechanism of Descartes and Boyle and the "magnetic philosophy" of Gilbert and Kepler, first attempted to quantitatively measure magnetic force and gravity, though he failed to discover their underlying principles. Then, out of the same milieu, came Isaac Newton, who proposed a pivotal new theoretical approach to the problem of gravity.

# Chapter 21

# Magnetism and Gravity: Hooke and Newton

## Section 1—John Wilkins and Magnetical Philosophy

A fact generally overlooked in science histories[1] is that, even while mechanical philosophy was dominating the intellectual scene on the Continent and also influencing British thinkers such as Power and Boyle, in Britain the "magnetical philosophy" initiated by Gilbert and Kepler remained equally strong.

Interest in magnets and magnetism seems to have been quite widespread in Britain around that time. In the preface to his *Magneticall Advertisements*, published in 1616, the archdeacon of Salisbury, William Barlowe, notes that "Many of our Nation …, not able to read Doctor Gilberts Booke in Latin, haue bin (euer since the first publishing thereof) exceeding desirous to haue it translated into English."[2] In fact, however, an English-language translation of Gilbert's *On the Magnet* did not appear until the nineteenth century, and in the meantime works such as *Magneticall Advertisements* and English physician Marke Ridley's *A Short*

*Treatise of Magneticall Bodies and Motions*, published in 1613, were written as alternatives for people who could not read Latin.

In 1630 the dramatist Ben Jonson came out with a comedy titled *The Magnetic Lady* whose heroine he named Lady Lodestone. Through the mid-seventeenth century, such interest in things magnetic fueled widespread admiration for Gilbert. In 1646, for example, Thomas Browne lauded Gilbert's achievements, asserting that

> although some assume the invention of [the magnet's] direction, and other have had the glory of the [compass] Card; yet in the experiments, grounds, and causes thereof, England produced the Father Philosopher, and discovered more in [the magnet] then [sic] Columbus or Americus did ever by it.[3]

Such sentiments smack of nationalistic pride in a British empire that was by then eclipsing Spain, Portugal, and other powers in ambitious endeavors around the world. But there was more to it than just that. Gilbert was considered the pioneer not merely of the specific discipline of magnetic studies but rather of a comprehensive philosophy—magnetical philosophy—on which natural science as a whole was to be refounded.

Magnetical philosophy was not really prescribed or codified as an established theory. Nonetheless, once purged of minor details and hylozoistic impurities, it can be characterized in two ways: narrowly, as a conception of nature that attributed the principle of the Earth's self-motion to its magnetic nature; and in a broader sense, as a cosmology that sought the origins of the entire solar system's activity and order in magnetic interactions among astronomical bodies themselves regarded as magnetic. Kepler had expanded Gilbert's narrower conception, and in Britain it was mostly the resulting broader understanding that was subsequently taken up and disseminated, primarily by professors at Gresham College.

Founded in London in 1597 in accordance with the will of prominent merchant Thomas Gresham, Gresham College was an institution of higher learning mainly for engineers and mariners. Accordingly, from its inception Gresham College was, unlike both Oxford and Cambridge, a hub of keen interest in such practical fields as navigation, geography, cartography, astronomy, and geomagnetism. Its first professor of geometry, Henry Briggs, was an expert in navigation who associated closely with such figures as Gilbert himself and the mathematician-cartographer Edward Wright. Henry Gellibrand, who in 1634 discovered the secular variation of magnetic declination, served as the college's fourth professor of astronomy; and Samuel Foster, its fifth (1636–1637) and seventh (1641–1652) professor of astronomy, had by then already embraced Kepler's theory of elliptical orbits.[4]

Then came a pivotal development: gatherings of a group of science enthusiasts at Foster's rooms in Gresham College from around 1645 evolved into a movement that led to the founding of the Royal Society. In an account written much later, mathematician John Wallis recalls that the group's purpose was "to discourse and consider of *Philosophical Enquiries*, and such as related thereunto: as *Physick*, *Anatomy*, *Geometry*, *Astronomy*, *Navigation*, *Staticks*, *Magneticks*, *Chymicks*, *Mechanicks*, and natural *Experiments*."[5] This list of subjects is intriguing: whereas today we might reasonably expect physics to be mentioned alongside astronomy, statics, and chemistry, in Wallis's list we find instead "magneticks." In Britain at that time, the study of magnetics was thus considered less a branch of physics than a discipline that subsumed physics as a whole. That is, as the true source of motion in all the universe, magnetic force was regarded as the key to clarifying all manner of questions about the natural world. Such was the intellectual milieu in which Christopher Wren, in his 1657 inaugural lecture as Gresham College professor of astronomy, remarked that he venerated Gilbert "not only as the sole Inventor of Magneticks, a new Science to be added to the Bulk of Learning, but as the Father of the new Philosophy"[6]

But let us return to the main thread of our story. The central figure in the group that gathered at Foster's was John Wilkins (1614–1672), who had graduated with an M.A. from Oxford in 1634. Wilkins is also known to history as the brother-in-law of Oliver Cromwell, and during the Puritan Revolution, after staunch Royalist scholars were ousted from Oxford by Parliamentarian forces, some members of Wilkins's group were installed in their place. Wilkins was made warden of Oxford's Wadham College in 1648, and the following year saw the Oxford appointments of Wallis as professor of geometry and Seth Ward—a Royalist but an acceptably moderate one, whose expertise included a deep knowledge of Keplerian theory—as professor of astronomy. Thus was formed, by around 1649, the circle of scientists that came to be known as the Oxford Group. Including among its members such figures as economist William Petty-known both as the father of modern statistics and as the founder of political economy-this group, too, was centered on Wilkins.

Around 1652 the group was joined by astronomer Lawrence Rooke; by Christopher Wren, who had learned Keplerian theory from Ward; and by Thomas Sprat, who later wrote a history of the Royal Society. Robert Boyle also became a member sometime in the late 1650s. Following Cromwell's death in 1658, Wilkins took up a post at Cambridge. With the Restoration of 1660, however, Wilkins and Ward found themselves back in London, having been removed from their respective positions at Cambridge and Oxford. Meanwhile, Rooke, who had already succeeded Foster as Gresham College's professor of astronomy, switched to professor of geometry in 1657, whereupon Wren took over the astronomy chair. Thus, several of the Oxford Group members relocated to London and began holding regular gatherings at Gresham College. This association was the immediate precursor to the Royal Society. At a November 1660 gathering that included Boyle, Wilkins, Petty, Rooke, and Wren, it was proposed that, "according to the manner in other countryes, where there were voluntary associations of men in academies, for the advancement of various parts of learning," they should similarly institutionalize their own meetings "for the promoting of experimentall philosophy."[7] The use of the term "experimental philosophy" suggests Francis Bacon's considerable influence on the group. It was decided that the meetings would be held weekly at Rooke's chamber in Gresham College. Out of these developments, the Royal Society was officially launched in July 1662 under a formal royal charter conferred by Charles II. Upon the Society's formal establishment, William Brouncker was appointed as its president and Wilkins and Henry Oldenburg as its secretaries. Wilkins was the greatest driving force behind the Society's founding, and we can reasonably assume that he had considerable influence in British intellectual circles at that time.

Although not a scientist known for any particular discoveries, Wilkins has been applauded as "a man who knew the technicalities of science, yet who … had the ability to explain those technicalities to the general reader, even to the 'meaner capacities' of ladies." In that role, he is remembered for a series of books explaining the latest scientific developments, such as Galileo's telescopic discoveries, in nonspecialist terms, thereby also playing a major part in the spread of heliocentric theory. His *The Discovery of a World in the Moone* (hereafter *Discovery*), published in 1638, is regarded as "one of the first important books of modern 'popular science.'"[8]

In the second edition of *Discovery*, published in 1640, Wilkins describes the "true nature of gravity" as

> a respective mutual desire of union, whereby condensed bodies, when they come within the sphere of their own vigour, do naturally apply themselves one to another by attraction or coition.[9]

In places, his phrasing suggests Gilbert's influence, but most noticeable of all is that Wilkins focuses on Kepler's virtual identification of magnetism with gravity and expresses a similar view himself. In *Discourse concerning a New Planet*, a 1640 work written in defense of heliocentric theory, Wilkins presents the Keplerian view that the sun causes the planets to orbit around it "by sending forth a kind of magnetic virtue in strait lines, from each part of its body"; and he proceeds to account for the Earth's gravity in the same way. That is, in explaining why a body cast up from the Earth's surface into the air is not left behind by the Earth's rotation, he posits that, just as the sun's rotation on its own axis makes the planets orbit around it, so "an arrow or bullet [is] carried round by the magnetical motion of our earth."[10]

English churchman Francis Godwin's *The Man in the Moone*, published in 1638, a few months after the first edition of Wilkins's *Discovery*, similarly asserts that

> those things which wee call heavie, do not sinke toward the Center of the Earth, as their naturall place, but as drawen by a secret property of the Globe of the Earth, or rather some thing within the same, in like sort as the Loadstone draweth Iron, being within the compasse of the beames attractive.[11]

In the second edition of *Discovery*, we can see the influence of Godwin. Aristotelian theory was thus being superseded on the basis of magnetical philosophy and the analogy that the latter drew between magnetism and gravity.

The idea that gravity was a force acting at a distance and therefore a kind of magnetic force had already been put forward by Francis Bacon. In *The New Organon* Bacon enumerates the kind of motion that must be involved if the moon raises seawater, if the fixed stars draw the planets to their apogees, or if the sun attracts Venus and Mercury in such a way that prevents them from moving any more than a certain distance from it. Regarding such motion, he reasons that "if it operates at great distances and on great masses of things," then it deserves special attention as "magnetic motion," "especially if it does not start from contact as most motions do." Bacon, who is thought not to have read Kepler's *New Astronomy*, not only knew nothing of elliptical orbits but even rejected heliocentric theory itself. Nonetheless, as in the passage quoted above, Bacon admits the existence of gravity as a force acting at a distance between astronomical bodies and interprets it simply as "magnetic action." Elsewhere in *The New Organon* he likewise refers to "the magnetic action by which iron is drawn towards a magnet, heavy objects towards the globe of the earth."[12] The idea of seeking the origin of the Earth's gravity in its magnetism derives less from Gilbert's *On the Magnet* than from misreadings of it, but it was nonetheless a fairly widespread view in Britain around that time.

Wilkins himself does not exactly claim that gravity *is* magnetism. On the contrary, in fact, he clearly distinguishes between the two forces, stating that

> it is another kind of affection which causes the union betwixt the iron and loadstone, from that which makes bodies move unto the earth.

"The former," he explains

> is some kind of nearness and similitude in their natures, for which philosophy, as yet, has not found a particular name. The latter does arise from that peculiar quality whereby the earth is properly distinguished from the other elements, which is its condensity.

It is clear, however, that Wilkins considers the operation of gravity to be similar to that of magnetic force as action at a distance. And indeed, several paragraphs after the passage just quoted, he even refers to gravity as "this magnetical virtue."[13]

On the topic of terrestrial gravity equated with magnetism, Wilkins also states in *Discovery* that

> you must not conceive, as if the orb of magnetical vigour were bounded in an exact superficies, or as if it did equally hold out just to such a determinate line, and no farther.... [It is] probable, that this magnetical vigour does remit of its degrees proportionally to its distance from the earth, which is the cause of it.[14]

Thus, not only are gravity and magnetism regarded as similar; both forces are also supposed to decrease in strength with distance. Here the influence of magnetical philosophy is clear.

Naturally we can infer Kepler's influence underlying Wilkins's references to gravity's attenuation with distance, the reason for which Wilkins takes to be "the weakness of the desire of union in dense bodies at a distance." Wilkins also recounts the belief that "a lump of ore or stone, as being on the ground, cannot be moved by less than six men, [but] being in the bottom of a deep mine, may be stirred by two"[15] Francis Bacon likewise records that this phenomenon "is affirmed … by many" and that accordingly

> it is very probable, that the *Motion of Gravitie* worketh weakly, both farre from the Earth, and also within the Earth.[16]

It is uncertain whether or not such a belief was in fact widely held among mine laborers of that time, but there is no doubt that the claim did have a direct influence on Robert Hooke, who, as we shall soon see, attempted to measure gravity's supposed attenuation at locations deep underground.

## Section 2—Robert Hooke and Mechanism

An account of the circumstances of the Royal Society's inception appears in Thomas Birch's *The History of the Royal Society of London* (hereafter *History of the Royal Society*), published in 1756. According to this chronicle, the group spent 1660 entirely in discussion of organizational matters, with the result that the record of its actual experiments does not begin until 2 January 1661. An entry dated 16 January that year reads: "The king sent two load-stones … with a message, that he expected an account from the society of some of the most considerable experiments upon them"; and another for the same day records that "a committe [sic] for magnetical experiments" was formed, led by members who also belonged to Gresham College.[17] From the very outset of the Royal

Society's activities, magnets seem to have been a topic of keen interest and magnetic force one of the Society's central research themes.

Seven years younger than Robert Boyle, Robert Hooke (1635–1703) entered Oxford in 1653. He became Boyle's assistant around 1658 and began participating in the Oxford Group around the same time. On Boyle's recommendation Hooke was appointed the Royal Society's curator of experiments in 1662, and he was made a fellow the following year. As curator of experiments, Hooke's duty was to carry out three or four suitable experiments per week on themes including those discussed at the Society's meetings. Given that the fellows' interests ranged from mathematics, astronomy, and meteorology to physics, chemistry, geology, zoology, and botany, this presumably meant quite a heavy workload. Nonetheless, Hooke not only fulfilled that role with great vigor and success but even served as a research adviser to the other members. Considering, also, the skill he displayed in creating, at Boyle's request, a vacuum pump that for its day represented the very cutting edge of advanced technology, it seems that Hooke had exceptional talents as an experimenter. Thus, he conducted the early Society's experimental activities under his sole management. A record entry of July 1664 confirms that he continued as curator of experiments after the Society received its royal charter. Around the same time, he was provided with lodgings at Gresham College, the Society's headquarters, where he lived for the rest of his life. In 1665, Society secretary Henry Oldenburg began publishing *Philosophical Transactions*, essentially a bulletin of the Society's activities, through which the now familiar system of open publication and peer review of research papers gradually took shape and priority claims to inventions and discoveries became publicly guaranteed. In the same year, Hooke was appointed Gresham College's professor of geometry; and upon taking over Oldenburg's executive duties after the latter's death in 1677, Hooke came to personify the Royal Society as a whole in both name and function.

Well known among Hooke's early accomplishments are his construction of the vacuum pumps central to Boyle's research on air and atmospheric pressure and his contributions to the establishment of Boyle's law concerning the relationship between gas pressure and gas volume. In 1665 came the publication of the work that brought Hooke himself widespread celebrity, his *Micrographia*. A deliberation mainly on microscopic observations of nature, this book is illustrated with Hooke's own intricate engravings of fleas, lice, and other microscopic subjects—images of such renown and exceptional quality that they still appear in biology textbooks today. It is also in this work that Hooke, having discovered the structural units of plant tissue under his microscope, coined the term "cell" to signify them. In that sense, *Micrographia* marks the birth of cytobiology. But the book also includes discussion of astronomical observations and natural philosophy in general, making it an invaluable window onto both the overall features of natural science at that time and Hooke's own particular views thereof.

In the preface to *Micrographia* Hooke writes that with the aid of microscopes "we may perhaps be inabled [sic] to discern all the secret workings of Nature, almost in the same manner as we do those that are the productions of Art, and are manag'd by Wheels, and Engines, and Springs, that were devised by humane Wit." He notes, further, that by means of experiments using such new observational instruments the Royal Society members "find some reason to suspect, that those effects of Bodies, which have been commonly attributed to *Qualities*, and those confess'd to be *occult*, are perform'd by the small *Machines* of Nature." In this respect, the young Hooke was an adherent of mechanism.[18]

Hooke's mechanistic views of nature are clearly seen in his *Lectures de potentia restitutiva, or, Of Spring* (hereafter *Of Spring*), published in 1678:

> I suppose then the sensible Universe to consist of body and motion.
>
> By Body I mean somewhat receptive and communicative of [vibrational] motion or progression. Nor can I have any other Idea thereof, for neither Extension nor Quantity, hardness nor softness, fluidity nor fixedness, Rarefaction nor Densation are the properties of Body, but of Motion or somewhat moved.
> ...
> These two [i.e. body and motion] do always counterballance each other in all the effects, appearances, and operations of Nature, and therefore it is not impossible but that they may be one and the same; for a little body with great motion is equivalent to a great body with little motion as to all its sensible effects in Nature.
>
> I do further suppose then that all things in the Universe that become the objects of our senses are compounded of these two ... namely, *Body*, and *Motion*. And that there is no one sensible Particle of matter but owes the greatest part of its sensible Extension to Motion whatever part thereof it ows [sic] to Body according to the common notion thereof.[19]

The position Hooke elaborates here is entirely mechanistic, and his argument that all objective reality consists of body and motion, of which motion is of greater effect, clearly reflects Boyle's influence.

Hooke's opinion is that the particles that make up any given matter have a specific oscillatory motion that determines the sensible properties of that matter. Particularly concerning forces acting between bodies, he stipulates in *Micrographia* that "the Congruity, in the Vibrative motions, may be the cause of all kind of attraction, not only Electrical, but Magnetical also, and therefore it may be also of Tenacity and Glutinousness."[20] In *Of Spring* he explains this "congruity" as a kind of resonance of specific types of vibration:

> By *Congruity* and *Incongruity* then I understand nothing else but an agreement or disagreement of Bodys [sic] as to their Magnitudes and motions.
>
> Those Bodies then I suppose congruous whose particles have the same Magnitude, and the same degree of Velocity, or else an harmonical proportion of Magnitude, and harmonical degree of Velocity. And those I suppose incongruous which have neither the same Magnitude, nor the same degree of Velocity, nor an harmonical proportion of Magnitude nor of Velocity.

In this connection, Hooke, following Descartes, posits the existence of a "subtil matter that incompasseth and pervades all other bodies."[21] On this basis he supposes that

> from a perfect congruity of the motions of two distant bodies, the intermediate fluid particles are separated and droven away from between them, and thereby those congruous bodies are, by the incompassing mediums, compell'd and forced neerer together.[22]

In other words, when two bodies are mutually congruous, the resonance of their vibratory motions forces out the medium between them, resulting in their being thrust together by the medium surrounding them. Here, the force between the bodies is understood mechanistically as action through a medium. The following is from *Micrographia*:

> We see therefore what is the reason of the *sympathy* or uniting of some bodies together, and of the *antipathy* or flight of others from each other: For *Congruity* seems nothing else but a *Sympathy*, and *Incongruity* an *Antipathy* of bodies; hence *similar* bodies once *united* will not *easily part*, and *dissimilar* bodies once *disjoyn'd* will not *easily unite* again.[23]

It is fair to say that by this argument Hooke attempts to provide, from the standpoint of mechanical philosophy, a grounding for the earlier magical notions of "sympathy" and "antipathy" and for the view, widely accepted since the age of Democritus and Plato, that like attracts like.

## Section 3—Hooke and Gravity: The Break from Mechanism

Hooke's adherence to mechanism was far from steadfast, however. According to science historian J. A. Bennett, the kind of mechanistic explanation we have just seen from Hooke was in fact "largely symbolic," aimed simply at maintaining "Hooke's credentials as a mechanist."[24] Indeed, in *Lampas*, printed in 1677, the year before *Of Spring* was published, Hooke states clearly: "I confess, I suppose Gravity to be otherwise performed than as *Des Cartes* has supposed."[25] And when actually discussing magnetism, gravity, and so on, he by no means limits himself to mechanistic expressions. On the contrary, in a report to the Royal Society dated 21 March 1666, he remarks of gravity that "it seems to be one of the most universal active principles in the world."[26] Here he takes gravity to be a fundamental—and irreducible—operation inherent in bodies, a view that clearly departs from the maxim of mechanical philosophy that all matter is essentially inert, passive, and devoid of quality.

Hooke's true position regarding gravity is apparent in the following excerpt from a paper he presented to the Royal Society on 23 May 1666, which, though rather long, I quote here in full because of its importance in connection with his concept of force:

> I have often wondered, why the planets should move about the sun according to Copernicus's supposition, being not included in any solid orbs … nor tied to it, as their center, by any visible strings; and neither depart from it beyond such a degree, nor yet move in a strait line, as all bodies, that have but one single impulse, ought to do: For a solid body, moved in a fluid, towards any part, (unless it be protruded aside by some near impulse, or be impeded in that motion by some other obviating body; or that the medium, through which it is moved, be supposed not equally penetrable every way), must persevere in its motion in a right line, and neither deflect this way nor that way from it. But all the celestial bodies, being regular solid bodies, and moved in a fluid, and yet moved in circular or elliptical lines, and not strait, must have some other cause, besides the first impressed impulse, that must bend their motion into that curve. And for the performance of this effect I cannot imagine any other likely cause besides these two: The first may be from an unequal density of the medium, thro' which the planetary body is to be moved; that is, if we suppose that part of the medium, which is farthest from the center, or sun, to be more dense outward, than that which is more near, it will follow, that the direct motion will be always deflected inwards, by the easier yielding of the inward, and the greater resistance of the outward part of that medium….
>
> … the second cause of inflecting a direct motion into a curve may be from an attractive property of the body placed in the

> center; whereby it continually endeavours to attract or draw it to itself. For if such a principle be supposed, all the phaenomena of the planets seem possible to be explained by the common principle of mechanic motions.[27]

Since Gassendi had conceived and Descartes had correctly formulated the law of inertia, it had become clear that what was needed to explain a planet's orbital motion was not, as Kepler had thought, a propulsive force tangential to the orbit but rather a force that curved the orbit toward the central body (i.e., the sun)—that is, a force that kept the planet moving around the sun instead of flying off on a tangent. In the passage just quoted, Hooke mentions as candidates for that required deflecting force both a mediated-action model—involving gradations of density of a fluid medium permeating cosmic space—and an action-at-a-distance model—involving an attractive force exerted by the central body. From a strict mechanist standpoint, the latter model should be rejected as a matter of course, but Hooke does not take that view. In this respect, Bennett aptly sums up Hooke's transitional, ambivalent position concerning the concept of force with the observation that "Hooke,… inheriting a different tradition as well as accepting the mechanical, could base his whole philosophy on Cartesian presuppositions, yet … feel quite at ease with the concept of attractive forces acting between heavenly bodies"[28]. This ambivalence in fact marked the watershed by which the inherent limits of naive mechanism were overcome and celestial mechanics began to take shape as a discipline of modern physics.

That Hooke was thus able to think flexibly beyond the fundamental rules of Cartesianism and mechanism is yet another fact attributable to the influence of Gilbert's "magnetical philosophy" and Bacon's "experimental philosophy." Hooke referred to his own scientific approach as "the *mechanical*, the *experimental* Philosophy,"[29] an expression acknowledging both Descartes's influence and Bacon's. In practice, however, the path Hooke actually took was not that of naive mechanism, with its plausible models contrived to explain the transmission of gravitational and magnetic force, but rather that of empirical science, accepting gravity and magnetism as empirical facts and exploring their operation and properties by means of experimentation and measurement.

Particularly evident in Hooke's thinking is his straightforward acceptance of the magnetic theory of gravity, an attitude that points to the considerable influence of magnetical philosophy. In a report to the Royal Society dated 21 March 1666, Hooke notes that

> Gilbert began to imagine [gravity] a magnetical attractive power, inherent in the parts of the terrestrial globe: the noble Verulam [i.e., Francis Bacon] also, in part, embraced this opinion; and Kepler (not without good reason) makes it a property inherent in all celestial bodies, sun, stars, planets.[30]

Regarding the opinion, accepted by the predecessors named, that gravity is a magnetic attractive force and also an inherent attribute of astronomical bodies, Hooke makes no negative comment at this point in time. Indeed, at a Royal Society meeting held on 14 December 1664, Hooke himself had raised the possibility of gravity's being magnetic. At a time when it was already known that the Earth's magnetism increased in strength with proximity to either pole, and, as Gellibrand had discovered, that magnetic declination changed over time (secular variation), Hooke thus considered the possibility that gravity, being a force derived from the Earth's magnetism, likewise increased in strength near the poles and even underwent secular variation.[31] It seems that Hooke at that time regarded gravitational force and magnetic force as virtually identical. Similarly, in a 1674 lecture titled *An Attempt to Prove the Motion of the Earth by Observations* (hereafter *Attempt to Prove the Motion of the Earth*) Hooke gives an account of Tycho Brahe's model of the celestial system as one in which

> the Planets make their Revolutions about [the sun] whilst the Earth stands still, and by its magnetism attracts the Sun, and so keeps him moving about it.[32]

This description of the interrelationships among the orbits of the sun, the planets, and Earth certainly accords with Tycho's system, but the notion that the Earth "by its magnetism attracts the Sun" is Hooke's interpretation.

Hooke's view of gravity as magnetic was based on the magnetical philosophy espoused since Gilbert's time, but in Hooke's case such thinking was particularly influenced by his experience of observing the comet of 1664–1665. He claimed that the power exerted by the sun on comets included not only an attractive force but a repulsive one as well, an assertion born of the need to explain "rationally" the cometary motions actually observed. This is why he supposed an even stronger correlation between gravity and magnetism than previous magnetical philosophy had suggested.

Hooke's theory of magnetic gravity appears in his 1678 work *Cometa* as follows:

> I suppose the gravitating power of the Sun in the center of this part of the Heaven in which we are, hath an attractive power upon all the bodies of the Planets, and of the Earth that move about it, and that each of those again have a respect answerable, whereby they may be said to attract the Sun in the same manner as the Load-stone hath to Iron, and the Iron hath to the

> Load-stone. I conceive also that this attractive virtue [i.e., of the sun] may act likewise upon several other bodies that come within the center of its sphere of activity, though 'tis not improbable also but that as on some bodies it may have no effect at all, no more than the Load-stone which acts on Iron, hath upon a bar of Tin, Lead, Glass, Wood, &c. so on other bodies, it may have a clean contrary effect, that is, of protrusion, thrusting off, or driving away, as we find one Pole of the Magnet doth the end of a Needle touched on the opposite part.[33]

Drawn mostly from observations of the 1664–1665 comet, Hooke's treatment of the topic here is adapted from a presentation he made to the Royal Society in August 1666. While rather incomprehensible when read without considering that background, his inquiry into the power exerted by the sun is intended to explain the orbit of an actually observed comet—one that approaches the sun in an almost straight line and then, at a point near the sun, turns sharply and hurtles away from it, again in an almost straight line, as if batted away by the sun itself. Although orbits are actually elliptical, in the case of comets they appear to move this way because their orbits are very elongated ellipses, with a high eccentricity and extremely distant aphelion, and also because what we on Earth can observe of their motion is only the small part relatively close to the sun.

From the fact that comets approach the sun in virtually a straight line and then end up receding from it also in an almost straight line, Hooke first of all infers that the sun's effect on comets has a finite sphere of activity. That is, it is only within the sun's sphere of activity, he surmises, that a comet veers in its motion under the influence of that force; outside the sphere, it travels in a straight line. He then argues that the reason a comet moves away from the sun after approaching it is that the sun's power is polar, comprising not only an attractive but also a repulsive force. As the comet passes near the sun, the latter's heat causes the comet's constituent parts to "become of other natures than they were before," so that, after reaching a point sufficiently close to the sun, they "are rather expelled from, than attracted towards it," with the result that the comet then travels away from the sun.[34] Needless to say, it is in magnetism that Hooke finds the prototype of such a force with both attracting and repelling aspects. In his conjectures about the polarity and selectivity of gravity, and also in his conception of its "sphere of activity," Hooke presumably drew heavily from Della Porta, Gilbert, and another of their successors, Wilkins.

Speculation that the force exerted by the sun on the planets was a magnetic force began with Kepler, and from it arose the view that the sun, too, was a massive magnet. Though rather difficult to imagine today, in the seventeenth century this opinion was quite widespread. German philosopher Gottfried Leibniz (1646–1716), who had a high regard for Kepler, wrote in 1689 that the sun could be thought of as a magnet ("sol tanquam magnes concipi potest").[35] John Flamsteed (1646–1719), who was born in the same year as Leibniz and became Britain's foremost authority in astronomical observation, serving as the first director of the Royal Greenwich Observatory, expressed similar views. In 1681 Flamsteed described the phenomenon of a comet's speeding toward and then immediately away from the sun with the words "as ye North pole of ye loadstone attracts ye one end of ye Magnetick needle but repells ye other"; and in 1685 he put it more plainly, referring to the sun as "the largest & most vigorous Magnet of our susteme [sic]."[36] The idea that the sun was a magnet and therefore exerted not only an attractive force but also, depending on the affected body and its position relative to the sun, a repulsive force seems to have been fairly widely believed at that time, at least in Britain.

## Section 4—Measuring Gravitational Force and Magnetic Force

In the abovementioned treatise *Of Spring* Hooke writes: "The Power of any Spring is in the same proportion with the Tension thereof: That is, if one power stretch or bend it one space, two will bend it two, and three will bend it three, and so forward."[37] This principle that a spring's extension or displacement is directly proportional to the force applied to the spring was discovered by Hooke and is known as Hooke's law. But what he accomplished here was more than simply the discovery of an empirical law about springs: the law's formulation is also significant for exemplifying the Keplerian approach of expressing force as a mathematical function and for providing the first concrete example thereof. In this respect, Hooke, while advocating "experimental philosophy," in fact clearly surpassed Bacon: Hooke sought to measure the strength of force quantitatively and to comprehend it in terms of mathematical laws.

Hooke followed that same approach in his studies on gravity and magnetism. His primarily purpose was to determine whether or not the Earth's gravity really was magnetic in origin. He declares this intention in a paper presented to the Royal Society on 21 March 1666, stating that

> first it will be requisite to consider, whether this gravitating or attracting power be inherent in the parts of the earth; and, if so, whether it be magnetical, electrical, or of some other nature distant from either

and furthermore that

> if it be magnetical, any body attracted by it ought to gravitate more, when nearer to its surface, than when farther off.[38]

Measuring change in gravitational force with distance was thus proposed as a means to corroborate the theory of magnetic gravity. It is important to note here that by "near the surface" and "farther off" Hooke refers not only to space above the Earth's surface (up in air) but also to that below it (underground).

Hooke made the first of such measurements using a balance scale set up at a high place such as the top of Westminster Abbey or the tower of Saint Paul's cathedral. On one pan of the scale he placed a heavy object attached with a long wire and, setting the pans of the scale in balance, measured the combined weight of the object and the wire. He then lowered the object by the wire almost to the ground and measured them again to determine any change in weight. Due to the effects of vibration, wind, and so on, however, he found that "nothing of certainty could be collected from these trials." His next task was to ascertain whether or not gravity decreased in strength underground. The motivation behind this undertaking presumably included the aforementioned reports by Bacon and Wilkins about large stones becoming lighter at the bottom of deep mine shafts. For Hooke, however, there was the more important question of whether or not gravity was in a straightforward sense magnetic, since it was conceivable that

> if all the parts of the terrestrial globe be magnetical, then a body at a considerable depth, below the surface of the earth, should lose somewhat of its gravitation, or endeavour downwards, by the attraction of the parts of the earth placed above it.

Underlying this conjecture, of course, was Gilbert's conception of the terrestrial globe, according to which "earth-substance," the main constituent of the Earth, was a magnetic kind of matter. To test the idea, Hooke measured weights of brass, wood, and flint attached to 80 feet (approx. 25 m) of rope inside a well near Oxford that was roughly 90 feet (approx. 30 m) deep. He found that "each of those bodies seemed to keep exactly the same gravity at the bottom of the well, that they had at the top." Even after repeating the experiment in a well almost 300 feet (approx. 90 m) deep, he failed to ascertain any significant difference in weight. Nonetheless, reasoning that the ratio of the portion of the Earth below the objects in the well to that above them was too great to make any drop in gravitational force detectable by such crude measurements, he concluded that these findings did not rule out the possibility of gravity's attenuation inside the Earth.[39]

In any case, this was only half of the program of experiments Hooke had planned; he still had to measure change in magnetic force with distance and to compare the findings with those on gravity. In his report of 21 March he gives the following description of experiments "to judge, whether there be any such variation of gravity; and, if such there be, whether it be analogous to that of a load-stone":

> Which experiments, as they are wholly new, being not attempted hitherto (that I know) by any; and as they may afford many helps toward the finding out the true nature of the magnet, and the laws and reasons of divers other motions; so if this analogy between the decrease of the attraction of the one [i.e., magnetic force], and of the gravity of the other, be found real, we may perhaps by the help of the load-stone, as it were, epitomise all the experiments of gravity, and determine, to what distance the gravitating power of the earth acts; and explicate perhaps divers other phaenomena of nature by ways not yet thought of.[40]

Presumably, Hooke, reflecting on the precedent of Gilbert's

studies of the geomagnetic field using a "terrella," was considering the possibility of using magnets to conduct experiments on the attenuation of gravity. Hooke thus shifted his approach from the measurement of gravity to the measurement of magnetic force.

In a record dated a week later (28 March), Hooke notes that he has built a device for the purpose of making

> experiments with upon a good loadstone for the finding out of the decrease of its attractive force upon a body, according as it is placed at greater and greater distances, in order to find out, whether gravitation be somewhat magnetical.

In explaining his decision to conduct such experiments he asserts that it is probable

> that if [the decrease in gravitational force and the decrease in magnetic force] hold the same proportion, they have the same cause.[41]

This represents the first expression of the crucial view that, not only should the strength of gravitational force and magnetic force be expressible as a function of distance, but also that if the forms of the two mathematical functions are equal then the two forces can be regarded as having the same source—or in other words, that the form of the mathematical function for a force is essentially what defines it and gives it its identity.

The entry of 4 April the same year states that "Mr. Hooke presented a table of the degrees of the loadstone's attraction of a little square oblong piece of iron at several distances from the pole of the magnet, as he had found it in making trials privately himself." The entry also provides the table as shown (in slightly adapted form) below.[42]

Hooke's method in this experiment was to place a piece of iron on one pan of a scale, balance it with weight in the other pan, then bring a magnet underneath the iron's pan up to a specific distance from it and record how much additional weight had to be added to the other pan to rebalance the scale. Essentially, this was the same method that had been previously suggested by Cusanus and Della Porta. In any case, the measurements yielded no mathematically simple functional relation between the force and the distance. Hooke seems to have continued making such measurements intermittently thereafter, but without achieving any clear results, and eventually he abandoned the notion of magnetical gravity. In 1682 he declared that "the *Magnetism* of *Gilbert* or *Kepler*...is a Propriety distinct from Gravity, and of quite another nature," and with this, magnetic philosophy quietly passed into history.[43]

Nonetheless, the fact that Hooke's work initiated an actual program to search for the mathematical law for magnetic force holds great significance. With that effort, knowledge about magnetic force, which had previously been associated with magical thought, took a great stride toward assuming the character of modern science.

Hooke's Measurements on Variation of Magnetic Force with Distance

| Distance (inches) | Weight (grains) |
|---|---|
| 6 | 0 |
| 4 | 0 1/8 |
| 2 | 2 13/16 (3 3/4)* |
| 1 | 17 6/8 (18 7/8) |
| 1/2 | 57 6/8 |
| 1/4 | 104 5/16 |
| 1/8 | 197 4/8 |

Note: Figures in parentheses are revised measurements provided by the Royal Society.

## Section 5—Hooke's "System of the World"

In the history surrounding the discovery of universal gravitation, Hooke is well known for his fierce dispute with Newton over prior claim to the discovery of the inverse-square law governing that force. In fact, however, there was already a widespread awareness at that time that, when some kind of physical influence or operation emanates isotropically from a central body, its strength decreases in inverse proportion to the square of the distance from that body.

French astronomer Ismaël Boulliau (Ismaël Bullialdus) proposed in 1638 that the intensity of light decreased in inverse proportion to the square of the distance from its source; and in a 1645 work he arrived at the inverse-square law for gravitational force by extending to three dimensions the law of efflux, which Kepler had considered only two dimensionally (see equation 18.1). Similarly, William Petty, in a 1674 work titled *A Discourse concerning the Use of Duplicate Proportion*, observes that sound also weakens in intensity in inverse proportion to the square of the distance from its source, and that to attain the same brightness of candlelight at different distances from the light source, it is necessary to increase the number of candles in proportion to the square of the distance.[44]

Accordingly, more important at this juncture than the question of prior claim to the discovery of the inverse-square law of force is the fact that Hooke put forward a new perspective on the orbital motions of planets and comets. In fact, it was by following the framework devised by Hooke that Newton was eventually able to provide a rigorous formulation of the inverse-square law of gravitational force.

The abovequoted paper that Hooke presented to the Royal Society on 23 May 1666 is referred to in Birch's *History of the Royal Society* as a "paper of Mr. Hooke concerning the inflection of a direct motion into a curve by a supervening attractive principle." As this description intimates, the main assertion of the discourse is that a planet's curvilinear orbital motion can be regarded as the superposition of the planet's rectilinear motion along the tangent to the orbit, which is due to inertia, and its deflection toward the center, which is due to the attractive force of the central body. This dispelled for the first time the misunderstanding and confusion inherent in Kepler's notions of rest due to inertia and continual motion due to force, thereby clearing the way toward a theoretically correct analysis of planetary motion. Hooke himself concludes: "By this hypothesis, the phaenomena of the comets as well as of the planets may be solved; and the motions of the secondary, as well as of the primary planets."[45] It would seem that Hooke arrived at this generalized view of the curved orbits of all planets around the sun from observation of the pattern of cometary motion—the almost straight progress toward the sun, sharp turn near it, then almost straight departure from it.

Much later, in a letter to Newton dated 24 November 1679, Hooke sought Newton's opinion regarding what Hooke called the "hypothesis or opinion of mine" about "compounding the celestiall motions of the planetts of a direct motion by the tangent & an attractive motion towards the centrall body."[46] To Newton, who until then had regarded planetary motion narrowly in terms of a balance between centripetal and centrifugal forces, this proposition by Hooke pointed toward a much more effective and general method of analysis. In the subsequent formulation of Newton's theory of planetary motion, therefore, "Hooke's contribution cannot be ignored."[47]

Working from such a theory of motion and from the idea of an attracting force operating on astronomical bodies, Hooke also put forward a grand vision of the "system of the world." This idea appears in his *Attempt to Prove the Motion of the Earth* (1674). Although the main focus of the treatise is to document astronomical observations made by telescope, let us focus here on a passage near the end of that work where he comments on "a System of the World differing in many particulars from any yet known." Hooke asserts there that the explanation of this "system of the world" relies on three suppositions:

> First, That all Coelestial Bodies whatsoever, have an attraction or gravitating power towards their own Centers, whereby they attract not only their own parts, and keep them from flying from them, as we may observe the Earth to do, but that they do also attract all the other Coelestial Bodies that are within the sphere of their activity; and consequently that not only the Sun and Moon have an influence upon the body and motion of the Earth, and the Earth upon them, but that ☿ [Mercury] also ♀ [Venus], ♂ [Mars], ♄ [Saturn], and ♃ [Jupiter] by their attractive powers, have a considerable influence upon its motion as in the same manner the corresponding attractive power of the Earth hath a considerable influence upon every one of their motions also. The second supposition is this, That all bodies whatsoever that are put into a direct and simple motion, will so continue to move forward in a streight line, till they are by some other effectual powers deflected and bent into a Motion, describing a Circle, Ellipsis, or some other more compounded Curve Line. The third supposition is, That these attractive powers are so much the more powerful in operating, by how much the nearer the body wrought upon is to their own Centers. Now what these several degrees are I have not yet experimentally verified; but it is a notion, which if fully prosecuted as it ought to be, will mightily assist the Astronomer to reduce all the Coelestial

> Motions to a certain rule, which I doubt will never be done true without it.[48]

We can summarize these suppositions as follows. First, the motions of all celestial bodies in the solar system are governed by an attractive force that operates reciprocally among them and over distance and is also the cause of the force (weight) by which bodies on or near each celestial body are pulled to it. Second, all curved motion is the result of deviation from rectilinear inertial motion, the cause of which deviation is the abovementioned attracting force. And third, the attracting force is characterized by the fact that it decreases with distance. Notice that all descriptions of mechanistic action through a medium—of the curvature of orbits being due to the varying density of an intervening fluid medium—have been abandoned. The central issue concerning force is no longer one of contriving plausible models to explain the origin or propagation of force in terms of a supposed material medium, but rather that of determining the quantitative behavior—that is, the mathematical laws—of a remotely acting force on the assumption that such a force does exist. Here we see the completion of the break from Cartesian mechanism. Moreover, Hooke's three hypotheses identified in broad outline what Newton went on to elaborate in finer detail in *Mathematical Principles of Natural Philosophy* (hereafter *Mathematical Principles*). It is in this sense that prominent Newton scholar Richard S. Westfall describes the abovequoted portion of Hooke's treatise as "this remarkable passage."[49]

Regarding gravity, Hooke did not stop at the merely qualitative description that it decreases with distance. In a letter to Newton dated 6 January 1680, he clearly specifies in quantitative terms that

> my supposition is that the Attraction always is in a duplicate proportion to the Distance from the Center Reciprocall.[50]

And finally, in 1682, Hooke dispenses with Descartes's vortex-model explanation of gravity by pointing out its fatal defect:

> The *Cartesian* Doctrine, and that of Mr. *Hobbs*, are both insufficient, because they do not give any reason why Bodies should descend towards the Center under or near the Poles.[51]

But despite coming so close to the crux of the problem and even pointing precisely in the direction of its solution, Hooke was unable to go any further, and left the task of actually providing that solution to Newton. At the end of his *Attempt to Prove the Motion of the Earth* (1674), Hooke writes:

> This I only hint at present to such as have ability and opportunity of prosecuting this Inquiry, and are not wanting of Industry for observing and calculating, wishing heartily such may be found, having my self many other things in hand which I would first compleat and therefore cannot so well attend it.[52]

Though he intuitively foresaw in general outline the kind of theory that was to come, he was too busy with his Royal Society duties to develop it himself.

There is also the fact that Hooke lacked sufficient mathematical skill to accomplish that task; he seems not to have had the same degree of talent as a mathematician that he had as an experimenter. Hence it was Newton, seven years Hooke's junior, who rigorously derived the law of universal gravitation, thereby refining Hooke's vision of the "system of the world" into an elaborate and grand mathematical system. Nonetheless, given that in physics it is almost always the correct formulation of a problem that brings its solution within reach, it must be acknowledged that Hooke's achievement was of enormous value.

## Section 6—Newton and his Mechanics

From Kepler's three laws of planetary motion, Newton deduced with mathematical rigor that an attractive force operated between the sun and each planet at a strength inversely proportional to the square of the distance between them. Hypothesizing that this attractive force was universal—operating among all bodies everywhere, namely both celestial bodies and terrestrial bodies—he was able to explain not only the motions of planets and their satellites but a number of other phenomenon as well, from the shape of the Earth to the ebb and flow of the tides. Providing thus the final solution to the problem of attraction among celestial bodies, a problem that had plagued scientists since Kepler, Newton elucidated the underlying order of the sun-centered universe (solar system). In deriving mathematical formulas for force from the motions of planets, Newton relied entirely on the approach to this problem suggested by Hooke, whereby orbital motion was taken to be the superposition of rectilinear inertial motion and deflection due to a central force. This reliance on Hooke's method is clearly evident in Newton's formulations of the principles of mechanics.

In *Mathematical Principles* Newton presents the following laws of motion as principles of mechanics:

> LAW I
>
> *Every body continues in its state of rest, or of uniform motion in a right line, unless it is compelled to change that state by forces impressed upon it.*
>
> . . .
>
> LAW II
>
> *The change of motion is proportional to the motive force impressed; and is made in the direction of the right line in which that force is impressed.*[53]

(The third law is the law of reciprocal actions.)

If we render the second law as a differential equation, as it is in modern texts on mechanics, then there is no need to formulate two separate laws, because the first law simply states what is derived when the external force in the second law is zero. In *Mathematical Principles*, however, Newton does not use differential equations. Rather, his method throughout the book is to find the relation for an infinitesimal time interval ($\Delta t$). As noted by the nineteenth-century physicist James Clerk Maxwell in his commentary on the second law, "By impressed force [Newton] means what is now called Impulse, in which the time during which the force acts is taken into account as well as the intensity of the force."[54] Hence, expressing the above two laws in present-day terms using vectors $\boldsymbol{v}(t+\Delta t) = \boldsymbol{v}(t) + (\boldsymbol{F}/m)\Delta t$, we can say that the infinitesimal displacement for minute time $\Delta t$ in the first law is $\Delta \boldsymbol{r}_1 = \boldsymbol{v}(t)\Delta t$ (displacement by inertial motion) and in the second law is $\Delta \boldsymbol{r}_1 = (\boldsymbol{F}/2m)(\Delta t)^2$ (displacement by external force). On this basis, these two displacements are independent.

As indicated in "Definition III" and "Definition IV" at the beginning of *Mathematical Principles*, on these grounds Newton counts inertia among the kinds of force, ranking it alongside "impressed force" as the "inertial force" or "innate force" of matter. With this, the two displacements are synthesized:

COROLLARY I

> *A body, acted on by two forces simultaneously, will describe the diagonal of a parallelogram in the same time as it would describe the sides by those forces separately.*[55]

In other words, the total displacement for the given time interval can be assigned as the vector sum of the displacements caused by the two forces ("innate" and "external"), that is, $\Delta \boldsymbol{r}_1 + \Delta \boldsymbol{r}_2$. This method for the analysis and synthesis of motion closely matches the method of analysis for planetary motion that Hooke proposed.

What Newton then did in *Mathematical Principles* was to obtain the law of force between celestial bodies from the observed planetary motions based on these laws of motion, and then to explain the system of the world. Thus, from Kepler's three laws of planetary motion, themselves derived from observation, Newton succeeded in deriving a law of the gravitational force operating between the sun and the planets (i.e, the formula thereof, which states that the gravitational force between two bodies, which exert that force upon each other, is inversely proportional to the square of the distance between them and directly proportional to the mass of both). His actual procedure was essentially as follows. First, he proved that Kepler's second law—according to which each planet's orbits lies on a plane that includes the sun (a proposition now also known as Kepler's zeroth law), and the radius vector of each orbit (the straight line between the planet and the sun) always sweeps out the same area in the same amount of time—means that the external force applied to a planet is a central force, that is, is always directed toward a central body. Next, from Kepler's first law, which states that each planet's orbit describes an ellipsis with the sun at one focus, he derived the relationship between the strength of the force and the planet's distance from the sun. Then, taking into account Kepler's third law, he showed that the force was proportional to the mass of the bodies involved. Newton demonstrated that the force thus derived was the same as that acting between the Earth and the moon and between the Earth and terrestrial bodies—or in other words, that all such force could be explained with a single mathematical function.

By showing that the force of gravity was truly universal, acting between all bodies everywhere, Newton brilliantly unraveled the problem that Kepler had first addressed: the dynamic order of the "system of the world." The adoption of this concept of gravity (universal gravitation) as a force acting at a distance and expressed as a mathematical function provided the impetus that took scientific thinking beyond the limits of seventeenth-century naive mechanism and launched modern physics as a mathematical science. The first edition of *Mathematical Principles* was published in 1687, the year before the Glorious Revolution and thirteen years after Hooke had first discoursed on the "system of the world." Fittingly, book 3 of *Mathematical Principles* bears the title "The System of the World."

Looking back over these developments, one can see that Hooke did more than lay the groundwork for acceptance of the anti-mechanistic notion of force acting at a distance: he also pointed the way toward the explication of planetary orbits. It was by those guidelines set by Hooke that Newton was able to complete his theory of universal gravitation and the world system. The overall process of mathematical reasoning that Newton carried out certainly constitutes a masterful and unrivaled work. Nonetheless, for their importance in providing the essential starting point for that process, we must acknowledge the key roles played by Kepler and later Hooke, and, by extension, the considerable influence of the theory of magnetic gravity.

## Section 7—Newton and Gravitational Force

Under the influence of figures such as Gilbert and Wilkins, in Britain the notion of force acting at a distance was to some extent already accepted, and when Hooke referred to gravity as such a force, he seems not to have aroused much criticism or objection.[56] Magnetical philosophy, which attributed the Earth's activity to its magnetism and its magnetism to its soul-like nature, was essentially unconnected to mechanism, and for this reason both mechanism and the idea of action at a distance loosely coexisted in the minds of British and especially London scientists. But then came Newton's *Mathematical Principles*, which, being written in Latin, immediately received great attention not only in Britain but also on the Continent, where his notion of "attraction" was severely criticized. As one scholar notes, "ideas of attraction ... hardly constituted orthodox mechanical philosophy in 1687."[57] The mechanistic view was more than just that matter was inert and passive. To mechanists, who prided themselves on having rid philosophy of teleological and hylozoistic views of nature, the idea that the sun exerted force upon planets across vast, utterly empty space suggested that the sun was *aware* of the existence and position of those distant bodies upon which it was supposed to act. For this reason, the idea was extremely difficult for mechanists to swallow.

The difference between the mechanistic view of action and the Newtonian view of force is well illustrated by the divergent opinions on acid put forward on the one hand by Newton and on the other by French Cartesian chemist Nicolas Lemery (1645–1715), a close contemporary of Newton's (two years his junior). In his *Cours de chymie* (Course on chemistry), first published in 1675 and thereafter in several further editions, Lemery proposed that the tart, prickling taste of acidic substances was due to their particles having sharp "spicules" or "acid points" that stimulate the nerves of the tongue, and that neutralization of acidic with alkaline substances occurred because the latter were composed of porous particles whose pores the "acid points" fit into.[58] This is a typical naive-mechanist explanation of matter's perceptible qualities and chemical reactions.

The contrasting explanation of acids that Newton put forward in 1692 is as follows:

> The Particles of Acids … are endued with a great Attractive Force; in which Force their Activity consists; and thereby also they affect and stimulate the Organ of Taste …
>
> The Particles of *Sal Alkali*, do consist of *Earthy* and *Acid* united together, after the same manner: But these Acids have so great an Attractive Force, that they can't be separated from the Salt by Fire.[59]

From a present-day perspective, expressions such as "spicules of acid particles" and "pores of alkaline particles" seem quite crude and even absurd, whereas Newton's critique of such notions is far more levelheaded and sound—as when he argues that the "Parts of all homogeneal hard Bodies which fully touch one another, stick together very strongly" and that "for explaining how this may be, some have invented hooked Atoms, which is begging the Question."[60]

Newton expresses his basic position on such questions thus:

> There are … Agents in Nature able to make the Particles of Bodies stick together by very strong Attractions. And it is the Business of experimental Philosophy to find them out.

This attitude derives from his overall view of nature, which in short is that

> Nature will be very comfortable to her self and very simple, performing all the great Motions of the heavenly Bodies by the Attraction of Gravity which intercedes those Bodies, and almost

> all the small ones of their Particles by some other attractive and repelling Powers which intercede the Particles.[61]

To Newton, the concept of attractive force was essential to the formulation of a valid theory of matter.

At the time, however, few shared that view except in England. Newton's idea of "attraction" drew a chorus of criticism, especially from Continental thinkers, for being a reversion to the notion of "occult qualities" and a rehashing of the theory of "sympathy" and "antipathy." A statement by Leibniz in 1716 summed up the prevailing view on the Continent: "All the natural forces of bodies, are subject to mechanical laws."[62] This was far from an unreasonable view to adopt in the context of the times. Consider, for example, what had been learned about the apparent "attractive" effect of the vacuum pump. According to previous natural philosophy, nature's "abhorrence of a vacuum" ensured that the water in the pump's pipe was *sucked* up by a mysterious power of attraction; but the seventeenth-century experiments by scientists from Torricelli to Boyle showed that the effect actually occurred because the atmosphere, which was in direct contact with the surface of the water outside the tube, *pushed* the water up. One can readily visualize and grasp the latter explanation, while the former is somewhat dubious. In that context, it was theorists persisting with the notion of "attraction," rather, who were seen as rigid conservatives rejecting new scientific knowledge. For all its prodigious mathematical armor, the essence of Newton's argument was thus regarded as setting scientific progress back an entire century.

However, that was not a fair criticism of the actual treatment of gravity in *Mathematical Principles.* At the beginning of the treatise, after presenting definitions of different kinds of force, Newton duly gives notice that

> I here design only to give a mathematical notion of those forces, without considering their physical causes and seats ... and use the words attraction, impulse, or propensity of any sort towards a centre, promiscuously, and indifferently, one for another; considering those forces not physically, but mathematically.[63]

These lines were retained unchanged from the book's first edition of 1687 through the third edition of 1726. And the thrust is the same in the following passage from Newton's *Opticks*:

> What I call Attraction may be perform'd by impulse, or by some other means unknown to me. I use that Word here to signify only in general any Force by which Bodies tend towards one another, whatsoever be the Cause.[64]

Newton thus remained entirely consistent on this point.

Nonetheless, his intent was not well understood, especially on the Continent. In a letter to Leibniz written just after the publication of *Mathematical Principles*, the Dutch physicist Huygens called Newton's idea of attractive force "absurd."[65] And soon after Newton's death, the French Cartesian Fontenelle, referring to Newton's idea of gravity having unknown causes but manifest effects, dismisses such thinking with the rhetorical question, "are they not properly causes which the Schoolmen call *occult qualities* ...?" Newton's "continual use of the word Attraction," Fontenelle argues,

> at least makes the Reader familiar with a notion exploded by the Cartesians, and whose condemnation had been ratified by all the rest of the Philosophers; and we must now be upon

> our guard, lest we imagine that there is any reality in it, and so expose our selves to the danger of believing that we comprehend it.[66]

Considering that this was part of a eulogy given on behalf of the French Academy on the occasion of Newton's death, one senses the depth of the rift between the Continental and Newtonian positions.

Even Leibniz, one of the few thinkers of Newton's era who could match him in intellectual talent, grumbles in his 1710 work *Theodicy* about the seeming reappearance of action at a distance:

> It is true that modern philosophers for some time now have denied the immediate natural operation of one body upon another remote from it, and I confess that I am of their opinion. Meanwhile remote operation has just been revived in England by the admirable Mr. Newton.[67]

Leibniz states his position explicitly in a letter of 1715:

> His [i.e., Newton's] philosophy appears to me rather strange and I cannot believe it can be justified. If every body is heavy it follows necessarily (whatever his supporters may say and however passionately they deny it) that gravity will be a scholastic occult quality or else the effect of a miracle.[68]

Thus it was widely thought that Newton had revived a specter that the new philosophy had already put to rest.

In Continental Europe, Newton was assailed from all sides. Eventually, it seems, even he could not let such criticism go unanswered. In the "General Scholium" at the end of the (1713) second edition of *Mathematical Principles* he offers the following rebuttal, essentially a statement of the difference between his and his critics' views on the methodological maxims and therefore the purpose of natural philosophy:

> Hitherto we have explained the phenomena of the heavens and of our sea by the power of gravity, but have not yet assigned the cause of this power. This is certain, that it must proceed from a cause .... But hitherto I have not been able to discover the cause of those properties of gravity from phenomena, and I frame no hypotheses; for whatever is not deduced from the phenomena is to be called an hypothesis; and hypotheses, whether metaphysical or physical, whether of occult qualities or mechanical, have no place in experimental philosophy.[69]

The "hypotheses" Newton rejects here meant any narrow mechanistic models—or what Alexandre Koyré calls "those that cannot be proved or disproved by mathematically treated experiment, specifically,... global qualitative explanations such as were attempted by Descartes."[70]

Moreover, instead of attempting to explain the transmission of force, as mechanism sought to do, Newton proposed determining the mathematical laws of force. At the end of *Mathematical Principles*, a few lines after the passage just quoted, Newton remarks that

> to us it is enough that gravity does really exist, and act according to the laws which we have explained, and abundantly serves to account for all the motions of the celestial bodies, and of our sea.[71]

Newton had thus abandoned the search for the cause or origin of gravitational force.

Indeed, as stated in his preface to the first edition of *Mathematical Principles*, from the outset Newton defines the purpose of "rational mechanics" as "the science of motions resulting from any forces whatsoever, and of the forces required to produce any motions, accurately proposed and demonstrated." On that basis, he limits the program and scope of his own natural philosophy (physics) to investigating "from the phenomena of motions … the forces of nature, and then from these forces to demonstrate the other phenomena." He spells out the steps of this approach as it applies to astronomy:

> I derive from the celestial phenomena the forces of gravity with which bodies tend to the sun and the several planets. Then from these forces, by other propositions which are also mathematical, I deduce the motions of the planets, the comets, the moon, and the sea.[72]

Specifically, from Kepler's three laws Newton deduces the mathematical law of universal gravitation as $F = Gm_1m_2/r^2$. Then, working back from this law, he explains the fall of terrestrial bodies, the orbiting of the moon around the Earth, and the phenomenon of the tides, and furthermore calculates the shape of the Earth; and the success of these attempts itself proves the validity of the law of universal gravitation. He regards the further ontological question of the "true nature" of universal gravitation as beyond the province of natural philosophy (physics), and likewise views related questions—such as how it is propagated through space or by what medium it reaches the bodies under its power—as matters with which there is no need to bother. That was Newton's official position. He thus extended to dynamics the stance of mathematical phenomenalism that Galileo had adopted in kinematics.

In her study of Newton's original handwritten manuscripts on alchemy, Betty Dobbs maintains that the point at which Newton definitively dispelled the mechanistic hypothesis of a great vortex of minute particles that fills space and carries the planets along in their orbits was when he mathematically proved Kepler's second law (the areal law).[73] If there were such minute particles operating on motions of planets—exerting force in a direction perpendicular to their radius vectors—then Kepler's second law would not hold. As Newton rigorously demonstrated, Kepler's second law is strictly true only when the force is a central force. It was with this step that Newton, though influenced by Cartesian mechanism in his youth, can be said to have finally parted company with impact models of force. That was also the point when Newton recognized that Tycho Brahe's precise quantitative observations and Kepler's rigorous mathematical laws were far more credible that any philosophical position.

Having rejected soul-invoking and magical action-at-a-distance explanations of magnetic and similar powers, mechanism thought it could dismantle magic by elucidating the mechanical operations of such forces. Ultimately, however, mechanism failed to explain gravity. By contrast, Newton rationalized the magical notion of force acting at a distance by conforming gravity to rigorous mathematical laws based on precise observation.

## Section 8—Magic Sanctified

Newton himself was in fact a quite mystical and religious person—at heart not as strictly modern a thinker as his achievements suggest. Indeed, economist John Maynard Keynes once called Newton "the last of the magicians."[74] But perhaps it was precisely that attitude that enabled Newton to accept the mysterious, miraculous operation of gravity as a force acting at a distance. Bear in mind that, while on the one hand conceiving his "system of the world" based on universal gravitation, on the other Newton devoted himself to alchemy. In fact, whereas he spent at most just over three years of his life in concentrated study of mechanics, including the writing of *Mathematical Principles*, he remained absorbed in his alchemical investigations for almost ten times as long. Dobbs, Westfall, and others who have analyzed the vast body of manuscripts on alchemy that Newton left identify the alchemical notion of "active principles" as the source of his concept of force.[75]

Gilbert's assertion that the Earth was a soul-endowed magnet had essentially been made to underpin the idea of the Earth's active nature. As we saw in the previous chapter, by Newton's time further nonmechanical agents, such as Power's "subtle spirits" and Boyle's "effluvia" endowed with "cosmical qualities," had been introduced into the debate in Britain, unlike on the Continent. As one scholar puts it, "the use of active principles in pre-Newtonian matter theory represents a clear and undeniable tradition in English mechanical philosophy and cannot be dismissed as nothing more than a series of minor aberrations by the writers involved."[76]

Meanwhile, in 1659, even Henry More, the Cambridge Platonist who had a considerable influence on Newton during the latter's student days, declared it certain

> that the phenomenon of gravity is repugnant to mechanical laws; and further that its explanation cannot be resolved into causes purely mechanical and corporeal; but that it is necessary here to admit certain additional causes which are immaterial and incorporeal.[77]

In any case, Newton believed that this dynamic world could not be explained if matter were the inert, passive thing that mechanism had claimed it to be. He makes the point in tentative terms in query 31, added for the 1706 Latin edition of his *Opticks*:

> It seems to me farther, that these Particles have not only a *Vis inertiae*, accompanied with such passive Laws of Motion as naturally result from that Force, but also that they are moved by certain active Principles, such as is that of Gravity, and that which causes Fermentation, and the Cohesion of Bodies.

He also declares there that

> we meet with very little Motion in the World, besides what is owing to these active Principles. And if it were not for these Principles, the Bodies of the Earth, Planets, Comets, Sun, and all things in them, would grow cold and freeze, and become inactive Masses; and all Putrefaction, Generation, Vegetation and Life would cease, and the Planets and Comets would not remain in their Orbs.[78]

By this reasoning, Newton had no doubt about the existence of active principles; for him, rather, the question was how to reconcile the passivity of matter with the activity of force. Part of his effort to do so is evident in *An Hypothesis Explaining the Properties of Light*, a discourse he presented to the Royal Society in 1675 and in which he proposes how "perhaps may all things be originated from aether."[79] Although this reintroduction of "aether" may appear to be a return to mechanism, that was not Newton's intention. In a letter of 1693 he does indeed write:

> It is inconceivable, that inanimate brute Matter should, without the Mediation of something else, which is not material, operate upon, and affect other Matter without mutual Contact.

But he then adds:

> Gravity must be caused by an Agent acting constantly according to certain Laws; but whether this Agent be material or immaterial, I have left to the Consideration of my Readers.[80]

Gravitational force is conveyed by "something," but that "something" to which Newton refers is not necessarily what we today think of as matter.

In query 28 of the 1707 edition of *Opticks*, Newton states more clearly that "the main Business of natural Philosophy is to argue from Phaenomena without feigning Hypotheses, and to deduce Causes from Effects, till we come to the very first Cause, which certainly is not mechanical." Via such thinking, the answer at which he ultimately arrives on the question of gravity's cause—the "first cause, which certainly is not mechanical"—is "a Being incorporeal, living, intelligent, omnipresent," or in other words, God.[81]

Richard Bentley, a Newtonian and contemporary of Newton's, puts this view plainly when he writes that

> Universal Gravitation, a thing certainly existent in Nature, is above all Mechanism and material Causes, and proceeds from a higher principle, a Divine energy and impression.[82]

There is also Newton's own statement in the "General Scholium" of the second edition of *Mathematical Principles* that "to discourse of [God] from the appearances of things, does certainly belong to Natural Philosophy."[83] As this suggests, in his view the mathematical principles of natural philosophy could be considered complete only when complemented by the theological principles of natural philosophy.

In the social and cultural context of Newton's time, in order to introduce into physics the magical notion of force acting at a distance, perhaps it was necessary not only to rationalize but also to sanctify that force. Whatever the metaphysical or theological underpinnings concerning force, the concept of universal gravitation expressed as a mathematical function proved to be extraordinarily effective, and came to play a pivotal role in the formation of modern physics and modern cosmology. Nonetheless, until they were finally purged by the French Enlightenment some time later, such metaphysical and theological "impurities" in ideas about force persisted, as we shall now see more clearly in the concurrent evolution of the debate on magnetic force.

## Section 9—Newton and Magnetic Force

Newton's *Mathematical Principles* provided a solution to the question of gravitational force, but there still remained the problem of magnetic force.

Although there are several places scattered throughout Newton's oeuvre where he touches on the subject of magnets, there are none in which he addresses the magnet or magnetic power itself as his main theme. One noteworthy experiment he conducted on the subject was that in which he showed that a magnet and a piece of iron floating in still water attracted each other until they joined at the midpoint, and thereafter remained at rest.[84] Newton regarded this as proof that the magnitude of a magnet's power to attract iron and that of iron's power to attract a magnet were equal. However, although the experiment was important for directly showing that the law of action and reaction holds true for magnetic force,[§§] it did not otherwise add to understanding of that power. It appears that Newton had comparatively little interest in magnetic force itself.

Newton became professor of mathematics at Cambridge at the age of 26 in 1669, but prior to that he had remained in quiet pursuit of his studies there, and it was not until the end of 1671 that he finally started to interact with the Royal Society in London. Perhaps because of this early isolation from the Royal Society, the formative process of Newton's scientific thought shows almost no influence from Gilbert's magnetical philosophy. The mechanical philosophy of Descartes and Boyle had a far greater impact on him. Indeed, Newton's notes from his student years include references to "magneticall rays" as the cause of magnetism and similarly to "rays of gravity" as the cause of gravity; there are even notes on his ideas for creating a perpetual-motion machine using currents of such rays.[85] Such thinking is entirely mechanistic. Furthermore, whereas the notes include frequent mention of Descartes and Boyle, and even show the influence of atomist Walter Charleton, they contain nothing about Gilbert, Kepler, or Wilkins. The young Newton dedicated himself to mechanism and atomism on the one hand and to Cambridge Platonism on the other.

Accordingly, even in later years Newton refused to accept certain views held by members of the Oxford and London groups, such as

[§§]That the law of action and reaction holds true for magnetic force and also for the other attractions was explicitly stated by one of Newton's pupils, John Theophilus Desaguliers, in 1719: "If the *Magnet* or *Load-Stone* attract *Iron*, the *Iron* will likewise equally attract the *Load-Stone*; this may be seen if you hold the *Iron* in a fixed Point, and the *Load-Stone* hang in a *Scale*, or be suspended by a String. The same thing is true in all other Attraction." Desaguliers, *A System of Experimental Philosophy*, 63.

that the sun was a magnet. In response to Hooke's and Flamsteed's discourse about the sun exerting a magnetic force on comets, Newton identified a flaw in the theory and flatly rejected it:

> I am the lesse inclined to beleive this attraction to be of a magnetick nature because ye [sun] is a vehemently hot body & magnetick bodies when made red hot lose their vertue.

In Newton's view, the

> whole body of ye Sun therefore must be red hot & consequently voyd of magnetism unless we suppose its magnetism of another kind from any we know.[86]

Similarly, corollary 5 of proposition 6 in the third book of *Mathematical Principles* reads as follows:

> The power of gravity is of a different nature from the power of magnetism; for the magnetic attraction is not as the matter attracted. Some bodies are attracted more by the magnet; others less; most bodies not at all. The power of magnetism in one and the same body may be increased and diminished; and is sometimes far stronger, for the quantity of matter, than the power of gravity; and in receding from the magnet decreases not as the square but almost as the cube of the distance, as nearly as I could judge from some rude observations.[87]

This passage appears in the second (1713) and third (1726) editions of the work, but in the first edition, published in 1687, the corresponding text appears as corollary 4 and with the final part stating more simply that magnetic power decreases by a ratio greater than the square of the distance. In either case, magnetic force was regarded as not conforming to the inverse-square law. Newton's grounds for considering magnetic force to be different from gravitational force thus seem to have been that (a) the latter was universal while the former was selective, and (b) the measured variation of magnetic force differed from that of gravity.

In Birch's *History of the Royal Society*, an entry dated 23 February 1687 relates that "Mr. Halley read a paper of his concerning an experiment for finding the comparative force of a loadstone at several distances"; and one for 2 March the same year reads: "An account of an experiment made by Mr. Halley for finding the proportion of the decrease of the magnetical virtue was read."[88] Unlike the one Hooke had applied, the method used in this case was to measure the angle of a magnetic needle's deviation under the effect of a magnet placed on the horizontal plane some distance away from the needle. The experimenter, Edmond Halley (1656–1742), is known not only for having urged the self-effacing Newton to write *Mathematical Principles* in the first place but also for contributing tremendous effort and even financial assistance toward getting it published. Considering this relationship between the two men, and also the fact that Newton was writing book 3 of the first edition of *Mathematical Principles* at the time of the experiment, it is conceivable that Halley's measurements were made at Newton's suggestion and behest. If so, then Newton's assertion in the first edition that the rate of attenuation of magnetic force was greater than the square of the distance was probably based on these measurements.

Much later, while preparing for the publication of the second edition of *Mathematical Principles*, Newton proposed, this time in his capacity as president of the Royal Society, that a similar set of measurements be made. According to science historian Robert Palter, Newton made this request on 20 March 1712, whereupon the young mathematician Brook Taylor (1685–1731), with assistance from Francis Hauksbee (c. 1670–1713), then the Royal Society's curator of experiments, endeavored to measure the distance dependence of magnetic force.[89] Hauksbee and Taylor announced the results of their measurements separately, Hauksbee first in the 1712 issue of the Society's *Philosophical Transactions* and Taylor via a letter, dated 25 June 1712, that he wrote to Society secretary Hans Sloane and that was published in edited form in the same journal (the 1715 and 1721 issues) after Hauksbee's death.

Taylor describes the method of their measurements as follows:

> We placed the great Loadstone belonging to the Royal Society so, that it's [sic] two Poles lay in the Plane of the Horizon, and were in a line exactly at right Angles with the natural Direction of the Needle we made use of, (which was that Dr. *Halley* had made to observe the Variations with). And by means of a Carriage contrived for that purpose, the Stone was easily moved to and fro, the Poles continuing always in the same Line. The Needle was so placed, that the Center it play'd upon was in the same Line with the Poles of the Stone; the North Pole being towards the Needle. We measured the Distances from the Center of the Needle to the Extremity of the Stone; and we found the Variations of the Needle from its natural Position to be as in the following Table.[90]

In other words, they aimed to determine the functional relation between the angle of the needle's deviation ($\varphi$) and its distance from the end of the magnet ($r$). In fact, the deviation of a magnetic needle is an effect of the interaction between the magnetic dipoles, making the relation between the force and the needle angle quite complex, but here it was tacitly assumed that angle $\varphi$ provided a direct measure of the strength of the magnetic force. Taylor gives the following account of the results:

> If it were known what point within the Stone, and what point in the Needle are the Centers of the Magnetical power, it would be easy to find the true powers of the Magnet at all the distances observed. For want of that Knowledge, I have computed the Forces from the Center of the Needle, and the Extremity of the Loadstone, and find, that at the distance of nine Feet, the Power alters faster, than as the Cubes of the distances, whereas at the distances of one and two Feet, the Power alters nearly as their Squares.[91]

The concern about the "center of the magnetical power" arose from the fact that, according to Hauksbee, the magnet used had "an irregular Figure."[92] In an effort to "try whether the Law, by which the Magnetism alters, could be reduced at all distances to any one certain power of those distances," Taylor then worked backward to identify the points that could be considered the centers of the magnetic force, only to find that this put the center of the magnet outside the magnet itself. From this he concludes that

> it seems to appear, that the power of Magnetism does not alter according to any particular power of the distances, but decreases much faster in the greater distances, than it does in the near ones.[93]

Taylor thus failed to identify any simple mathematical function regarding magnetic force. Although his measurement data differed somewhat from Hauksbee's, the two men's conclusions were the same.

Since it appeared in any case that magnetic force did not follow the same laws as did gravitational force, Newton adopted a different approach to the former, one in which he unwaveringly pursued a mechanistic and materialistic model of mediated action based on the propagation of "magnetic effluvia."

In a manuscript fragment written around 1673 and titled "De aere et aethere" (On air and ether), Newton writes:

> I believe everyone who sees iron filings arranged into curved lines like meridians by effluvia circulating from pole to pole of the [load-]stone will acknowledge that these magnetic effluvia are of this kind.[94]

Furthermore, in *An Hypothesis Explaining the Properties of Light*, a paper submitted to the Royal Society in 1675, he counterposes "magnetic effluvia" to the "gravitating principle." And in a draft revision to *Mathematical Principles* that is thought to have been written in the 1690s, he writes that gold, since it does not shield magnetic force, must have minute pores allowing the "magnet's effluvia" to pass through it.[95] And even in later writings, such as number 22 of the "queries" that he added for the 1717 and other revised editions of his 1704 treatise *Opticks*, Newton refers to "Effluvia of a Magnet" and "magnetick Effluvia" as if their existence were self-evident.[96]

There is also the following passage in the notes that mathematician David Gregory (1661–1708) of Edinburgh took from what Newton told him when the two met in Cambridge in March 1694:

> Magnetic virtue is destroyed by a flame, and by heat: a rod of iron, either by standing long in a perpendicular position, or by cooling in an erect position, acquires magnetic virtue from the Earth. But it gets magnetic virtue too with a strong blow of a hammer at either extremity. If it is struck hard at one or other

> end the poles of the iron rod are interchanged: if it is struck in the middle (say with hammering at an anvil) it quite loses its magnetism. And so this virtue seems to be produced by mechanical means.[97]

This is the same reasoning that Henry Power had employed earlier in his mechanistic theory of magnetic force. Newton no doubt supposed that a property such as magnetic power, which acted only upon iron and furthermore could be conferred and taken away at will by heating and hammering, had to be fundamentally different from what he had identified as the *universal* force of gravitation.

Newton moved to London in 1696 and assumed the presidency of the Royal Society in November 1703, following Hooke's death in March that year. In that position, which he held until his own death in 1726, Newton came to wield tremendous influence over the Society and by extension the entire British scientific community. The Royal Society became stacked with devoted Newtonians under Newton's paternalistic rule. Accordingly, it is fair to say, Royal Society research on natural philosophy (physics) during the first third of the eighteenth century was conducted along lines generally either prescribed or approved by Newton. In a review of magnetism studies that was included in the 1730 issue of *Philosophical Transactions*, Servington Savery asserts that mutual attraction between magnets and the attraction of iron by magnets were due to "an invisible Force which differs from that of Gravitation, and also of Electricity."[98] That magnetic force and gravitational force were of fundamentally different natures seems to have been the common understanding in the Newtonian circle at that time.

Research by the Royal Society into magnetic force did in fact proceed from the assumption of "magnetical matter" and "magnetic effluvia." In 1716, Halley, a loyal Newtonian, attempted to explain the aurora—the "strange and surprizing *Phaenomena*" observed not only in London but throughout Britain in March that year—in terms of "magnetic effluvia" flowing out from the North Pole. In this, the first ever attempt to link the aurora phenomenon to geomagnetism, Halley notes: "I assume the *Effluvia* of the Magnetical Matter for this purpose."[99] The passage in question shows clearly that the existence of "magnetic effluvia" was endorsed by the members of the Newtonian circle. And as evidence that streams of such effluvia exist outside the Earth's surface, Halley, following Newton, points to the pattern of curved lines that iron filings form around a magnet.

Even after Newton's death, his influence over the Royal Society and the London scientific community was sustained until the final demise of the Newtonian circle itself with the deaths, in 1742 and 1744, respectively, of Halley and of John Theophilus Desaguliers, a natural philosopher hailed as a "master of the art of exposition of Newtonian philosophy: without mathematics."[100] Nonetheless, the entry "Magnetism" in the fifth edition of Ephraim Chambers' *Cyclopedia*, published in London around that time (1741–1743), expressly states that the "opinion that principally prevails among the moderns is that of Des Cartes."[101] On the subject of magnetic force, Newton did not, after all, put forward a theory that could supersede Descartes's.

Not long after that, in 1747, a report on experiments with magnets was published in *Philosophical Transactions* by Royal Society fellow Gowin Knight (1713–1772), a physicist known for developing a superior new method of manufacturing artificial magnets and for improving nautical compasses. At the end of the report, Knight presents a consideration of the "Cause of the surprizing

*Phaenomena* of the Loadstone" that includes the following propositions:

> *Proposition* I*st.*
> *The magnetic Matter of a Loadstone moves in a Stream from one Pole to the other internally, and is then carried back in curve Lines externally, till it arrives again at the Pole where it first entered, to be again admitted.*
>
> ...
>
> *Prop.* 2.
> *The immediate Cause, why two or more magnetical Bodies attract each other, is the Flux of one and the same Stream of magnetical Matter through them.*
>
> ...
>
> *Prop.* III.
> *The immediate Cause of magnetic Repulsion is the Conflux and Accumulation of the magnetic Matter.*

Although no mention is made of "grooved particles," this is in effect a revival of Descartes's vortex theory. Among the evidence Knight cites for this view are the following experiments:

> *Experiment* I.
> If we lay a magnetical Body under a Piece of Paper or Glass that is strewed over with steel Filings or magnetical Sand, and by striking the Table put the Filings in Motion, they will readily dispose themselves in such a manner as to represent, with great Exactness, the Course of the magnetic Matter.... By this Experiment the curve Lines in which the magnetical Matter returns back to the Pole where it first enter'd are accurately expressed by the Arrangement of the Filings....
>
> ...

> *Exp.* III.
>
> If the South Pole of one Magnet be opposed to the North of another, most of the magnetic Matter is carried directly out of one into the other: and does not return back in curve Lines till after having passed through both Magnets… .
>
> …
>
> *Exp.* V.
>
> If the North or South Poles of two Magnets be opposed to each other, the Filings will exhibit the Appearance of two Streams meeting; and the Curves of each will all be turned towards the opposite Pole of the same Magnet….[102]

These passages are from descriptions of only some of the experiments Knight cites, but here, too, as in Newton's own writings, the curved-line pattern formed by iron filings in a magnetic field is given as solid evidence for the existence of "magnetic matter" and its "flux."

Thus, in Britain, as one historian points out, "in regard to magnetism, Newton's thinking remained throughout almost all his life very close to that of his supposed arch-rival, Descartes!" and "most of our early 'Newtonian' group adopted the same approach." Unlike in the case of gravitation, notes the same historian, "in the magnetic case, even the most convinced Newtonians continued to uphold orthodox Cartesian opinions throughout the first half of the eighteenth century."[103] However, since these Newtonians considered "magnetic effluvia," the key agent of vortical motion, to be a specific kind of matter, their position was closer to Boyle's than to the "orthodox Cartesian" view of magnetism.

* * *

In the seventeenth century, the emergent philosophy of mechanism and the revived philosophy of atomism were extolled in Continental Europe as the cornerstones of the new science. According to that science, material bodies were essentially inert and force could be propagated only by direct contact. This thinking came to exert considerable influence in Britain as well.

At the same time, however, British thinkers, particularly the group that formed around Wilkins and eventually evolved into the Royal Society, carried on, along with Bacon's "experimental philosophy," the "magnetical philosophy" of Gilbert and Kepler, which held that astronomical bodies exerted a seemingly magnetic influence upon one another. Britain therefore enjoyed a cultural climate more receptive to the idea that the force acting between astronomical bodies was a kind of action at a distance similar to magnetic force. Hence, even the mechanist Hooke explored two different positions concerning the force acting on planets and making them orbit the sun: a Cartesian view that attributed the force to the varying density of the fluid medium filling the intervening space; and the supposition that the sun pulled the planets toward it by some form of action at a distance. Then, regarding actual planetary motion, Hooke put forward the view that what scientists should investigate was not how a force diverting planets toward a central body could possibly arise or be transmitted, but simply how the strength of that force changed (decreased) with increasing distance from the central body. This marked a watershed in mechanistic thinking—as is illustrated in simplified form in Hooke's fundamental opinion that whether or not magnetic force and gravitational force are the same depends on whether or not they can be represented by the same form of mathematical function.

Hooke further proposed a method of analysis by which, taking planetary motion to be a superposition of rectilinear inertial motion tangential to the orbit and deflected motion toward the center due to an external force, one could then derive the attractive force from the sun. This analysis was then realized mathematically in Newton's *Mathematical Principles.*

In this connection it is worth noting Newton's candid statement of his basic perspective on natural philosophy (physics) in query 31 of his *Opticks*. First he asserts that

> the *Aristotelians* gave the Name of occult Qualities ... to such Qualities only as they supposed to lie hid in Bodies, and to be the unknown Causes of manifest Effects: Such as would be the Causes of Gravity, and of magnetick and electrick Attractions, and of Fermentations, if we should suppose that these Forces or Actions arose from Qualities unknown to us, and uncapable of being discovered and made manifest. Such occult Qualities put a stop to the Improvement of natural Philosophy, and therefore of late Years have been rejected. To tell us that every Species of Things is endow'd with an occult specifick Quality by which it acts and produces manifest Effects, is to tell us nothing.

Up to this point, Newton's view is identical to that of mechanist thinkers from Galileo and Descartes to Power. But in the words that immediately follow the above we find the originality and novelty that set Newton apart:

> But to derive two or three general Principles of Motion from Phaenomena, and afterwards to tell us how the Properties and Actions of all corporeal Things follow from those manifest Principles, would be a very great step in Philosophy, though the Causes of those Principles were not yet discover'd.[104]

Newton's position on the question of force was that it was enough to derive a mathematical law from phenomena and then apply that law to explain various other phenomena.

In this manner, Newton advocated a philosophy whereby, once one derives from Kepler's laws the fact that an attractive force (universal gravitation) acts between astronomical bodies in inverse proportion to the square of their distance from each other and in direct proportion to their mass, and furthermore thereby fully and quantitatively explains the motions of planets, comets, moons, and so on, then there is little point probing further into the nature of that force or its mechanism of transmission. And indeed, the success of mechanics rested on the fact that the law of universal gravitation that Newton thus derived did in fact explain the ebb and flow of tides, account for the shape of the Earth, and enable accurate prediction of the return of Halley's comet. Whereas mechanism had sought to explode magic by revealing the apparatus by which force was transmitted, Newton, by elucidating a law of force, rationalized magic and incorporated it into the mathematical science of physics. But Newton did more than rationalize a seemingly magical force; he also had to sanctify it by identifying its ultimate source as the omnipresent God of Christianity. Such were the constraints of his age.

Regarding the relationship between gravitational force and magnetic force, however, Newton, having avoided the influence of magnetical philosophy, had always distanced himself from the magnetic-gravity theories advanced by thinkers in the lineage from Kepler to Hooke. Moreover, Newton, like Hooke, considered the gravity-magnetism relationship in terms of whether or not the relevant mathematical functions matched one another. Since the inverse-square law could not be shown to apply in the case of magnetic force, he reasoned that it must be fundamentally different from the force of gravitation, and conceived of the former in mechanistic and materialistic terms as a type of mediated action involving "magnetic effluvia," "magneticall rays," or "magnetic matter." This was the state of thought about gravitational and magnetic force at the end of the first third of the eighteenth century, when Newton died.

It was not until after his death that the laws of magnetic force were likewise mathematically ascertained, the constraints of seventeenth-century naive mechanism thereby finally thrown off, and the study of magnetism properly incorporated into the mathematical sciences. Those developments were also a process of purging from the concept of force the metaphysical and theological impurities that Newton had attached to it.

Chapter 22

# Epilogue: Ascertaining the Laws of Magnetic Force

## Section 1—Musschenbroek's and Helsham's Measurements

In the period following Newton's death, prevailing opinion in Britain about magnetic force, unlike that about gravitational force, was mechanistic and materialistic, accounting for magnetic phenomena in terms of vortices of fluid matter circulating within and around magnets. But in the eighteenth century there appeared—notably among Continental and Irish scholars rather than among the British—those who investigated magnetism in much the same way that Newton had approached gravity, namely, by seeking the mathematical laws of magnetic force.

The first steps along that path were recorded by Dutch scientist Pieter van Musschenbroek (1692–1761). After studying medicine at Leiden (Leyden) University, in 1717 Musschenbroek visited England to deepen his knowledge of British experimental natural philosophy. Thereafter, while holding successive professorships at the University of Duisburg, Utrecht University, and Leiden University, he played a central role in disseminating Newtonian thought in the Dutch Republic during the first half of the eighteenth century.

Musschenbroek's efforts to measure magnetic force began in 1724. The results of those first measurements were recorded in a letter he sent to John Theophilus Desaguliers. The letter was published in *Philosophical Transactions* in 1725. In it, Musschenbroek denies the existence of any "magnetic effluvium" or "magnetic fluid" on the grounds that, even when an object is placed between a magnet and a piece of iron, the force with which the magnet attracts the iron is the same as when no such object is interposed. That is, while admitting that he did not know exactly what caused magnetic power, he thus reasoned that the cause could not possibly be something material.[1]

Musschenbroek's reasoning here is essentially the same as that of William Gilbert, who had likewise considered magnetic force to be nonmaterial because of the absence of a shielding effect. Later, in a work titled *Dissertatio physica experimentalis de magnete* (Physical experimental treatise on the magnet) and published in 1729, Musschenbroek declares that "I am of the opinion that there does not exist a magnetic Fluid save in the mind of one imagining it."[2]

Musschenbroek therefore directed his magnetic studies toward measuring how the strength of magnetic force changes with distance, and for that purpose he devised a more precise version of the balance-scale experiments suggested by Cusanus and Della Porta. In his letter to Desaguliers, Musschenbroek includes the raw data of various measurements, remarks that he does not think it possible to draw any conclusions from the experiment about the relation between force and distance, and ultimately reaches the verdict that there is no distinct relation between the two. He failed to discover any law for magnetic force.[3]

The first assertion of an inverse-square law for magnetic force was made by Richard Helsham (c. 1682–1738), who used virtually the same method as Musschenbroek's to measure attractive force between a magnet and a piece of iron. Helsham was born in Ireland, studied at Trinity College, and taught mathematics and natural philosophy in Dublin. His measurements appear in *A Course of Lectures in Natural Philosophy*, published posthumously in 1739.

Helsham starts out by clearly delineating the objective of his magnetic research:

> Many and surprising are the properties both of the loadstone and magnetical needle, which however I shall not here consider; my intent at present being only to shew from experiment the law of magnetical attraction; or in other words, to shew in what proportion the attractive power of the loadstone varies according to the different distances of the iron which it attracts.[4]

The device he used for his experiment was simple: a balance with a magnet suspended at one end, counterbalancing weights at the other end, and a flat plate of iron under the magnet:

> ... let a flat piece of iron be placed beneath [the suspended loadstone] at the distance of four tenth parts of an inch, the stone will immediately descend, and adhere to the iron: let the stone again be removed to the same distance, and a weight of four grains and four tenth parts of a grain be thrown into the scale at the other end of the balance; this weight will be an exact counterbalance to the attractive force, and prevent the descent of the stone; but if any part of the weight be taken out,

> the attraction will prevail, and carry the stone down. If the stone be placed at half the former distance, that is to say, at the distance of two tenth parts of an inch above the iron, the weight necessary to hinder its descent will be about seventeen grains and a half, that is four times as much as before. Consequently, the attractive force of the stone at the single distance from the iron, is to the same at the double distance as four to one, that is reciprocally as the squares of the distances.

Helsham's description of the experiment ends there. He then refers to Newton's suggestion, in corollary 5 of proposition 6 in the third book of *Mathematical Principles of Natural Philosophy* (*Mathematical Principles*), that magnetic force varies in inverse proportion almost to the cube of the distance, and remarks that "notwithstanding this objection I shall still venture to affirm the law of magnetical attraction to be such as makes it act with forces which are in the reciprocal duplicate ratio of the distance."[5]

Helsham's conclusion should really be called a conviction, one that he had held prior to conducting any experiment on the subject. His consistent reference to what *will* be or *will* happen makes one wonder whether he did in fact carry out the experiment and take the measurements. At the very least, the fact that he arrived at such a sweeping conclusion after only one round of measurements is enough to imply that he had accepted the conclusion beforehand.

Helsham's *A Course of Lectures in Natural Philosophy* was subsequently reprinted a number of times as a textbook on Newtonianism and the new experimental physics. Nonetheless, apart from its mention by Royal Society fellow James Ferguson (1710–1770) in his 1770 work *Introduction or Lectures on Electricity*,[6] Helsham's magnetic research seems to have been more or less disregarded in histories about inquiry into magnetic phenomena.

No reference whatsoever is made to it in Johann Christian Poggendorff's *Geschichte der Physik* (History of Physics), Ferdinand Rosenberger's *Die Geschichte der Physik* (The History of Physics), Edmund Hoppe's *Geschichte der Elektrizität* (History of Electricity), Park Benjamin's *The Intellectual Rise in Electricity*, Edmund Whittaker's *A History of the Theories of Aether and Electricity*, or Max Jammer's *Concepts of Force*. In his biographical study of Charles-Augustin de Coulomb, C. Stewart Gillmor mentions a number of figures who preceded Coulomb in formulating an inverse-square law for magnetic force, but Helsham's name is not among them.[7] An exception to this trend of disinterest is Robert E. Schofield's *Mechanism and Materialism*, but even there it is maintained that "Helsham's assertion, in 1739, of its inverse-square nature was rightly submerged" because it conflicted with measurements made by Brook Taylor, Musschenbroek, and Desaguliers.[8]

On the other hand, both the needle-deflection measurements by Taylor and Hauksbee and the balance-scale measurements by Musschenbroek failed to identify any clear functional relation between the force and the distance. One reason for this failure is that (to put it in modern terms) their experiments measured force acting between magnetic dipoles—that is, the resultant force between two pairs of magnetic poles—and they were unaware of the fact that such force varies in complex ways depending on the shape of the magnets used, their relative positions, and the ratio of their size. Particularly in regard to Musschenbroek's measurements, it hardly requires actual measurement to realize that, since the distances between the poles facing each other range from almost 0 up to 13.5 inches, the ratios of the four distances between each pole would vary widely and therefore preclude the resultant force from being a simple function of distance. Another reason concerns the measurements based on needle deflection: contrary to Taylor's and Hauksbee's naive assumption, the angle of the needle's deflection is not in itself a direct indication of—that is, it is not proportional to—the magnitude of the force.

In short, though we speak of measuring simply the strength of the force, in fact such experiments measure certain mechanical effects produced by the force. Accordingly, it is necessary to identify in advance a quantity to be measured that corresponds in a one-to-one ratio with the strength of force, and to prepare a theoretical framework of the relationship between that quantity and the magnitude of the force. At the same time, the measuring equipment must be suitably designed so as to magnify the desired effect, inhibit factors causing disturbance, and make measurement of the target physical quantity as easy and accurate as possible. For these reasons, accurate measurement of magnetic force required, on the one hand, further development of the theory of mechanics, particularly more precise theory of rigid-body equilibrium and motion; and, on the other, technical advances enabling the manufacture of more powerful and suitably shaped magnets. It was after these advances, both achieved around the middle of the eighteenth century, that magnetic research was elevated to the level of mathematical physics.

## Section 2—Calandrini's Measurements

In the field of measurements based on the deflection of a magnetic needle, Swiss scientist Jean-Louis Calandrini (1703–1758), a professor at the Academy of Geneva, derived the theoretical relationship between magnetic force and the angle of the needle's deflection, and by applying that knowledge he became the first to measure magnetic force precisely.

From 1739 to 1742, an edition of Newton's *Mathematical Principles*, scrupulously annotated by Thomas Le Seur and Francois Jacquier, was published in Geneva. In a note to Newton's remark that "the power of magnetism decreases not as the square but almost as the cube of the distance" (in corollary 5 of proposition 6 of book 3), this edition records that the hypothesis had been experimentally confirmed. The note in question is thought to have been written by Calandrini. Let us look more closely at this note, drawing from the full English translation of it provided by science historian Robert Palter.[9]

An outline of the proposed experimental apparatus is shown in Figure 22.1. The figure shows a magnetic needle's horizontal deviation as seen from directly above, and is explained as follows:

> Let *ACB* be the magnetic meridian; let *NCS* be a magnetic needle drawn by the action of magnet *M* out of the magnetic meridian. Let the line *CM*, drawn from the center of the needle [*C*] to the center of the magnet [*M*], be perpendicular to the magnetic meridian. Let it be supposed that the distance $\overline{CM}$ from the center of the needle to the center of the magnet is physically infinite. (553–554)

In this context, "physically infinite" means that the length ($2a$) of the needle *SN* is negligible in comparison to the distance ($r$) from the center ($C$) of the needle to the magnet's center $M$ (i.e., $r \gg 2a$).

The rationale behind the experiment is that it should measure the deflection of the needle in a state of equilibrium between the force (couple) exerted by the Earth's magnetic field and the force exerted by the magnet. To begin with, Calandrini provides the following explanation of the influenc e of the Earth's magnetic field on the needle's deflection:

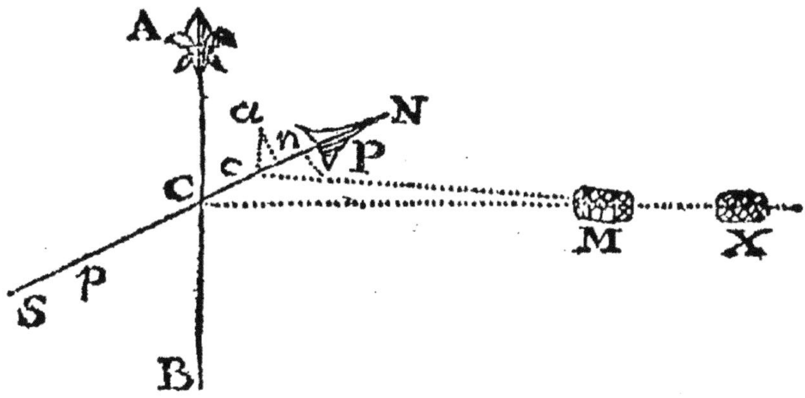


Figure 22.1: Arrangement used for Calandrini's measurements.
*Source*: Palter, "Early Measurements of Magnetic Force."

> The magnetic force of the Earth draws the needle back from position *SCN* to *BCA*, but as the force lies obliquely to the latter position, it must be resolved into two forces, one perpendicular to the line *SCN*, one parallel to it. The latter acts to no effect against the resistance of center *C*, the former causes gyration of the needle. Therefore, if at a certain point *c*, $\overline{ac}$ should represent the total magnetic force [exerted on that point by the Earth], $\overline{an}$ [a component of the force perpendicular to the needle] will represent the force with which the needle is turned. This latter [i.e., $\overline{an}$] is to the total magnetic force at that point [i.e., $\overline{ac}$] as the sine of the angle *acn* [$\angle acn = ACN = \varphi$] … is to the radius [ $\overline{an} = \overline{ac} \sin \varphi$]. It may be supposed that an equal force is being exercised at all points on *CN*; but in the part *CS* that force acts repulsively, and therefore conspires with the force which turns the part *CN* and doubles its efficacy. (554)

This "force which turns" the needle around *C* is what in modern terminology we would call a force contributing to the moment of the force around *C*—that is, a component of the force perpendicular to *C*. Furthermore, a "point" on *CN* essentially means an infinitesimal part of *CN*. The note continues as follows:

> It is to be noted that if equal forces at all points on *CN* should act equally and perpendicularly to turn that line [i.e., *CN*], the efficacy of them all will be the same as if the sum of all the forces were acting perpendicularly at a point *P*, distant by $2/3\,\overline{CN}$ [$= 2a/3$] from center *C*. The total magnetic force turning the part *CN* may therefore be reckoned as concentrated at this point; and, by the same reasoning, the repulsive force turning the part *CS* may be reckoned as concentrated at a point *p*, $2/3\,\overline{CS}$ [$= 2a/3$] distant from center *C*. On account of the equality of the lines $\overline{CN}$, $\overline{CS}$, and therefore of the segments $\overline{CP}$, $\overline{Cp}$, the total magnetic force, both attractive [between *C* and *N*] and repulsive [between *C* and *S*], can be reckoned as applied [as attraction in the same direction] at point *P*. (554)

Note, however, that there is an error here in the condition "if equal forces at all points on *CN* should act equally and perpendicularly to turn that line," since this does not lead to the subsequent description of *P* as the point at a distance of $2a/3$ from the center. A more correct description is as follows.

Let *H* be the strength of the geomagnetic field, *X* a given point on *CN*, and *x* the distance from center *C* to point *X*. If at point *X* the ratio of the strength of the force per unit length of the needle that responds to the unit magnetic field is $\mu(x)$, then, at that point, the force exerted on the needle by the geomagnetic field in the infinitesimal interval of width $dx$ is $\mu(x)Hdx$.

Thus, the aggregate moment of the force about point *C* (torque) that the geomagnetic field exerts on the *CN* segment can be expressed, by denoting $\angle acn = ACN = \varphi$, as

$$T_{\mathrm{E}} = \int_0^a \mu(x)H \times x \sin\varphi\, dx = H\left\{\int_0^a \mu(x)x dx\right\} \sin\varphi. \qquad (22.1)$$

Here, with $P$ as the point where the resultant force exerted by the geomagnetic field is applied and $\langle x\rangle$ as the distance from $C$ to $P$, we can write

$$\int_0^a \mu(x)xdx = \langle x\rangle \int_0^a \mu(x)dx. \tag{22.2}$$

If the intensity of the force is, as posited in the quotation above, the same at all points on $CN$ [that is, if $\mu(x) = \text{const.}$], then $\langle x\rangle = a/2$. Conversely, in order to derive $\langle x\rangle = 2a/3$, the equation would have to take the form $\mu(x) = mx$, with $m$ as a constant. This error, however, has no effect on the subsequent results. Indeed, whatever function form $\mu(x)$ takes, if equation 22.3 defines $\langle x\rangle$, then the subsequent argument holds true.

On this basis, if the distance to magnet $M$ is sufficiently far in comparison to the needle's length ($2a$), then the force applied by the magnet to all infinitesimal intervals along the needle can likewise be expressed as $\mu(x)Fdx$. Then the aggregate moment of the force about point $C$ (torque) applied by magnet $M$ to segment $CN$ acts in the direction opposite to the torque $T_{\mathrm{E}}$; and, denoting $\angle NcM = \theta$, we can likewise express its magnitude as

$$T_{\mathrm{M}} = \int_0^a \mu(x)F \times x \sin\theta\, dx = F\left\{\int_0^a \mu(x)xdx\right\} \sin\theta. \tag{22.3}$$

Here, since $F$ can be regarded as the same at all points along the needle, its point of action can also be regarded as the same as $P$, the point of application of geomagnetic force, regardless of the function form of $\mu(x)$.

The principle behind this measurement is that "in a case where the needle is at rest, the magnetic force of the Earth turning the needle is equal to the force of the magnet turning the needle" (555); or, in other words, that the two moments are in balance. In that case,

> the total magnetic force of the Earth is to the total force of the magnet $M$ as the sine of the deviation [$\theta$] of the needle from the magnet is to the sine of the declination [$\varphi$] of the needle from the magnetic meridian [$H : F = \sin\theta : \sin\varphi$]. (555)

Accordingly, if the position of the magnet is changed, then $F$, the force exerted by the magnet, will change with the distance ($r$) from magnet $M$ to the *centrum rotationis acus*, that is, to $P$, the center of the moment of force. Thus, $F$ can be considered a function of $r$, namely, $F(r)$. Therefore,

> the force of the magnet at various distances (infinite with respect to the magnitude of the needle) is as the sine of the declination [$\sin\varphi$] of the needle from the magnetic meridian divided by the sine of its deviation [$\sin\theta$] from the magnet. (555)

This can be expressed more clearly as follows:

$$T_E = T_M \qquad \text{i.e.} \qquad H\sin\varphi = F(r)\sin\theta$$

$$\therefore \qquad F(r) = H\frac{\sin\varphi}{\sin\theta} \propto \frac{\sin\varphi}{\sin\theta}. \tag{22.4}$$

The results of the measurements were given in a table reproduced here in simplified form as Table 22.1. On the basis of the table, for all the measurement values we find that $r^3 \times (\sin\varphi \div \sin\theta) = 1.9 \times 10^5$. From this, and taking into account equation 22.4, we get

$$F(r) \propto \frac{\sin\varphi}{\sin\theta} \propto \frac{1}{r^3}. \tag{22.5}$$

Table 22.1: Calandrini's Measurements.

| r | $\varphi$ | $\theta$ | $r^3 \times (\sin\varphi \div \sin\theta)$ |
|---|---|---|---|
| 40 | 75° | 19°27′ | $1.856 \times 10^5$ |
| 50 | 61° | 35°41′ | $1.874 \times 10^5$ |
| 60 | 44°30′ | 53°42′ | $1.878 \times 10^5$ |
| 80 | 21° | 77°06′ | $1.882 \times 10^5$ |
| 100 | 11° | 85°46′ | $1.913 \times 10^5$ |
| 120 | 6°20′ | 89°22′ | $1.906 \times 10^5$ |
| 150 | 3°20′ | 91°15′ | $1.963 \times 10^5$ |
| 160 | 2°40′ | 91°39′ | $1.906 \times 10^5$ |

In this manner, Calandrini arrived at the following conclusion, endorsing the suggestion Newton had put forward in *Mathematical Principles*:

> In receding from the magnet, magnetic force decreases in the ratio almost of the cube, so far at least as could be noticed from these rough observations. (558)

## Section 3—John Michell and the Inverse-square Law

In the needle-deflection measurements of magnetic force that had previously been conducted by a succession of researchers from Halley to Taylor, Hauksbee, and others, the angle of the needle's deflection had been thought to provide in itself a rough indication of the strength of the magnetic force at work. In place of such loose reasoning, Calandrini theoretically determined, from the balance of the moments of force, the precise relation between the magnetic force and the needle's angle of deflection; and on that basis, he derived a function for magnetic force from measurements of the angle of the needle's deflection. This constituted a major advance in understanding about how to measure magnetic force and was the first illustration of the notion, subsequently expressed by French physicist and philosopher of science Pierre Duhem, that "a physical experiment is not simply the observation of a group of facts but also the translation of these facts into a symbolic language with the aid of rules borrowed from physical theories."[10]

Because Calandrini's only purpose was to support the proposition Newton had made in *Mathematical Principles*, his argument concludes at the point where he professes to have confirmed Newton's assertion that magnetic force is in inverse proportion to the cube of distance. Actually, what Calandrini derived was the behavior of the resultant force among the four poles of the magnet and needle; the task of seeking the separate force acting between any two poles was left to later thinkers.

In Calandrini's measurements, it is assumed, though not explicitly stated, that the dimensions of the magnet itself are negligibly small in comparison with $r$, the distance between the magnet and the needle. That is, $r \gg D$, where $2D$ represents the magnet's length. Let us suppose, then, that the force between the magnetic poles takes the following form, where $C$ is constant:

$$f(r) = \frac{C}{r^n}. \tag{22.6}$$

In this case, the resultant force exerted on $Q$ (any given point on the needle) by both poles of the magnet would be as follows (given $\angle QMC = \alpha$):

$$\begin{aligned} F(r) &= f(r - D\cos\alpha) - f(r + D\cos\alpha) \\ &= \frac{C}{(r - D\cos\alpha)^n} - \frac{C}{(r + D\cos\alpha)^n} \\ &= \frac{2nCD\cos\alpha}{r^{n+1}}\left\{1 + O\left(\frac{D}{r}\right)\right\}. \end{aligned}$$

Hence, in order for this force to be in inverse proportion to the cube of $r$, it is necessary that $n = 2$, which means that the force between the magnetic poles must be inversely proportional to the square of the distance.

The first person to point this out was English natural philosopher John Michell (1724–1793), in 1750. Michell was a versatile scholar who studied at Cambridge and subsequently lectured there on Hebrew, Greek, algebra, and geometry. He is known for his early research in electromagnetics, for introducing statistics into such fields as seismology and astronomy, and for developing, toward the end of his career, a torsion balance in collaboration with renowned English physicist Henry Cavendish on the occasion of the latter's direct measurement of the force of gravity between two bodies.[11]

Michell's early research on magnetism is recorded in his *A Treatise of Artificial Magnets*, published in 1750. In this work, Michell remarks on a number of principles that he had experimentally verified concerning magnets, including the following:

> *Thirdly*, Each pole attracts or repels exactly equally, at equal distances, in every direction. This is a property, which perhaps those, who imagine Magnetism to depend upon a subtle fluid, may not be very willing to admit, as being utterly inconsistent with such an Hypothesis; but it is capable of being proved by a great variety of experiments....
>
> *Fourthly*, The Magnetical Attraction and Repulsion are exactly equal to each other. Most people, who have mentioned anything relating to this property of the Magnet, have agreed, not only that the Attraction and Repulsion of Magnets are not equal to each other, but that also they do not observe the same rule of increase and decrease. Their mistake in this matter arose from

> their not attending to the different degrees of strength, that magnets have, in different circumstances....[12]

Michell realized that these two points were at the heart of the problem with measurements conducted up to that time. What others had failed to fully grasp was that, whereas the force exerted by each pole of the magnet had been expressed as a simple function of distance, in fact the force exerted between two magnets with north and south poles is the combined effect of the forces at work among two pairs of poles, and so the force acting between two magnets depends not simply on the distance between them but also on their respective shapes, sizes, and positions.

Michell then concludes as follows:

> *Sixthly*, The attraction and Repulsion of magnets decreases, as the squares of the distances from the respective poles increase. There have been some, who have imagined, that the decrease of the Magnetic attraction and Repulsion is inversely as the cubes of the distances; others as the squares; and others, that it follows no certain *ratio* at all, but that it is much quicker at greater distances, than at small ones, and that it is different in different stones. Among these last is Dr. Brook [*sic*] Taylor, and P. Muschenbroek [*sic*], who seem to have been pretty accurate in their experiments.... The conclusions of these gentlemen were drawn from their experiments, without their being aware of the *third* property of magnets, just mentioned. If they had made proper allowances for that, together with the increase and diminuation of power in the magnets they tried their experiments with, all the irregularities, they complained of ... might very well be accounted for, and the whole of their experiments coincide with the squares of the distances inversely.

Michell thus follows Helsham in declaring that the force between magnetic poles obeys the inverse-square law. Regrettably, it is not made clear what experiments Michell actually conducted or precisely how he reached this conclusion. He writes:

> This property, from some experiments I have made myself, and from those I have seen of others, seems very probable; but I do not pretend to lay it down as certain, not having made experiments enough yet, to determine it with sufficient exactness.

It seems that Michell, like Helsham, had already adopted the hypothesis of the inverse-square law *a priori.*

Michell's *A Treatise of Artificial Magnets* was primarily a technical manual for the manufacture of magnets, as is indicated in the work's subtitle, "In which is shewn An easy and expeditious Method of making them, Superior to the best *Natural* Ones." In Britain, successive advances in the manufacture of magnets had been made by Servington Savery in 1730, Gowen Knight in 1746 and 1747, and John Canton in 1750,[13] and the "double-touch" method developed by Michell himself subsequently came into wide use. With these improvements, it was now relatively easy to make magnets that not only had strong magnetic power but were also well shaped with uniform width. Particularly important was the new ability to make strong magnets that were also sufficiently elongated to be suitable for investigating the effect of each individual pole separately, an ability that in turn made possible the precise and refined measurements of magnetic force that came later.

In his *Tentamen theoriae electricitatis et magnetismi* (*Aepinus's Essay on the Theory of Electricity and Magnetism*), published in 1759, German physicist Franz Aepinus (1724–1802) touches on Musschenbroek's and others' failed attempts to discover a function form for the relationship between distance and magnetic force. He then writes: "Although I have found this matter difficult enough, I am still not without hope that, if an experiment is properly set up with the help of artificial magnets,... there is a possibility that the law itself will finally be unearthed."[14]

By that time, it was clear that the key to measuring the force between magnetic poles was the use of artificial magnets suitably formed for that purpose. And indeed, the measurements conducted (as we will soon see) by Tobias Mayer and Charles-Augustin de Coulomb in the later part of the eighteenth century were made possible by the development of elongated artificial magnets of uniform width. Mayer noted that "artificial magnets tend to be much more regular in their action than natural magnets";[15] and it was only by using artificial magnets manufactured using the "double-touch" method that Coulomb succeeded in directly measuring the attractive force between magnetic poles.

## Section 4—Tobias Mayer and the Demise of the Vortex Theory

The next scholar to arrive at the inverse-square law for magnetic force on the basis of refined theoretical considerations and meticulously planned experiments was Tobias Mayer (1723–1762) of the University of Göttingen. Mayer is known primarily as an astronomer, in particular for having devised lunar tables accurate enough for determining longitude even at sea, but his interests spanned all areas of science.

Mayer's research into magnetism began in 1757, and in 1760 he wrote *Theoria magnetis* (*The Theory of the Magnet and Its Application to Terrestrial Magnetism*). The treatise was published in Göttingen in 1972, along with an English translation, as part of volume 3 of *The Unpublished Writings of Tobias Mayer*, but in its own day it was not made public. Eric G. Forbes, editor of the *Unpublished Writings*, notes that this is why Mayer's "achievements in the physical sciences are much less appreciated than they deserve to be."[16]

Much of the historical significance of Mayer's treatise lies in his deducing that the force between magnetic poles varies as the inverse square of the distance, but also important is the sweeping criticism of the Cartesian vortex theory that he sets out in the first chapter, titled "An Examination of the commonly held view of magnetic theory." Let us begin our consideration of Mayer's contribution by looking at that critique.

As we have seen, Newton's theory of gravitation was not widely accepted on the Continent in the mid-eighteenth century. This was because the view of science Newton put forward in *Mathematical Principles*—that it was sufficient for science to derive the laws of force from phenomena and to quantitatively prove further phenomena predicted on the basis of those laws—was very different not only from the earlier Scholastic view that one could not understand anything without first understanding its essence but also from the mechanistic quest for explanations of *causes*. Descartes had spoken of inquiring into the causes of nature's workings just as someone trained in handling machinery surmises the inner workings of a machine from observation of its outward appearance. And the atomist Pierre Gassendi had expressed a similar view:

> Wherever possible in the study of nature we make use of anatomy, chemistry and other aids so as to understand, by

> breaking down the bodies as far as possible and dividing them as it were into their component parts, what these elements are and what manner of criteria helped in their composition.[17]

At that time, the approach represented by such mechanistic or atomistic reductionism was regarded as the new direction of science.

As it turned out, however, the mechanistic view failed to explain gravitational force. Newton's theory of universal gravitation, on the other hand, was extremely persuasive, and after Newton's death the Newtonian position regarding force gradually came to be accepted on the Continent as well as in Britain. In 1738, Voltaire, who was instrumental in disseminating Newton's ideas in France, declared in *The Elements of Sir Isaac Newton's Philosophy* that "Descartes's Cause of Gravity and Motion is a Chimera,"[18] and it was from around that time that Cartesian mechanism rapidly fell out of favor even on the Continent.

The situation with regard to magnetic force, however, was quite different. In 1742, the Académie royale des sciences de Paris (Paris Academy) held a competition on the subject of the "explanation of the attraction of the magnet and iron, the pointing of the magnetized needle toward the north, and its declination and inclination."[19] Four years later, the Academy presented three winning essays, one by the great Swiss mathematician Leonhard Euler, another by the Swiss mathematician Daniel Bernoulli and his father Johann Bernoulli, and the third by French natural philosopher Etienne-Francois Du Tour; and the essays appeared in print in 1752. What is notable here is that all three essays are essentially based on the Cartesian vortex theory of magnetic force as arising from the action of "minute matter" circulating within and around magnets.

It is not surprising that Euler adopted the Cartesian view. As one of the editors of *Leonhardi Euleri opera omnia* (The complete works of Leonhard Euler) aptly puts it, Euler was an "outdated and fairly orthodox Cartesian"[20] who, even at that late juncture, yielded to the vortex theory even for explaining gravitation.[21] Little wonder, then, that he attempted to account for magnetic power as well in terms of vortices of minute matter. Indeed, in a later work titled *Lettres à une Princesse d'Allemagne* (*Letters of Euler to a German Princess*), Euler states:

> As you see nothing that impels [small bits of iron and steel] toward the loadstone, we say that the loadstone attracts them, and this phenomenon we call *attraction*. It cannot be doubted, however, that there is a very subtile, though invisible, matter, which produces this effect, by actually impelling the iron toward the loadstone.

Surmising, furthermore, that there are "canals" inside magnets, he goes on at great length to offer an account of the magnet's attraction and repulsion by means of a vortical motion of magnetic particles circulating in and around the magnet via the canals.[22]

Matters were different in the case of Daniel Bernoulli. Considering himself a thoroughgoing Newtonian ("ein völliger Newtonianer"), on the question of gravitational force Daniel Bernoulli naturally took Newton's side, advocated opposition to Descartes, and rejected the theory of vortices. But when it came to magnetic force, even Daniel Bernoulli embraced the vortex theory.[23] As we saw in the previous chapter, however, Newton himself accepted the vortex theory concerning magnetism, so it is not surprising that Daniel Bernoulli did the same.

Such was the milieu in which Tobias Mayer, in his 1760 treatise *The Theory of the Magnet*, took Newton's methodology regarding gravitational force and applied it wholesale to magnetic force as well. The treatise is thus very important not only in connection with the laws of magnetic force but also in terms of the purpose of physical theory and the methodology of physics itself, and as such deserves careful consideration in the present context.[24]

The first chapter begins as follows:

> §1
>
> Among those effects of nature which the collective efforts of scientists have so far failed to explain satisfactorily, an important place is held by the remarkable phenomena of magnetism. Despite the researches of many eminent scientists in this field, despite the theories and hypotheses they have elaborated to explain the nature of magnetism, not one of them has been able to claim that he has given a clear and complete account of even one of the phenomena in question; nor does any of their theories satisfy this important condition of truth, namely that of enabling us to express the force and action of the magnet in mathematical terms, and to define geometrically the extent of its observed operation. (TM 1.1:63–64)

Mayer's view is that what matters is mathematical laws, and that there is no point probing beyond them for the true nature of forces or the mechanism of their propagation. This is precisely the position that Newton adopted with regard to gravitational force.

In section 2, Mayer criticizes previous inquiries that claimed to have identified the general principle of magnetism from what was really only limited experience of a few particular cases. Then, in section 3, he likewise dismisses both mystics who proclaim that "the law of magnetism ... is quite different in kind from the common laws of mechanics which are universally obeyed by the rest of the physical world," and agnostics who claim that "it is not given to man to penetrate these mysteries of nature" (TM 1.3:64).

The next target of Mayer's criticism is Cartesian theory. Regarding the distressingly backward state of magnetic studies at the time, whose explanations, he remarks, would satisfy only "the credulous and those content with vague arguments," Mayer writes that

> physicists are generally agreed that there exists some rare form of matter—call it aether, a vortex or a magnetic material—which flows around every magnet, rushing away from one pole towards the other; and that it is by reason of this ebb and flow that a magnet attracts iron or another magnet, governs the direction of a magnetic needle, etc. They see the existence of such a vortex confirmed by the regular shape and position of iron filings brought within a magnetic field. And this may indeed be said to be the sole result to date of the labour expended by our physicists in the study of magnetism. (TM 1.4:64)

In section 5 he makes it clear what he thinks of the vortex theory: "This very theory,... which has received such wide acceptance, has been in my opinion a major obstacle to farther progress in this field" (TM 1.5:65). In explaining his view, he elaborates the following all-out attack (at every conceivable level) against the vortex hypothesis.

First, he points out that

> very little is known about the motion of fluid bodies at all, and their internal movement, such as we have in vortices, is particularly obscure. Despite much research into this topic, no general formula has yet been found to fit every individual

> case, as with the movement of solid bodies. This being so, the movement of so rare a form of matter as is supposed to surround a magnet will be all the more difficult to explain. (TM 1.5:65)

In short, the problem was that the science of fluid mechanics itself was not yet well established. Underlying this was the fact that, with figures such as Gottfried Leibniz, Pierre Varignon, Jacob Hermann, the Bernoullis, and Leonhard Euler having refined and analytically rewritten the theory of mechanics since the publication of Newton's *Mathematical Principles*, particle and rigid-body mechanics had, unlike in Descartes' time, already acquired a theoretical framework in the form of mathematically rigorous physics. Nonetheless, research in refined fluid mechanics had begun only in the mid-eighteenth century, and the formulation of a general-purpose theory had not yet succeeded. In any case, considering the standards of mechanistic theory in the eighteenth century, it is hardly surprising that thinkers of that time discounted the vortex theory as no longer even deserving to be called a theory.

Mayer then points out that the vortex theory, being nothing more than an *ad hoc* explanation—an arbitrary redescription of the phenomena in question—failed to provide insight into any underlying principles:

> But let us concede for the moment that the phenomena of magnetism are produced by a vortex of some rare form of matter in perpetual motion. Are we any farther forward? Do we not simply start asking more questions? How did this movement originate, what set the matter in motion? Are we not then forced to postulate a second vortex surrounding the first—or something even less likely? (TM 1.6:65)

Accordingly, Mayer's conclusion about and evaluation of the vortex theory is this: "Instead of getting closer to the truth by this enquiry (which is altogether too Cartesian) into the causes of the phenomena in question, we become increasingly enmeshed in difficulties which are quite insuperable" (TM 1.6:65).

Mayer's third point of contention—which he broaches by stating that the "inadequacy of the vortex theory in explaining natural phenomena is further illustrated by the case of the heavenly bodies" (TM 1.7:65)—is that the theory has already been disproved in the field of astronomy:

> Descartes used this theory [i.e., his vortex theory] to explain the movement of the planets around the sun. Throughout the half century or so that this hypothesis found favour with philosophers, its adherents were unable to explain properly one single phenomenon; not one of the many inequalities of planetary motion was accounted for. But if the theory satisfied the physicists of that period, it certainly did not satisfy the mathematicians and astronomers, who are sceptical of vague explanations and fanciful generalities ..., and demand a theory which enables them to produce accurate computations of the motions of the planets which agree with their observations. That is why throughout this period they preferred the hypotheses of circular and elliptical motions, which were at least more fruitful. Finally, of course, Newton, with his brilliant theory of universal gravitation, once and for all rescued astronomical physics from the mythical vortices as well as from the many difficulties of purely geometric hypotheses. (TM 1.7:65–66)

Fourth, Mayer turns his critical gaze on the curved patterns that iron filings form around a magnet, a phenomenon previously often pointed to as evidence of the vortical motion of a supposed magnetic fluid—a proposition that even Newton had put forward. Mayer points out that the patterns do not in fact corroborate the supposed vortical motion. That is, the behavior of the individual filings does not match the vortical movement predicted by the theory, since

> if one considers more closely the phenomenon produced by iron filings scattered around a magnet, a phenomenon which, as I said before, is commonly thought to provide visible proof of the existence of a vortex, one will see that it rather tells against the idea of the ebb and flow of matter in a vortex. For if this matter, as it flows away from one pole, is deflected on to the side of the magnet and then proceeds on to the other pole, why does the fluid not break up and carry all the filings which are in its way to the same pole? How does it come about that the filings which are nearer to the pole from which the vortex matter is flowing away do not follow its movement to the other pole, but, if they are made a little freer by shaking the table, actually go in the opposite direction towards their nearer pole? Finally why do the filings which are equidistant from each pole make for neither but for a point midway between them? It is my view that every filing whose nearest point on the magnet is other than the pole towards which the vortex matter is moving testifies in this experiment against the vortex hypothesis. (TM 1.9:66)

This argument that the patterns of iron filings fail to show the existence of any vortical motion of magnetic matter—indeed, that examining the movements of individual filings reveals that they clearly violate the hypothesis of vortical motion—was very persuasive. In short, Mayer dismisses the idea of vortices as no more than a "figment of the imagination" and declares that "there is no need to spend any more time reviewing the errors of this theory [i.e., the vortex theory] which can best be refuted by the new theory of magnetism I am about to propose" (TM 1.11:67).

Thus ends the first chapter of *The Theory of the Magnet*, a chapter in which Mayer demolishes the hypothesis of vortical motion so decisively as to lay it to rest once and for all.

Despite the important role it thus played in the demise of Cartesian mechanism, Mayer's treatise has received scant attention, even in Germany. Three decades after Mayer's treatise, Immanuel Kant states uncritically in his *Critique of Pure Reason* that "we cognize the existence of a magnetic matter penetrating all bodies from the perception of attracted iron filings."[25] Nor is Mayer's work on magnetism mentioned in nineteenth-century German histories of physics and magnetic studies, such as those by Poggendorff, Rosenberger, and Hoppe; or even in twentieth-century histories of physics, such as those by the physicists Max von Laue and Friedrich Hund.

## Section 5—Mayer's Method of Magnetic Research

On the basis of his critique as outlined above, Mayer proposed that the proper approach to magnetic research must entail accurate measurement of the dependence of magnetic force on distance as well as formulation of a rigorous quantitative law of magnetic force. He begins the second chapter of *The Theory of the Magnet* by proposing as a fundamental stance that the science of magnetic force should identify the *law* of magnetic force rather than seek its *cause*:

> The failure of the physicists' explanations of magnetic phenomena seems to me on closer inspection attributable to ignorance not so much of the cause of magnetic attraction as of the law governing the increase or decrease in its intensity at different distances from the magnet. For it is certain that if there were agreement about the magnitude of the force which

> is expected by a magnet on iron or by one magnet on another, whatever the position and distance were, then each individual phenomenon could be clearly and completely explained by the principles of mechanics alone, without the aid of any fictitious or vague hypotheses, even although the question of causality remained unresolved. The phenomena of universal gravitation, elasticity and hardness of bodies have been explained, although their causes remain mysteries; and they could not be better or more simply explained even if their ultimate causes were known. (TM 2.12:67)

This leads Mayer to the following declaration:

> This being so, we shall confine our discussion to the laws according to which the magnetic force operates from a given distance, and ignore any discussion of why this force exists, leaving that for others to decide if they can. (TM 2.13:67)

This was an unequivocal reformulation of the program that Newton had conducted with regard to gravitational force and a declaration of intent to apply it to magnetic force as well. In other words, it proclaimed the banishment of ontology and metaphysics from magnetic theory.

In implementing this program, Mayer begins with the observation that the force of a magnet

> is strongest at [the magnet's] poles, but the intervening space is not altogether without the power of attraction, although the latter decreases the farther from the pole and the nearer to the middle one goes, so that at the exact mid-point no attraction is experienced at all. (TM 2.14:67–68)

The basis for this reasoning is the following experiment.

> *Experiment I*
>
> Spread a quantity of iron filings uniformly on a flat surface, and place on them an artificial magnet of uniform thickness ... You will observe that at the extremities of the magnet many more filings adhere more firmly than in the space between the poles, and that at the exact mid-point there are very few filings if any at all. (TM 2.14:68)

Mayer notes here that a piece of iron "is a magnet for at least so long as it is exposed to the action of a magnet" (TM 2.17:68–69). Since the subject of his inquiry is the law of magnetic force rather than the nature of magnets, he proceeds on the basis that there is no longer any reason to distinguish magnets from pieces of magnetized iron, and therefore regards natural magnets and artificial magnets in exactly the same light.

Mayer infers from the results of the experiment that

> the unlike poles of a magnet or a piece of iron always attract each other, and the like poles always repel ... But it is not only unlike poles that attract one another, but also the particles of a magnet or piece of iron which happen to be nearer to those poles ... In defining, therefore, the area of action of magnetic force,... we must pay attention to the action of individual magnetic particles on the individual particles of a magnetic body and vice versa. (TM 2.17:69)

(By "particles of a magnetic body" he means simply minute parts of a magnet; the expression does not imply anything along the lines of atomism or particle theory).

Mayer offers a more detailed explanation by supposing—as shown in Figure 22.2a—two bar magnets of uniform thickness (*ACB* and

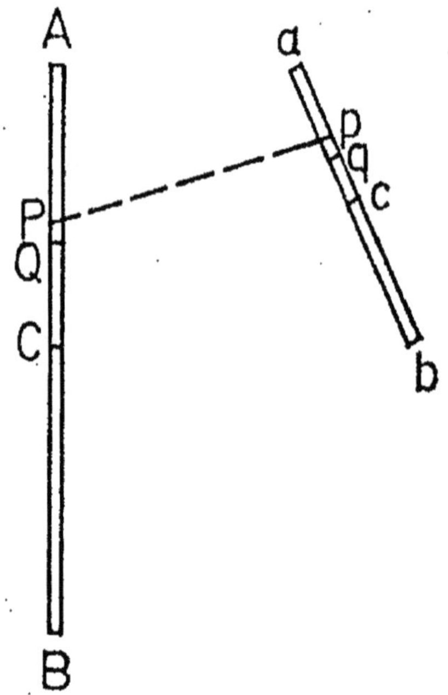


Figures 22.2a–22.2d: Arrangements used for Mayer's measurements.
*Source*: Mayer, *Unpublished Writings*.

*acb*) with centers at *C* and *c*, respectively, and with poles *A* and *a* and *B* and *b* attracting one another.

> The magnet *AB* will then act so that the individual particles between *A* and *C* will attract the individual particles between *a* and *c* and repel those between *c* and *b*. Similarly, the particles between *B* and *C* will attract those between *b* and *c* and repel those between *a* and *c*, there being always a repelling force equal to any attracting force. The farther any particle is from its centre *C* or *c*, or the nearer it is to any pole, the greater the attraction or repulsion will be. It will also be further strengthened the nearer any particle of the magnetic body *ab* is to the particles of the magnet *AB*. (TM 2.18:69)

> We can see from this how complex the complete action of the magnet *AB* is upon the magnetic body *ab*; and how difficult it is to derive the laws of the operation and extent of this action directly from observation. For it was seen by experiment that any one particle of a magnet can act on a magnetic body with a different force from all the others and in a different direction; these experiments serve to record only the sum of all the forces and the mean direction, so that it can scarcely be determined how much of this is attributable to any particular particle. (TM 2.19:69)

The method Mayer adopts is to estimate the magnetic force acting between "particles of the magnetic body" by a simple experiment; theoretically calculate on that basis the resultant force between magnets positioned in specific ways; and then see whether or not the estimates are confirmed by actual measurements. This is precisely the modern hypothetico-deductive method of proof and experimental verification.

Mayer begins with the arrangement shown in Figure 22.2b—a magnet (*AB*) and a magnetic needle (*ab*) placed in a straight line. While varying the distance *r* between *AB* and *ab* (*r* being the distance between points *A* and *a*), at each distance he allows *ab* to oscillate around its center point (*c*) and measures the period of oscillation (*T*). With the length of *ab* given here as 1.2 inches and the length of *AB* as 1 foot, the results are as follows.

| | | | | |
|---|---|---|---|---|
| $r$ | 1.0, | 2.2, | 4.7 | (inch) |
| $T$ | 0.2, | 0.4, | 0.8 | (second) |
| $r/T$ | 5.0, | 5.5, | 5.875 | (inch/second) |

Theory of mechanics tells us that in such a case the applied force $F$ is inversely proportional to the square of period $T$. In the above

Figure 22.2b

measurements, however, the distance $r$ between $a$ and $A$ increases at a slightly faster rate than does the period $T$. From this Mayer surmises that

> it is clear that the magnetic forces acting upon the needle in this case decrease at a slightly slower rate than the square of the distance increases. (TM 2.22:70)

This argument can be put into modern terms as follows. If $F$ is the force acting upon half of the needle, the moment of force around point $c$ (the center of magnetic needle $ab$) can be expressed as $F\langle x\rangle \sin u$, where $\langle x\rangle$ means the distance from $c$ to the point of action and $u$ is the angle of the needle's deflection from the equilibrium position. Then, by denoting the moment of inertia around $c$ as $I$, we can formulate the equation of motion for the needle's oscillations ($|u| \ll 1$) as

$$I\frac{d^2u}{dt^2} = -2F\langle x\rangle \sin u \cong -2F\langle x\rangle u.$$

Therefore, if $T$ is the period of oscillation, then

$$T^2 = (2\pi)^2 \frac{I}{2F\langle x\rangle} \qquad \therefore \qquad F \propto \frac{1}{T^2}$$

Note that Euler had already deduced essentially the same relationship between the magnitude of force and the period of oscillation in 1743.[26] Then, from the third row of the above table showing Mayer's results, we can surmise that $T \propto r^{(1-\varepsilon)}$ (where $\varepsilon$ is within the range $0 < \varepsilon \ll 1$). Therefore,

$$F \propto r^{-2(1-\varepsilon)}.$$

On the basis of this simple inquiry, Mayer provisionally concludes as follows:

> From this we can confidently conjecture that the force with which any single particle of a magnet acts, corresponds exactly to the square of the distance. For we can show that if the force exerted by each individual particle varies exactly with the square of the distance their combined force should decrease at a somewhat slower rate. Moreover, it can scarcely be doubted, if we compare other similar laws, that this one which states that the force of the individual particles of a magnet varies in inverse proportion to the square of the distance, is one which is consistent with nature. For she observes the same law in producing all her effects of this type e.g. light, heat, universal gravitation etc. (TM 2.23:70–71)

Here too one gets the impression that the conclusion had in fact preceded the investigation. Helsham had done the same, and indeed it seems that many thinkers of that time who embraced Newtonian natural philosophy took the inverse-square law to be true not only for gravitation but as a universal law of nature. We see this again in the 1765 assertion by French natural philosopher Georges-Louis Leclerc Comte de Buffon, that the notion of attractive force acting in accordance with the inverse-square law could be applied even to chemical affinities.[27]

In section 24, Mayer states that the change in magnetic force from the center of a magnet to either pole—that is, the variation in the magnetic force of the magnet's constituent particles depending on how far they are from its center—depends on the magnet's shape. But he goes on to hypothesize that, in the case of a "straight artificial magnet of uniform thickness," "the force of the particles increases in direct simple proportion to the distance from the centre." This is equivalent to Calandrini's hypothesis that $\mu(x) = mx$. In the same passage, Mayer declares that "we shall regard [such a straight, uniform magnet] as infinitely thin, according it a thickness of zero" (TM 2.24:71). What made this theoretical step possible was the recent advent of artificial magnets that more or less fit that description.

## Section 6—Mayer's Theory: Hypotheses and Deductive Process

In chapter 3 of *The Theory of the Magnet*, Mayer calculates on the basis of the above hypothesis the magnetic force acting between two magnets. For his first problem, he posits a formula for magnetic force between "magnetic particles," or minute segments, of two magnets—$AB$ (with length $2a$) and $ab$ (with length $2\alpha$)—positioned in any way, as shown in Figure 22.2a.

Mayer considers the force exerted between the infinitely small segments *PQ* and *pq*, where *PQ* has a length of $dx$ and is at point *P* at distance $x$ from *C*, the center of *AB*, and *pq* has a length of $dy$ and is at point *p* at distance $y$ from *c*, the center of *ab*. According to the hypothesis, the strength of the magnetic force of segment *PQ* is proportional to $\mu(x)dx = mxdx$, and that of segment *pq* is proportional to $\mu^*(y)dy = m^*ydy$. If *R* is the distance between *P* and *p*, then the force can be given as

$$\mu(x)\mu^*(y)\frac{1}{R^2}dxdy = \frac{mm^*}{R^2}xydxdy. \qquad (22.7)$$

The resultant force acting between the two magnets can then be expressed by integrating the equation with respect to $x$ and $y$.

Particularly in the second problem addressed in chapter 3, Mayer in fact uses integral calculation to derive the resultant force between two bar magnets positioned in a straight line, as shown in Figure 22.2c. Let *A* and *a* be two unlike, mutually attractive poles. If $r$ is the distance between *C* (the center of *AB*) and *c* (the center of *ab*), then the distance between *P* and *p* is $R = r - x - y$, and the force between the two magnets can be derived by integrating equation 22.7 with respect to $x$ and $y$. Mayer makes the calculation according to two hypotheses, one for the force exerted on *p* by all points of magnet *AB* (hypothesis 1); and the other for the force exerted only by the corresponding point on *AB*, that is, the point that satisfies the condition $\pm x/a = y/\alpha$ (hypothesis 2). Mayer's description of hypothesis 2 is that

> the parts which act thus on one another are those which in similar magnets are the same distance from their centres or whose distance from their centres is proportional to the distance

Figure 22.2c

> between their poles. All other parts are in respect of those indifferent i.e. they neither attract them like those of unlike nature, nor repel them like those of like nature. (TM 3.34:76)

Mayer then describes an experiment to measure the magnetic force between two bar magnets positioned in a straight line as indicated in Figure 22.2d. *Aa* is a nonmagnetic body attached at point *A* to prevent the large magnet *AB* and the small magnet *ab* from coming into direct contact with one another. The small magnet *ab* is suspended at *O* by a thread of length *L* so that it can pivot freely within the plane of the magnetic meridian, where $\alpha\beta$ in the figure represents the lowest position of *ab*, and $O\gamma$ and *Oc* represent the *L*-long thread when vertical and when inclined, respectively.

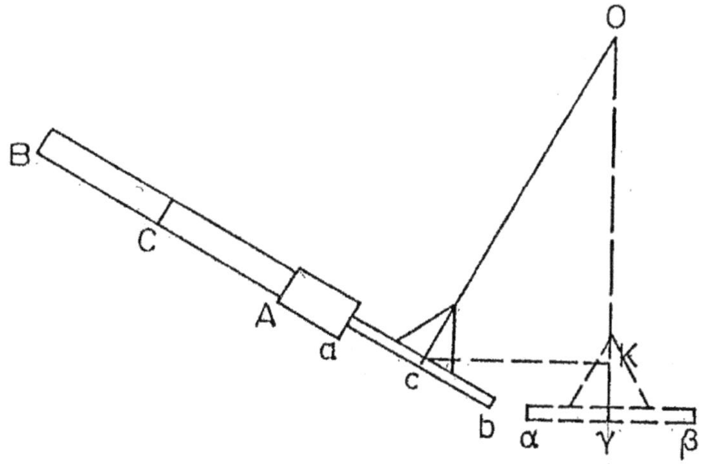


Figure 22.2d

Beginning in a state whereby the small magnet is touching *Aa* due to *AB*'s attractive force, if *AB* is slowly raised, *ab* rises with *AB*, remaining at a right angle to the thread, until it reaches the position where the thread *Oc* reaches the angle $\angle cOK = \theta$. Beyond that position, the equilibrium of the attractive force between the two magnets ($F$) and the weight ($W$) of the small magnet is broken and the small magnet detaches from *Aa*. By measuring the distance $x$ between $c$ and $K$ at that point, we find that the attractive force between the magnets when $A$ and $a$ are at distance $\overline{Aa}$ is $F = W \sin\theta = Wx \div L$.

By comparing the force thus measured between two magnets positioned in a straight line with the force he had calculated on the basis of hypotheses 1 and 2, Mayer finds that "it is clear that the experiment definitely favours the second hypothesis" (TM 4.45:84). From there, he eventually arrives at the following conclusion:

> I have found that the second hypothesis always satisfies the phenomena, and I have no hesitation in offering it as a true

> account of magnetism and one consistent with nature. If anyone is still inclined to doubt that, I rather think he must consider doubtful and uncertain all the proofs and explanations that physicists have ever given. For since the truth of a physical system is proved only by experience and its ability to satisfy the phenomena, anyone who denies a system thus established seems to me to be denying the phenomena themselves and the evidence of the senses. (TM 4.50:86)

Although I have not provided the numerical values of Mayer's measurements here, it is certain that all his measurements using a number of magnets matched his calculations based on hypothesis 2. However, not only the postulate $\mu(x) \propto x$ but also the idea in hypothesis 2 that only the corresponding points of the two magnets exert force on one another seem highly arbitrary, and it is difficult to grasp what physical significance they could hold. Indeed, one could easily suspect Mayer of coming up with hypothesis 2 as an afterthought when his measurements failed to match the calculations of hypothesis 1. Even if it would be going too far to make such an accusation, the presence of hypothesis 2 makes it difficult to accept that Mayer's measurements directly demonstrated that an attractive force operating between magnetic poles is inversely proportional to the square of the distance between them.

Nonetheless, Mayer's great contributions to magnetic inquiry were that he rid magnetic theory of the fanciful vortex hypothesis that had been in vogue since Descartes and steered magnetic research toward discovering the laws rather than the causes of magnetic force.

## Section 7—Coulomb's Confirmation of the Inverse-square Law

The fundamental weakness in Calandrini's and Mayer's measurements was that the hypothesis that the strength of the force exerted or received by particles (minute segments) of magnets is proportional to their distance from the center, namely, $\mu(x) \propto x$, was not itself directly proved. French engineer Charles-Augustin de Coulomb (1736–1806) resolved this problem by using artificial magnets with magnetic poles localized to as small a point as possible at each end. By directly measuring the attractive force between the poles of such magnets, he showed, virtually beyond doubt, that the force obeyed the inverse-square law. Coulomb presented a series of research results on electrical and magnetic force to the Académie royale des sciences de Paris (Paris Academy) in seven installments between 1785 and 1789. Let us examine the second of those papers, presented in 1785 and known as the Second Memoir, which covers his measurements of magnetic force.[28]

The straight artificial magnets Coulomb employed were needles magnetized by the double-touch method developed by Michell, 25 *pouce* in length (approximately 64 cm, as 1 *pouce* is roughly equivalent to 1 inch) and 1.5 *ligne* in width (approximately 0.32 cm, as 1 *ligne* is 1/12 *pouce*). He used such long, thin magnets so as to isolate and measure the effect of each single pole. Coulomb's work thus bears out how crucial the development of the technique to manufacture artificial magnets was to progress in the measurement of magnetic force.

Although Coulomb employs the term "magnetic fluid," this is not the force-propagating medium that Descartes theorized to be flowing around magnetic bodies. Rather, Coulomb's concept is something like that of a magnetic charge existing within the magnet as the source of its magnetism. Coulomb supposes that this "fluid" is made up of "magnetic molecules" (corresponding to Mayer's "magnetic particles"). This mirrors his notion that "electric fluid" is an agglomeration of "electric molecules," or in other words charged particles. In Coulomb's model, that is, electric force operates between electric molecules as action at a distance, and magnetic force operates between magnetic molecules likewise as action at a distance. Coulomb was thus a staunch proponent of the idea of action at a distance. In the first experiment, he determines that the "magnetic fluid" is "concentrated at 10 *lignes* [5/6 *pouce*, or approximately 2 cm] from the ends of the needle." This finding obviates the need for such questionable hypotheses as $\mu(x) \propto x$.

Before describing his actual experiments, Coulomb emphasizes that the "magnetic fluid acts by attraction or repulsion according to the ratio compounded directly of the density of the fluid and inversely of the square of the distances between its molecules." In other words, if the respective densities of magnetic fluid are $\mu_1$ and $\mu_2$ and the distance is $r$, then the magnetic force acting between the magnetic molecules is proportional to $\mu_1\mu_2/r^2$. This clearly resembles the proposition that the force of universal gravitation acting on two mass points of mass $m_1$ and $m_2$ separated by distance $r$ takes the form $Gm_1m_2/r^2$. Coulomb had arrived at a law of the same form for electrostatic force in his First Memoir, and here too his conviction about the universality of the inverse-square law seems to have preceded his actual inquiry. It is therefore fair to say that the purpose of his experiments and measurements was not to discover a still unknown formula for force but rather to confirm this presupposition.

Coulomb's first trial was of a dynamical nature, measuring the frequency of oscillation of a magnetic needle subjected to a magnet's force. This is almost identical to the preliminary experiment conducted by Mayer. A small magnetic needle roughly 1 *pouce* (just under 3 cm) in length is suspended from a silk thread and brought to rest horizontally along the magnetic meridian. The abovementioned long, slender magnet is then set vertically with one pole on the same horizontal plane as the magnetic needle and at a given distance from the needle's center along the same magnetic meridian, and the needle is made to oscillate horizontally, the deflecting angle of the needle from the magnetic meridian being denoted by $u$. Although the moment of force due to the Earth's magnetic field is proportional to $\sin u$, at small angles $\sin u$ is approximately equal to $u$, so the moment of restoring force can be expressed as $Ku$. Similarly, if the distance $r$ to the pole of the magnet is sufficiently large compared to the length of the needle, then that pole exerts almost the same degree of force on every point on the needle, and the resulting couple can be expressed as $F\langle x\rangle \sin u \cong F\langle x\rangle u$ (where $\langle x\rangle$ is the length from the center of the needle to the point of action of that force).

Accordingly, the equation for the needle's rotation, where $I$ is the inertial moment, is

$$I\frac{d^2u}{dt^2} = -(F\langle x\rangle + K)u.$$

From this, for $\nu$ as the needle's oscillation frequency

$$\nu^2 = \frac{1}{(2\pi)^2}\frac{F\langle x\rangle + K}{I} \propto F\langle x\rangle + K.$$

Specifically, if the force is measured when the magnet is not near the needle (i.e., with an infinitely large distance between the magnetic pole and the magnetic needle)—that is, when the oscillation frequency due solely to the Earth's magnetic field is $\nu_0$—then $\nu_0^2 \propto K$ and hence

$$F \propto \nu^2 - \nu_0^2.$$

Measuring the oscillation frequencies ($v$) at various distances ($r$) between the needle and the pole thus gives the ratios for the distance variation of $F$. The following table shows Coulomb's measurements.

| $r$ (pole-to-needle distance) | $v$(1/minute) | $v^2 - v_0^2 \propto F$ | $r^2 \times (v^2 - v_0^2)$ |
|---|---|---|---|
| $\infty$ (no magnet) | $15 = v_0$ | | |
| 4 (pouce) | 41 | $41^2 - 15^2 = 1456$ | 23,296 |
| 8 (pouce) | 24 | $24^2 - 15^2 = 351$ | 22,464 |
| 16 (pouce) | 17 | $17^2 - 15^2 = 64$ | 16,384 |

Finding thus that $1456 : 351 : 64 \cong 1 : 1/4 : 1/16 = 1/4^2 : 1/8^2 : 1/16^2$, Coulomb concludes that $F \propto r^{-2}$. One could also say that $F \times r^2 \propto (v^2 - v_0^2) \times r^2$ is constant. (Note that the value for the distance of 16 *pouce* does not accord well with the others. Thinking that in that case the force of the upper pole of the vertical magnet affected the measurement because of the great distance between the pole and the needle, Coulomb gives a corrected value of 79 for $v^2 - v_0^2$, which makes the third value 20,224.)

One must admit that this alone cannot be said to prove unquestionably the inverse-square law for magnetic force. In particular, the considerable discrepancy between the final value and the previous two is hard to overlook.

However, we must also take into account other measurements that Coulomb presents in the same memoir, this time in the area of statics. The horizontal arrow in Figure 22.3 is a slender artificial magnet 24 *pouce* (approximately 61 cm) long. It is suspended from a torsion wire running through a 30-*pouce* (approximately 76-cm) tube (*id*) that rises vertically from the center of the magnet. The

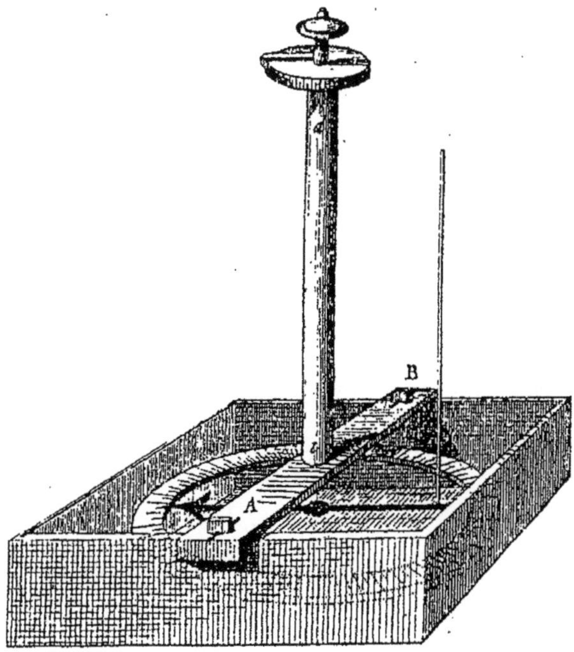


Figure 22.3: Coulomb's magnetic torsion balance. *Source*: Coulomb, *Vier Abhandlungen*.

wire can be twisted by turning the knob on the top of the tube. When the wire is in its normal, untwisted state, the horizontal magnet is aligned in the direction of the magnetic meridian. The rod parallel to *id* is also an elongated artificial magnet. Its lower pole (*P* in Figures 22.4a and 22.4b) is positioned at the initial static position of one of the horizontal magnet's poles (*Q* in Figures 22.4a and 22.4b), and then the knob at the top of the torsion wire is turned so as to bring the horizontal magnet to rest. In this case, both magnets are sufficiently long that it can be assumed that the only force exerted is that between the two poles (*P* and *Q*). The force acting between the poles can thus be ascertained from the angle of rotation of the knob.

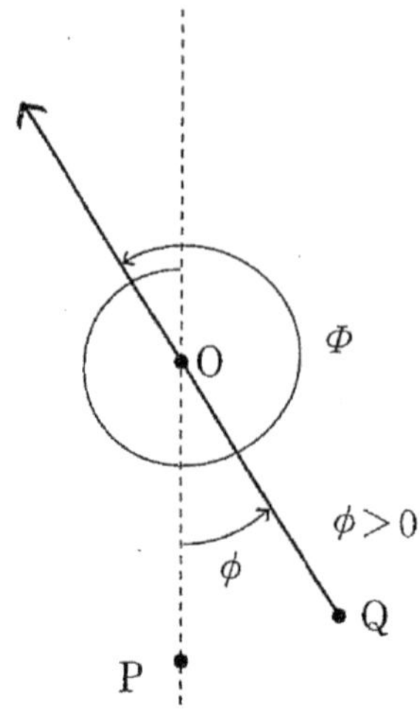


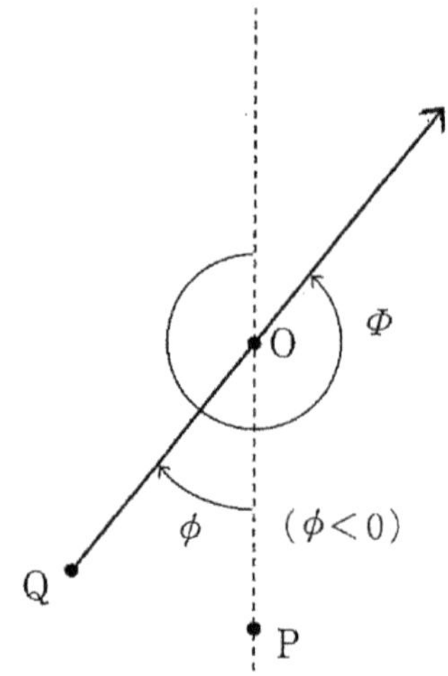


Figure 22.4a Overview of Coulomb's measurements. Figure 22.4b

Let the angle of rotation of the knob be $\Phi$ and the horizontal magnet's angle of deflection from the magnetic meridian be $\varphi$. (Given that the knob can be turned further than one complete rotation, $\Phi = \varphi$ does not necessarily hold, and in general $\Phi - \varphi =$ an integer $\times\, 360°$.) In this state, since the wire's torsion angle is $\Phi - \varphi$, the moment of restoring force due to the elasticity of the wire acting on the horizontal magnet is $N(\Phi - \varphi)$. Meanwhile, since the moment of the couple due to the Earth's magnetic field is proportional to $\sin \varphi \cong \varphi$, it can be expressed as $M\varphi$ (where $\varphi$ is relatively small). If the length of the horizontal magnet is $2a$, the moment of force due to force $F$ between the horizontal magnet and the vertical magnet is $Fa \cos(\varphi/2) \cong Fa$, and consequently the equilibrium of the horizontal magnet is

$$Fa = N(\Phi - \varphi) + M\varphi$$

Meanwhile, for small $\varphi$ values, the distance between the two poles exerting force on each other ($P$ and $Q$ in Figures 22.5a and 22.5b) can be approximated to $r = 2a\,\sin(\varphi/2) \cong a\varphi$.

First, the knob is rotated without the vertical magnet in place, whereupon the horizontal magnet seeks the position of equilibrium resulting solely from the terrestrial magnetic field and the elastic force of the wire (Figure 22.5b). Note that, if we take the angles in the figure with the counterclockwise as the positive, in this case we get $\varphi < 0$; and in the actual measurements, the angles were $\Phi = 700°$ and $\varphi = -20°$ . Hence,

$$0 = N \times 720° + M \times (-20°) \qquad \therefore \qquad M = 36N$$

This brings us to the measurements then taken with the vertical magnet in place, which were as follows.

| $\Phi$ | $\varphi$ | $N(\Phi - \varphi) + M\varphi = N(\Phi + 35\varphi)$ | $r = a\varphi$ |
|---|---|---|---|
| 24° | 24° | $N \times 864°$ | $a \times 24°$ |
| 3 turns + 17° = 1097° | 17° | $N \times 1692°$ | $a \times 17°$ |
| 8 turns + 12° = 2892° | 12° | $N \times 3312°$ | $a \times 12°$ |

This gives $864 : 1692 : 3312 \cong 1/4 : 1/2 : 1 \cong 1/24^2 : 1/17^2 : 1/12^2$ and hence $F \propto r^{-2}$. Alternatively, if we formulate it as

$$r^2 \times F = (a\varphi)^2 \times \{N(\Phi - \varphi) + M\varphi\} = a^2 N\varphi^2 \times (\Phi + 35\varphi),$$

then, for the measurements in order from the top of the above table (and eliminating the constant factor $a^2N$), we get

$$24^2 \times 864 = 497664,\ 17^2 \times 1692 = 488988,\ 12^2 \times 3312 = 476928.$$

Since these three values can be regarded as in effect equal at a high degree of probability, we can posit that $r^2 \times F = \text{const.}$

Toward the end of the memoir, Coulomb summarizes his findings (including for his measurements of electrostatic force) as follows:

1. The electric action, whether repulsive or attractive, of two electrified spheres, and therefore of two electrified molecules, is in the ratio compounded of the densities of the electric fluid of the two electrified molecules and inversely as the square of the distances;
2. In a needle 20 to 25 inches in length, magnetized by the double-touch method, the magnetic fluid can be supposed to be concentrated at 10 *lignes* from the ends of the needle;
3. When a needle is magnetized, in whatever position it is placed, it is always attracted back to this meridian by a force, constant and parallel to the meridian, of which the resultant passes always through the same point of the suspended needle:
4. The attractive and repulsive force of the magnetic fluid, as of the electric fluid, is exactly in the ratio directly of the densities and inversely of the square of the distances between the magnetic molecules.

In other words, at the separation distance $r$ in either case, the magnitude of electrostatic force ($f_E$) between two electric charges ($e_1$ and $e_2$) and the magnitude of magnetic force ($f_M$) between two magnetic poles ($m_1$ and $m_2$) can be represented as

$$f_E \propto \frac{e_1 e_2}{r^2}, \qquad f_M \propto \frac{m_1 m_2}{r^2}$$

This is what we now know as Coulomb's law, a formulation that ranks with Newton's law of universal gravitation as one of the cornerstones of physical science.

Writing around half a century later, in 1825, French physicist André-Marie Ampère, who had studied at the École Polytechnique, the training school for engineers that was established in Paris during the Revolutionary Period, makes the following remarks regarding the end point of this long investigative process initiated by Cusanus and Della Porta, continued by the members of the Royal Society of London and others, then completed by Mayer and Coulomb:

> To observe first the facts, varying the conditions as much as possible, to accompany this with precise measurement, in order to deduce general laws based solely on experience, and to deduce therefrom, independently of all hypothesis regarding the nature of the forces which produce the phenomena, the mathematical value of these forces, that is to say, to derive the formula which represents them, such was the road which Newton followed. This was the approach generally adopted by the learned men of France to whom physics owes the immense progress which has been made in recent times, and similarly it has guided me in all my research into electrodynamic phenomena. I have relied solely on experimentation to establish the laws of the phenomena and from them I have derived the formula which alone can represent the forces which are produced; I have not investigated the possible cause of these forces, convinced that all research of this nature must proceed from pure experimental knowledge of the laws and from the value, determined solely by deduction from these laws, of the individual forces in the direction which is, of necessity, that of a straight line drawn through the material points between which the forces act.[29]

As Newtonian mechanics spread on the Continent in the second half of the eighteenth century, there was a growing recognition that the key to understanding gravity was the determination of the mathematical laws governing it, as well as a gradual acceptance that the questions philosophy had posed about gravity's "essential nature" or "cause" were not central to physics. The research conducted by Coulomb extended that program of investigation to the issue of electromagnetic force. As theory about the forces of electricity and magnetism thus also came to be rationalized under the rubric of action at a distance expressed as mathematical functions, such thinking became indispensable to mathematical physics and thus an integral element of modern physical science as a whole.

With this gradual rationalization of our view of the physical world, magnetic force, which from ancient times had been regarded as an exemplar of "occult forces," was systematically divested of its magical character. Scientists no longer even spoke of the kind of theological underpinnings that Newton himself inferred about gravitational force. Physics abandoned attempts to gain an ontological understanding of the essences of things and satisfied itself with seeking mathematical certainty regarding the laws that governed phenomena. This was the birth of mathematical physics.

* * *

From the late medieval period to the dawn of the modern era, magic, alchemy, and astrology were the intellectual traditions that seriously addressed questions of force and natural action, particularly action at a distance—which Scholastic philosophy could not explain rationally. Magic, especially the natural magic of the late Renaissance, described electric force (the amber effect), magnetic force, and other actions of the natural world in terms of "sympathy and antipathy" and "occult power." Moreover, rather than probing the essential nature of such forces, magic took the approach of investigating, through experiment and observation, how such forces behaved so as make use of them. Alchemy was likewise intellectually progressive in its simlar emphasis on experiment and observation. This approach ran directly counter to the Scholastic method of logical deduction from the "true natures" of things—which implied that nothing could be understood without first knowing its ontological essence.

Meanwhile, mechanism, such as was advocated by Descartes, Gassendi, and others, which emerged in the early modern period as a philosophy to supersede Scholasticism, attempted to dismantle magic by providing mechanical explanations of forces, or rather of their causes and mechanisms of propagation. However, mechanism failed to achieve that self-imposed goal.

Ultimately, what deepened our understanding of force and opened the way to new scientific advances was not the invention of models for the propagation of force, as mechanism had attempted, much less the purely rational deduction of the nature of force from supposedly absolute first principles. The breakthrough came, rather, from an approach that laid aside such questions of essence and cause in favor of ascertaining the mathematical laws underlying force through experiment and observation and particularly through precise measurement. The final steps in that process—achieved by Hooke and Newton concerning gravity and by Mayer and Coulomb concerning magnetism—marked the birth of the concept of force as it is understood in modern physics.

We thus observe that the progress of inquiry into magnetic force and force in general unfolded not along the path of reductionism charted by Plato, Epicurus, and Lucretius in antiquity and by Descartes and Gassendi in the modern era but rather in the direction anticipated by such thinkers as the mystic Cusanus and the magician Della Porta. Thus the notion of action at a distance—having originated in a hylozoistic and magical world view, having persisted even in the ideas of Gilbert and Kepler, and despite being rejected by mechanism precisely for those hylozoistic and magical aspects—was ultimately affirmed in the form of laws expressed as mathematical functions. With that step, action at a distance secured its place in natural science, the Copernican system was given a truly dynamical foundation, and thus modern physics was born.

Accordingly, any simplistic schema suggesting that modern mechanism supplanted medieval Scholasticism and thus brought about a scientific revolution is inaccurate, at least with respect to the question of force, the core issue of modern physics. The mechanist Galileo, for example, considered the problem of the solar system to be a matter beyond the province of physical science; Cartesian mechanism was mere fancy with no connection to Tycho Brahe's observations or Kepler's theories; and neither Galileo nor Descartes was able to understand gravity, especially gravitational force between celestial bodies. By contrast, it was figures whose ideas retained the influence of hylozoism—Gilbert, Kepler, and so on—that incorporated the concept of force into the very heart of natural science in the early years of the modern age; and it was their ideas that continued to have a major impact in the seventeenth century, particularly in England, in the form of "magnetic philosophy." Under that influence, thinkers such as Hooke—a mechanist who was nonetheless heavily influenced by magnetic philosophy—and Newton—who devoted himself to alchemy and was influenced by the Cambridge Platonists—built up the theory of gravitational force. And among the core ideas that have in fact survived far beyond those early stages of the modern era we find Kepler's laws of planetary motion, Boyle's law of air pressure, Newton's law of gravitation, and Coulomb's law of electromagnetic force.

We see, then, that there was no tearing down of magic in the way mechanism had intended. Once the magical concept of action at a distance had been grasped and rationalized in terms of mathematical laws, the general principle was established that natural philosophy (physics) was a science dealing with verifiable laws that could also be expressed with mathematical rigor. As a result, natural philosophers simply lost interest in the problems peculiar to magic. To be sure, in the formative process that gave rise to modern science, science was inspired by natural magic; particularly in the formulation of the concept of force, science inherited from magic and astrology the notion of action at a distance and from alchemy the notion of active principle. But once natural philosophy had established its own method, magic was no longer a subject of discourse among the natural philosophers and no longer had any significant impact on people's view of the natural world. It would be wrong to underestimate in hindsight the contribution that magic made at the infancy of modern science; but, of course, it would be an even greater error to suggest that magic has retained some kind of persisting significance simply because it played that role at one stage in the evolution of modern science.

And with that, our long story of the protoscience of magnetism and gravity comes to an end. Before long, the Newtonian and Coulombian concept of action at a distance was revised in a process leading from Faraday to Einstein. But that is a later story.

# Notes

## Introduction

1. Jammer, *Concepts of Force*, 14.
2. Plato, *Timaeus* 80C.
3. Aristotle, *Physics* 7.1.242b25, 7.2.243a3.
4. R. Bacon, *On the Multiplication of Species*, 62.
5. Gilbert, *On the Magnet* 2.2:57.
6. Charleton, *Physiologia*, 343, 345.
7. William Foster, quoted in Debus, *Chemical Philosophy*, 280.
8. Leibniz, *Leibniz-Clarke Correspondences*, 66.
9. Quoted in Crombie, *Augustine to Galileo* 2:59; idem, *Robert Grosseteste*, 212 n. 2 (Latin). See also Hesse, *Forces and Fields*, 102.
10. Paracelsus, *Diseases that Deprive Man of his Reason*, 153.
11. F. Bacon, *New Organon* 2.37:168, 2.48:198.
12. Agricola, *De natura fossilium*, 84.
13. Gilbert, *On the Magnet* 2.2:46.
14. A. Smith, *Essays on Philosophical Subjects*, 42.
15. Einstein, "Autobiographical Notes," 9.
16. Thomas Aquinas, *Summa theologica* 2.2, q. 96, art. 2.
17. Pomponazzi, *De naturalium effectuum causis*, 22–23 and idem, *Les causes des merveilles de la nature*, 121; my translation.

18. Thomas Aquinas, *Soul*, 10.

19. Ficino, *Three Books on Life* 3:315–16.

20. See for example White, *Medieval Technology and Social Change*, 177; Johnson, *Astronomical Thought*, 215–16.

21. Stevin, *Principal Works of Simon Stevin* 3:129.

22. Kepler, *Epitome of Copernican Astronomy*, 919.

23. Quoted in Kelly, *The "DE MUNDO" of William Gilbert*, 70.

24. Quoted in Birch, *History of the Royal Society of London* 2:70.

25. Mitchell, "Chapters in the History of Terrestrial Magnetism, Ch. 4," 324.

26. Schmitt, preface to *Chusei no meishin* (*Les superstitions*), 4; my translation.

**Chapter 1—Ancient Greece: The Science of Magnetism is Born**

1. Aristotle, *On the Soul* 1.2.405a19.

2. Diogenes Laertius, *Lives of Eminent Philosophers* 1.1:24.

3. See Lloyd, *Aristotle*, 181; T. S. Hall, *Ideas of Life and Matter*, 14, 108; Hesse, *Forces and Fields*, 37.

4. Barnes, ed., *Early Greek Philosophy*, 24. See also Freeman, *Pre-Socratic Philosophers*, 66.

5. Alexander of Aphrodisias, *Quaestiones 2.16–3.15*, 28 (Bruns's edition: p. 72 lines 10–18); emphasis added; brackets in source.

6. Freeman, *Pre-Socratic Philosophers*, 198. See also Aristotle, *On Generation and Corruption* 1.8.324b27, 325b1.

7. Diogenes Laertius, *Lives of Eminent Philosophers* 9.9:57.

8. Alexander of Aphrodisias, *Quaestiones 2.16–3.15*, 29–30 (Bruns's edition: p. 73 lines 14–25); brackets in source.

9. Barnes, ed., *Early Greek Philosophy*, 208–9. See also Diogenes Laertius, *Lives of Eminent Philosophers* 9.7:44.

10. Aristotle, *On Sense and the Sensible* 4.442b11.

11. Diogenes Laertius, *Lives of Eminent Philosophers* 9.7:47.

12. Barnes, ed., *Early Greek Philosophy*, 208.

13. Homer, *Odyssey* 17.218.

14. Barnes, ed., *Early Greek Philosophy*, 214.

15. Diogenes Laertius, *Lives of Eminent Philosophers* 9.6:31.

16. Plato, *Timaeus* 81A. *Timaeus* subsequently cited in this chapter in parentheses in the text, with the abbreviation "Tim." followed by the Stephanus reference.

17. Galen, *On the Natural Faculties* 3.15:319.

18. Quoted in J. A. Smith, "Precursors to Peregrinus," 36 n. 92.

19. R. Bacon, *Opus majus*, 631.

20. Paracelsus, *Astronomia magna*, 108; my translation. The original reads: "Gleiches in Gleiches, das gehört zusammen. Gleiches und Ungleiches vereingt sich nichit."

21. *Pirotechnia of Vannoccio Biringuccio*, 115.

22. Kepler, *New Astronomy*, 55.

23. Alexander of Aphrodisias, *Quaestiones 2.16–3.15*, 29 (Bruns's edition: p. 72 line 30–p. 73 line 8); brackets in source.

24. Plato, *Ion* 533DE.

25. Dictionary of Scientific Biography, s.v. "Empedocles of Acraga," 367–69; Freeman, *Pre-Socratic Philosophers*, 195; Farrington, *Greek Science*, 58; B. Russell, *History of Western Philosophy*, 72.

26. Roller and Roller, "Prenatal History of Electrical Science," 345; idem, "Development of the Concept of Electric Charge," 541; D. H. D. Roller, *"DE MAGNETE" of William Gilbert*, 22.

27. Plutarch, *Moralia* 13:69–73 (Stephanus's edition: p. 1005).

28. Aristotle, *Physics* 8.10.267a2.

29. Aristotle, *Meteorology* 4.1.378a13.

30. Aristotle, *On Generation and Corruption* 2.2, 2.3.

31. Aristotle, *Physics* 1.5.188b23.

32. Aristotle, *On Generation and Corruption* 2.8, 2.9.

33. Aristotle, *Physics* 4.8.215a3.

34. Aristotle, *On the Heavens* 1.3.270b13, 1.2.269a7, 1.3.270a5, 1.2.269a31.

35. Aristotle, *Physics* 7.1.241b24; idem, *On the Heavens* 2.6.288a28.

36. Aristotle, *Physics* 8.4.255b33.

37. Ibid. 8.4.255a29.

38. Aristotle, *Metaphysics* 12.7.1072b6, 1072b29, 1074a30.

39. Aristotle, *Physics* 7.1.242b27.

40. Ibid. 8.5.256b24.

41. Aristotle, *On the Soul* 2.4.415b9, 1.3.406a30, 1.4.408b30.

42. Farrington, *Greek Science*, 166–67; Lloyd, *Greek Science after Aristotle*, 9–10.

43. Diogenes Laertius, *Lives of Eminent Philosophers* 5.2:36, 37.

44. Theophrastus, *On Stones*; subsequently cited in this chapter in parentheses in the text, with the abbreviation "OS" followed by the paragraph number.

45. Pliny, *Natural History* 37.13.

46. Aristotle, *Meteorology* 3.6.378a20f.

47. Boyle, The Works of the Honourable Robert Boyle, 4: 343.

48. Gilbert, *On the Magnet* 5.12:208, 210.

49. Aristotle, *On the Soul* 2.2.413a21, 2.1.412b16.

50. Farrington, *Greek Science*, 31 The periodization by Bertrand Russell (*History of Western Philosophy*, 229) is nearly the same.

## Chapter 2—The Hellenistic Age

1. The full text of *Letter to Herodotus* is printed in Diogenes Laertius, *Lives of Eminent Philosophers* 10. In this chapter, the *Letter* is cited in parentheses in the text, with the abbreviation "LH" followed by the volume number and section number as given in the *Lives*.

2. Quoted in the introduction to Lucretius, *De rerum natura* (*On the Nature of Things*), p. X.

3. In this chapter, *On the Nature of Things* is subsequently cited in parentheses in the text, with the abbreviation "ONT" followed by the volume number and line number.

4. Boyle, The Works of Honourable Robert Boyle, 4, 68. On Gassendi, see Brett, *Philosophy of Gassendi*, 102–3, 222–23.

5. Brett, *Philosophy of Gassendi*, 73; Boas, "Establishment of the Mechanical Philosophy," 430; Duhem, *Aim and Structure of Physical Theory*, 122–23; Cyrano de Bergerac, *L'autre monde*, 972.

6. Cardano, *Book of My Life*, 59.

7. Heiberg, *Mathematics and Physical Science in Classical Antiquity*, 99.

8. In this chapter, *On the Natural Faculties* is cited in parentheses in the text, with the abbreviation "ONF" followed by the book number, the chapter number, and the page number of *Claudii Galeni opera omnia*.

9. Aristotle, *Metaphysics* 4.1014b17; idem, *On the Soul* 2.3.414a31. See also idem, *On the Soul* 2.2.413a25.

10. See *Hippocrates* 4, "Nature of Man."

11. Lloyd, *Greek Science after Aristotle*, 141.

12. Gilbert, *On the Magnet* 2.3:62.

13. Alexander of Aphrodisias, *Quaestiones 2.16–3.15*, 29 (Bruns's edition: p. 72 lines 21–27); brackets in source.

14. Ibid., 30 (Bruns's edition: p. 73 lines 26–30); brackets in source.

15. Ibid., 29 (Bruns's edition: p. 73 lines 8–12); brackets in source.

16. Ibid., 30 (Bruns's edition: p. 74 lines 6–8); brackets in source.

17. Ibid., 30–31 (Bruns's edition: p. 74 lines 8–14); brackets in source.

18. Ibid., 31 (Bruns's edition: p. 74 lines 21–28); brackets in source.

19. Ibid., 31 (Bruns's edition: p. 74 lines 28–30).

20. Porphyry, *On Abstinence from Killing Animals*, 117.

21. Acosta, *Natural and Moral History of the Indies*, 54–55.

22. For an abridged version of Bertelli's report, see Hazard, "Early History of the Mariner's Compass," 179. See also Mitchell, "Chapters in the History of Terrestrial Magnetism, Ch. 1," 114; and Mottelay, *Bibliographical History of Electricity and Magnetism*, 7.

## Chapter 3—The Days of the Roman Empire

1. Wolff, *Cultural Awakening*, 125. See also Fontana, *Distorted Past*, 11.

2. B. Russell, *History of Western Philosophy*, 283–84.

3. Aelian, *Historical Miscellany* 4.21:207, 12.19:369.

4. R. French, *Ancient Natural History*, 262.

5. In this chapter, *The Greek Herbal of Dioscorides* is cited in parentheses in the text, with the abbreviation "GH" followed by the volume number and the page number of the edition used.

6. Thorndike, *History of Magic and Experimental Science* 1:154; Singer, *From Magic to Science*, 185–86; Crombie, *Augustine to Galileo* 1:38.

7. William of Conches, *Philosophia mundi* 4.6; R. Bacon, *Opus majus*, 620. See also Haskins, *Renaissance of the Twelfth Century*, 329.

8. Vives, *On Education*, 169.

9. Jones, *Ancients and Moderns*, 5–6.

10. Singer, *History of Biology*, 94.

11. Helmont, *Van Helmont's Works*, 13. See also Debus, *Chemical Philosophy*, 347.

12. Arber, *Herbals*, 10.

13. Dodds, *Greeks and the Irrational*, 293.

14. Sigerist, *History of Medicine* 1:343, 486.

15. Thorndike, *History of Magic and Experimental Science* 1:581.

16. Gilbert, *On the Magnet* 1.1:2.

17. Pliny the Younger, *Letters of the Younger Pliny*, 88.

18. Ibid., 166–68.

19. In this chapter, *Natural History* is cited in parentheses in the text, with the abbreviation "NH" followed by the book number and chapter number.

20. Ward, *Venerable Bede*, 35; Thorndike, *History of Magic and Experimental Science* 1:635.

21. William of Conches, *Philosophia mundi* 3.6; *Metalogicon of John of Salisbury*, 145; *Ruodlieb* 2:16.

22. Southern, *Making of the Middle Ages*, 71.

23. Hugh of Saint Victor, *Didasalicon* 3.2.

24. References to Pliny's *Natural History* are particularly numerous in the two works by Albert.

25. Richard de Bury, *Philobiblon*, 64.

26. Oresme, *Nicole Oresme and the Kinetics of Circular Motion*, 287, 315, 319.

27. Febvre and Martin, *Coming of the Book*, 276.

28. Oviedo, *Natural History of the West Indies*, 3.

29. Agricola, *De natura fossilium*, 2.

30. Acosta, *Natural and Moral History of the Indies*, 53.

31. R. Bacon, *Opus majus*, 155.

32. Haskins, *Renaissance of the Twelfth Century*, 112.

33. Tacitus, *The Agricola and the Germania*, 139.

34. Agricola, *De natura fossilium*, 85–86; Gilbert, *On the Magnet* 1.6:18.

35. Grant, *Physical Science in the Middle Ages*, 8. See also D. H. D. Roller, *"DE MAGNETE" of William Gilbert*, 25.

36. Stahl, *Roman Science*, 106.

37. Jammer, *Concepts of Force*, 45.

39. Plutarch, *Moralia* 8:175 (Stephanus's edition: p. 641).

40. *Claudian* 2:234–37.

41. Bromehead, "Geology in Embryo," 115.
42. R. French, *Ancient Natural History*, 263.
43. Aelian, *On the Characteristics of Animals* 2.10.14:304–305.
44. Kunz, *Curious Lore of Precious Stones*, 96.
45. B. Russell, *History of Western Philosophy*, 257–58.
46. Dodds, *Greeks and the Irrational*, 246. See also ibid., 292.
47. Stahl, *Roman Science*, 119.

## Chapter 4—Christianity in the Middle Ages

1. Augustine, *City of God* 21.4:970–71.
2. Ibid. 21.7:978.
3. Ibid. 21.5:971.
4. Augustine, *Confessions* 10.35:241–42; italics in source.
5. Ibid. 10.35:243.
6. Gervase, OTIA IMPERIALIA : *Recreation for an Emperor*, 564–65.
7. Augustine, *City of God* 21.4:970, 971.
8. See D. H. D. Roller, *"DE MAGNETE" of William Gilbert*, 26.
9. Mandeville, *Mandeville's Travels*, 116.
10. See section 6 of chapter 4 (Albert the Great), section 4 of chapter 9 (Cusanus), section 5 of chapter 13 (Agricola), section 2 of chapter 15 (Pomponazzi), and chapter 16 (Della Porta).
11. Isidore, *Etymologies*, 254.
12. Marbode, *De lapidibus*, line 30 (p. 36).
13. *Metalogicon of John of Salisbury*, 188. John states here that "the diamond was finally cut by means of lead with goat's blood." Among Pliny, Augustine, and Isidore, however, none makes any reference to lead in this connection.
14. Hartmann von Awe, *Erec*, line 8438 (p. 472); Wolfram, *Parzival* 2.105, lines 20–21 (vol. 1, pp. 180–81).
15. Albert the Great, *Book of Minerals*, 70.
16. R. Bacon, *Opus majus*, 584.
17. Paracelsus, *Hermetic and Alchemical Writings* 1:17.
18. Augustine, *On Christian Doctrine* 2.9.14:42.
19. Ibid. 2.16.24:50, 2.16.25:51, 2.29.45:65. See also Grant, *Physical Science in the Middle Ages*, 4–5.

20. Pliny, *Natural History* 30.1.

21. Basil, *Regulae fusius tractatae*, 1047, 1050; my translation. See also, Thomas Aquinas, *Summa theologica* 2.1, q. 85, art. 5; Alford, "Medicine in the Middle Ages," 388–89.

22. Sigerist, *Civilization and Disease*, 140.

23. See C. Hill, *Intellectual Origins of the English Revolution*, 25.

24. Milis, "Introduction," 6. See also Seibt, *Glanz und Elend des Mittelalters*, 528.

25. Fontana, *Distorted Past*, 29.

26. Augustine, *On Christian Doctrine* 2.23.36:59, 2.29.45:65.

27. Borst, *Alltagsleben im Mittelalter*, 488; my translation.

28. Southern, *Making of the Middle Ages*, 166.

29. Bede, *Ecclesiastical History of the English People*, 45.

30. *Marbode's Christian Symbolic Lapidary*, 125.

31. Marbode, *De lapidibus*, lines 22–23 (p. 34).

32. Sarton, *Introduction to the History of Science* 1:741; see also Klein-Franke, "Knowledge of Aristotle's Lapidary," 141.

33. Jammer, *Concepts of Force*, 46.

34. Dioscorides, *Greek Herbal* 5:146.

35. Pliny, *Natural History* 36.34; Isidore, *Etymologies*, 253.

36. Marbode, *De lapidibus*, lines 270–83 (p. 56).

37. Pliny, *Natural History* 37.60. See also Thorndike, *History of Magic and Experimental Science* 1:80–81.

38. Isidore, *Etymologies*, 255.

39. Marbode, *De lapidibus*, line 594 (p. 81).

40. Ibid., lines 284–311 (p. 58), lines 369–370 (p. 64).

41. Wolfram, *Parzival* 16.792, lines 1–5 (vol. 2, p. 351); my translation.

42. *La quête du graal*, 225; my translation. The original French reads: "le vertus des pierres, la force des herbes."

43. *King Horn*, line 571.

44. Guillaume de Lorris and Jean de Meun, *Romance of the Rose*, lines 1067–1070 (p. 45).

45. Jacobus de Voragine, *Golden Legend*, 346.

46. *Reinke de Vos*, lines 4881–82, 4902–4; my translation.

47. Greene, *Pandosto*, pp. xxvii–xxviii.

48. Marlowe, *Tragical History of Doctor Faustus*, lines 135–37 (pp. 8–9).

49. Spenser, *Faerie Queene*, 4.3.24 (vol. 2, p. 34).

50. Charon, "Knowledge of Herbs," 122–23.

51. *Hildegard von Bingen's "PHYSICA"*, 138.

52. Ibid., 152.

53. Ibid., 151.

54. D. H. D. Roller, *"DE MAGNETE" of William Gilbert*, 28.

55. Albert the Great, *Book of Minerals*, 103–4; brackets in source.

56. Thorndike, *History of Magic and Experimental Science* 2:531–32.

57. Thomas Aquinas, *On Spiritual Creatures*, 36.

58. Thomas Aquinas, *Commentary on Aristotle's Physics* 7.3.903:461.

59. Albert the Great, *Book of Minerals*, 93.

60. Ibid., 55–56; brackets in source.

61. Bartholomew, *On the Properties of Things*, 368.

62. See Schipperges, *Der Garten der Gesundheit*, 86, 125.

63. Quoted in Bromehead, "Geology in Embryo," 124.

64. Boyle, The Works of the Honourable Robert Boyle, 3: 512–561, esp. 517.

## Chapter 5—The Discovery of Magnetic Directionality

1. Seibt, *Glanz und Elend des Mittelalters*, 528; my translation.

2. Fontana, *Distorted Past*, 52.

3. See White, *Medieval Technology and Social Change*, ch. 2; idem, *Medieval Religion and Technology*, 77; Gimpel, *Medieval Machine*, ch. 3; Morrall, *Medieval Imprint*, 129–34.

4. Seibt, *Glanz und Elend des Mittelalters*, 204–5; Gimpel, *Medieval Machine*, 87–89.

5. Verger, *Les universités au moyen âge*, 45–46.

6. Marenbon, *Later Medieval Philosophy*, 15.

7. Wolff, *Cultural Awakening*, 183–84. See also, Lindberg, "Transmission of Greek and Arabic Learning," 60–61.

8. *Dictionary of Scientific Biography*, s. v. "Gerbert"; Southern, *Making of the Middle Ages*, 170, 180; Wolff, *Cultural Awakening*, 193–94; Haskins, *Renaissance of the Twelfth Century*, 311.

9. Dufourcq, *La vie quotidienne dans l'Europe médiévale sous domination arabe*, 241–42.

10. Ibid., 70–71; my translation.

11. Libera, *La philosophie médiévale*, 387; my translation.

12. Sarton, *Introduction to the History of Science* 2:283–84; Marenbon, *Later Medieval Philosophy*, 53; Haskins, *Renaissance of the Twelfth Century*, 292–93; Lindberg, "Transmission of Greek and Arabic Learning," 72.

13. Grant, *Physical Science in the Middle Ages*, 20. See also Haskins, *Renaissance of the Twelfth Century*, ch. 12.

14. Needham, *Science and Civilisation in China* 4:249; Mitchell, "Chapters in the History of Terrestrial Magnetism, Ch. 1," 110, 130; Sarton, *Introduction to the History of Science* 1:740.

15. Gilbert, *On the Magnet* 1.1:4. See also Mitchell, "Chapters in the History of Terrestrial Magnetism, Ch. 1," 121.

16. Sarton, *Introduction to the History of Science* 1:741. See also ibid. 2:509; Turner, *Science in Medieval Islam*, 119; D. R. Hill, *Islamic Science and Engineering*, 59.

17. E. G. R. Taylor, *Haven-finding Art*, 96. See also ibid., 126–27; Mitchell, "Chapters in the History of Terrestrial Magnetism, Ch. 1," 123, 130; J. A. Smith, "Precursors to Peregrinus," 24.

18. *Dictionary of the Middle Ages*, s. v. "Compass, magnetic."

19. Aczel, *Riddle of the Compass*, ch. 5; Della Porta, *Natural Magick* 7.32; Gilbert, *On the Magnet* 1.1:4.

20. Braudel, *La Mediterranée: L'espace et l'histoire*, 176; my translation.

21. Winter, "Who Invented the Compass?" 99.

22. This quotation is from the English translation by Harradon. A number of Latin codices of the *Letter* are extant, among them one in which the cited passage has "in all parts (*in omnibus partibus*)" instead of "in all ports (*in omnibus portibus*)" and "Normannie, Flandrie" instead of "Normannie, Picardie et Flandrie." For further details, see Thompson, "Petrus Peregrinus," 15.

23. Quoted in J. A. Smith, "Precursors to Peregrinus," 37 (Latin and English). The same passage appears in Hellmann, "Die Anfänge der magnetischen Beobachtungen," 126 (Latin); D. H. D. Roller, *"DE MAGNETE" of William Gilbert*, 35 (Latin); Benjamin, *Intellectual Rise in Electricity*, 128–29 (English); Mottelay, *Bibliographical History of Electricity and Magnetism*, 31 (English); and Needham, *Science and Civilisation in China* 4:246 (English). See also Bromehead, "Alexander Neckam on the

Compass-needle," 139–40; and May, "Alexander Neckam and the pivoted Compass Needle," 283–84.

24. Humboldt, *Cosmos* 2:656n.

25. Quoted in Needham, *Science and Civilisation in China* 4:246–47. Versions of the French-language original of this passage, with some differences among them, appear in Needham's work; in Mottelay, *Bibliographical History of Electricity and Magnetism*, photograph of codex, facing p. 30; in J. A. Smith, "Precursors to Peregrinus," 39 n. 108; and in Aczel, *Riddle of the Compass*, 30–31. Other English translations appear in *Dictionary of the Middle Ages*, s. v. "Compass, magnetic"; J. A. Smith, "Precursors to Peregrinus," 39; Aczel, *Riddle of the Compass*, 31; and E. G. R. Taylor, *Haven-finding Art*, 95. A German translation appears in Lipmann, "Geschichte der Magnetnadel bis zur Erfindung des Kompasses (gegen 1300)," 22.

26. Quoted in J. A. Smith, "Precursors to Peregrinus," 41 (Latin and English). The passage also appears in D. H. D. Roller, *"DE MAGNETE" of William Gilbert*, 36 (Latin); Benjamin, *Intellectual Rise in Electricity*, 154 (English); and Mottelay, *Bibliographical History of Electricity and Magnetism*, 31 (English).

27. Thorndike, *History of Magic and Experimental Science* 2:388.

28. Mitchell, "Chapters in the History of Terrestrial Magnetism, Ch. 1," 128.

29. Sarton, *Introduction to the History of Science* 2:629.

30. Benjamin, *Intellectual Rise in Electricity*, 131.

31. Needham, *Science and Civilisation in China* 4:246.

32. This passage and those of Scot's cited in the next paragraph are quoted in J. A. Smith, "Precursors to Peregrinus," 45 n. 131 (Latin and English). They also appear in Haskins, "Michael Scot and Frederick II," 270–71 n. 2 (Latin).

33. Quoted in Mottelay, *Bibliographical History of Electricity and Magnetism*, 44. See also Haskins, "Michael Scot and Frederick II," 270 n. 2; J. A. Smith, "Precursors to Peregrinus," 45, 52 n. 165; and Gilbert, *On the Magnet* 2.2:11.

34. Quoted in J. A. Smith, "Precursors to Peregrinus," 45 (English), 45 n. 130 (Latin). See also Haskins, "Michael Scot and Frederick II," 270 n. 2.

35. Haskins, "Michael Scot and Frederick II," 264.

36. Thorndike, *Michael Scot*, 7. See also Haskins, *Renaissance of Twelfth Century*, 333; and Thorndike, *History of Magic and Experimental Science* 2:316.

37. Thorndike, *Michael Scot*, 33.

38. See Haskins, "Michael Scot and Frederick II," 255, 259; Crombie, *Augustine to Galileo* 1:93, 160; idem, *Robert Grossteste*, 42, 46; and Lindberg, "Transmission of Greek and Arabic Learning," 67.

39. R. Bacon, *Opus majus*, 63. See also, Haskins, "Michael Scot and Frederick II," 259 n. 6.

40. Marenbon, *Later Medieval Philosophy*, 52. See also Johnson, *Astronomical Thought*, 60; and Libera, *La philosophie médiévale*, 383.

41. Sarton, *Introduction to the History of Science* 2:491.

42. See *Dictionary of the Middle Ages*, s. v. "Frederick II of the Holy Roman Empire," s. v. "Melfi, Constitutions of."

43. Haskins, "The *De arte venandi cum avibus* of the Emperor Frederick II," 334.

44. Frederick II, *Art of Falconry*, 3–4.

45. Ibid., 79.

46. Albert the Great, *On Animals* 23.2:1546.

47. Crombie, *Augustine to Galileo* 2:17.

**Chapter 6—Thomas Aquinas and his Understanding of Magnetism**

1. Origen, *On First Principles* 4.1:266.

2. Seel, *Der Physiologus*, esp. items 37 ("Feuerstein"), 42 ("Starker Diamantstein"), and 38 ("Magnetstein").

3. Bonaventura, *Soul's Journey into God*, 84–85.

4. Augustine, *Confessions* 7.9.

5. Quoted in Wolff, *Cultural Awakening*, 239.

6. Aristotle, *Metaphysics* 1.1.980a.

7. Aristotle, *On the Heavens* 2.1238b8; idem, *On the Generation of Animals* 4.4.770b11.

8. Aristotle, *Physics* 2.7.198b3.

9. William of Conches, *Philosophia mundi* 1.13; my translation.

10. See "The Condemnation of Aristotle's Books on Natural Philosophy in 1210 at Paris" and "The Natural Books of Aristotle in the

Arts Curriculum at the University of Paris in 1255," in Grant, ed., *A Source Book in Medieval Science*, 42–44.

11. On the history of translations of Aristotle's works in the Middle Ages, see Libera, *La philosophie médiévale*, 358–63.

12. Thomas Aquinas, *Summa theologica* 1, q. 46, art. 2.

13. Aristotle, *Physics* 7.2.243a7; see also 8.3,254a13 and 8.7.260a28.

14. Aristotle, *Metaphysics* 7.7.1033a3.

15. See Thomas Aquinas, *Commentary on the Metaphysics of Aristotle* 12.2:856, 1031.

16. Thomas Aquinas, *Principles of Nature*, 8.

17. Thomas Aquinas, *On Being and Essence*, 65, 63.

18. Jean of Saint-Amand, *Expositio super antidotarium Nicolai*, 156.

19. Duhem, *Aim and Structure of Physical Theory*, 11.

20. Aristotle, *Physics* 8.3.253b5.

21. Thomas Aquinas, *Commentary on Aristotle's Physics* 2.1.145:77, 2.2.145:76.

22. Thomas Aquinas, *Commentary on the Metaphysics of Aristotle* 12.3:863–64.

23. Thomas Aquinas, *On Spiritual Creatures*, 36.

24. Thomas Aquinas, *Soul*, 10; brackets in source.

25. Aristotle, *History of Animals*, 8.1.588b4. See also Lovejoy, *Great Chain of Being*, 56–57.

26. Thomas Aquinas, *Treatise on Separate Substances*, 26.

27. Aristotle, *Physics* 7.2.243a45, b19.

28. Thomas Aquinas, *Commentary on Aristotle's Physics* 7.3.903:460–61.

29. Ibid. 2.6.189:100.

30. Thomas Aquinas, *On the Unity of the Intellect Against the Averroists*, 33.

31. Thomas Aquinas, *Summa theologica*, 2.2, q. 96, art. 2.

32. Aristotle, *Meteorology* 1.2.339a22.

33. Thomas Aquinas, *Commentary on the Metaphysics of Aristotle* 12.9:899.

34. Thomas Aquinas, *Treatise on Separate Substances*, 105, 88.

35. Ibid., 28.

36. Ibid., 31, 34, 27.

37. Thomas Aquinas, *On the Truth of the Catholic Faith* 3.2:45–46.

38. Thomas Aquinas, *Disputed Questions on Truth* 1:251.

39. Thomas Aquinas, *Summa theologica* 2.2, q. 96, art. 2.

40. See F. S. Taylor, *Alchemists*, 99.

41. Buridan, *Quaestiones* 2, q. 5 (pp. 145, 148); my translation.

42. Thomas Aquinas, *On Being and Essence*, 35, 37.

43. Thomas Aquinas, *Super evangelium S. Ioannis lectura* 1.1.26:8; my translation. The original Latin reads: "cum volo concipere rationem lapidis, oportet quod ad ipsam ratiocinando perveniam."

44. Thomas Aquinas, *Commentary on the Metaphysics of Aristotle* 12.12:925.

45. Thomas Aquinas, *Treatise on Separate Substances*, 88, 86.

**Chapter 7—Roger Bacon and the Propagation of Magnetic Force**

1. R. Bacon, *Opus majus*, 112. *Opus majus* subsequently cited in this chapter in parentheses in the text, with the abbreviation "OMj" followed by the page number of the edition used.

2. Michelet, *La sorcière*, 412; my translation.

3. On the broad implications of Bacon's "scientia experimentalis" see Thorndike, *History of Magic and Experimental Science* 2:649–59; Lindberg, "Science as Handmaiden"; and Hackett, "Roger Bacon on *Scientia experimentalis*." On the fact that "experientia" or "experimentum" as used by Bacon do not designate experiment in the modern sense, see Dijksterhuis, *Mechanization of the World Picture*, 138; and Wallace, "The Philosophical Setting of Medieval Science," 98–99.

4. Plato, *Timaeus* 27D–28A, 29BC.

5. Aristotle, *Posterior Analytics* 1.2.71b12, 19, 72a25.

6. Ibid. 2.19.100b5.

7. Ibid. 2.19.99b35, 100a4, 1.18.81b8.

8. Aristotle, *Physics* 1.1; idem *Metaphysics* 5.11.

9. Aristotle, *Posterior Analytics* 1.13.79a2.

10. Grosseteste, *On Light*, 10–11.

11. On the details and background of Grosseteste's *Concerning Lines, Angles, and Figures* (*De lineis, angulis et figures*), see McEvoy, *Philosophy of Robert Grosseteste*, part 3, chapter 1.

12. Grosseteste, *Concerning Lines, Angles, and Figures*, 385–86.

13. Ibid., 387.

14. Ibid.

15. Ibid., 386.

16. Ibid., 385.

17. Galileo, *Assayer*, 241–42.

18. In this chapter, *On the Multiplication of Species* is cited in parentheses in the text, with the abbreviation "OMS" followed by the page number of the edition used.

19. Quoted in Crombie, *Robert Grosseteste*, 206–7; brackets in source. The Latin original of this part of *Opus minus* appears in Mitchell, "Chapters in the History of Terrestrial Magnetism, Ch. 2," 272 n. 14, and in D. H. D. Roller, *"DE MAGNETE" of William Gilbert*, 38–39.

20. Mitchell, "Chapters in the History of Terrestrial Magnetism, Ch. 2," 243–44.

21. Aristotle, *On the Heavens* 4.3.310b3.

22. Buridan, *Questions on the Four Books on the Heavens and the World of Aristotle*, 558.

23. Aristotle, *Physics* 4.1.208b10.

24. Quoted in Wolfson, *Crescas' Critique of Aristotle*, 563.

25. Quoted in Mottelay, *Bibliographical History of Electricity and Magnetism*, 44; italics in source. See also J. A. Smith, "Precursors to Peregrinus," 52, 52 n. 165; and Benjamin, *Intellectual Rise in Electricity*, 156.

26. Crathorn, *On the Possibility of Infallible Knowledge*, 280.

### Chapter 8—Petrus Peregrinus and his *Letter concerning the Magnet*

1. For details of the thirty or so confirmed Latin manuscripts of Peregrinus's *Letter concerning the Magnet* (*Epistola de magnete*) currently extant in Europe, see Thompson, "Petrus Peregrinus de Maricourt and His *Epistola DE MAGNETE*"; and Schlund, "Petrus Peregrinus von Maricourt," *Archivum Franciscanum Historicum* 5 (1912): 22–40. There are at least three English-language translations, one by Silvanus Thompson, first printed in 1902 in a limited run of 250 copies for private circulation, and one each by Brother Arnold (Joseph Charles Mertens) and H. D. Harradon. The Arnold translation was first published in 1904 and was

reproduced in Grant, ed., *A Source Book in Medieval Science*, 368–76. The Harradon translation appears in *Terrestrial Magnetism and Atmospheric Electricity* 48 (1943): 3–17. A French-language translation appears along with the Latin original in *Revue d'histoire des sciences* 28 (1975): 193–234. In this chapter, all quotations from the *Letter concerning the Magnet* are from the Harradon translation and are cited in parentheses in the text, with the abbreviation "LCM" followed by the part number and chapter number.

2. Schlund, "Petrus Peregrinus von Maricourt," *Archivum Franciscanum Historicum* 4 (1911): 636 n. 5.

3. Fleming, "The Earth, a Great Magnet," 47.

4. Kant, *Critique of Pure Reason*, 109.

5. Zilsel, "The Origins of William Gilbert's Scientific Method," 29–30.

6. Schlund, "Petrus Peregrinus von Maricourt," *Archivum Franciscanum Historicum* 4 (1911): 437; my translation of the original German, which reads: "die erste Arbeit nach der reinexakten, induktiv-empirischen Methode der Naturwissenschaft"; Mottelay, *Bibliographical History of Electricity and Magnetism*, 45; Wightman, *Science and the Renaissance* 1:63.

7. Quoted in Crombie, *Robert Grosseteste*, 205; and in idem, *Science, Optics and Music*, 53. The Latin original of this passage appears in J. A. Smith, "Precursors to Peregrinus," 69 n. 249.

8. Vesalius, *Fabrica*, 520. See also O'Malley, *Andreas Vesalius of Brussels, 1514–1564*, 318.

9. Stock, "Science, Technology and Economic Progress in the Early Middle Ages," 45. On Gerbert, see Grant, *Physical Science in the Middle Ages*, 13–14; and Wolff, *Cultural Awakening*, 188.

10. *The "Didascalicon" of Hugo of Saint Victor* 3.1 and appendix A. See also White, *Medieval Religion and Technology*, 247.

11. Benjamin, *Intellectual Rise in Electricity*, 167.

12. White, *Machina ex Deo*, 70.

13. Crombie, *Augustine to Galileo* 2:25.

14. Ibid. 1:67.

15. Wolff, *Cultural Awakening*, 135.

16. White, *Medieval Technology and Social Change*, 119; idem, *Medieval Religion and Technology*, 54, 220.

17. Quoted in Klemm, *History of Western Technology*, 87. See also White, *Medieval Technology and Social Change*, 131.

18. *Roger Bacon's Letter*, 15, 26.

19. White, *Machina ex Deo*, 70. See also idem, *Medieval Technology and Social Change*, 134.

20. Reynolds, *Stronger Than a Hundred Men*, 54. See also Gimpel, *Medieval Machine*, 10–11.

21. White, *Medieval Technology and Social Change*, 83–84; Reynolds, *Stronger Than a Hundred Men*, 69–79; J. Harvey, *Medieval Craftsmen*, 97–99.

22. Klemm, *History of Western Technology*, 77, 85; White, *Medieval Technology and Social Change*, 85–89; J. Harvey, *Medieval Craftsmen*, 100.

23. Johannsen, *Geschichte des Eisens*, 28.

24. Le Goff, *Intellectuals in the Middle Ages*, 6.

25. For the discussion of and quotations from Jean's *Expositio*, I have relied on the translation and commentary in Thorndike, "John of St. Amand on the Magnet."

26. Crombie, *Augustine to Galileo* 1:133.

27. Libera, *La philosophie médiévale*, 418.

## Chapter 9—Nicolaus Cusanus and the Quantification of Magnetic Force

1. Cassirer, *The Individual and the Cosmos in Renaissance Philosophy*, 7.

2. Kristeller, *Renaissance Thought*, 58.

3. Le Goff, *Intellectuals in the Middle Ages*, 135.

4. Thorndike, *Science and Thought*, 134.

5. On the life of Nicolaus Cusanus, see Meuthen, *Nikolaus von Kues*.

6. Cusanus, *On Peaceful Unity of Faith*, 33.

7. *Of Learned Ignorance* is cited in the present chapter in parentheses in the text, with the abbreviation "OLI" followed by the book number, chapter number, and page number.

8. Bruno, *Infinite Universe and Worlds*, 311; brackets in source.

9. Oresme, *Book of the Heavens*, book 2, chapter 28 (pp. 600–609). See also Crombie, *Augustine to Galileo* 2:88–96; Grant, "Late Medieval Thought, Copernicus, and the Scientific Revolution," 202–11; idem, *Physical Science in the Middle Ages*, 64–69.

10. Bruno, *Infinite Universe and Worlds*, 307.

11. Ibid., 311.

12. Gilbert, *On the Magnet* 5.5:227.

13. Augustine, *Literal Meaning of Genesis* 1.4:107–111.

14. Cassiodorus, *Divine and Human Readings*, 142–43.

15. Cusanus, *Layman on Wisdom*, 74.

16. Cusanus, *On Actualized-possibility*, 81.

17. Cusanus, *Layman on Wisdom*, 22.

18. Cusanus, *De staticis experimentis*, 241–42.

19. Ibid., 250.

20. Ibid., 247.

21. Ibid., 242.

22. Ibid., 244.

23. Ibid., 242.

24. Dee, *Mathematicall Praeface*, biiijv.

25. Petty, preface to *Political Arithmetick*. See also Kargon, "William Petty's Mechanical Philosophy," 64.

26. Hales, introduction to *Vegetable Staticks*, xxxi.

27. Cusanus, *De staticis experimentis*, 249.

28. Cassirer, *Das Erkenntnisproblem* 1:314–18, 352–59.

29. Jammer, *Concepts of Force*, chapter 4.

30. Cusanus, *De staticis experimentis*, 249.

## Chapter 10—The Rediscovery of Things Ancient: Magic in the Early Renaissance Period

1. On Salutati and Bruni, see Garin, *Italian Humanism*, chapters 1 and 2, and idem, *Science and Civic Life*, chapter 1.

2. Ficino, *Philebus Commentary*, 272–73.

3. Quoted in Gurevich, *Categories of Medieval Culture*, 125.

4. R. Bacon, *Opus majus* 2.9:52; 4.2:129. See also ibid. 6.12:621; 2.9:53.

5. Ficino, *Commentary on Plato's Symposium*, 58 (English translation p. 155).

6. Copernicus, *On the Revolutions*, 5, 12, 22.

7. Hermes Trismegistus, *Hermetica* 1:194–95, 202–3. See also Boas, *Scientific Renaissance*, 69.

8. Hermes Trismegistus, *Hermetica* 1:204–5.

9. Kristeller, *Eight Philosophers*, 66.

10. Pico, *Dignity of Man*, 5.

11. Quoted in Shumaker, *Occult Sciences in the Renaissance*, 16–27. See also Walker, *Spiritual and Demonic Magic*, 58.

12. Febvre and Martin, *Coming of the Book*, 180–86.

13. Ibid., 112–13; Rossi, *Birth of Modern Science*, 42.

14. Eisenstein, *Printing Revolution*, 112.

15. See, for example, Yates, *Giordano Bruno*, 17–18, 80–81; P. French, *John Dee*, 83.

16. Thomas, *Religion and the Decline of Magic*, 228. See also Hansen, "Science and Magic," 483–84.

17. Pico, *Dignity of Man*, 26, 28.

18. Ficino, "Apologia," in *Three Books on Life*, 398.

19. See also Walker, *Spiritual and Demonic Magic*, part 1, chapter 2; part 2, chapter 4.

20. See Thorndike, *History of Magic and Experimental Science* 2:347.

21. Oresme, *Nicole Oresme and the Medieval Geometry of Qualities and Motions*, 337.

22. Ficino, "Apologia," in *Three Books on Life*, 396.

23. Ficino, *De vita*, book 3 ("De vita coelitus comparanda") is subsequently cited in the present chapter in parentheses in the text, with the abbreviation "TBL" followed by the chapter numbers and line numbers of *Three Books on Life*.

24. See Walker, *Spiritual and Demonic Magic*, 43, 222; Copenhaver, "Scholastic Philosophy and Renaissance Magic in the *De vita* of Marsilio Ficino," 523–54.

25. Ficino, *Commentary on Plato's Symposium*, 91 (English translation p. 199).

26. Ibid., 78 (English translation, p. 183).

27. On the life of Agrippa, see Seligmann, *History of Magic*, chapter 13 and Zilboorg, *History of Medical Psychology*, 201–6.

28. Nashe, *Unfortunate Traveller*, 299; Marlowe, *Tragical History of Doctor Faust*, scene 1, line 114.

29. See, for example, Zilboorg, *History of Medical Psychology*, 201–2; Sarton, *Six Wings*, 214–15; Michelet, *La sorcière* 2:50; Easlea, *Witch Hunting, Magic, and the New Philosophy*, 39.

30. *Three Books of Occult Philosophy* is cites in the present chapter in parentheses in the text, with the abbreviation “TBOP” followed by the chapter numbers and page numbers.

31. F. Bacon, *New Organon* 2.51:219.

32. Alberti, *Art of Building* 10.15:365.

33. Bruno, *On Magic*, 105.

34. Butterfield, *Origins of Modern Science*, 35.

35. Fontana, *Distorted Past*, 107.

## Chapter 11—The Age of Exploration and the Discovery of Magnetic Declination

1. *Carmina Burana*, 19, 6b.

2. De Lorris and De Meun, *Romance of the Rose*, 36, lines 1156–60.

3. De Bury, *Philobiblon*, 23.

4. Chaucer, *Parliament of Birds*, 132, lines 148–50.

5. Petrarch, *Canzoniere*, 129.

6. Pliny, *Natural History* 36.25, 7.56. See also ibid. 4.12.

7. Herodotus, *History* 7.42 (p. 224).

8. Cortes, *The Sphere and Art of Navigating*, 86; Norman, *Newe Attractive*, 1.

9. Haskins, *Renaissance of the Twelfth Century*, 316.

10. Polo, *Travels*, 239.

11. Pliny, *Natural History* 7.2.

12. Whitfield, *Image of the World*, 21.

13. Le Goff, *Time, Work, and Culture*, 189–200.

14. Ptolemy, *Ancient India as Described by Ptolemy*, 239.

15. Mandeville, *Mandeville’s Travels*, 188, 118.

16. *Sanct Brandan Seefahrt*, quoted in Fraunberger, *Elektrizität in Barock*, 17–18.

17. Spenser, *Faerie Queene*, book 2, cantos 12–14 (vol. 1, p. 314).

18. Goethe, *Sorrows of Young Werther*, 56.

19. Quoted in Benjamin, *Intellectual Rise in Electricity*, 155; J. Smith, “Precursors to Peregrinus,” 53, note 168.

20. Nordenskiöld, *Facsimile-Atlas*, 65, figure XXXII. See also Crone, introduction to Stevin, *Principal Works of Simon Stevin* 3:396, note 68.

21. Magnus, *Historia de gentibus septentrionalibus*, 126. The map is printed in Nordenskiöld, *Facsimile-Atlas*, 59, figure 32.

22. Da Orta, *Simples and Drugs of India*, 351.

23. Gilbert, *On the Magnet* 1.2:5; 4.1:153.

24. F. Bacon, *New Organon* 1.129:100.

25. Campanella, *City of the Sun*, 121.

26. Cardano, *Book of My Life*, 189–90.

27. Sherrington, *Endeavour of Jean Fernel*, 17. See also Sarton, *Six Wings*, 193.

28. F. Bacon, *New Organon* 1.72:60.

29. Aristotle, *Meteorology*, 362b7.

30. Ovid, *Metamorphoses*, 30.

31. Pliny, *Natural History* 2.67.

32. Origen, *On First Principles* 2.3:90.

33. Macrobius, *Dream of Scipio*, 201–2.

34. Pigafetta, *Magellan's Voyage*, 41.

35. Oviedo, *Natural History of the West Indies*, 14; my translation.

36. Gómara, *Historia General de las Indias* 1:16; my translation.

37. Da Orta, *Simples and Drugs of India*, 349.

38. Acosta, *Natural and Moral History of the Indies*, 89.

39. Vives, *On Education*, 8–9.

40. Sherrington, *Endeavour of Jean Fernel*, 17.

41. F. Bacon, *New Organon* 1.71:59.

42. Ibid. 1.93:78; 1.84:69. See also idem, *Advancement of Learning* 2.2:14.

43. Acosta, *Natural and Moral History of the Indies*, 53.

44. Poggendorff, *Geschichte der Physik*, 270; my translation.

45. On Bertelli's paper, see Mitchell, "Chapters in the History of Terrestrial Magnetism, Ch. 1," 138, note 103; Benjamin, *Intellectual Rise in Electricity*, 202; Krafft, "Sphaera activitatis–orbis virtuitis," 121; Singer, *Scientific Ideas to 1900*, 324.

46. Humboldt, *Cosmos* 2:656–57.

47. Mottelay, *Bibliographical History of Electricity and Magnetism*, 65–66.

48. Mitchell, "Chapters in the History of Terrestrial Magnetism, Ch. 2," 241–80; Heathcote, "Early Nautical Charts," 13–29.

49. Hellmann, "Die Anfänge der magnetischen Beobachatungen," 113; idem, "The Beginnings of Magnetic Observations," 74.

50. Motteley, *Bibliographical History of Electricity and Magnetism*, 65–66; Humboldt, *Cosmos* 2:657; Irving, *Christopher Columbus* 1:198.

51. Quoted in Mitchell, "Chapters in the History of Terrestrial Magnetism, Ch. 2," 253.

52. Quoted in ibid., 252.

53. Quoted in ibid., 253.

54. Quoted in ibid,. 254.

55. Falero, *The Sphere and the Art of Navigation*, 80–84.

56. The oldest known document containing a treatise on determining longitude by means of declination of a magnetic needle is said to be that written by Joãn de Lisboa in 1514, which I have not been able to consult. See Crone, introduction to Stevin, *Principal Works of Simon Stevin* 3:393.

57. Castro, *Magnetic Observations*, 187–98.

58. Hellmann, "Die Anfänge der magnetischen Beobachatungen," 125; idem, "The Beginnings of Magnetic Observations," 84.

59. Castro, *Magnetic Observations*, 197. See also Crone, introduction to Stevin, *Principal Works of Simon Stevin* 3:396.

60. Castro *Magnetic Observations*, 194.

61. G. Hartmann, "The Letter of Georg Hartmann to Duke Albrecht of Prussia," 128–30.

62. Cortes, *The Sphere and Art of Navigating*, 84–91.

63. Mercator, "The Letter of Gerhard Mercator to Antonius Perrenotus," 201–2.

64. Ibid., 201. On the prime meridian, see Mercator, "Text and Translation of the Legends of the Original Chart of the World by Gerhard Mercator, issued in 1569," 20–21.

65. Hellmann, "Die Anfänge der magnetischen Beobachatungen," 119; idem, "The Beginnings of Magnetic Observations," 80.

66. Crone, introduction to Stevin, *Principal Works of Simon Stevin* 3:397.

67. Nordenskiöld, *Facsimile-Atlas*, 90, figure 60, figure XLVII; Campbell, *Early Maps*, 22–23.

68. Quoted in Crone, introduction to Stevin, *Principal Works of Simon Stevin* 3:396.

69. Bruno, *On Magic*, 121–22.

## Chapter 12—Robert Norman and *The Newe Attractive*

1. G. Hartmann, "The Letter of Georg Hartmann to Duke Albrecht of Prussia," 128.

2. *The Newe Attractive* is cited in the present chapter in parentheses in the text, with the abbreviation "TNA" followed by numbers referring to chapters and pages.

3. See figure 12.1.

4. Mitchell, "Chapters in the History of Terrestrial Magnetism, Ch. 3," 77; 79, note 2.

5. Montaigne, *Complete Essays*, 644.

6. Thevet, *Newe Found Worlde*, 138.

7. Cartier, *Navigations to Newe France*.

8. Gilbert, *On the Magnet* 4.6:162.

9. Pumfrey, "O tempora, O magnes! A Sociological Analysis of the Discovery of Secular Magnetic Variation in 1634," 191–92; E. G. R. Taylor, *Tudor Geography*, 157–58.

10. Stevin, *Principal Works of Simon Stevin* 3:422–23; idem, *Haven-finding Art*, 1.

11. C. Hill, *Intellectual Origins of the English Revolution*, 15.

12. Nef, *La naissance de la civilisation industrialle et le monde contemporain*, chapter 2; Farrington, *Francis Bacon*, 13.

13. Merton, *Social Theory and Social Structure*, 666.

14. Sombart, *Krieg und Kapitalismus*, 47–49.

15. Borough, preface to *Variation of the Cumpas*.

16. Simpkins, "Early Editions of Euclid in England," 230; Singer, *Scientific Ideas to 1900*, 194.

17. H. Gilbert, *Queene Elizabethes Academy*, 10.

18. Rossi, *Francis Bacon*, 6–7; idem, *Philosophy, Technology, and the Arts in the Early Modern Era*, 10; E. G. R. Taylor, *Mathematical Practitioners*, 321.

19. D. E. Smith, "New Information respecting Robert Recorde," 296.

20. Easton, "The Early Editions of Robert Recorde's *Ground of Artes*," 515–32; Johnson and Larkey, "Robert Recorde's Mathematical Teaching and the Anti-Aristotelian Movement," 59–87.

21. Recorde, *Castle of Knowledge*, 164–65. See also Patterson, "Recorde's Cosmology, 1556 ," 208–18; Johnson and Larkey, "Robert Recorde's Mathematical Teaching and the Anti-Aristotelian Movement."

22. P. French, *John Dee*, 172.

23. Ibid., 40. On Dee's association with Mercator, see Crane, *Mercator*,164–69.

24. C. Smith, *John Dee*, 1.

25. P. French, *John Dee*, 1.

26. Yates, *Theatre of the World*, 18.

27. Rossi, *Francis Bacon*, 40; C. Hill, *Intellectual Origins of the English Revolution*, 37.

28. See, for example, Verger, *Les universités au moyan âge*, chapter 6, section 2; C. Hill, *Intellectual Origins of the English Revolution*, chapter 2, section 4, and Appendix.

29. Quoted in Bennett, "The Mechanics' Philosophy and Mechanical Philosophy," 10.

30. Dee, *Mathematicall Praeface*, Aiiiv, Aiiiir.

31. E. G. R. Taylor, *Mathematical Practitioners*, 170; C. Hill, *Intellectual Origins of the English Revolution*, 16.

**Chapter 13—Mining and the Continued Peculiarity of Magnetism**

1. Palissy, *Admirable Discourse*, 26. See also ibid., 23–24.
2. Stevin, *Principal Works of Simon Stevin* 3:608.
3. Febvre and Martin, *Coming of the Book*, 105, 108.
4. Gilmont, "Protestant Reformations and Reading," 214.
5. Febvre and Martin, *Coming of the Book*, 109, 249.
6. Gilmont, "Protestant Reformations and Reading," 214.
7. From Stone, "The French Language in Renaissance Medicine," 322.
8. Packard, *Ambroise Pare*, 106, 110; C. Hill, *Intellectual Origins of the English Revolution*, 75, note 2; 28–31.
9. Sarton, *Six Wings*, 110. See also O'Malley, *Andreas Vesalius of Brussels*, 44; Febvre and Martin, *Coming of the Book*, 329–30.
10. Kaplan, *Robert Recorde*, 92.
11. Davis, *Society and Culture in Early Modern France*, 220.
12. Cassirer, *The Individual and the Cosmos in Renaissance Philosophy*, 56–57.

13. Biringuccio, *The PIROTECHNIA of Vannoccio Biringuccio*, 21–22. In the present chapter, this work is subsequently cited in parentheses in the text, with the abbreviation "PVB" followed by page numbers.

14. Vives, *On Education*, 209.

15. Paracelsus, *Selected Writings*, 57, 61.

16. See Nef, *La naissance de la civilisation industrielle et le monde contemporain*, 115. Nef relates that he was informed of this episode by Herbert Clark Hoover, one of the English translators of *De re metallica* (and later president of the United States).

17. *De re metallica* was translated from the first Latin edition (1556) into English in 1912 by Herbert Clark Hoover (later president of the United States) and his wife Lou Henry Hoover, an eminent Latin scholar. Their translation was republished in 1950 by Dover Publishing. The copious notes appended to the Hoovers' translation, which cover all the previous literature and history, are highly valued by geologists (Bromehead, "Geology in Embryo," 113). In 1928, following a proposal by Conrad Matschoss, a science historian and the director of the Verein Deutscher Ingenieure (Association of German Engineers), a German translation of the 12th book of *De re metallica* was published through the collaboration of many scholars. And in 1959, a facsimile reprint of a 1561 Latin edition was published in Italy. In the present chapter, subsequent citations of *De re metallica* are based on the English translation and are given in parentheses in the text, with the abbreviation "DRM" followed by page numbers.

18. Pliny, *Natural History* 34.49.

19. Origen, *On First Principles* 3.1:158.

20. D'Anghiera, *Decades of the Newe Worlde*, 133r.

21. Paracelsus, *Concerning the Nature of Things*, 182.

22. Ercker, *Ores and Assaying*, 10–11.

23. Boyle, The Works of the Honourable Robert Boyle, 4: 79–84.

24. Aepinus, *Essay on the Electricity and Magnetism*, 430.

25. Quoted in Holmyard, *Alchemy*, 135.

26. Biringuccio, *The PIROTECHNIA of Vannoccio Biringuccio*, 19; Agricola, *De re metallica*, 141–48, 247; Ercker, *Ores and Assaying*, 287.

27. In the present chapter, citations of *De natura fossilium* are based on the English translation in The Geological Society of America, Special

Paper No. 63 (1955) and are given in parentheses in the text with the abbreviation "DNF" followed by page numbers of this translation.

28. Rossi, *Francis Bacon*, 4.

## Chapter 14—Paracelsus and Magnet Therapy

1. O'Malley, *Andreas Vesalius*, 50.

2. From Paracelsus's *Programm der Baseler Vorlesung*. This English translation is from Pachter, *Magic into Science*, 152–53.

3. Bruno, *Cause, Principle and Unity*, 52–53.

4. Paracelsus, *Seven Defensiones*, 34.

5. Bullough, "Population and the Study and Practice of Medieval Medicine," 63.

6. This English translation is from Paracelsus, *Selected Writings*, 4.

7. Paracelsus, *Sämtliche Werke* 10:115.

8. This English translation is from Stillman, *Paracelsus*, 76.

9. From Paracelsus's *Von den tartarischen Krankheiten*. This English translation is from Pachter, *Magic into Science*, 7.

10. This English translation is from Paracelsus, *Selected Writings*, 55.

11. Paracelsus, *On the Miners' Sickness*, 57, 91.

12. Paracelsus, *Seven Defensiones*, 29; the passage from *The Doctor's Labyrinth* is from idem, *Selected Writings*, 52.

13. F. Bacon, *Advancement of Learning* 2.10:115.

14. This English translation is from Paracelsus, *Selected Writings*, 59–60.

15. C. Webster, *From Paracelsus to Newton*, 26. See also Paracelsus, *The Prophecies of Paracelsus*.

16. Cassirer, *The Individual and the Cosmos in Renaissance Philosophy*, 112. See also Pagel, *Paracelsus*, 218–23; Pagel, "Paracelsus and the Neoplatonic and Gnostic Tradition," 125–66; and Garin, "The Philosopher and the Magus," 148.

17. This English translation is from Paracelsus, *Selected Writings*, 39.

18. This English translation is from Paracelsus, *Selected Writings*, 21.

19. Paracelsus, *Volumen paramirum*,in *Paracelsus Sämtliche Werke*, 1: 211–12.

20. From Paracelsus's *Volumen paramirum*. This English translation is from Paracelsus, *Selected Writings*, 40.

21. This English translation is from Paracelsus, *Selected Writings*, 61.

22. From Paracelsus's *The Doctor's Labyrinth*. This English translation is from Paracelsus, *Selected Writings*, 143.

23. From Paracelsus's *Paragranum*. This English translation is from Paracelsus, *Hermetic and Alchemical Writings of Paracelsus* 2:148.

24. These English translations are from Stillman, *Paracelsus*, 52, and Pachter, *Magic into Science*, 136. See also Paracelsus's *Paragranum* in *Hermetic and Alchemical Writings of Paracelsus* 2:151.

25. Paracelsus, *Seven Defensiones*, 14.

26. These English translations are from Pachter, *Magic into Science*, 125, and Paracelsus, *Selected Writings*, 60–61.

27. From Paracelsus's *Paragranum*. This English translation is from Paracelsus, *Hermetic and Alchemical Writings of Paracelsus* 2:151.

28. Thomas, *Religion and the Decline of Magic*, 228. See also Sarton, *Six Wings*, 180.

29. Paracelsus, *Seven Defensiones*, 26–27.

30. From Paracelsus's *Caelum philosophorum*. This English translation is from Paracelsus, *Hermetic and Alchemical Writings of Paracelsus* 1:17.

31. Paracelsus, *Astronomia magna*, 107–8.

32. Plato, *Timaeus*, 81A.

33. From Paracelsus's *Von den natürlichen Dingen* (*Concerning the Nature of Things*). This English translation is from Paracelsus, *Hermetic and Alchemical Writings of Paracelsus* 2:132.

34. Chaucer, *Canterbury Tales*, 476; Gower, *Confessio amantis*, volume 4, line 247.

35. See Paracelsus, *Volumen paramirum*, in *Paracelsus Sämtliche Werke* 1:211–12, 214–15; idem, *Hermetic and Alchemical Writings of Paracelsus* 1:186–87; 2:150; and Holmyard, *Alchemy*, 156.

36. Paracelsus, *Herbarius*, 123; brackets in source.

37. Pachter, *Magic into Science*, 104.

38. Paracelsus, *Herbarius*, 123.

39. Ibid., 123–24; brackets in source.

40. Ibid., 125; brackets in source.

41. From Bartholomew's *On the Properties of Things*. This English translation is from Grant. ed., *A Source Book in Medieval Science*, 368.

42. Beckmann, *Inventions, Discoveries and Origins* 1:43–47.

43. From Paracelsus's *Von den natürlichen Dingen* (*Concerning the Nature of Things*). This English translation is from Paracelsus, *Hermetic and Alchemical Writings of Paracelsus*, 158, 140.

44. Ibid., 131–32. See also ibid., 145.

45. Ibid., 132–3.

46. See, for example, Singer, *Scientific Ideas to 1900*, chapters 4–7, especially p. 214.

47. Montaigne, *Complete Essays* 2.12:642–43. See also Debus, *Man and Nature in the Renaissance*, 20–33.

48. Quoted in Debus, *Chemical Philosophy*, 149

49. Quoted in ibid., 178.

50. Quoted in F. Hartmann, *Paracelsus*, 28.

51. See Pagel, *Paracelsus*, 95, note 251; and idem, "The 'Wild Spirit' (Gas) of John Baptist van Helmont (1579–1644) and Paracelsus," 1–13.

52. J. Webster, *Duchess of Malfi*, act 5, scene 2; Jonson, *Alchemist*, act 2, scene 3.

53. Rossi, *Birth of Modern Science*, 140.

54. W.Harvey, *Anatomical Disquisition*, 286; Debus, *Man and Nature in the Renaissance*, 69–73.

55. Haggard, *Devils, Drugs, and Doctors*, 330.

56. Paracelsus, *Diseases that Deprive Man of his Reason*, 153.

57. Quoted in Duhem, *Aim and Structure of Physical Theory*, 234.

58. Quoted in Debus, *English Paracelsians*, 121. See also Pagel, *Paracelsus*, 343.

59. Quoted in Debus, *Chemical Philosophy*, 280.

60. This paragraph is based mainly on Debus, *Chemical Philosophy*, 1.

61. Boas, "Review of *Natural Magick*," 76.

## Chapter 15—Changes in Magical Thought during the Late Renaissance

1. Hugh of Saint Victor, *Didascalicon*, 154.

2. Walker, *Spiritual and Demonic Magic*, 36, 83–84; Burke, *Renaissance Italy 1420–1540*, 179.

3. Thomas, *Religion and the Decline of Magic*, 255, 49.

4. R. Bacon, *Opus majus* 6:631.

5. Ibid. 6:632–33. See also ibid. 6:587.

6. *Roger Bacon's Letter*, 38.

7. R. Bacon, *Opus majus* 6:630.

8. Yates, *Giordano Bruno*, 131. See also Shumaker, *Occult Sciences in the Renaissance*, 134–35.

9. Quoted in Rossi, *Francis Bacon*, 18–19.

10. Kristeller, *Eight Philosophers*, 87–88.

11. Grafton, *Cardano's Cosmos*, 73.

12. My translations, based on Pomponazzi, *De naturalium effectuum causis sive de Incantationibus*, 230 (French translation, p. 225) and 98 (French translation, p. 162).

13. My translation, based on ibid., 22–23 (French translation, p. 121).

14. Ibid., 242, 23, 42 (French translation, pp. 231, 122, 133).

15. My translation, based on ibid., 73–74 (French translation, p. 151).

16. Walker, *Spiritual and Demonic Magic*, 164.

17. Garin, "The Philosopher and the Magus," 123. See also Copenhaver, "Did Science Have a Renaissance?"

18. Cassirer, *The Individual and the Cosmos in Renaissance Philosophy*, 81.

19. Thomas, *Religion and the Decline of Magic*, 54.

20. Easlea, *Witch Hunting, Magic, and the New Philosophy*, 110.

21. Scot, *Discoverie of Witchcraft*, 236–37. See also ibid., 234–35.

22. Ibid., 237–38.

23. R. Bacon, *Opus majus* 4:129.

24. Crombie, *Augustine to Galileo* 1:68.

25. Thomas Aquinas, *Summa theologica*, 2.2, question 96, article 2.

26. Ficino, *Three Books on Life* 3.12:301.

27. Biringuccio, *The PIROTECHNIA of Vannoccio Biringuccio*, 114.

28. Crombie, *Styles of Scientific Thinking* 1:332. See also Wolff, *Cultural Awakening*, 284.

29. P. French, *John Dee*, 162.

30. Quoted in Thorndike, *History of Magic and Experimental Science* 4:501.

31. R. Bacon, *Opus majus* 4:133.

32. Agrippa, *Occult Philosophy*, 71, 60.

33. Albert the Great, *The Book of Secrets of Albertus Magnus*, 82.

34. Boas, *Scientific Renaissance*, 21.

35. Thorndike, *History of Magic and Experimental Science* 1:469; ibid. 2:977.

36. C. Smith, *John Dee*, 244. See also Johnson, *Astronomical Thought*, 80.

37. Clulee, "John Dee's Mathematics and the Grading of Compound Qualities," 197; idem, "Astrology, Magic, and Optics," 642, 670–73, 678. See also idem, *John Dee's Natural Philosophy*, 64–70.

38. *Propaedeumata aphoristica* is cited in the present chapter with roman numerals identifying the relevant aphorisms.

39. C. Smith, *John Dee*, 243.

40. Clulee, "Astrology, Magic, and Optics," 677. See also idem, *John Dee's Natural Philosophy*, 67.

41. *The Mathematicall Praeface* is cited in the present chapter in parentheses in the text, with the abbreviation "TMP" followed by page references.

42. Crombie, *Styles of Scientific Thinking* 1:507.

43. Clulee, "At the Crossroads of Magic and Science," 59 and note 9. See also idem, *John Dee's Natural Philosophy*, 170–76 and 286, note 72.

44. Debus, *Man and Nature in the Renaissance*, 8; idem, introduction to John Dee, *Mathematicall Praeface*, 21–22.

45. Boas, *Scientific Renaissance*, 185.

46. P. French, *John Dee*, 142.

47. Yates, *Theatre of the World*, 5.

48. *The Book of My Life* is cited in the present chapter in parentheses in the text, with the abbreviation "BML" followed by page numbers.

49. Thorndike, *History of Magic and Experimental Science* 5:573.

50. Cardano, *De subtilitate rerum*, 360–61.

51. From Cardano's *De subtilitate rerum*, 443–44. This English translation is from Benjamin, *Intellectual Rise in Electricity*, 248.

52. From Cardano's *De subtilitate rerum*, 444. This English translation is from Benjamin, *Intellectual Rise in Electricity*, 249.

53. From Cardano's *De subtilitate rerum*, 357. This English translation is from Cass, *The First Book of Jerome Cardan's* De subtilitate, 75.

54. Bruno, *Infinite Universe and Worlds*, 285.

55. Bruno, *On Magic*, 105.

56. Ibid., 120.
57. Ibid., 121.
58. Bruno, *Cause, Principle and Unity*, 5.
59. Bruno, *Infinite Universe and Worlds*, 372.
60. Rossi, *Birth of Modern Science*, 26.

## Chapter 16—Della Porta's Investigations into Magnetism

1. Febvre and Martin, *Coming of the Book*, 115, 249.
2. De Villamil, *Newton*, 92.
3. Febvre and Martin, *Coming of the Book*, 180.
4. Della Porta, *Natural Magick*, 160. Unless otherwise indicated, in the present chapter the title *Natural Magick* denotes the second edition, which is subsequently cited in parentheses in the text, with the abbreviation "NM" followed by book and chapter numbers.
5. Eamon, "Books of Secrets in Medieval and Early Modern Science," 37; idem, *Science and the Secrets of Nature*, chapter 9. See also Nicolson, *Dictionary of the History of the Ideas*, s.v. "Virtuoso."
6. Burckhardt, *Civilization of the Renaissance in Italy*, 319.
7. Cassirer, *The Individual and the Cosmos in Renaissance Philosophy*, 63.
8. Burckhardt, *Civilization of the Renaissance in Italy*, 319.
9. Shumaker, *Occult Sciences in the Renaissance*, 111.
10. Ibid., 118.
11. Crombie, *Science, Optics and Music*, 176.
12. Rosenberger, *Die Geschichte der Physik* 1:139.
13. Cassirer, *The Individual and the Cosmos in Renaissance Philosophy*, 152. See also Boas, *Scientific Renaissance*, 189.
14. Lindberg, *Theory of Vision*, 184–85; Wolf, *History of Science, Technology and Philosophy*, 249; Crombie, *Science, Optics and Music*, 218–19, 226, 290; Poggendorff, *Geschichte der Physik*, 169.
15. Galileo, *Sidereal Messenger*, 36.
16. Kepler, *Kepler's Conversation with Galileo's Sidereal Messenger*, 15.
17. Crombie, *Styles of Scientific Thinking* 1:655.
18. Kay, *William Gilbert's Renaissance Philosophy of the Magnet*, 50.
19. Boas, "Review of *Natural Magick*," 76.

20. Gilbert, *On the Magnet* 1.1:6. On the English translation of this book, see note 4 chapter 17.

21. The English translations from the first edition of *Natural Magick* are mine, based on the 1572 Italian edition, *De i miracoli et maravigliosi effetti dalla natura prodotti*, 80–82, with some wording based on book 7 of the English translation of the second edition used throughout the present chapter.

22. Gilbert, *On the Magnet* 2.3:64.

23. Della Porta seems to confuse Thales with Anaxagoras.

24. Mottelay, *Bibliographical History of Electricity and Magnetism*, 74; Thorndike, *History of Magic and Experimental Science* 6:420; Boas, *Scientific Renaissance*, 188–89; idem, "Review of *Natural Magick*," 76.

25. Elliott, *Electromagnetics*, 398.

26. Gilbert, *On the Magnet* 1.1:2.

27. Ibid. 3.13:143.

28. Ibid. 1.12:30.

29. Ibid. 3.13:143.

30. Ibid. 2.4:66–67.

31. Ibid. 2.23:91.

32. Ibid. 2.25:93.

33. Barlow, *Magneticall Advertisements*, 6.

34. Quoted in Kay, *William Gilbert's Renaissance Philosophy of the Magnet*, 137.

35. Abromitis, *William Gilbert as Scientist*, Appendix I.

36. King, "The Natural Philosophy of William Gilbert and His Predecessors," 124.

37. Browne, *The Works of Sir Thomas Browne* 2:58–59.

38. Jammer, *Concepts of Force*, 76.

39. Sarton, *Six Wings*, 96,

40. Ibid., 86.

41. Rossi, *Francis Bacon*, 28; idem, *Birth of Modern Science*, chapter 2; Yates, *Rosicrucian Enlightenment*, 120.

42. Kepler to Mästlin, 8 February 1601, in Baumgardt, *Johannes Kepler: Life and Letters*, 64.

43. Stevin, *Principal Works of Simon Stevin* 3:610–11.

44. C. Webster, *From Paracelsus to Newton*, 60.

45. Mach, *Principles of Physical Optics*, 12.

46. C. Webster, *From Paracelsus to Newton*, 59.

47. Eamon, "Books of Secrets in Medieval and Early Modern Science," 40–41.

48. Thorndike, *History of Magic and Experimental Science* 2:267.

49. Quoted in Eamon, "Books of Secrets in Medieval and Early Modern Science," 30–31; idem, *Science and the Secrets of Nature*, 47.

50. *Roger Bacon's Letter*, 38–41. See also idem, *Opus majus* 6:621.

51. Pico, *Dignity of Man*, 30.

52. Eamon, "Books of Secrets in Medieval and Early Modern Science," 30–31. See also Rossi, *Birth of Modern Science*, 18–22.

53. Boas, *Scientific Renaissance*, 183.

## Chapter 17—William Gilbert's *On the Magnet*

1. Such accounts are uncritically retold in, for example, Harré, "William Gilbert" and Suter, "A Biographical Sketch of Dr. William Gilbert of Colchester."

2. Langdon-Brown, "William Gilbert: His Place in the Medical World," 136–39.

3. Thompson, "The Family and Arms of Gilbert of Colchester," 200; D. H. D. Roller, *"De magnete" of William Gilbert*, 64.

4. The first (1600) edition of *De magnete* (*On the Magnet*) was reprinted in 1892 in Berlin and again in 1967 in Brussels, the latter by the publisher Culture et Civilisation. It was first translated into English by Paul Fleury Mottelay as *De Magnete*, published in New York in 1893. Several years later, a second English translation, by Silvanus Thompson and titled *On the Magnet*, was distributed in a limited-number edition of 250 copies to celebrate the 300th anniversary of the original publication. The Thompson translation is so meticulous that it reproduces even the pagination of the original Latin edition. Although the title page has "ANNO MCM" (1900) in imitation of the "ANNO MDC" (1600) on the original edition, in reality the Thompson translation appeared in 1901. It was published to commemorate the founding of the Gilbert Club in 1889. Mottelay's translation was republished in 1952 by Encyclopaedia Britannica and in 1958 by Dover Publications. Thompson's translation was

also reprinted in 1958 by Basic Books in New York. In the present chapter, subsequent citations of *On the Magnet* are from the Thompson translation and, where not specified in the text, are given in parentheses, with the abbreviation "OM" followed by the relevant chapter and page numbers.

5. F. Bacon, *Advancement of Learning* 1.3.9:23.

6. Chapman, "William Gilbert and the Science of his Time," 134.

7. Zilsel, "Origins of William Gilbert's Scientific Method," 19.

8. Benjamin, *Intellectual Rise in Electricity*, 279.

9. Whewell, *History of the Inductive Sciences*, part 3, book 12, 1:37; Tait, *Lectures on Some Recent Advances in Physical Science*, 12; Wolf, *History of Science, Technology and Philosophy* 1:293. See also Suter, "A Biographical Sketch of Dr. William Gilbert of Colchester," 255.

10. King, "The Natural Philosophy of William Gilbert and His Predecessors," 124.

11. Jones, *Ancients and Moderns*, 16.

12. Boas, *Scientific Renaissance*, 191.

13. Rossi, *Birth of Modern Science*, 150.

14. Kay, *William Gilbert's Renaissance Philosophy of the Magnet*, 87.

15. F. Bacon, *Advancement of Learning* 1.5.7:35; idem, *New Organon* 1.54:46.

16. Hall, *Revolution in Science*, 257.

17. From Cardano's *De subtilitate*, 444. This English translation is from Benjamin, *Intellectual Rise in Electricity*, 249.

18. Priestley, *History and Present State of Electricity*, 5.

19. Kay, *William Gilbert's Renaissance Philosophy of the Magnet*, 91.

20. Johannsen, *Geschichte des Eisens*, chapter 7, section 8.

21. Quoted in Mottelay, *Bibliographical History of Electricity and Magnetism*, 90.

22. Hesse, "Gilbert and the Historians," 130.

23. Mitchell, "Chapters in the History of Terrestrial Magnetism, Ch. 1," 130, note 1; Sarton, *Six Wings*, 96–97; Fleming, "The Earth, a Great Magnet," 49; Chapman, "William Gilbert and the Science of His Time," 135.

24. Burtt relates that "The power of this magnetic soul to act at a distance, he [Gilbert] explained by the conception of a magnetic effluvium emitted by the loadstone"; and Butterfield gives a similar account. See

Burtt, *Metaphysical Foundations*, 158; and Butterfield, *Origins of Modern Science*, 142. Gilbert, however, not only did not offer such an explanation but indeed explicitly negated that view. See Gilbert, *On the Magnet*, 66.

25. Digges, *A Perfit Description of the Caelestiall Orbes*, 78. See also Johnson, *Astronomical Thought*, 164–70.

26. Quoted in Boas and Hall, "Tycho Brahe's System of the World," 258.

27. Bruno, *Infinite Universe and Worlds*, 302.

28. Freudenthal, "Theory of Matter and Cosmology in William Gilbert's *De magnete*," 22–37.

29. White, *Medieval Religion and Technology*, 56; Johnson, *Astronomical Thought*, 216.

30. Quoted in Bennett, "Cosmology and the Magnetical Philosophy," 171, and in idem, *Mathematical Science of Christopher Wren*, 59.

31. Stevin, *Principal Works of Simon Stevin* 3:129.

32. Ridley, *Magneticall Bodies and Motions*, 13–14.

33. Johnson, *Astronomical Thought*, 274.

34. Aristotle, *On the Soul* 2.1.412b17.

35. Bruno, *Infinite Universe and Worlds*, 328.

36. Lindsay, "William Gilbert and Magnetism in 1600," 281.

37. Chalmers, "The Lodestone and the Understanding of Matter in Seventeenth Century England," 86.

38. Boas, *Scientific Renaissance*, 190, 194.

39. Kearney, *Science and Change*, 110. See also ibid., 21–22.

40. Feingold, *Mathematicians' Apprenticeship*, 10.

41. Galileo, *Sidereal Messenger*, 57.

42. Zilsel, "The Origins of William Gilbert's Scientific Method," 9.

43. Galileo, *Dialogue*, 406.

44. Kay, *William Gilbert's Renaissance Philosophy of the Magnet*, Preface.

45. See Ridley, *Magneticall Animadversions*, 10, and Abromitis, *William Gilbert as Scientist*, 113.

## Chapter 18—Johannes Kepler and the Magnetical Philosophy

1. Nicolson, *Voyages to the Moon*, 23.

2. The edition of this work that I consulted, titled *Mysterium Cosmographicum: The Secret of the Universe*, presents an English translation alongside the Latin text of the original 1621 edition. In the present chapter, this work is cited in parentheses in the text, with the abbreviation "SU" followed by the relevant chapter numbers and page numbers of this English translation.

3. Copernicus, *On the Revolutions*, preface, 5.

4. Ibid. 1.10:22.

5. Baumgardt, *Johannes Kepler: Life and Letters*, 41.

6. Quoted in Duhem, *To Save the Phenomena*, 5.

7. Plato, *Timaeus* 1.11, 38C–39D.

8. Aristotle, *On the Heavens*, book 2, chapter 3.

9. Ptolemy, *Almagest* 9.2:420; brackets in source. See also ibid. 3.1:140.

10. Quoted in Holton, "Johannes Kepler's Universe," 345.

11. Copernicus, *On the Revolutions* 1.10:22, 1.9:18.

12. Ibid. 1.4:10–11, 4.2:176.

13. Osiander, "To the Reader Concerning the Hypotheses of this Work," in ibid., XX.

14. Campanella, *City of the Sun*, 109.

15. This English translation is from Jardine, *Birth of History and Philosophy of Science*, 144–45 (English). For the corresponding passage from the original Latin work (*Apologia pro Tychone contra Ursum*), see ibid., 92–93.

16. *New Astronomy* is the title of the English translation of Kepler's *Astronomia nova*. The English edition is subsequently cited in the present chapter in parentheses in the text, with the abbreviation "NA" followed by the relevant chapter and page numbers.

17. Quoted in Jardine, *Birth of History and Philosophy of Science*, 250. In the present chapter, subsequent citations of books 4 and 5 of Kepler's *Epitome of Copernican Astronomy* refer to the English translation in *Great Books of the Western World* 16 and are given in parentheses in the text, with the abbreviation "ECA" followed by the relevant page numbers.

18. Applebaum, "Keplerian Astronomy after Kepler," 460; Duhem, *Aim and Structure of Physical Theory*, 246.

19. Gilbert, *On the Magnet* 4:230.

20. Ibid. 4:219, 227.

21. Ibid. 4:232.

22. This English translation is from Kelly, *The "De mundo" of William Gilbert*, 67. See also Thoren, *Lord of Uraniborg*, 86.

23. Gilbert, *On the Magnet* 4:224.

24. Jones, *Ancients and Moderns*, 20.

25. Gilbert, *On the Magnet* 4:231.

26. Kelly, *The "De mundo" of William Gilbert*, 70.

27. F. Bacon, *New Organon* 2:36, 158.

28. Godwin, *Man in the Moone*, 65.

29. Quoted in Birch, *History of the Royal Society of London* 2:70. See also the letter of Hooke to Boyle, March 21, 1666, in Boyle, *Works* 4:505–8.

30. Butterfield, *Origins of Modern Science*, 141.

31. This English translation is from Jammer, *Concepts of Force*, 84–85. For the original, see *Johannes Kepler Gesammelte Werke*, book 16, number 508.

32. This English translation is from Jardine, *Birth of History and Philosophy of Science*, 146. For the corresponding passage from the original Latin work (*Apologia pro Tychone contra Ursum*), see ibid., 94.

33. This English translation is from Pumfrey, "Magnetical Philosophy and Astronomy, 1600-1650," 48; brackets in source. For the original, see *Johannes Kepler Gesammelte Werke*, book 14, number 242.

34. *Johannes Kepler Gesammelte Werke*, book 15, number 357.

35. Leibniz, *Mathematische Schriften* 6:147.

36. Gingerich, "Kepler's Place in Astronomy," 264; Barbour, *Discovery of Dynamics*, 118, 242, 283–86.

37. This English translation is from Kepler, *Optics*, 19–20. For the original, see *Astronomiae pars optica*, in *Johannes Kepler Gesammelte Werke*, book 2, page 19.

38. This English translation is from Holton, "Johannes Kepler's Universe," 342; brackets in source. For the original, see *Johannes Kepler Gesammelte Werke*, book 15, number 325.

39. *Johannes Kepler Gesammelte Werke*, book 15, number 335, lines 93–95; my translation. The original reads: "Solis corpus est circulariter magneticum et convertitur in suo spacio, transferens orbem virtuitis suae, quae non est attractoria sed promotoria."

40. This English translation is from Baldwin, "Magnetism and the Anti-Copernican Polemic," 156. For the original, see *Johannes Kepler Gesammelte Werke*, book 16, number 463.

41. Bennett, "Hooke and Wren and the System of the World," 36; idem, "Cosmology and the Magnetical Philosophy," 171; idem, *Mathematical Science of Christopher Wren*, 59–60.

42. Euler, *Mechanica*, in *Leonhardi Euleri Opera Omnia*, second series, 1:31–32.

43. *Johannes Kepler Gesammelte Werke*, book 15, number 358, lines 49–51.

44. Leibniz, *Mathematatische Schriften* 6:175.

45. This English translation is from Jammer, *Concepts of Force*, 82. For the original, see *Johannes Kepler Gesammelte Werke*, book 15, number 340.

46. This English translation is from Rosen, *Kepler's Somnium*, 221 (Appendix H); brackets in source. For the original, see *Johannes Kepler Gesammelte Werke*, book 15, number 358.

47. Kepler's original notes to *Somnium* are subsequently cited in parentheses in the text, with the abbreviation "N" followed by the number of the relevant note.

48. Quoted in Koestler, *Sleepwalkers*, 243. See also Caspar, *Kepler*, 181–85.

49. See Rosen, *Kepler's Conversation with Galileo's Sidereal Messenger*, 140.

50. Dee, *Mathematicall Praeface*, bjv, biijr.

51. See Jammer, *Concepts of Force*, chapter 4, and Thorndike, "The True Place of Astrology in the History of Science."

52. Dee, *Mathematicall Praeface*, biiijr.

53. Duhem, *Aim and Structure of Physical Theory*, 240.

54. Crombie, *Augustine to Galileo* 2:36, 140.

## Chapter 19—Seventeenth-century Mechanism and Notions of Force

1. Mendelssohn, *Science and Western Domination*, 74; Kearney, *Science and Change*, 153.

2. Galileo, *Assayer*, 276.

3. Descartes, *Principles of Philosophy* 1.71:32–33, 2.4:40.

4. Quoted in Duhem, *Aim and Structure of Physical Theory*, 122.

5. Pyle, *Atomism and Its Critics*, 142.

6. Galileo, *Assayer*, 241.

7. Descartes, *Principles of Philosophy* 4.187:275.

8. Pliny, *Natural History* 2.99; Dante, *Divine Comedy: Paradise*, canto 16, lines 82–83; Gilbert, *On the Magnet* 2.16:86; F. Bacon, *New Organon* 2.48:198; Della Porta, *Natural Magick* 1.88. See also Darwin, *Tides and Kindred Phenomena*, chapter 4.

9. Stevin, *Principal Works of Simon Stevin* 3:332.

10. Gower, *Confessio amantis*, volume 7, lines 721–24.

11. Shakespeare, *The Winter's Tale*, act 1, scene 2, line 428.

12. Galileo, *Dialogue*, 442.

13. Wallace, *Galileo and His Sources*, 294; Drake, *Telescopes, Tides, and Tactics*, 206–7.

14. F. Bacon, *New Organon* 2.36:159.

15. Galileo, *Dialogue*, 462, 445.

16. Jammer, *Concepts of Force*, 46, 56; Duhem, *Aim and Structure of Physical Theory*, 234; Shumaker, *Occult Sciences in the Renaissance*, 30.

17. Ficino, *Commentary on Plato's Symposium*, 200.

18. P. French, *John Dee*, 92.

19. Galileo, *Dialogue*, 19, 242.

20. Casper, *Kepler*, 109.

21. Mach, *Science of Mechanics*, 231.

22. Galileo, *On Motion*, chapter 3; idem, *Letters on Sunspots*, 113.

23. Galileo, *Dialogue*, 234. See also ibid., 410, and Duhem, *Aim and Structure of Physical Theory*, 121–22.

24. Galileo, *Dialogue*, 34. See also Crombie, *Augustine to Galileo* 2:165.

25. Galileo, *Two New Sciences*, 77.

26. Ibid., 158–59.

27. Galileo, *Letters on Sunspots*, 123–24.

28. Galileo, *Assayer*, 237.

29. Casper, *Kepler*, 136.

30. Descartes, *Philosophical Works of Descartes* 1:299. Note, however, that subsequent citations of the *Principles of Philosophy* in the present chapter are to the Miller and Miller translation and are given

in parentheses in the text, with the abbreviation "PP" followed by the relevant part and article.

31. Descartes, *Discourse on the Method* 6:121.

32. Ibid. 5:108.

33. Descartes, *The World*, 339.

34. Descartes, *Discourse on the Method* 6:121.

35. Quoted in Eamon, "Technology as Magic," 171 (English), 204 (Italian).

36. Wilkins, *Mathematicall Magick*, 143.

37. Fontenelle, *Plurality of Worlds*, 18, 16; italics in source.

38. Westfall, *Construction of Modern Science*, 42.

39. *Physiologia* is subsequently cited in parentheses in the text, with the abbreviation "PEGC" followed by the relevant page number.

40. Kargon, "Walter Charleton, Robert Boyle, and the Acceptance of Epicurean Atomism in England," 186.

41. Westfall, *Never at Rest*, 89, 96; Deason, "Reformation Theology and the Mechanistic Conception of Nature," 181.

42. The passage quoted is an almost verbatim reproduction of Gassendi's own statement, which is quoted in Duhem, *Aim and Structure of Physical Theory*, 88.

43. Molière, *The Learned Women*, act 3, scene 2.

44. Quoted in Meli, *Equivalence and Priority*, 49.

45. Deason, "Reformation Theology and the Mechanistic Conception of Nature," 168.

## Chapter 20—Robert Boyle and the Transformation of Mechanism in Britain

1. *The Advancement of Learning* and *The New Organon* are cited in the present chapter in parentheses in the text, with the abbreviation "AL" followed by the relevant book number, article number, and subarticle number for the former, and the abbreviation "NO" followed by the relevant book number and article number for the latter.

2. F. Bacon, *New Organon* 1.74 and "Preface to the Great Renewal," 13.

3. Ibid., "Preface to the Great Renewal," 13.

4. Aristotle, *Posterior Analytics* 99b20, 100b5, 88a18.

5. F. Bacon, *New Organon*, "Preface to the Great Renewal," 18. See also ibid. 1.50.

6. Whitehead, *Science and the Modern World*, 66. See also Losee, *Philosophy of Science*, 66–67.

7. Spinoza to Oldenburg, April 1662, *Correspondence of Spinoza*, 91.

8. Butterfield, *Origins of Modern Science*, 121; Rossi, *Francis Bacon*, 14; Farrington, *Francis Bacon*, 79.

9. Whittaker, *Theories of Aether and Electricity* 1:35, note 4; Browne, *Pseudodoxia epidemica*, 118.

10. Chalmers, "Sir Thomas Browne, True Scientist," 38.

11. Willey, *Seventeenth Century Background*, 44.

12. Browne, *Religio Medici*, 41.

13. Browne, *Pseudodoxia epidemica*, 90.

14. Hooke, *Micrographia*, 46. See also Jones, *Ancients and Moderns*, 185.

15. C. Webster, "Discovery of Boyle's Law"; Cohen, "Newton, Hooke, and 'Boyle's Law.'"

16. In the present chapter, *Experimental Philosophy* is cited in parentheses in the text, with the abbreviations "EP" followed by the relevant page number.

17. Boas, "Introduction," in ibid., xvii. On Power's *Experimental Philosophy*, see also C. Webster, "Henry Power's Experimental Philosophy."

18. Huygens, *Treatise on Light*, 12.

19. Thorndike, *History of Magic and Experimental Science* 8:211.

20. See Boas, "Introduction," in Experimental Philosophy, xxii.

21. Westfall, *Construction of Modern Science*, 77.

22. *The Origin of Forms and Qualities according to the Corpuscular Philosophy* is contained in *The Works of the Honourable Robert Boyle* and is cited in the present chapter in parentheses in the text as "Works" followed by the relevant volume number and page number.

23. Boyle, *The Sceptical Chymist*, 182.

24. Boyle, *Works* 3:48.

25. Kargon, "Epicurean Atomism in England," 188.

26. Boyle, *The Sceptical Chymist*, 200; italics in source.

27. Boas, "Establishment of the Mechanical Philosophy," 467.

28. *About the Excellency and Grounds of the Mechanical Hypothesis* is contained in *Works* 4:67–78.

29. *Experiments, Notes, &c.* is contained in *Works* 4:230–353. The main contents of *Experiments, Notes, &c.* are "Of the Mechanical Origin of Heat and Cold," 236–59; "Experiments and Notes about the Mechanical Production of Magnetism," 340–45; and "Experiments and Notes about the Mechanical Production of Electricity," 345–54.

30. *Of the Cause of Attraction by Suction* is contained in *Works* 4:128–44.

31. *Of the Systematical or Cosmical Qualities of Things* is contained in *Works* 3:306–25.

32. Boyle, *Sceptical Chymist*, 104–5.

33. *Suspicions about Some Hidden Qualities in the Air* is contained in *Works* 4:85–103.

34. Schofield, *Mechanism and Materialism*, 15–16.

35. Quoted in Kargon, "William Petty's Mechanical Philosophy," 64.

36. See Henry, "Occult Qualities and the Experimental Philosophy."

37. Maddison, *The Honourable Robert Boyle*, 261.

**Chapter 21—Magnetism and Gravity: Hooke and Newton**

1. In *The Construction of Modern Science*, for example, Richard S. Westfall treats this period but completely neglects the influence of Gilbert.

2. Barlow, *Magneticall Advertisements*, B2r.

3. Browne, *Works of Sir Thomas Browne* 2:98.

4. J. L. Russell, "Kepler's Laws of Planetary Motion," 19.

5. Quoted in Weld, *History of the Royal Society* 1:31.

6. Quoted in Bennett, "Cosmology and the Magnetical Philosophy," 171; idem, *Mathematical Science of Christopher Wren*, 59.

7. Quoted in Weld, *History of the Royal Society* 1:65.

8. Nicolson, *Voyages to the Moon*, 93.

9. Wilkins, *Mathematical and Philosophical Works* 1:114.

10. Ibid. 1:244, 219.

11. Godwin, *Man in the Moone*, 46–47.

12. F. Bacon, *New Organon* 2.48:198, 2.37:169.

13. Wilkins, *Mathematical and Philosophical Works* 1:115, 117.

14. Ibid. 1:124.

15. Ibid. 1:117, 116.

16. F. Bacon, *Sylva Sylvarum*, 10.

17. Birch, *History of the Royal Society* 1:10, 12.

18. Hooke, *Micrographia*, "The Preface," a2v, g1r.

19. Gunther, *Early Science in Oxford* 8:339–40.

20. Hooke, *Micrographia*, 31.

21. Gunther, *Early Science in Oxford* 8:339, 341.

22. Hooke, *Micrographia*, 31.

23. Ibid., 16.

24. Bennett, "Magnetical Philosophy and Astronomy," 228.

25. Gunther, *Early Science in Oxford* 8:183.

26. Quoted in Birch, *History of the Royal Society* 2:70. This report also appears in full in Boyle, *Works* 4:506–8.

27. Quoted in Birch, *History of the Royal Society* 2:91.

28. Bennett, "Cosmology and the Magnetical Philosophy," 173.

29. Hooke, *Micrographia*, "The Preface," a2r.

30. Quoted in Birch, *History of the Royal Society* 2:70.

31. Ibid. 1:507.

32. Gunther, *Early Science in Oxford* 8:3.

33. Ibid. 8:228–29.

34. Ibid. 8:229.

35. Leibniz, *Mathematische Schriften* 6:152.

36. Flamsteed to Halley, 17 February 1680/1, in Newton, *Correspondence of Isaac Newton* 2:338; Flamsteed to Newton, 5 January 1684/5, in ibid. 2:409.

37. Gunther, *Early Science in Oxford* 8:333.

38. Quoted in Birch, *History of the Royal Society* 2:70.

39. Quoted in ibid. 2:70–71.

40. Quoted in ibid. 2:72.

41. Quoted in ibid. 2:75.

42. Quoted in ibid. 2:77.

43. Hooke, *Nature of Comets*, 183. See also Pumfrey, "Mechanizing Magnetism in Restoration England."

44. Duhem, *Aim and Structure of Physical Theory*, 246; Applebaum, "Keplerian Astronomy after Krepler," 460; Wilson, "From Kepler's Laws to Universal Gravitation," 107; Kargon, "William Petty's Mechanical Philosophy," 65; Hatch, "Boulliau, Ismaël (1605–1694)" in *Encyclopedia of the Scientific Revolution*.

45. Quoted in Birch, *History of the Royal Society* 2:90, 92.

46. Hooke to Newton, 24 November 1679, in Newton, *Correspondence of Isaac Newton*, 2:297.

47. Westfall, *Construction of Modern Science*, 152. See also Westfall, *Never at Rest*, 382–92; idem, "Hooke and Universal Gravitation," 260; Wilson, "Newtonian Achievement in Astronomy," 240; Lohne, "Hooke *versus* Newton"; Patterson, "Hooke's Gravitation Theory: I"; and idem, "Hooke's Gravitation Theory: II."

48. Gunther, *Early Science in Oxford* 8:27–28.

49. Westfall, *Never at Rest*, 382.

50. Hooke to Newton, 6 January 1679/80, in Newton, *Correspondence of Isaac Newton* 2:309.

51. Hooke, *Nature of Comets*, 183.

52. Gunther, *Early Science in Oxford* 8:28.

53. Newton, *Mathematical Principles*, 13 (Cajori edition).

54. Maxwell, *Matter and Motion*, 32.

55. Newton, *Mathematical Principles*, 14 (Cajori edition).

56. See Bennett, "Cosmology and the Magnetical Philosophy," 175.

57. Dobbs, *Janus Faces of Genius*, 4.

58. Quoted in Partington, *History of Chemistry* 3:33, and in Westfall, *Construction of Modern Science*, 70, 72. See also Thackray, *Atoms and Powers*, 200.

59. Newton, "Nature of Acids," *Isaac Newton's Papers & Letters*, 257. See also idem, *Opticks*, 385–86.

60. Newton, *Opticks*, 388.

61. Ibid., 394, 397.

62. Leibniz to Clarke, 18 August 1716, *Leibnitz-Clarke Correspondence*, 95.

63. Newton, *Mathematical Principles*, 5 (Cajori edition).

64. Newton, *Opticks*, 376.

65. Huygens to Leibniz, 18 November 1690, quoted in Duhem, *Aim and Structure of Physical Theory*, 15.

66. Fontenelle, "Elogium of Sir Isaac Newton," 463, 454.

67. Leibniz, *Theodicy*, 85.

68. Leibniz to Conti, November or December 1715, *Leibniz-Clarke Correspondence*, 184; brackets in source.

69. Newton, *Mathematical Principles*, 546–47 (Cajori edition).

70. Koyré, *Newtonian Studies*, 16, note 3. See also Cohen, *Franklin and Newton*, 125–26, and Duhem, *Aim and Structure of Physical Theory*, 321.

71. Newton, *Mathematical Principles*, 547 (Cajori edition).

72. Newton, *Mathematical Principles*, XVII–XVIII (Cajori edition).

73. Dobbs, *Janus Faces of Genius*, 131–32, 167. See also Westfall, *Never at Rest*, 454–58, and McGuire, "Newton's Invisible Realm," 157.

74. Keynes, "Newton, the Man," 27.

75. Dobbs, *Janus Faces of Genius*, 4–5, 91, 252; Westfall, *Never at Rest*, 299-300, 304; idem, "Alchemy in Newton's Career," 224, 229–30; idem, "Newton and Alchemy," 324.

76. Henry, "Occult Qualities and the Experimental Philosophy," 338.

77. Quoted in Burtt, *Metaphysical Foundations*, 133. See also C. Webster, *From Paracelsus to Newton*, 69; Westfall, *Never at Rest*, 304; idem, "Alchemy in Newton's Career," 216; idem, "Newton and Alchemy," 331; McGuire, "Newton's Invisible Realm," 184–85.

78. Newton, *Opticks*, 401, 399–400.

79. Newton, *An Hypothesis Explaining the Properties of Light*, 180.

80. Newton to Bentley, 25 February 1693, in Newton, *Isaac Newton's Papers & Letters*, 302, 303.

81. Newton, *Opticks*, 369, 370.

82. Bentley, *Confutation of Atheism*, in Newton, *Isaac Newton's Papers & Letters*, 344.

83. Newton, *Mathematical Principles*, 546 (Cajori edition).

84. Newton, *Mathematical Principles*, 25 (Cajori edition).

85. Newton, *Certain Philosophical Questions*, 376–77, 430–31. See also Westfall, *Never at Rest*, 91.

86. Newton to [?Crompton], 28 February 1680/1 and ? April 1681, in Newton, *Correspondence of Isaac Newton* 2:341, 2:360; brackets and question marks in source.

87. Newton, *Mathematical Principles*, 414 (Cajori edition).

88. Birch, *History of the Royal Society* 4:526–27.

89. Palter, "Early Measurements of Magnetic Force," 547.

90. B. Taylor, "Account of an Experiment," 295.

91. B. Taylor, "Extract of a Letter," 204.

92. Hauksbee, "Power of the Load-stone," 507.

93. B. Taylor, "Extract of a Letter," 204–5.

94. Newton, "De aere et aethere," 228 (Latin text on p. 220); brackets in source.

95. Newton, *An Hypothesis Explaining the Properties of Light*, 180; idem, "Draft addition to the *Principia*," 314 (Latin), 316 (English).

96. Newton, *Opticks*, 267, 353.

97. Gregory, "Memoranda," in Newton, *Correspondence of Isaac Newton* 3:338 (Latin text at 3:335).

98. Savery, "Magnetical Observations and Experiments," 300.

99. Halley, "Appearance of the Lights," 406, 427.

100. Cohen, *Franklin and Newton*, 244.

101. Quoted in Home, introduction to *Aepinus's Essay*, by Franz Aepinus, 158.

102. Knight, "Magnetical Experiments," 665–69.

103. Home, "'Newtonianism' and the Magnet," 256, 263. See also idem, introduction to *Aepinus's Essay*, by Franz Aepinus, 139, 151–55.

104. Newton, *Opticks*, 401–2.

## Chapter 22—Epilogue: Ascertaining the Laws of Magnetic Force

1. Musschenbroek, "De viribus magneticis," 376, 370.

2. Quoted in Home, introduction to *Aepinus's Essay*, by Franz Aepinus, 161–62.

3. Musschenbroek, "De viribus magneticis," 374.

4. Helsham, *Lectures in Natural Philosophy*, 19. The reprinted edition I consulted is based on the fourth edition, but the relevant passages do not seem to differ from those in the first edition.

5. Ibid., 19–20.

6. On Ferguson's comment, see Mottelay, *Bibliographical History of Electricity and Magnetism*, 232.

7. Gillmor, *Coulomb*, 193.

8. Schofield, *Mechanism and Materialism*, 174. On Desaguliers's experiment, see Desaguliers, "Account of Some Magnetical Experiments."

9. Palter, "Early Measurements of Magnetic Force," 544–58. In this section, Palter's translation of the note is cited in parentheses in the text, with the relevant page numbers of Palter's paper.

10. Duhem, *Aim and Structure of Physical Theory*, 156. See also ibid., 147.

11. See McCormmach, "John Michell and Henry Cavendish."

12. This and subsequent citations in this section from *A Treatise of Artificial Magnets* are quoted in Hardin, "Scientific Work of John Michell," 29.

13. Weld, *History of the Royal Society* 1:509–10.

14. Aepinus, *Aepinus's Essay*, 327.

15. Mayer, *Unpublished Writings* 3:68 (Latin text at 3:36).

16. Forbes, preface to ibid. 3:1.

17. Quoted in Rossi, "Hermeticism, Rationality and the Scientific Revolution," 252.

18. Voltaire, *Isaac Newton's Philosophy*, 185.

19. Home, introduction to *Aepinus's Essay*, by Franz Aepinus, 144.

20. Fleckenstein, "Vorwort des Herausgeber," in *Leonhardi Euleri opera omnia*, series secunda, volumen quantum, XI; my translation. The original of the phrase quoted is "ein verspäteter und ziemlich orthodoxer Cartesianer."

21. Euler, *To a German Princess* 1:243. See also idem, *Anleitung zur Naturlehre*, chapter 19.

22. Euler, *To a German Princess*, volume 1, letter 55, and volume 2, letters 62–65.

23. D. Bernoulli to Euler, 21 January 1742, and 4 February 1744. Cited in Boss, *Newton and Russia*, 136; Forbes, introduction to *The Unpublished Writings of Tobias Mayer* 3:8.

24. *The Theory of the Magnet*, volume 3 of *The Unpublished Writings of Tobias Mayer*, is subsequently cited in parentheses in the text, with the abbreviation "TM" followed by the relevant chapter, section, and page numbers.

25. Kant, *Critique of Pure Reason*, 325–26.

26. Roche, *Mathematics of Measurement*, 139.

27. Thackray, *Atoms and Powers*, 158–60.

28. Subsequent quotations from Coulomb's memoir are from Gillmor, *Coulomb*, 182–92.

29. Ampère, *Mathematical Theory of Electrodynamic Phenomena*, 156.

# Bibliography

## Abbreviations

| | |
|---|---|
| ***AHES*** | *Archive for History of Exact Sciences* |
| ***AJP*** | *American Journal of Physics* |
| ***BJHS*** | *The British Journal for the History of Science* |
| ***BJPS*** | *The British Journal for the Philosophy of Science* |
| ***GBWW*** | *Great Books of the Western World (Encyclopaedia Britannica)* |
| ***JHA*** | *Journal for the History of Astronomy* |
| ***JHI*** | *Journal of the History of Ideas* |
| ***JKGW*** | *Johannes Kepler Gesammelte Werke* |
| ***JMH*** | *Journal of Medieval History* |
| ***SBMS*** | *A Source Book in Medieval Science* |
| ***TMAE*** | *Terrestrial Magnetism and Atmospheric Electricity* |

## Reference Works

**Applebaum, Wilbur, ed.** *Encyclopedia of the Scientific Revolution: From Copernicus to Newton*. New York: Garland, 2000.

**Gillispie, Charles Coulston, ed.** *Dictionary of Scientific Biography.* 18 vols. New York: Scribner, 1970–1981.

**Grant, Edward, ed.** *A Source Book in Medieval Science.* Harvard University Press, 1974.

**Strayer, Joseph R., ed.** *Dictionary of the Middle Ages.* 15 vols. New York: Scribner, 1982–1989.

**Wiener, Philip P., ed.** *Dictionary of the History of Ideas: Studies of Selected Pivotal Ideas.* 5 vols. New York: Charles Scribner's Sons, 1968–1974.

## Primary Sources

Note: Where appropriate, the original year of publication is given in parenthesis after the title of the edition consulted.

**Acosta, José de.** *Natural and Moral History of the Indies* (1590). Translated by F. M. López-Morillas. Edited by J. E. Mangan. Introduction and commentary by W. Mignolo. Duke University Press, 2002.

**Aelian.** *Historical Miscellany.* Translated by N. G. Wilson. Harvard University Press, 1997.

_________. *On the Characteristics of Animals.* Translated by A. F. Scholfield. 3 vols. Harvard University Press, 1958–1959.

**Aepinus, Franz Ulrich Theodosius.** *Tentamen theoriae electricitatis et magnetismi* (1759). Translated by P. J. Connor as *Aepinus's Essay on the Theory of Electricity and Magnetism.* Introduction and notes by R. W. Home. Princeton University Press, 1979.

**Agricola, Georgius.** *De natura fossilium* (1546). Translated by M. C. Bandy and J. A. Bandy as *Textbook of Mineralogy.* Special Paper 63 of the Geological Society of America, 1955.

_________. *De re metallica* (1556). Translated by H. C. Hoover and L. H. Hoover. New York: Dover, 1950.

**Agrippa, Cornelius.** *Three Books of Occult Philosophy or Magic* (1533). Translated and edited by W. F. Whitehead. New York: AMS Press, 1982.

**Alberti, Leon Battista.** *On the Art of Building in Ten Books* (1485). Translated by J. Rykwert, N. Leach, and R. Tavernor. Cambridge, MA: MIT Press, 1988.

**Albert the Great.** *Book of Minerals.* Translated by D. Wyckoff. Oxford: Clarendon Press, 1967.

_________. *The Book of Secrets of Albertus Magnus: Of the Virtues of Herbs, Stones, and Certain Beasts, Also a Book of the Marvels of the World.* Translated and edited by M. R. Best and F. H. Brightman. York Beach: Weiser, 1999.

_________. *On Animals: A Medieval Summa Zoologica.* Translated by K. F. Kitchell and I. M. Resnick. 2 vols. Johns Hopkins University Press, 1999.

**Alexander of Aphrodisias.** *Quaestiones 2.16–3.15.* Translated by R. W. Sharples. Cornell University Press, 1994.

**Ampère, André Marie.** *On the Mathematical Theory of Electrodynamic Phenomena* (1825). Translated by O. M. Blunn. In *Early Electrodynamics*, edited by R. A. R. Tricker, 155–200. Oxford: Pergamon Press, 1965.

**Anonymous.** *Carmina Burana.* In *Bibliothek des Mittelalters* 13. Frankfurt: Deutsche Klassiker, 1978.

_________. *Der Physiologus: Tiere und ihre Symbolik.* Translated by O. Seel. 2nd ed. Zurich: Artemis, 1967.

_________. *King Horn.* In *The Chief Middle English Poets.* Edited by J. L. Weston, 93–109. New York: Phaeton Press, 1968.

_________. *La quete du graal: Édition présentée et établie par Albert Béguin et Yves Bonnefoy.* Paris: Édition du Seuil, 1965.

_________. *The Love Songs of the Carmina Burana*. Translated by E. D. Blodgett and R. A. Swanson. New York: Garland, 1987.

_________. *Reinke de Vos und satirisch-didaktische Dichtung*. Edited by E. Wolff. Tokyo: Sanshusya, Tokyo, 1973.

_________. *The Ruodlieb: The First Medieval Epic of Chivalry from Eleventh-century Germany*. Translated by G. B. Ford, Jr. Leiden: Brill, 1965.

**Aristotle.** *The History of Animals*. Translated by D. W. Thompson. In GBWW 9: 3–158.

_________. *Metaphysics*. Translated by W. D. Ross. In GBWW 8:495–626.

_________. *Meteorology*. Translated by E. W. Webster. In GBWW 8:443–494.

_________. *On Generation and Corruption*. Translated by H. H. Joachim. In GBWW 8:407–441.

_________. *On Sense and the Sensible*. Translated by J. I. Beare. In GBWW 8:673–689.

_________. *On the Generation of Animals*. Translated by A. Platt. In GBWW 9:253–331.

_________. *On the Heavens*. Translated by J. L. Stocks. In GBWW 8:357–405.

_________. *On the Soul*. Translated by J. A. Smith. In GBWW 8:627–668.

_________. *Physics*. Translated by R. P. Hardie and R. K. Gaye. In GBWW 8:257–355.

_________. *Posterior Analytics*. Translated by C. R. G. Mure. In GBWW 8:95–137.

**Augustine.** *The City of God*. Translated by H. Bettenson. Introduction by J. O'Meara. London: Penguin Books, 1984.

_________. *Confessions*. Translated by R. S. Pine-Coffin. London: Penguin Books, 1961.

_________. *The Literal Meaning of Genesis*. Translated by J. H. Taylor. 2 vols. New York: Newman Press, 1982.

_________. *On Christian Doctrine*. Translated by D. W. Robertson, Jr. Indianapolis/New York: Bobbs-Merrill, 1958.

**Bacon, Francis.** *The Advancement of Learning* (1605). New York: Modern Library, 2001.

_________. *The New Organon* (1620). Translated and edited by L. Jardine and M. Silverthorne. Cambridge University Press, 2000.

_________. *Sylva Sylvarum: Or, A Naturall Historie*. London: Haviland, 1639.

**Bacon, Roger.** *De multiplicatione specierum (On the Multiplication of Species)*. In *Roger Bacon's Philosophy of Nature*, edited by D. C. Lindberg, 1–269. South Bend: St. Augustine's Press, 1998.

_________. *The Opus majus of Roger Bacon*. Translated by R. B. Burke. Bristol: Thoemmes Press, 2000.

_________. *Roger Bacon's Letter Concerning the Marvelous Power of Art and of Nature and Concerning the Nullity of Magic*. Translated by T. L. Davis. New York: AMS Press, 1982.

**Barlow, William.** *Magneticall Advertisements* (1616). New York: Da Capo Press, 1968.

**Bartholomew the Englishman.** Excerpt from *De rerum proprietatibus* (*On the Properties of Things*). Translated by E. Grant. In SBMS, 367–68.

**Basil.** *Regulae fusius tractatae*. In *Patrologiae cursus completus* 31, *S. P. N. Basilii Opera omnia quae extant*, edited by J-P. Migne, 890–1051. Turnhout: Brespols, 1977.

**Bede.** *Ecclesiastical History of the English People*. Translated by Leo Sherley-Price. Revised by R. E. Latham. London: Penguin Books, 1990.

**Bentley, Richard.** *A Confutation of Atheism from the Origin and Frame of the World* (1693). In *Isaac Newton's Papers & Letters on Natural Philosophy*, 2nd edition, edited by I. B. Cohen, 313–94. Harvard University Press, 1978.

**Bernard le Bovier de Fontenelle.** *See* **Fontenelle, Bernard le Bovier de.**

**Birch, Thomas.** *The History of the Royal Society of London* (1756–1757). 4 vols. Brussels: Culture et Civilisation, 1968.

**Biringuccio, Vannoccio.** *The Pirotechnia of Vannoccio Biringuccio* (1540). Translated by C. S. Smith. Cambridge: M.I.T. Press, 1966.

**Bonaventure.** *The Soul's Journey into God.* In *Bonaventure: The Soul's Journey into God, The Tree of Life, The Life of St. Francis*, translated by E. Cousins. New York: Paulist Press, 1978.

**Borough, William.** *A Discourse of the Variation of the Cumpas, or Magneticall Needle* (1581). Norwood, NJ: Walter J. Johnson, 1974.

**Boyle, Robert.** *The Sceptical Chymist* (1661). London: Dent, 1991.

_________. *The Works of the Honourable Robert Boyle* (1772). Edited by T. Birch. 6 vols. Hildesheim: Georg Olms, 1965–1966.

**Browne, Thomas.** *The Works of Sir Thomas Browne.* Edited by G. Keynes. 4 vols. University of Chicago Press, 1964.

**Bruno, Giordano.** *Cause, Principle and Unity: And Essays on Magic.* Translated and edited by R. de Lucca and R. J. Blackwell. Cambridge University Press, 1998.

_________. *On the Infinite Universe and Worlds* (1584). In *Giordano Bruno: His Life and Thought: With Annotated Translation of His Work On the Infinite Universe and Worlds*, translated by D. W. Singer. New York: Greenwood Press, 1968.

**Buridan, Jean.** *Iohannis Buridani Quaestiones super libris quattuor de caelo et mundo*. Edited by E. A. Moody. New York: Kraus, 1970.

_________. *Questions on the Four Books on the Heavens and the World of Aristotle*. In M. Glagett, *The Science of Mechanics in the Middle Ages*, 557–64 and 594–99. University of Wisconsin Press, 1959.

**Campanella, Tommaso.** *The City of the Sun: A Poetical Dialogue* (1602). Translated by D. J. Donno. University of California Press, 1981.

**Cardano, Girolamo (Jerome Cardan).** *The Book of My Life: De vita propria liber* (1576). Translated by J. Stoner. New York: Dover, 1962.

_________. *De subtilitate rerum* (1550). In Girolamo Cardano, *Opera omnia*, vol. 3 (1662). New York: Johnson Reprint, 1967.

_________. *The First Book of Jerome Cardan's De Subtilitate.* Translated by M. M. Cass. Williamsport: Bayard Press, 1934.

**Cartier, Jacques.** *Navigations to Newe Fraunce* (1580). Translated by J. Florio. Ann Arbor: University Microfilms, 1966.

**Cassiodorus.** *An Introduction to Divine and Human Readings.* Translated by L. W. Jones. New York: Octagon Books, 1966.

**Castro, João de.** *Extracts on Magnetic Observations from Logbooks of João de Castro, 1538–1539 and 1541.* Translated by J. de Sampaio Ferraz. In H. D. Harradon, "Some Early Contributions to the History of Geomagnetism—VII," TMAE 49 (1944), 187–98.

**Charleton, Walter.** *Physiologia Epicuro-Gassendo-Charltoniana: Or, A Fabrick of Science Natural, Upon the Hypothesis of Atoms* (1654). Introduction by R. Kargon. New York: Johnson Reprint, 1966.

**Chaucer, Geoffrey.** *The Canterbury Tales.* Translated by J. U. Nicolson. In GBWW 22:157–550.

_________. *The Parliament of Birds.* Translated by B. Stone. In *Love Visions*, 123–49. Harmondsworth: Penguin Books, 1983.

**Claudian.** *Claudian: With an English Translation by Maurice Platnauer.* 2 vols. Harvard University Press, 1963.

**Copernicus, Nicolaus.** *On the Revolutions.* Translated by E. Rosen. Johns Hopkins University Press, 1978.

**Cortes, Martin.** *Brief Compendium on the Sphere and Art of Navigating.* In H. D. Harradon, "Some Early Contributions to the History of Geomagnetism—II & III," TMAE 48 (1943), 84–91.

**Coulomb, Charles Augustin.** "Second Mémoire sur l'Électricite et le Magnetisme," *Memoires de l'Acadamie Royale des Sciences* (1785), 578–611.

_________. *Vier Abhandlungen über die Elektricität und den Magnetismus.* In *Ostwald's Klassiker der Exakten Wissenschaften* 13:12–42. Leipzig: Engelmann, 1890.

**Crathorn, William.** *On the Possibility of Infallible Knowledge.* In *Mind and Knowledge*, edited by R. Pasnau, 245–301. Cambridge University Press, 2002.

**Cusanus, Nicolaus (Nicholas de Cusa).** *De staticis experimentis* (1450). In *Unity and Reform: Selected Writings of Nicholas de Cusa*, edited by J. P. Dolan, 241–60. University of Notre Dame Press, 1962.

_________. *The Layman on Wisdom and the Mind* (1450). Translated by M. L. Führer. Ottawa: Dovehouse, 1989.

_________. *Of Learned Ignorance* (1440). Translated by G. Heron. Introduction by D. J. B. Hawkins. Westport: Hyperion, 1979.

_________. *On Actualized-possibility* (1460). In *A Concise Introduction to the Philosophy of Nicholas of Cusa*, by J. Hopkins, 60–153. University of Minnesota Press, 1978.

_________. *On Peaceful Unity of Faith, Nicholas of Cusa's De pace fidei and Cribratio alkorani* (1437). Translated by J. Hopkins. Minneapolis: A. J. Banning Press, 1990.

**Cyrano de Bergerac.** *L'autre mond in Les Etats et Empires de la Luna.* In *Libertins du XIIe siècle*, edited by J. Prévot, 903–90. Paris: Gallimard, 1998.

**D'Anghiera, Pietro Martire.** *The Decades of the Newe Worlde or West India* (1555). Translated by R. Eden. Ann Arbor: University Microfilms, 1966.

**Dante Alighieri.** *The Divine Comedy.* Translated by D. L. Sayers and B. Reynolds. 3 vols. Harmondsworth: Penguin Books, 1949, 1955, 1962.

**da Orta, Garcia.** *Colloquies on the Simples and Drugs of India.* Translated by C. Markham. London: Henry Sotheran, 1913.

**Darwin, George Howard.** *The Tides and Kindred Phenomena in the Solar System.* London: Murray, 1901.

**de Acosta, José.** *See* **Acosta, José de.**

**de Bury, Richard.** *Philobiblon.* Introduction by A. Taylor. University of California Press, 1948.

**de Castro, João.** *See* **Castro, João de.**

**Dee, John.** *John Dee on Astronomy: Propaedeumata aphoristica (1558 and 1568).* Edited and translated by W. Shumaker. Introduction by J. L. Heilbron. University of California Press, 1978.

_________. *The Mathematicall Praeface to the Elements of Geometrie of Euclid of Megara* (1570). Introduction by A. G. Debus. New York: Science History Publications, 1975.

_________. *Propaedeumata aphoristica.* See *John Dee on Astronomy*

**de Fontenelle, Bernard le Bovier.** *See* **Fontenelle, Bernard le Bovier de.**

**de Gómara, Francisco López de.** *See* **Gómara, Francisco López de.**

**della Mirandola, Pico.** *See* **Pico della Mirandola.**

**della Porta, Giambattista (John Baptista Porta).** *De i miracoli et maravigliosi effetti dalla natura prodotti.* Venice: Salicato, 1572. Originally published in Latin as *Magia naturalis libris IV* (1558).

_________. *Natural Magick.* Translated by T. Young. New York: Basic Books, 1957. Originally published in Latin as *Magia naturalis libris XX* (1589).

**de Lorris, Guillaume, and Jean de Meun.** *The Romance of the Rose.* Translated by C. Dahlberg. Princeton University Press, 1971 [1995].

**de Oviedo.** *See* **Oviedo, Gonzalo Fernández de.**

**Desaguliers, John Theophilis.** "Account of Some Magnetical Experiments," *Philosophical Transactions* 40 (1737/1738), 384–87.

**Descartes, René.** *Discourse on the Method of Rightly Conducting the Reason* (1637). In *The Philosophical Works of Descartes*, translated by E. S. Haldane and G. R. T. Ross, vol. 1, 79–130. Cambridge University Press, 1931.

_________. *The Philosophical Works of Descartes.* Translated by E. S. Haldane and G. R. T. Ross. Cambridge University Press, 1931.

_________. *Principles of Philosophy* (1644). Translated by V. R. Miller and R. P. Miller. Dortrecht: D. Reidel, 1983.

_________. *The World* (1633). In *Descartes: Selections*, translated and edited by R. M. Eaton, 312–49. New York: Charles Scribner's Sons, 1955.

**Digges, Thomas.** *A Perfit Description of the Caelestiall Orbes according to the Most Aunciente Doctrine of the Pythagoreans, Lately Reuiued by Copernicvs and by Geometricall Demonstrations Approued* (1576). In "Thomas Digges, the Copernican System, and the Idea of the Infinity of the Universe in 1576," by F. R. Johnson and S. V. Larkey, *Huntington Library Bulletin* 5 (1934), 78–95.

**Diogenes Laertius.** *Lives of Eminent Philosophers.* Translated by R. D. Hicks. 2 vols. Harvard University Press, 2000.

**Dioscorides.** *The Greek Herbal of Dioscorides* (1655). Translated by J. Goodyer. Edited by R. T. Gunther. New York: Hafner, 1968.

**Einstein, Albert.** "Autobiographical Notes." In *Albert Einstein: Philosopher-Scientist*, edited by P. A. Schilpp, 2–94. La Salle: Open Court, 1970.

**Ercker, Lazarus.** *Lazarus Ercker's Treatise on Ores and Assaying* (1580). Translated by A. G. Sisco and C. S. Smith. University of Chicago Press, 1951.

**Euler, Leonhard.** *Anleitung zur Naturlehre.* In *Leonhardi Euleri Opera Omnia*, series 3, vol. 1. Leipzig/Berlin: Teubner, 1926.

_________. *Letters of Euler to a German Princess* (1795). 2 vols. Translated by H. Hunter. Bristol: Thoemmes Press, 1997.

_________. *Mechanica, sive motus scientia analytice exposita.* In *Leonhardi Euleri Opera Omnia*, series 2, vols. 1 and 2. Leipzig/Berlin: Teubner, 1912.

**Falero, Francisco.** *Treatise on the Sphere and the Art of Navigation* (1535). In H. D. Harradon, "Some Early Contributions to the History of Geomagnetism—II & III," TMAE 48 (1943), 80–84.

**Fernández de Oviedo, Gonzalo.** *See* **Oviedo, Gonzalo Fernández de.**

**Ficino, Marsilio.** *Marsilio Ficino's Commentary on Plato's Symposium* (1469). Translated by S. R. Jayne. Columbia: University of Missouri, 1944.

_________. *The Philebus Commentary* (1469). Translated by M. J. Allen. University of Calfornia Press, 1975.

_________. *Three Books on Life* (1489). Translated and annotated by C. V. Kaske and J. R. Clark. Binghamton: The Renaissance Society of America, 1998.

**Flamsteed, John.** Flamsteed to Halley, 17 February 1680/1. In *The Correspondence of Isaac Newton* 2:338. Cambridge University Press, 1959–1977.

_________. Flamsteed to Newton, 5 January 1684/5. In *The Correspondence of Isaac Newton* 2:409. Cambridge University Press, 1959–1977.

**Fontenelle, Bernard le Bovier de.** *A Conversation on the Plurality of Worlds* (1686). Translated from the French. London: Evans, 1758.

_________. "The Elogium of Sir Issac Newton" (1728). In *Isaac Newton's Papers & Letters on Natural Philosophy*, edited by I. B. Cohen and R. E. Schofield, 2nd ed., 444–74. Harvard University Press, 1978.

**Frederick II of Hohenstaufen.** *The Art of Falconry: being the De arte venandi cum avibus of Frederick II of Hohenstaufen.* Translated and edited by C. A. Wood and F. M. Fyfe. Stanford University Press, 1943 [1961].

**Galen.** *Galen, On the Natural Faculties.* Translated by A. J. Brock. Harvard University Press, 1991.

**Galilei, Galileo.** *Dialogue concerning the Two Chief World Systems* (1632). Translated and annotated by S. Drake. University of California Press, 1967.

_________. *Discoveries and Opinions of Galileo.* Translated and annotated by S. Drake. Garden City: Doubleday, 1957.

_________. "Excerpts from *The Assayer* (1623)." In *Discoveries and Opinions of Galileo*, 229–81. Garden City: Doubleday, 1957.

_________. *Letters on Sunspots* (1613). In *Discoveries and Opinions of Galileo*, 87–114. Garden City: Doubleday, 1957.

_________. *On Motion and On Mechanics*. Translated by I. E. Drabkin and S. Drake. University of Wisconsin Press, 1960.

_________. *Sidereus nuncius, or, The Sideral Messenger* (1610). Translated and annotated by A. van Helden. University of Chicago Press, 1989.

_________. *Two New Sciences* (1634). Translated and annotated by S. Drake. University of Wisconsin Press, 1974.

**Garcia da Orta.** *See* **da Orta, Garcia**

**Gervase of Tilbury.** *Gervase of Tilbury: Otia imperialia: Recreation for an Emperor*. Translated and edited by S. E. Banks and J. W. Binns. Oxford: Clarendon Press, 2002.

**Giambattista della Porta.** *See* **della Porta, Giambattista.**

**Gilbert, Humphrey.** *Queene Elizabethes Academy*. Edited by F. J. Furnivall. Early English Text Society, 1869.

**Gilbert, William (Guilielmi Gilberti).** *De magnete*. Translated by P. F. Mottelay. New York: Dover, 1958.

_________. *De magnete, Magneticisque corporibus, et de magno magnete tellure: Physiologia nova, plurimis & argumentis, & experimentis demonstrata* (1600). Brussels: Culture et Civilisation, 1967.

_________. *On the Magnet*. Translated by S. P. Thompson. New York: Basic Books, 1958.

**Godwin, Francis.** *The Man in the Moone: or, A Discourse of a Voyage Thither, by Domingo Gonsales, the Speedy Messenger* (1638). New York: Da Capo Press, 1972.

**Goethe, Johann Wolfgang von.** *The Sorrows of Young Werther* (1774). Translated and annotated by M. Hulse. Harmondsworth: Penguin Books, 1989.

**Gómara, Francisco López de.** *Historia General de las Indias* (1552). Barcelona: Obras Maestras, 1954.

**Gower, John.** *Confessio amantis*. Translated by T. Tiller. Harmondsworth: Penguin Books, 1963.

**Greene, Robert.** *Greene's "Pandosto" or "Dorastus and Fawnia": being the Original of Shakespeare's "Winter's Tale."* Edited by P. G. Thomas. London: Chatto & Windus, 1907.

**Gregory, David.** "Memoranda." In *The Correspondence of Isaac Newton* 3:338. Cambridge University Press, 1959–1977.

**Grosseteste, Robert.** *Concerning Lines, Angles, and Figures.* Translated by D. C. Lindberg. In SBMS, 385–88.

________. *On Light.* Translated by C. C. Riedl. Marquette University Press, 1978.

**Guillaume de Conches.** *See* **William of Conches.**

**Guillaume de Lorris and Jean de Meun.** *See* **de Lorris, Guillaume, and Jean de Meun.**

**Hales, Stephen.** *Vegetable Staticks* (1727). Edited by M. A. Hoskin. London: Macdonald, 1969.

**Halley, Edmund.** "An Account of the Late Surprizing Appearance of the Lights Seen in the Air." *Philosophical Transactions* 29 (1716), 406–28.

**Hartmann, Georg.** "The Letter of Georg Hartmann to Duke Albrecht of Prussia (1544)." In H. D. Harradon, "Some Early Contributions to the History of Geomagnetism—IV," TMAE 48 (1943), 128–30.

**Hartmann von Aue.** *Erec.* In *Bibliothek des Mittelalters* 5. Frankfurt: Deutsche Klassiker, 2004.

**Harvey, William.** *An Anatomical Disquisition on the Motion of the Heart and Blood in Animals* (1628). Translated by R. Willis. In GBWW 28:267–312.

**Hauksbee, Francis.** "An Account of Experiments concerning the Proportion of the Power of the Load-stone at Different Distances." *Philosophical Transactions* 27 (1712), 506–11.

**Helmont, Jean Baptiste van.** *Van Helmont's Works.* Translated by J. Chandler. London: 1664.

**Helsham, Richard.** *A Course of Lectures in Natural Philosophy* (1767). Dublin: Physics Department of Trinity College, 1999.

**Hermes Trismegistus.** *Hermetica: The Ancient Greek and Latin Writings Which Contain Religious or Philosophic Teachings Ascribed to Hermes Trismegistus.* Translated and edited by W. Scott. Oxford: Clarendon Press, 1924.

**Herodotus.** *The History of Herodotus.* Translated by G. Rawlinson. In GBWW 6:1-341.

**Hildegard von Bingen.** *Hildegard von Bingen's Physica: The Complete English Translation of Her Classic Work on Health and Healing.* Translated by P. Throop. Rochester: Healing Arts Press, 1998.

**Hippocrates.** "Nature of Man." In *Hippocrates*, vol. 4, translated by W. H. S. Jones, 1–41. London: Heinemann, 1967.

**Homer.** *The Odyssey.* Rendered into English prose by S. Butler. In GBWW 4:183-322.

**Hooke, Robert.** *An Attempt to Prove the Motion of the Earth by Observations* (1674). In *Early Science in Oxford*, edited by R. T. Gunther, vol. 8, 1–28.

_________. *Cometa, or Remarks about Comets* (1678). In *Early Science in Oxford*, edited by R. T. Gunther, vol. 8, 217–69.

_________. *A Discourse of the Nature of Comets.* In *The Posthumous Works of Robert Hooke*, 149–92. New York: Johnson Reprint, 1969.

_________. Hooke to Newton, 24 November 1679. In *The Correspondence of Isaac Newton* 2:297. Cambridge University Press, 1959–1977.

_________. Hooke to Newton, 6 January 1679/1680. In *The Correspondence of Isaac Newton* 2:309. Cambridge University Press, 1959–1977.

_________. *Lamp, or Descriptions of Some Mechanical Improvements of Lamps* (1677). In *Early Science in Oxford*, edited by R. T. Gunther, vol. 8, 155–209.

_________. *Micrographia* (1665). Mineola, NY: Dover, 2003.

_________. *Potentia Resistutiva, or Spring* (1678). In *Early Science in Oxford*, edited by R. T. Gunther, vol. 8, 333–56.

**Hugh of Saint Victor.** *The "Didascalicon" of Hugh of Saint Victor: A Medieval Guide to the Arts.* Translated and annotated by J. Taylor. Columbia University Press, 1991.

**Huygens, Cristiaan.** *Treatise on Light.* Translated by S. P. Thompson. University of Chicago Press, 1945.

**Isidore of Seville.** *The Etymologies.* Translated by E. Brehaut in "An Encyclopedist of the Dark Ages: Isidore of Seville," *Studies in History, Economics and Public Law* 48 (1912), 1–274.

_________. "Isidore of Sevilla, On Man and Monsters: A Translation of 'Isidori Etymologiarum sive originum liber XI: De homine et portentis.'" In "Isidore of Seville: The Medical Writings," by W. D. Sharpe, *Transactions of the American Philosophical Society*, n.s., 54 (1964), 5–75.

**Jacobus de Voragine.** *The Golden Legend: Selections.* Translated and selected by C. Stage. London: Penguin Books, 1998.

**Jean of Saint-Amand.** *Expositio super antidorum Nicolai.* Translated by L. Thorndike in "John of St. Amand on the Magnet," *Isis*, 36 (1946), 156–57.

**John of Salisbury.** *The Metalogicon of John of Salisbury: A Twelfth-Century Defense of the Verbal and Logical Arts of the Trivium.* Translated and annotated by D. D. McGarry. University of California Press, 1955.

**Jonson, Ben.** *The Alchemist* (1612). In *Ben Jonson's Plays*, vol. 2, 1–90. London: Dent, 1957.

**Kant, Immanuel.** *Critique of Pure Reason* (1781). Translated and edited by P. Guyer and A. W. Wood. Cambridge University Press, 1998.

**Kepler, Johannes.** *Astronomia nova* (1609). In JKGW 3:1–424.

_________. *Astronomiae pars optica* (1604). In JKGW, 2:5–391.

_________. *A Defense of Tycho against Ursus/Apologia pro Tychone contra Ursum.* In *The Birth of History and Philosophy of Science*, by N. Jardine, 85–207. Cambridge University Press, 1984.

_________. *Epitome astronomiae Copernicanae* (1618–1621). In JKGW 7:1–537.

_________. *Epitome of Copernican Astronomy.* Books 4 and 5. Translated and selected by C. G. Wallis. In GBWW 16: 839–1004.

_________. *Kepler's Conversation with Galileo's Sidereal Messenger* (1610). Translated and annotated by E. Rosen. New York, Johnson Reprint, 1965.

_________. *Kepler's Somnium: The Dream, or Posthumous Work on Lunar Astronomy.* Translated by E. Rosen. University of Wisconsin Press, 1967.

_________. Letters, in JKGW, Bd.14–Bd.16.

_________. Letters, in *Johannes Kepler: Life and Letters*, translated and edited by C. Baumgardt. New York: Philosophical Library, 1951.

_________. *Mysterium cosmographicum* (1596). In JKGW 1:1–80; ed. altera (1621) in JKGW 8:5–128.

_________. *Mysterium cosmographicum: The Secret of the Universe.* Translated by A. M. Duncan. Introduction and commentary by E. J. Aiton. Preface by I. B. Cohen. New York: Abaris Books, 1981.

_________. *New Astronomy.* Translated by W. H. Donahue. Cambridge University Press, 1992.

_________. *Optics: Paralipomena to Witelo, & Optical Part of Astronomy.* Translated by W. H. Donahue. Santa Fe: Green Lion Press, 2000.

**Knight, Gowin.** "A Collection of the Magnetical Experiments Communicated to the Royal Society." In *Philosophical Transactions* 44 (1746/1747), 656–72.

**Leibniz, Gottfried Wilhelm.** *The Leibniz-Clarke Correspondence.* Translated, edited, and annotated by H. G. Alexander. Manchester University Press, 1956.

_________. *Mathematische Schriften.* Edited by G. I. Gerhardt. Hildesheim: Georg Olms Verlag, 1971.

_________. *Theodicy: Essays on the Goodness of God, the Freedom of Man, and the Origin of Evil* (1710). Translated by E. M. Huggard. Edited by A. M. Farrer. London: Routledge & Kegan Paul, 1952.

**Lorris, Guillaume de.** *See* **de Lorris, Guillaume, and Jean de Meun.**

**Lucretius.** *De rerum natura.* Translated by W. H. D. Rouse. Revised by M. F. Smith. Harvard University Press, 1975.

**Macrobius, Ambrosius Aurelius.** *Commentary on the Dream of Scipio.* Translated and annotated by W. H. Stahl. Columbia University Press, 1952.

**Magnus, Olaus.** *Historia de gentibus septentrionalibus: Romae 1555/A Description of the Northern Peoples: Rome 1555.* Translated by P. Fisher and H. Higgens. Edited by P. Foote. 3 vols. London: Hakluyt Society, 1996–1998.

**Mandeville, John.** *Mandeville's Travels.* Translated by M. Letts. Nendeln: Kraus Reprint, 1967.

**Marbode (Bishop of Rennes).** *De lapidibus.* With English translation by J. M. Riddle in "Marbode of Rennes' De lapidibus," *Sudhoffs Archiv: Zeitschrift für Wissenschaftsgeschichte*, Beihefte 20 (1977), 34–118.

________. *Marbode's Christian Symbolic Lapidary in Prose.* With English translation by J. M. Riddle in "Marbode of Rennes' De lapidibus," *Sudhoffs Archiv: Zeitschrift für Wissenschaftsgeschichte*, Beihefte 20 (1977), 125–29.

**Marco Polo.** *See* **Polo, Marco.**

**Marlowe, Christopher.** *The Tragical History of Doctor Faust* (1592–1593). In *Eight Famous Elizabethan Plays.* New York: Modern Library, 1932.

**Maxwell, James Clerk.** *Matter and Motion* (1877). London: Sheldon Press, 1925.

**Mayer, Tobias.** *The Unpublished Writings of Tobias Mayer.* Translated and edited by E. G. Forbes. 3 vols. Vol. 3, *The Theory of the Magnet and Its Application to Terrestial Magnetism.* Göttinsgen: Vandenhoeck and Ruprecht, 1972.

**Mercator, Gerhard.** "Gerhard Mercator of Rupelmonde to Antonius Perrenotus, Most Venerable Bishop of Arras, A.D. 1546." In H. D. Harradon, "Some Early Contributions to the History of Geomagnetism—VI," TMAE 48 (1943), 201–2.

________. "Nova et aucta orbis terrae descriptio ad usum navigantium emendate accommodate." In "Text and Translation of the Legends of the Original Chart of the World by Gerhard Mercator, issued in 1569," *Hydrographic Review* 9 (1932), 7–45.

**Molière.** *The Learned Women.* In *The Dramatic Works of Molière*, translated into English prose by C. H. Wall, vol. 3. London: George Bell, 1908.

**Montaigne, Michel de.** *The Complete Essays.* Translated and edited by M. A. Screech. Harmondsworth: Penguin Books, 1991.

**Musschenbroek, Pieter van.** "De Viribus Magneticis. Epistola Viri Doctissimi P. Muschenbroek, M. D. ad J. Th. Desaguliers, LL. D. R. S. S.," *Philosophical Transactions* 33 (1725), 370–78.

**Nashe, Thomas.** *The Unfortunate Traveller* (1594). In *Shorter Novels*, vol. 1, 261–356. London: Dent, 1949.

**Newton, Isaac.** *Certain Philosophical Questions: Newton's Trinity Notebook*. Edited by J. E. McGuire and M. Tamny. Cambridge University Press, 1983.

_________. *The Correspondence of Isaac Newton*. 7 vols. Cambridge University Press, 1959–1977.

_________. "De aere et aethere." In *Unpublished Scientific Papers of Isaac Newton*, 214–20 (Latin) and 221–28 (English).

_________. "Draft Addition to the Principia Book III, Proposition VI, Corollaries 4 and 5." In *Unpublished Scientific Papers of Isaac Newton*, 314 (Latin) and 316 (English).

_________. *An Hypothesis Explaining the Properties of Light* (1675). In *Isaac Newton's Papers & Letters on Natural Philosophy*, 178–90.

_________. *Isaac Newton's Papers & Letters on Natural Philosophy*. Edited by I. B. Cohen and R. E. Schofield. 2nd ed. Harvard University Press, 1978.

_________. *Isaac Newton's Philosophiae naturalis principia mathematica*. Edited by A. Koyré and I. B. Cohen. 2 vols. Harvard University Press, 1972.

_________. *Mathematical Principles*. See *Philosophiae naturalis principia mathematica* and *Sir Isaac Newton's Mathematical Principles of Natural Philosophy and His System of the World*.

_________. *Opticks* (1730). New York: Dover, 1952.

_________. *Philosophiae naturalis principia mathematica* (1687). 1st ed. Brussels: Culture et Civilisation, 1965.

_________. *Philosophiae naturalis principia mathematica* (1726). 3rd ed. See *Isaac Newton's Philosophiae naturalis principia mathematica*.

_________. *Sir Isaac Newton's Mathematical Principles of Natural Philosophy and His System of the World* (1729). Translated by

A. Mott. Revised by F. Cajori. University of California Press, 1947.

________. *Unpublished Scientific Papers of Isaac Newton: A Selection from the Portsmouth Collection in the University Library, Cambridge*. Translated and edited by A. R. Hall and M. B. Hall. Cambridge University Press, 1978.

**Nicolaus Cusanus.** *See* **Cusanus, Nicolaus.**

**Norman, Robert.** *The Newe Attractive* (1581). Norwood: Walther J. Johnson, 1974.

**Olaus Magnus.** *See* **Magnus, Olaus.**

**Oresme, Nicole.** *Nicole Oresme and the Kinematics of Circular Motion: Tractatus de commensurabilitate vel incommensurabilitate motuum celi.* Translated and edited by E. Grant. University of Wisconsin Press, 1971.

________. *Nicole Oresme and the Medieval Geometry of Qualities and Motions: A Treatise on the Uniformity and Difformity of Intensities Known as Tractatus de configutationibus qualitatum et motuum*. Translated and edited by M. Clagette. University of Wisconsin Press, 1968.

________. *On the Book of the Heavens and the World of Aristotle.* In *The Science of Mechanics in the Middle Ages*, edited by M. Clagget, 463–64, 570–71, and 600–609. University of Wisconsin Press, 1959.

**Origen.** *Origen, On First Principles.* Translated and annotated by G. W. Butterworth. Gloucester: Peter Smith, 1973.

**Orta, Garcia da.** *See* **da Orta, Garcia.**

**Osiander, Andreas.** "To the Reader concerning the Hypotheses of This Work." In Copernicus, *On the Revolutions*, XX. Johns Hopkins University Press, 1978.

**Ovid.** *The Metamorphoses of Ovid.* Translated by M. M. Innes. Harmondsworth: Penguin Books, 1955.

**Oviedo, Gonzalo Fernández de (Gonzalo Fernández de Oviedo y Valdés).** *Natural History of the West Indies.* Translated and edited by S. A. Stoudemire. University of North Carolina Press, 1959.

**Palissy, Bernard.** *The Admirable Discourses of Bernard Palissy.* Translated by A. La Rocque. University of Illinois Press, 1957.

**Paracelsus.** *Astronomia Magna, oder die ganze Philosophia sagax der grossen und kleinen Welt.* Frankfurt, Peter Lang, 1999.

_________. *Concerning the Nature of Things.* In *The Hermetic and Alchemical Writings of Paracelsus*, edited by A. W. Wait, 1:120–94. Berkeley: Shambhala, 1976.

_________. *Das Baseler Programm* (1527). In *Sämtliche Werke*, edited by K. Sudhoff and W. Matthiessen, 4:3–4. Hildesheim: Georg Olms, 1996.

_________. *Das erste Buch der Grossen Wundarzney* (1536). In *Sämtliche Werke*, edited by K. Sudhoff and W. Matthiessen, 10:7–200. Hildesheim: Georg Olms, 1996.

_________. *The Diseases that Deprive Man of His Reason.* Translated by G. Zilboorg. In *Four Treatises of Theophrastus von Hohenheim, Called Paracelsus*, 127–212. Johns Hopkins University Press, 1996.

———. *Four Treatises of Theophrastus von Hohenheim, Called Paracelsus.* Edited by H. E. Sigerist. Johns Hopkins University Press, 1996.

_________. "The Herbarius of Paracelsus." Translated by B. Moran in *Pharmacy in History* 35 (1993), 99–127.

_________. *The Hermetic and Alchemical Writings of Paracelsus.* Edited by A. W. Wait. 3 Vols. Vol. 1, *Hermetic Chemistry.* Berkeley: Shambhala, 1976.

_________. *On the Miners' Sickness and Other Miners' Diseases.* Translated by G. Rosen. In *Four Treatises of Theophrastus von Hohenheim, Called Paracelsus*, 43–126. Johns Hopkins University Press, 1996.

_________. *The Prophecies of Paracelsus*. London: Rider, 1974.

_________. *Sämtliche Werke*. Edited by K. Sudhoff and W. Matthiessen. Hildesheim: Georg Olms, 1996.

_________. *Selected Writings*. Translated by N. Guterman. Edited by J. Jacobi. Princeton University Press, 1979.

_________. *Seven Defensiones*. Translated by C. L. Temkin. In *Four Treatises of Theophrastus von Hohenheim, Called Paracelsus*, 1–41. Johns Hopkins University Press, 1996.

_________. *Volumen paramirum*. In *Sämtliche Werke*, edited by K. Sudhoff and W. Matthiessen, 1:163–240. Hildesheim: Georg Olms, 1996.

**Petrarch, Francesco.** *See* **Petrarca, Francesco.**

**Petrarca, Francesco.** *Canzoniere*. Translated by J. G. Nichols. Manchester: Carcanet, 2002.

**Petrus Peregrinus.** *Epistola Petri Peregrini de Maricourt ad Sygerum de Foucaucourt militem: De magnete* (1269). In "Le *De magnete* de Pierre de Maricourt: Traduction et commentaire," by P. Radelet-de Grave and D. Speiser, *Revue d'Histoire des Sciences* 28 (1975), 201–230 (Latin-French).

_________. *The Letter of Peregrinus on the Magnet*. Translated by Brother Arnold (J. C. Mertens). In SBMS, 368-376.

_________. *The Letter of Peter Peregrinus de Maricourt to Sygerus de Foucaucourt, Soldier, concerning the Magnet*. Translated by H. D. Harradon in "Some Early Contributions to the History of Geomagnetism—I," TMAE 48 (1943), 6–18.

**Petty, William.** *Political Arithmetick* (1690). Düsseldorf: Verlag Wirtschaft und Finanzen, 1992.

**Pico della Mirandola.** *On the Dignity of Man*. Translated by C. G. Wallis. Indianapolis: Bobbs-Merrill Company, 1965.

**Pietro Martire d'Anghiera.** *See* **D'Anghiera, Pietro Martire.**

**Pigafetta, Antonio.** *Magellan's Voyage: A Narrative Account of the First Circumnavigation.* Translated and edited by R. A. Skelton. 2 vols. Yale University Press, 1969.

**Plato.** *Ion.* Translated by B. Jowett. In GBWW 7:142–48.

_________. *Timaeus.* Translated by B. Jowett. In GBWW 7:442–77.

**Pliny (Pliny the Elder).** *Natural History.* Translated by H. Rackham. 10 vols. Harvard University Press, 1961–1968.

**Pliny (Pliny the Younger).** *The Letters of the Younger Pliny.* Translated by B. Radice. Harmondsworth: Penguin Books, 1969.

**Plutarch.** *Plutarch's Moralia.* Vol. 8, translated by P. A. Clement and H. B. Hoffleit. Vol. 13, part 1, translated by H. Cherniss. Harvard University Press. 1969, 1976.

**Polo, Marco.** *The Travels of Marco Polo.* Translated by R. Latham. Harmondsworth: Penguin Books, 1958.

**Pomponazzi, Pietro.** *De naturalium effectuum causis sive de Incantationibus.* (1567). Hildesheim: Georg Olms, 1970.

_________. *Les causes des merveilles de la nature: ou Les enchantements.* Translated and annotated by H. Busson. Paris: Les Éditions Rieder, 1930. (French)

**Porphyry.** *On Abstinence from Killing Animals.* Translated by G. Clark. Cornell University Press, 2000.

**Porta, Giambattista della.** *See* **della Porta, Giambattista.**

**Power, Henry.** *Experimental Philosophy, in Three Books, Containing New Experiments, Microscopical, Mercurial, Magnetical* (1664). Introduction by M. Boas Hall. New York: Johnson Reprint, 1966.

**Priestley, Joseph.** *The History and Present State of Electricity.* 4th edition. London, 1775.

**Ptolemy.** *Ancient India as Described by Ptolemy.* Translated by J. W. McCrindle. Edited by R. Jain. New Delhi: Today and Tomorrow's Printers and Publishers, 1974.

_________. *Ptolemy's Almagest*. Translated and annotated by G. J. Toomer. Princeton University Press, 1997.

**Recorde, Robert.** *The Castle of Knowledge* (1556). Amsterdam: Johnson Reprint, 1975.

**Richard de Bury.** *See* **de Bury, Richard.**

**Ridley, Marke.** *Magneticall Animadversions*. London. 1617.

_________. *A Short Treatise of Magneticall Bodies and Motions*. London, 1613.

**Savery, Servington.** "Magnetical Observations and Experiments." *Philosophical Transactions* 36 (1730), 295–340.

**Scot, Reginald.** *The Discoverie of Witchcraft* (1584). London, 1886.

**Shakespeare, William.** *The Winter's Tale*. In GBWW 27: 489–523.

**Smith, Adam.** *Essays on Philosophical Subjects*. Edited by W. P. D. Wightman and J. C. Bryce. Oxford: Clarendon Press, 1980.

**Spenser, Edmund.** *The Faerie Queene*. Introduction by J. W. Hales. 2 vols. London: Dent, 1917–1919.

**Spinoza, Baruch de.** *The Correspondence of Spinoza*. Translated, edited, and annotated by A. Wolf. New York: Russell & Russell, 1966.

**Stevin, Simon.** *The Haven-finding Art: or, The Way to Find Any Haven or Place at Sea, by the Latitude and Variation* (1599). Translated by E. Wright. New York: Da Capo Press, 1968.

_________. *The Principal Works of Simon Stevin*. Vol. 3, *Astronomy, Navigation*. Amsterdam: Swets & Zeitlinger, 1961.

**Tacitus, Cornelius.** *The Agricola and the Germania*. Translated by H. Mattingly. Revised by S. A. Handford. Harmondsworth: Penguin Books, 1970.

**Taylor, Brook.** "An Account of an Experiment Made by Dr. Brook Taylor Assisted by Mr. Hawkesbee, in Order to Discover the

Law of the Magnetical Atrraction." *Philosophical Transactions* 29 (1715), 294–95.

_________. "Extract of a Letter from Dr. Brook Taylor, F. R. S. to Sir Hans Sloan." *Philosophical Transactions* 31 (1721), 204–8.

**Theophrastus.** *On Stones.* Greek Text with English translation and commentary by E. R. Caley and J. F. C. Richards. Ohio State University, 195.

**Thevet, André (Thevet, Andrewe).** *The New Found Worlde, or Antarctike.* (1568). New York: Da Capo Press, 1971.

**Thomas Aquinas.** *Commentary on Aristotle's Physics.* Translated by R. J. Blackwell, R. J. Spath, and W. E. Thirlkel. Introduction by V. J. Bourke. Foreword by R. McInerny. Notre Dame: Dumb Ox Books, 1999.

_________. *Commentary on the Metaphysics of Aristotle.* Translated by J. P. Rowan. Chicago: Regnery, 1961.

_________. *The Disputed Questions on Truth.* Vol. 1. Translated by R. W. Mulligan. Chicago: Regnery, 1952.

_________. *On Being and Essence.* Translated by A. Maurer. Toronto: Pontificial Institute of Mediaeval Studies, 1968.

_________. *On Spiritual Creatures (De spiritualibus creaturis).* Translated by M. C. FitzPatrick and J. J. Wellmuth. Marquette University Press, 1949.

_________. *On the Truth of the Catholic Faith: Summa contra Gentiles.* Vol. 3, part 2. Translated by V. J. Bourke. New York: Hanover House, 1956.

_________. *On the Unity of the Intellect against the Averroists.* Translated by B. H. Zedler. Marquitte University Press, 1968.

_________. *The Principles of Nature.* In *Selected Writings of St. Thomas Aquinas*, 1–28. Translated and annotated by R. P. Goodwin. Indianapolis: Bobbs-Merrill, 1965.

_________. *The Soul: A Translation of St. Thomas Aquinas' De Anima.* Translated by J. P. Rowan. St. Louis: B. Herder, 1951.

_________. *Summa theologica.* 5 vols. Translated by Fathers of the English Dominican Province. Westminster: Christian Classics, 1981.

_________. (Tomás de Aquino). *Super evangelium S. Ioannis lectura.* Torino: Marietti, 1972.

_________. *Treatise on Separate Substances.* Translated and annotated by F. J. Lescoe. West Hartford: Saint Joseph College, 1959.

**van Helmont.** *See* **Helmont, Jean Baptiste van**

**van Musschenbroek, Pieter.** *See* **Musschenbroek, Pieter van.**

**Vesalius, Andreas.** *De Humani Corporis Fabrica.* In *The Development of Modern Science*, vol. 2 of *In Moments of Discovery*, edited by G. Schwartz and P. W. Bishop, 515–53. New York: Basic Books, 1958.

_________. *Fabrica.* In *Andreas Vesalius of Brussels, 1514–1564*, by C. D. O'Malley, appendix, 317–24. University of California Press, 1964.

**Vives, Juan Luis.** *On Education: A Translation of the De tradendis disciplinis of Juan Luis Vives* (1531). Introduction by F. Watson. Cambridge University Press, 1913.

**Voltaire.** *The Elements of Sir Isaac Newton's Philosophy* (1738). Translated by J. Hanna. London: Cass, 1967.

**von Eschenbach, Wolfram.** *See* **Wolfram von Eschenbach.**

**Webster, John.** *The Duchess of Malfi* (1614). In *Eight Famous Elizabethan Plays.* New York: Modern Library, 1932.

**Weld, Charles Richard.** *A History of the Royal Society, with Memoirs of the Presidents* (1848). 2 vols. Bristol: Thoemmes Press, 2000.

**William of Conches.** *Philosophia mundi.* Edited by G. Maurach. University of South Africa, 1974.

**Wilkins, John.** *The Mathematical and Philosophical Works of the Right Rev. John Wilkins* (1802). Vol. 1, book 1, *The Discovery of a New World: or, A Discourse Tending to Prove, That It Is Probable There May Be Another Habitable World in the Moone* (1640), 1–130. Vol. 1, book 2, *A Discourse concerning a New Planet: Tending to Prove, That Our Earth Is One of the Planets* (1640), 131–261. Vol. 2, book 1, *Mathematicall Magick, or the Wonders That May be Performed by Mechanicall Geometry* (1648), 88–246. London: Cass, 1970.

**Wolfram von Eschenbach.** *Parzival.* In *Bibliothek des Mittelalters* 8. Frankfurt: Deutsch Klassiker, 1994.

## Secondary Sources

**Abromitis, Lois Irene.** *William Gilbert as Scientist: The Portrait of a Renaissance Amateur.* Brown University PhD thesis, 1977.

**Aczel, Amir D.** *The Riddle of the Compass.* New York: Harcourt, 2001.

**Alford, John A.** "Medicine in the Middle Ages: The Theory of a Profession." *Centennial Review* 23 (1979), 377–96.

**Applebaum, Wilbur.** "Keplerian Astronomy after Kepler: Researches and Problems." *History of Science* 34 (1996), 451–504.

**Arber, Agnes.** *Herbals, Their Origin and Evolution: A Chapter in the History of Botany 1470–1670.* Cambridge Univeristy Press, 1938.

**Baldwin, Martha R.** "Magnetism and the Anti-Copernican Polemic." JHA 16 (1985), 155–74.

**Barbour, Julian B.** *The Discovery of Dynamics.* Oxford University Press, 2001.

**Barnes, Jonathan, ed.** *Early Greek Philosophy.* London: Penguin Books, 2001.

**Baumgardt, Carola.** *Johannes Kepler: Life and Letters.* New York: Philosophical Library, 1951.

**Beckmann, Johann.** *A History of Inventions, Discoveries and Origins.* Translated by W. Johnston. London: Bohn, 1846.

**Benjamin, Park.** *The Intellectual Rise in Electricity: A History.* London: Longmans, Green & Co., 1895.

**Bennett, James A.** "Cosmology and the Magnetical Philosophy, 1640–1680." JHA 12 (1981), 165–77.

_________. "Hooke and Wren and the System of the World: Some Points towards an Historical Account." BJHS 8 (1975), 32–61.

_________. "Magnetical Philosophy and Astronomy from Wilkins to Hooke." In *Planetary Astronomy from the Renaissance to the Rise of Astrophysics*, edited by R. Taton and C. Wilson (vol. 2 of *The General History of Astronomy*, edited by M. A. Hoskin), 222–30. Cambridge University Press, 1989.

_________. *The Mathematical Science of Christopher Wren.* Cambridge University Press, 1982.

_________. "The Mechanics' Philosophy and the Mechanical Philosophy." *History of Science* 24 (1986), 1–28.

**Bernal, John Desmond.** *Science in History.* London: Watts, 1954.

**Boas, Marie (Marie Boas Hall).** "The Establishment of the Mechanical Philosophy." *Osiris* 10 (1952), 412–541.

_________. "Introduction." In *Experimental Philosophy*, by H. Power, ix–xxvii. New York: Johnson Reprint, 1966.

_________. "Review of Natural Magick." *Isis* 50 (1959), 76.

_________. *The Scientific Renaissance 1450–1630.* London: Collins, 1962.

**Boas, Marie, and Rupert Hall.** "Tycho Brahe's System of the World." *Occasional Notes* 3 (1959), 253–63.

**Bonelli, Maria Luisa Righini.** *See* **Righini Bonelli, Maria Luisa.**

**Borst, Otto.** *Alltagsleben in Mittelalter.* Frankfurt: Insel, 1983.

**Boss, Valentin.** *Newton and Russia: The Early Influence, 1698–1796.* Harvard University Press, 1972.

**Braudel, Fernand.** *La Méditerranée: l'aspace et l'histoire.* Flammarion, 1985.

**Brett, George S.** *The Philosophy of Gassendi.* London: Macmillan, 1908.

**Bromehead, Cyril N.** "Alexander Neckam on the Compass-Needle." TMAE 50 (1945), 139–40.

_________. "Geology in Embryo (up to 1600 A.D.)." *Proceedings of the Geologists' Association* 56 (1945), 89–134.

**Brother Potamian.** *See* **Potamian (Brother Potamian).**

**Bruno, Leonard C.** *Landmarks of Science.* New York: Facts on File, 1989.

**Bullough, Vern L.** "Population and the Study and Practice of Medieval Medicine." *Bulletin of the History of Medicine* 36 (1962), 62–69.

**Burckhardt, Jacob.** *The Civilization of the Renaissance in Italy.* Translated by S. G. C. Middlemore. London: Penguin Books, 1990.

**Burke, Peter.** *Culture and Society in Renaissance Italy 1420–1540.* New York: Charles Scribner's Sons, 1972.

**Burtt, Edwin Arthur.** *The Metaphysical Foundations of Modern Physical Science.* London: Routledge & Kegan Paul, 1972.

**Butterfield, Herbert.** *The Origins of Modern Science 1300–1800.* London: G. Bell and Sons, 1957.

**Campbell, Tony.** *Early Maps.* New York: Cross River Press, 1981.

**Caspar, Max.** *Kepler.* Translated and edited by C. D. Hellman. Introduction and references by O. Gingerich. New York: Dover, 1993.

**Cassirer, Ernst.** *Das Erkenntnisproblem in der Philosophie und Wissenschaft der neueren Zeit.* Bd. 1. Dritte Aufl. Berlin: Bruno Cassirer, 1922.

_________. *The Individual and the Cosmos in Renaissance Philosophy.* Translated by M. Domandi. University of Pennsylvania Press, 1963.

**Chalmers, Gordon Keith.** "The Lodestone and the Understanding of Matter in Seventeenth Century England." *Philosophy of Science* 4 (1937), 75–95.

_________. "Sir Thomas Browne, True Scientist." *Osiris* 2 (1936), 28–79.

**Chapman, Sydney.** "William Gilbert and the Science of His Time." *Nature* 154 (1944), 132–36.

**Charon, Véronique.** "The Knowledge of Herbs." In *The Pagan Middle Ages*, edited by L. Milis, 109–28. Woodbridge: Boydell Press, 1998.

**Clulee, Nicholas H.** "Astrology, Magic, and Optics: Facets of John Dee's Early Natural Philosophy." *Renaissance Quarterly* 30 (1977), 632–80.

_________. "At the Crossroads of Magic and Science: John Dee's Archemastrie." In *Occult and Scientific Mentalities in the Renaissance*, edited by B. Vickers, 57–71. Cambridge University Press, 1984.

_________. "John Dee's Mathematics and the Grading of Compound Qualities." *Ambix* 18 (1971), 178–211.

_________. *John Dee's Natural Philosophy: Between Science and Religion*. Routledge, 1988.

**Cohen, I. Bernard.** *Franklin and Newton: An Inquiry into Speculative Newtonian Experimental Science and Franklin's Work in Electricity as an Example Thereof.* Philadelphia: American Philosophical Society, 1956.

_________. "Newton, Hooke, and 'Boyle's Law' (Discovered by Power and Towneley)." *Nature* 204 (1964), 618-621.

**Copenhaver, Brian P.** "Did Science Have a Renaissance?" *Isis* 83 (1992), 387–407.

_________. "Scholastic Philosophy and Renaissance Magic in the *De vita* of Marsilio Ficino." *Renaissance Quarterly* 37 (1984), 523–54.

**Crane, Nicholas.** *Mercator: The Man Who Mapped the Planet.* London: Phoenix, 2002.

**Crombie, Alistair Cameron.** *Augustine to Galileo.* London: Heinemann Educational Books, 1979.

_________. *Robert Grosseteste and the Origins of Experimental Science 1100–1700.* Oxford: Clarendon Press, 1953/1971.

_________. *Science, Optics and Music in Medieval and Early Modern Thought.* London: Hambledon Press, 1990.

_________. *Styles of Scientific Thinking in the European Tradition.* 3 vols. London: Duckworth, 1994.

**Crone, Ernst.** "Introduction." In *The Principal Works of Simon Stevin* 3:363–417. Amsterdam: Swets & Zeitlinger, 1961.

**Davis, Natalie Zemon.** *Society and Culture in Early Modern France: Eight Essays.* Stanford University Press, 1975.

**Deason, Gary.** "Reformation Theology and the Mechanistic Conception of Nature." In *God and Nature*, edited by D. C. Lindberg and R. L. Numbers, 167–91. University of California Press, 1986.

**Debus, Allen G.** *The Chemical Philosophy: Paracelsian Science and Medicine in the Sixteenth and Seventeenth Centuries.* 2 vols. New York: Science History Publications, 1977.

_________. *The English Paracelsians.* New York: Franklin Watts, 1966.

_________. "Introduction." In *The Mathematicall Praeface to the Elements of Geometrie of Euclid of Megara*, by John Dee, 1–30. New York: Science History Publications, 1975.

_________. *Man and Nature in the Renaissance.* Cambridge University Press, 1978.

**de Libera, Alain.** *See* **Libera, Alain de.**

**de Villamil, Richard.** *Newton: The Man.* New York: Johnson Reprint, 1972.

**Dijksterhuis, Eduard J.** *The Mechanization of the World Picture.* Translated by C. Dikshoorn. Oxford University Press, 1961.

**Dobbs, Betty J. T.** *The Janus Faces of Genius: The Role of Alchemy in Newton's Thought.* Cambridge University Press, 2002.

_________. "Newton's Alchemy and His Theory of Matter." *Isis* 73 (1982), 511–28.

**Dodds, Eric Robertson.** *The Greeks and the Irrational.* University of California Press, 1951.

**Drake, Stillman.** *Telescopes, Tides and Tactics.* University of Chicago Press, 1983.

**Dufourcq, Charles-Emmanuel.** *La vie quotidienne dans l'Europe médiévale sous domination Arabe.* Paris: Hachette, 1978.

**Duhem, Pierre.** *The Aim and Structure of Physical Theory.* Translated by P. P. Wiener. Princeton University Press, 1954.

_________. *To Save the Phenomena: An Essay on the Idea of Physical Theory from Plato to Galileo.* Translated by E. Doland and C. Maschler. Introductory essay by S. L. Jaki. University of Chicago Press, 1969.

**Eamon, William.** "Books of Secrets in Medieval and Early Modern Science." *Sudhoffs Archiv* 69 (1985), 26–49.

_________. *Science and the Secrets of Nature: Books of Secrets in Medieval and Early Modern Culture.* Princeton University Press, 1994.

_________. "Technology as Magic in the Late Middle Ages and the Renaissance." *Janus* 70 (1983), 171–212.

**Easlea, Brian.** *Witch Hunting, Magic, and the New Philosophy: An Introduction to Debates of the Scientific Revolution, 1450–1750.* Brighton: Harvester Press, 1980.

**Easton, Joy B.** "The Early Editions of Robert Recorde's *Ground of Artes.*" *Isis* 58 (1967), 515–32.

**Eisenstein, Elizabeth L.** *The Printing Revolution in Early Modern Europe.* Cambridge University Press, 1983.

**Elliott, Robert S.** *Electromagnetics: History, Theory, and Applications.* New York: IEEE Press, 1993.

**Farrington, Benjamin.** *Francis Bacon: Philosopher of Industrial Science.* London: Lawrence and Wishart, 1951.

_________. *Greek Science: Its Meaning for Us.* London: Penguin Books, 1953.

**Febvre, Lucien, and Henri-Jean Martin.** *The Coming of the Book: The Impact of Printing, 1450–1800.* Translated by D. Gerard. London: Verso, 1984.

**Feingold, Mordechai.** *The Mathematicians' Apprenticeship: Science, Universities and Society in England, 1560–1640.* Cambridge University Press, 1984.

**Fleckenstein, Joachim Otto.** "Vorwort des Herausgeber." In *Leonhardi Euleri Opera Omnia*, series 2, vol. 5, VII-L. Lausanne: Orell-Füssli, 1957.

**Fleming, John Ambrose.** "The Earth, a Great Magnet." TMAE 2 (1897), 45–60.

**Fontana, Josep.** *The Distorted Past: A Reinterpretation of Europe.* Translated by C. Smith. Oxford: Blackwell, 1995.

**Forbes, Eric Gray.** "Preface" and "Introduction." In *The Unpublished Writings of Tobias Mayer*, translated and edited by E. G. Forbes, vol. 3, 1–31. Göttinsgen: Vandenhoeck and Ruprecht, 1972.

**Fraunberger, Fritz**. *Elektrizität im Barock.* 2te Aufl. Köln: Aulis Verlag Deubner, n.d.

**Freeman, Kathleen.** *The Pre-Socratic Philosophers: A Companion to Diels, Fragmente der Vorsokratiker.* 2nd edition. Oxford: Blackwell, 1949.

**French, Peter.** *John Dee: The World of an Elizabethan Magus.* London: Routledge & Kegan Paul, 1972.

**French, Roger.** *Ancient Natural History: Histories of Nature.* London: Routledge, 1994.

**Freudenthal, Gad.** "Theory of Matter and Cosmology in William Gilbert's *De magnete.*" *Isis* 74 (1983), 22–37.

**Garin, Eugenio.** *Italian Humanism: Philosophy and Civic Life in the Renaissance.* Translated by P. Munz. Oxford: Blackwell, 1965.

_________. "The Philosopher and the Magus." In *Renaissance Characters,* edited by E. Garin and translated by L. G. Cochrane. University of Chicago Press, 1997.

_________. *Science and Civic Life in the Italian Renaissance.* Translated by P. Munz. Garden City: Anchor, 1969.

**Gillmor, Charles Stewart.** *Coulomb and the Evolution of Physics and Engineering in Eighteenth-Century France.* Princeton University Press, 1971.

**Gilmont, Jean-Francois.** "Protestant Reformations and Reading." In *A History on Reading in the West,* edited by G. Cavallo and R. Chartier and translated by L. G. Cochrane. Oxford: Polity Press, 1999.

**Gimpel, Jean.** *The Medieval Machine: The Industrial Revolution of the Middle Ages.* 2nd edition. Aldershot: Wildwood House, 1988.

**Gingerich, Owen.** "Kepler's Place in Astronomy." *Vistas in Astronomy* 18 (1975), 261–78.

**Goff, Jacques Le.** *See* **Le Goff, Jacques.**

**Grafton, Anthony.** *Cardano's Cosmos: The Worlds and Works of a Renaissance Astrologer.* Harvard University Press, 1999.

**Grant, Edward.** "Late Medieval Thought, Copernicus, and the Scientific Revolution." JHI 23 (1962), 197–220.

_________. *Physical Science in the Middle Ages.* Cambridge University Press, 1977).

**Gunther, Robert T., ed.** *The Cutler Lectures of Robert Hooke*. Vol. 8 of *Early Science in Oxford*, edited by R. T. Gunther and A. E. Gunther. London: Dawsons of Pall Mall, 1968.

**Gurevich, A. J.** *Categories of Medieval Culture*. Translated by G. L. Campbell. London: Routledge & Kegan Paul, 1985.

**Hackett, Jeremiah.** "Roger Bacon on *Scientia experimentalis*." In *Roger Bacon and the Sciences: Commemorative Essays*, edited by J. Hackett, 277–316. Leiden: Brill, 1997.

**Haggard, Howard W.** *Devils, Drugs, and Doctors*. London: Heinemann, 1929.

**Hall, A. Rupert.** *The Revolution in Science 1500–1750*. London: Longman, 1983.

**Hall, Marie Boas.** *See* **Boas, Marie.**

**Hall, Thomas S.** *Ideas of Life and Matter: Studies in the History of General Physiology, 600 B.C.–1900 A.D.* University of Chicago Press, 1969.

**Hansen, Bert.** "Science and Magic." In *Science in the Middle Ages*, edited by D. C. Lindberg, 483–506. University of Chicago Press, 1978.

**Hardin, Clyde L.** "The Scientific Work of the Reverend John Michell." *Annals of Science* 22 (1966) 27–47.

**Hartmann, Franz.** *The Life and Doctrines of Philippus Theophrastus, Bombast of Hohenheim, known by the Name of Paracelsus* (1891). Pomeroy, WA: Health Research, 1963.

**Harradon, H. D.** "Some Early Contributions to the History of Geomagnetism—I." TMAE 48 (1943), 3–17.

_________. "Some Early Contributions to the History of Geomagnetism—II & III." TMAE 48 (1943), 79–91.

_________. "Some Early Contributions to the History of Geomagnetism—IV." TMAE 48 (1943), 127–30.

_________. "Some Early Contributions to the History of Geomagnetism—V." TMAE 48 (1943), 197–99.

________. "Some Early Contributions to the History of Geomagnetism—VI." TMAE 48 (1943), 200-202.

________. "Some Early Contributions to the History of Geomagnetism—VII." TMAE 49 (1944), 185-99.

________. "Some Early Contributions to the History of Geomagnetism—VIII." TMAE 50 (1945), 63-68.

**Harré, Rom.** "William Gilbert." In *Early Seventeenth Century Scientists*, edited by R. Harré, 1-24. London: Pergamon Press, 1965.

**Harvey, John.** *Medieval Craftsmen.* New York: Drake, 1975.

**Haskins, Charles Homer.** "The 'De Arte Venandi cum Avibus' of the Emperor Frederick II." *English Historical Review*, July (1921), 334-55.

________. "Michael Scot and Frederick II." *Isis* 4 (1922), 250-75.

________. *The Renaissance of the Twelfth Century.* Cleveland: World Publishing, 1957.

**Hatch, Robert A.** "Bouliau, Ismaël (1605-1694)." In *Encyclopedia of the Scientific Revolution*, edited by W. Applebaum, 98-99. New York: Garland, 2000.

**Hazard, D. L.** "Early History of the Mariner's Compass and Earliest Knowledge of the Magnetic Declination according to Bertelli." TMAE 8 (1903), 179-83.

**Heathcote, Niels Hugh de Vaudrey.** "Early Nautical Charts." *Annals of Science* 1 (1936), 13-29.

**Heiberg, Johan L.** *Mathematics and Physical Science in Classical Antiquity.* Translated by D. C. Macgregor. Oxford University Press, 1922.

**Hellmann, Gustav.** "Die Anfänge der magnetischen Beobachtungen." *Zeitschrift der Gesellschaft für Erdkunde zu Berlin* 32 (1897), 112-36; published in English as "The Beginnings of Magnetic Observations," TMAE 4 (1899), 73-86.

**Henry, John.** "Occult Qualities and the Experimental Philosophy: Active Principles in Pre-Newtonian Matter Theory." *History of Science* 24 (1986), 335–81.

**Hesse, Mary B.** *Forces and Fields: The Concept of Action at a Distance in the History of Physics.* London: Thomas Nelson and Sons, 1961.

_________. "Gilbert and the Historians." BJPS 11 (1960), 1–10, 130–42.

**Hill, Christopher.** *Intellectual Origins of the English Revolution.* Oxford: Clarendon Press, 1965.

**Hill, Donald R.** *Islamic Science and Engineering.* Edinburgh University Press, 1993.

**Holmyard, Eric John.** *Alchemy.* New York: Dover, 1990.

**Holton, Gerald.** "Johannes Kepler's Universe: Its Physics and Metaphysics." AJP 24 (1956), 340–51.

**Home, Roderick Weir.** "Introduction." In *Aepinus's Essay on the Theory of Electricity and Magnetism*, by F. Aepinus, 1–224. Princeton University Press, 1979.

_________. "'Newtonianism' and the Theory of the Magnet." *History of Science* 15 (1977), 252–66.

**Hoppe, Edmund.** *Geschichte der Elektrizität.* Leipzig: Barth, 1884.

**Humboldt, Alexander von.** *Cosmos: A Sketch of a Physical Description of the Universe.* 4 vols. Translated by E. C. Otté et al. London: Bohn, 1848–1852.

**Huntley, Frank Livingstone.** *Sir Thomas Browne: A Biographical and Critical Study.* University of Michigan Press, 1962.

**Jammer, Max.** *Concepts of Force: A Study in the Foundations of Dynamics.* Harvard University Press, 1957.

**Jardine, Nicholas.** *The Birth of History and Philosophy of Science: Kepler's* A Defence of Tycho against Ursus *with Essays on Its Provenance and Significance.* Cambridge University Press, 1984.

**Johannsen, Otto.** *Geschichte des Eisens.* 2te Aufl. Düsseldorf: Verlag Stahleisen M.B.H., 1925.

**Johnson, Francis R.** *Astronomical Thought in Renaissance England: A Study of the English Scientific Writings from 1500 to 1645.* Johns Hopkins Press, 1937.

**Johnson, Francis R., and Sandford V. Larkey.** "Robert Recorde's Mathematical Teaching and Anti-Aristotelian Movement." *Huntington Library Bulletin* 7 (1935), 59–87.

_________. "Thomas Digges, the Copernican System, and the Idea of the Infinity of the Universe in 1576." *Huntington Library Bulletin* 5 (1934), 69–117.

**Jones, Richard Foster.** *Ancients and Moderns: A Study of the Rise of the Scientific Movement in Seventeenth-Century England.* 2nd edition. Peter Gloucester: Smith, 1961.

**Kaplan, Edward.** *Robert Recorde (c. 1510–1558): Studies in the Life and Works of a Tudor Scientist.* New York University PhD thesis, 1960.

**Kargon, Robert.** "Walter Charleton, Robert Boyle, and the Acceptance of Epicurean Atomism in England." *Isis* 55 (1964), 184–92.

_________. "Willam Petty's Mechanical Philosophy." *Isis* 56 (1965), 63-66.

**Kay, Charles D.** *William Gilbert's Renaissance Philosophy of the Magnet.* University of Pittsburgh PhD thesis, 1981.

**Kearney, Hugh.** *Science and Change: 1500–1700.* New York: McGraw-Hill, 1971.

**Kelly, Suzanne.** *The* De mundo *of William Gilbert.* Amsterdam: Menno Hertzberger, 1965.

**Keynes, John Maynard.** "Newton, the Man." In *Newton Tercentenary Celebrations: 15–19 July 1946*, 27–34. Cambridge University Press, 1947.

**King, W. James.** "The Natural Philosophy of William Gilbert and His Predecessors." *Bulletin: Contributions from the Museum of History and Technology* 218 (1959), 121–39.

**Klein-Franke, Felix.** "The Knowledge of Aristotle's Lapidary during the Latin Middle Ages." *Ambix* 17 (1970), 137–42.

**Klemm, Friedrich.** *A History of Western Technology.* Translated by D. W. Singer. London: Allen & Unwin, 1959.

**Koestler, Arthur.** *The Sleepwalkers: A History of Man's Changing Vision of the Universe.* New York: Macmillan, 1959.

**Koyré, Alexandre.** *Newtonian Studies.* London: Chapman & Hall, 1965.

**Krafft, Fritz.** "Sphaera activitatis—orbis virtutis. Das Entstehen der Vorstellung von Zentralkräften." *Sudhoffs Archiv* 54 (1970), 113–40.

**Kristeller, Paul Oskar.** *Eight Philosophers of the Italian Renaissance.* Stanford University Press, 1964.

__________. *Renaissance Thought: The Classic, Scholastic, and Humanistic Strains.* New York: Harper, 1961.

**Kunz, George Frederick.** *The Curious Lore of Precious Stones.* Philadelphia: Lippincott, 1913.

**Langdon-Brown, Walter.** "William Gilbert: His Place in the Medical World." *Nature* 154 (1944), 136–39.

**Le Goff, Jacques.** *Intellectuals in the Middle Ages.* Translated by T. L. Fagan. Cambridge: Blackwell, 1993.

__________. *Time, Work, and Culture in the Middle Ages.* Translated by A. Goldhammer. University of Chicago Press, 1980.

**Libera, Alain de.** *La philosophie médiévale.* Paris: Presses Universitaires de France, 1993.

**Lindberg, David C.** "Science as Handmaiden: Roger Bacon and the Patristic Tradition." *Isis* 78 (1987), 518–36.

_________. *Theories of Vision from al-Kindi to Kepler.* University of Chicago Press, 1976.

_________. "The Transmission of Greek and Arabic Learning to the West." In *Science in the Middle Ages*, edited by D. C. Lindberg, 52–90. University of Chicago Press, 1978.

**Lindberg, David C., ed.** *Science in the Middle Ages.* University of Chicago Press, 1978.

**Lindberg, David C., and Ronald L. Numbers, eds.** *God and Nature: Historical Essays on the Encounter between Christianity and Science.* University of California Press, 1986.

**Lindsay, Robert Bruce.** "William Gilbert and Magnetism in 1600." AJP 8 (1940), 271–82.

**Lipmann, Edmund O. von.** "Geschichte der Magnetnadel bis zur Erfindung des Kompasses (gegen 1300)." *Quellen und Studien zur Geschichte der Natur-wissenschaften und der Medizin* 3 (1932), 1–49.

**Lloyd, Geoffrey E. R.** *Aristotle: The Growth and Structure of His Thought.* Cambridge University Press, 1968.

_________. *Greek Science after Aristotle.* New York: Norton, 1973.

**Lohne, Johannes.** "Hooke *versus* Newton: An Analysis of the Documents in the Case on Free Fall and Planetary Motion." *Centaurus* 7 (1960), 6–52.

**Losee, John Price**. *A Historical Introduction to the Philosophy of Science.* Oxford University Press, 1972.

**Lovejoy, Arthur O.** *The Great Chain of Being: A Study of the History of an Idea.* Harvard University Press, 1976.

**Mach, Ernst.** *The Principles of Physical Optics: An Historical and Philosophical Treatment.* Translated by J. S. Anderson and A. F. A. Young. New York: Dover, 1926.

_________. *The Science of Mechanics: A Critical and Historical Account of Its Development.* Translated by T. J. McCormack. La Salle, IL: Open Court, 1960 [2006].

**Maddison, Robert E. W.** *The Life of the Honourable Robert Boyle F.R.S.* London: Taylor and Francis, 1969.

**Marenbon, John.** *Later Medieval Philosophy (1150–1350): An Introduction.* London/New York: Routledge, 1987.

**May, W. E.** "Alexander Neckam and the Pivoted Compass Needle." *Journal of the Institute of Navigation* 8 (1955), 283–84.

**McCormmach, Russell.** "John Michell and Henry Cavendish: Weighing the Stars." BJHS 4 (1968), 126–55.

**McEvoy, James.** *The Philosophy of Robert Grosseteste.* Oxford: Clarendon Press, 1982.

**McGuire, James E.** "Force, Active Principles, and Newton's Invisible Realm." *Ambix* 15 (1968), 154–208.

**Meli, Domenico Bertoloni.** *Equivalence and Priority: Newton versus Leibniz.* Oxford: Clarendon Press, 1993.

**Mendelssohn, Kurt.** *Science and Western Domination.* London: Thames and Hudson, 1976.

**Merton, Robert K.** *Social Theory and Social Structure.* Enlarged edition. New York: Free Press, 1968.

**Meuthen, Erich.** *Nikolaus von Kues, 1401–1464: Skizze einer Biographie.* Münster: Verlag Aschendorff, 1976.

**Michelet, Jules.** *La sorcière.* In *Oeuvres Complètes de J. Michelet* 37. Paris: Ernest Flammarion, 1898.

**Milis, Ludo J. R.** "Introduction: The Pagan Middle Ages: A Contradiction in Terms?" In *The Pagan Middle Ages*, edited by L. J. R. Milis, 1–12. Woodbridge: Boydell Press, 1998.

_________. **ed.** *The Pagan Middle Ages.* Translated by T. Guest. Woodbridge: Boydell Press, 1998.

**Mitchell, A. Crichton.** "Chapters in the History of Terrestrial Magnetism, Ch. 1, On the Directive Property of a Magnet in the Earth's Field and the Origin of the Nautical Compass." TMAE 37 (1932), 105–46.

_________. "Chapters in the History of Terrestrial Magnetism, Ch. 2, The Discovery of the Magnetic Declination." TMAE 42 (1937), 241–80.

_________. "Chapters in the History of Terrestrial Magnetism, Ch. 3, The Discovery of the Magnetic Inclination." TMAE 44 (1939), 77–80.

_________. "Chapters in the History of Terrestrial Magnetism, Ch. 4, The Development of Magnetic Science in Classical Antiquity." TMAE 51 (1946), 323–51.

**Morrall, John B.** *The Medieval Imprint.* London: Penguin Books, 1967.

**Mottelay, Paul Fleury.** *Bibliographical History of Electricity and Magnetism, Chronologically Arranged.* London: Griffin, 1922.

**Needham, Joseph.** *Science and Civilisation in China.* Vol. 4. Cambridge University Press, 1962.

**Nef, John.** *La naissance de la civilisation industrielle et le monde contemporain.* Paris: Colin, 1954.

**Nicolson, Marjorie Hope.** "Virtuoso." In *Dictionary of the History of Ideas* 4:486–90.

_________. *Voyages to the Moon.* New York: Macmillan, 1948.

**Nordenskiöld, Adolf Erik.** *Facsimile-Atlas to the Early History of Cartography* (1889). Translated by J. A. Ekelof and C. R. Markham. New York: Kraus Reprint, 1961.

**O'Malley, Charles D.** *Andreas Vesalius of Brussels, 1514–1564.* University of California Press, 1964.

**Pachter, Henry M.** *Magic into Science: The Story of Paracelsus.* New York: Schuman, 1951.

**Packard, Francis R.** *Life and Times of Ambroise Paré (1510–1590).* New York: Hoeber, 1921.

**Pagel, Walter.** *Paracelsus: An Introduction to Philosophical Medicine in the Era of the Renaissance.* Basel: Karger, 1958.

_________. "Paracelsus and the Neoplatonic and Gnostic Tradition." *Ambix* 8 (1960), 125–66.

_________. "The 'Wild Spirit' (Gas) of John Baptist van Helmont (1579–1644) and Paracelsus." *Ambix* 10 (1962), 1–13.

**Palter, Robert.** "Early Measurements of Magnetic Force." *Isis* 63 (1972), 544–58.

**Partington, James Riddick.** *A History of Chemistry.* 4 vols. London: Macmillan, 1969–1972.

**Patterson, Louise Diehl.** "Hooke's Gravitation Theory and Its Influence on Newton: I: Hooke's Gravitation Theory." *Isis* 40 (1949), 327–41.

_________. "Hooke's Gravitation Theory and Its Influence on Newton: II: The Insufficiency of the Traditional Estimate." *Isis* 41 (1950), 32–45.

_________. "Recorde's Cosmography, 1556." *Isis* 42 (1951), 208–18.

**Poggendorff, Johann Christian.** *Geschichte der Physik* (1879). Schaan, Liechtenstein: Sändig Reprint Verlag, 1983.

**Potamian (Brother Potamian).** "Gilbert of Colchester." *Popular Science Monthly* 59 (1901), 336–50.

**Pumfrey, Stephen.** "Magnetical Philosophy and Astronomy, 1600–1650." In *Planetary Astronomy from the Renaissance to the Rise of Astrophysics*, edited by R. Taton and C. Wilson (vol. 2 of *The General History of Astronomy*, edited by M. A. Hoskin), 45–53. Cambridge University Press, 1989.

_________. "Mechanizing Magnetism in Restoration England: The Decline of Magnetic Philosophy." *Annals of Science 44* (1987), 1–22.

_________. "'O tempora, O magnes!' A Sociological Analysis of the Discovery of Secular Magnetic Variation in 1634." BJHS 22 (1989), 181–214.

**Pyle, Andrew.** *Atomism and Its Critics*. Bristol: Thoemmes Press, 1995.

**Radelet-de Grave, Patricia, and David Speiser.** "Le *De magnete* de Pierre de Maricourt: Traduction et commentaire." *Revue d'Histoire des Sciences* 28 (1975), 193–234.

**Reynolds, Terry S.** *Stronger Than a Hundred Men: A History of the Vertical Water Wheel*. Johns Hopkins University Press, 1983.

**Riddle, John M.** "Marbode of Rennes' *De lapidibus*: Considered as a Medical Treatise with Text, Commentary, and C. W. King's Translation, together with Text and Translation of Marbode's Minor Works on Stones." *Sudhoffs Archiv: Zeitschrift für Wissenschaftsgeschichte*, Beihefte 20 (1977), 1–144.

**Righini Bonelli, Maria Luisa, and William R. Shea, eds.** *Reason, Experiment, and Mysticism in the Scientific Revolution*. New York: Science History Publications, 1975.

**Roche, John J.** *The Mathematics of Measurement: A Critical History*. London: Athlone Press, 1998.

**Roller, Duane H. D.** *The "De magnete" of William Gilbert*. Amsterdam: Hertzberger, 1959.

**Roller, Duane, and Duane H. D. Roller.** "The Development of the Concept of Electric Charge: Electricity from the Greeks to Coulomb." In *Harvard Case Histories in Experimental Science*, edited by J. B. Conant and L. K. Nash, 2:541–639. Harvard University Press, 1970.

_________. "The Prenatal History of Electrical Science." AJP 21 (1953), 343–56.

**Rosen, Edward.** "Appendix." In *Kepler's Somnium*, by J. Kepler, 177–239. University of Wisconsin Press, 1967.

_________. "Notes." In *Kepler's Conversation with Galileo's Sidereal Messenger*, by J. Kepler, 51–157. New York, Johnson Reprint, 1965.

**Rosenberger, Ferdinand.** *Die Geschichte der Physik.* 2 vols. Braunschweig: Vieweg, 1882.

**Rossi, Paolo.** *The Birth of Modern Science.* Translated by C. de N. Ipsen. Oxford: Blackwell, 2001.

_________. *Francis Bacon: From Magic to Science.* Translated by S. Rabinovitch. University of Chicago Press, 1968.

_________. "Hermeticism, Rationality, and the Scientific Revolution." In *Reason, Experiment, and Mysticism in the Scientific Revolution*, edited by M. L. Righini Bonelli and W. R. Shea, 247–73. New York: Science History Publications, 1975.

_________. *Philosophy, Technology, and the Arts in the Early Modern Era.* Translated by S. Attanasio and edited by B. Nelson. New York: Harper & Row, 1970.

**Russell, Bertrand.** *History of Western Philosophy.* London: Routledge, 2000.

**Russell, John L.** "Kepler's Laws of Planetary Motion: 1609–1666." BJHS 2 (1964), 1–24.

**Sarton, George.** *Introduction to the History of Science.* 3 vols. Robert E. Krieger, 1975.

_________. *Six Wings: Men of Science in the Renaissance.* Indiana University Press, 1957.

**Schipperges, Heinrich.** *Der Garten der Gesundheit: Medizin in Mittelalter.* Munich: Artemis, 1985.

**Schlund, Erhard.** "Petrus Peregrinus von Maricourt: Sein Leben und seine Schriften." *Archivum Franciscanum Historicum* 4 (1911), 436–55, 633–43; 5 (1912), 22–40.

**Schmitt, Jean-Claude.** "Les superstitions." Chapter 4 of *Histoire de la France religieuse*, vol. 1, edited by J. Le Goff and R. Rémond. Paris: Seuil, 1988.

**Schofield, Robert E.** *Mechanism and Materialism: British Natural Philosophy in an Age of Reason.* Princeton University Press, 1970.

**Schwartz, Gorge, and Philip W. Bishop, eds.** *Moments of Discovery.* 2 vols. New York: Basic Books, 1958.

**Seibt, Ferdinand.** *Glanz und Elend des Mittelalters.* Berlin: Goldmann, 1987.

**Seligman, Kurt.** *The History of Magic and the Occult.* New York: Random House, 1997.

**Sherrington, Charles.** *The Endeavour of Jean Fernel, with a List of the Editions of His Writings.* Cambridge University Press, 1946.

**Shumaker, Wayne.** *The Occult Sciences in the Renaissance. A Study in Intellectual Patterns.* University of California Press, 1972.

**Sigerist, Henry E.** *Civilization and Disease.* University of Chicago Press, 1943.

_________. *A History of Medicine.* Vol. 1. Oxford University Press, 1955.

**Simpkins, Diana M.** "Early Editions of Euclid in England." *Annals of Science* 22 (1966), 225–49.

**Singer, Charles.** *From Magic to Science: Essays on the Scientific Twilight.* New York: Dover, 1958.

_________. *A History of Biology to about the Year 1900: A General Introduction to the Study of Living Things.* London: Abelard-Shuman, 1959.

_________. *A Short History of Scientific Ideas to 1900.* Oxford: Clarendon Press, 1959.

**Smith, Charlotte Fell.** *John Dee (1527–1608).* London: Constable, 1909.

**Smith, David E.** "New Information respecting Robert Recorde." *American Mathematical Monthly* 28 (1921), 296–300.

_________. *Rara Arithmetica.* Boston: Ginn, 1908.

**Smith, Julian A.** "Precursors to Peregrinus: The Early History of Magnetism and the Mariner's Compass in Europe." JMH 18 (1992), 21–74.

**Sombart, Werner.** *Krieg und Kapitalismus.* New York: Arno Press, 1975.

**Southern, Richard W.** *The Making of the Middle Ages.* London: Hutchinson, 1953.

**Stahl, William H.** *Roman Science: Origins, Development, and Influence to the Later Middle Ages.* University of Wisconsin Press, 1962.

**Stillman, John Maxson**. *Theophrastus Bombastus von Hohenheim called Paracelsus: His Personality and Influence as Physician, Chemist and Reformer.* Chicago: Open Court, 1920.

**Stock, Brian.** "Science, Technology and Economic Progress in the Early Middle Ages." In *Science in the Middle Ages*, edited by D. C. Lindberg, 1–51. University of Chicago Press, 1978.

**Stone, Howard.** "The French Language in Renaissance Medicine." *Bibliothèque d'Humanisme et Renaissance* 15 (1953), 315–46.

**Suter, Rufus.** "A Biographical Sketch of Dr. William Gilbert of Colchester." *Osiris* 10 (1952), 368–84.

_________. "Dr. William Gilbert of Colchester." *Scientific Monthly* 70 (1950), 254–61.

**Tait, Peter Guthrie.** *Lectures on Some Recent Advances in Physical Science with a Special Lecture on Force.* London: Macmillan, 1876.

**Taton, René, and Curtis Wilson, eds.** *Planetary Astronomy from the Renaissance to the Rise of Astrophysics.* Vol. 2 of *The*

*General History of Astronomy*, edited by M. A. Hoskin. Cambridge University Press, 1989.

**Taylor, Eva Germain Rimington.** *The Haven-Finding Art: A History of Navigation from Odysseus to Captain Cook.* London: Hollis & Carter, 1956.

_________. *The Mathematical Practitioners of Tudor & Stuart England.* Cambridge University Press, 1954.

_________. *Tudor Geography, 1485–1583.* London: Methuen, 1930.

**Taylor, F. Sherwood.** *The Alchemists.* New York: Arno Press, 1974.

**Thackray, Arnold.** *Atoms and Powers: An Essay on Newtonian Matter-theory and the Development of Chemistry.* Harvard University Press, 1970.

**Thomas, Keith.** *Religion and the Decline of Magic.* London: Weidenfeld and Nicolson, 1971/1980.

**Thompson, Silvanus.** "The Family and Arms of Gilbert of Colchester." *Transactions of the Essex Archaeological Society*, n.s., 9 (1906), 197–211.

_________. "Petrus Peregrinus de Maricourt and His *Epistola de magnete*." *Proceedings of the British Academy 2* (1905/6), 1–23.

**Thoren, Victor E.** *The Lord of Uraniborg: A Biography of Tycho Brache.* Cambridge University Press, 1990.

**Thorndike, Lynn.** "De Lapidibus." *Ambix* 8 (1960), 6–23.

_________. *A History of Magic and Experimental Science.* 8 vols. Columbia University Press, 1923–1958.

_________. "John of St. Amand on the Magnet." *Isis* 36 (1946), 156–57.

_________. *Michael Scot.* London: Nelson, 1965.

_________. *Science and Thought in the Fifteenth Century.* New York: Hafner, 1963.

_________. "The True Place of Astrology in the History of Science." *Isis* 46 (1955), 273–78.

**Turner, Howard R.** *Science in Medieval Islam: An Illustrated Introduction*. University of Texas Press, 1995.

**Verger, Jacques.** *Les universités au moyen age*. Paris: Presses Universitaires de France, 1973.

**Vickers, Brian ed.** *Occult and Scientific Mentalities in the Renaissance*. Cambridge University Press, 1984.

**Viets, Henry.** "De staticis experimentis of Nicolaus Cusanus." *Annals of Medical History* 4 (1922), 115–35.

**Villamil, Richard de.** *See* **de Villamil, Richard.**

**von Humboldt, Alexander.** *See* **Humboldt, Alexander von.**

**von Lipmann, Edmund O.** *See* **Lipmann, Edmund O. von.**

**Walker, Daniel P.** *Spiritual and Demonic Magic: From Ficino to Campanella*. Nendeln: Kraus Reprint, 1976.

**Wallace, William A.** *Galileo and His Sources: The Heritage of the Collegio Romano in Galileo's Science*. Princeton University Press, 1984.

_________. "The Philosophical Setting of Medieval Science." In *Science in the Middle Ages*, edited by D. C. Lindberg, 91–119. University of Chicago Press, 1978.

**Ward, Benedicta.** *The Venerable Bede*. London: Chapman, 1998.

**Webster, Charles.** "The Discovery of Boyle's Law, and the Concept of the Elasticity of Air in the Seventeenth Century." AHES 2 (1965), 441–502.

_________. *From Paracelsus to Newton: Magic and the Making of Modern Science*. Cambridge University Press, 1982.

_________. "Henry Power's Experimental Philosophy." *Ambix* 14 (1967), 150–78.

**Westfall, Richard S.** *The Construction of Modern Science: Mechanisms and Mechanics*. Cambridge University Press, 1977.

_________. "Hooke and the Law of Universal Gravitation." BJHS 3 (1967), 245–61.

_________. *Never at Rest: A Biography of Isaac Newton.* Cambridge University Press, 1980.

_________. "Newton and Alchemy." In *Occult and Scientific Mentalities in the Renaissance*, edited by B. Vickers, 315–35. Cambridge University Press, 1984.

_________. "The Role of Alchemy in Newton's Career." In *Reason, Experiment, and Mysticism in the Scientific Revolution*, edited by M. L. Righini Bonelli and W. R. Shea, 189–232. New York: Science History Publications, 1975.

**Whewell, William.** *History of the Inductive Sciences.* 3 vols. (1837). London: Frank Cass, 1967.

**White, Lynn, Jr.** *Machina ex Deo: Essays in the Dynamism of Western Culture.* MIT Press, 1968.

_________. *Medieval Religion and Technology: Collected Essays.* University of California Press, 1978.

_________. *Medieval Technology and Social Change.* Oxford University Press, 1962.

**Whitehead, Alfred North.** *Science and the Modern World.* New York: Macmillan, 1926.

**Whitfield, Peter.** *The Image of the World: 20 Centuries of World Maps.* San Francisco: Pomegranate Artbooks, 1994.

**Whittaker, Edmund.** *A History of the Theories of Aether and Electricity.* Vol. 1, *The Classical Theories.* London: Nelson, 1951.

**Wightman, William P. D.** *Science and the Renaissance.* 2 vols. Edinburgh: Oliver and Boyd, 1962.

**Willey, Basil.** *The Seventeenth Century Background: Studies in the Thought of the Age in Relation to Poetry and Religion.* Harmondsworth: Penguin Books, 1972.

**Wilson, Curtis.** "From Kepler's Laws, So-called, to Universal Gravitation: Empirical Factors." AHES 6 (1970), 89–170.

_________. "The Newtonian Achievement in Astronomy." In *Planetary Astronomy from the Renaissance to the Rise of Astrophysics*, edited by R. Taton and C. Wilson (vol. 2 of *The General History of Astronomy*, edited by M. A. Hoskin), 233–74. Cambridge University Press, 1989.

**Winter, Heinrich.** "Who Invented the Compass?" *Mariner's Mirror* 23 (1937), 95–102.

**Wolf, Abraham.** *A History of Science, Technology, and Philosophy in the 16th & 17th Centuries.* 2 vols. Gloucester: Smith, 1968.

**Wolff, Philippe.** *The Cultural Awakening.* Translated by A. Carter. New York: Pantheon Books, 1968.

**Wolfson, Harry Austryn**. *Crescas' Critique of Aristotle: Problems of Aristotle's* Physics *in Jewish and Arabic Philosophy.* Harvard University Press, 1971.

**Yates, Frances A.** *Giordano Bruno and the Hermetic Tradition.* University of Chicago Press, 1964/1991.

_________. *The Rosicrucian Enlightenment.* London: Routledge & Kegan Paul, 1972.

_________. *Theater of the World.* University of Chicago Press, 1969.

**Zilboorg, Gregory.** *A History of Medical Psychology.* New York: Norton, 1967.

**Zilsel, Edgar.** "The Origins of William Gilbert's Scientific Method." JHI 2 (1941), 1–32.

Ingram Content Group UK Ltd.
Milton Keynes UK
UKHW021314070723
424728UK00009B/101

9 789813 223769